Lecture Notes in Computer Science　　10846

Commenced Publication in 1973
Founding and Former Series Editors:
Gerhard Goos, Juris Hartmanis, and Jan van Leeuwen

More information about this series at http://www.springer.com/series/7407

Fedor V. Fomin · Vladimir V. Podolskii (Eds.)

Computer Science – Theory and Applications

13th International Computer Science Symposium in Russia, CSR 2018
Moscow, Russia, June 6–10, 2018
Proceedings

 Springer

Editors
Fedor V. Fomin
Department of Informatics
University of Bergen
Bergen
Norway

Vladimir V. Podolskii
Steklov Mathematical Institute
Moscow
Russia

ISSN 0302-9743   ISSN 1611-3349   (electronic)
Lecture Notes in Computer Science
ISBN 978-3-319-90529-7   ISBN 978-3-319-90530-3   (eBook)
https://doi.org/10.1007/978-3-319-90530-3

Library of Congress Control Number: 2018941549

LNCS Sublibrary: SL1 – Theoretical Computer Science and General Issues

Printed on acid-free paper

This Springer imprint is published by the registered company Springer International Publishing AG
part of Springer Nature
The registered company address is: Gewerbestrasse 11, 6330 Cham, Switzerland

Preface

The 13th International Computer Science Symposium in Russia (CSR 2018) was held during June 6–10, 2018, in Moscow, Russia. The symposium was organized by the Higher School of Economics. It was the 13th event in the CSR series of regular international meetings, following St. Petersburg (2006), Ekaterinburg (2007), Moscow (2008), Novosibirsk (2009), Kazan (2010), St. Petersburg (2011), Nizhny Novgorod (2012), Ekaterinburg (2013), Moscow (2014), Listvyanka (2015), St. Petersburg (2011), and Kazan (2017). CSR covers a wide range of areas in theoretical computer science and its applications.

The opening lecture was given by Noga Alon (Tel Aviv University and Princeton). Seven other invited plenary lectures were given by Markus Bläser (Saarland University), Vladimir Gurvich (Rutgers University), Alexander Kulikov (St. Petersburg Department of Steklov Institute of Mathematics), Kurt Mehlhorn (Max-Planck-Institut für Informatik), Michael Saks (Rutgers University), Rahul Santhanam (University of Oxford), and László A. Végh (London School of Economics).

This volume contains the accepted papers and those sent by the invited speakers. We received 42 submissions in total, and out of these the Program Committee selected 24 papers for presentation at the symposium and for publication in the proceedings. Each submission was reviewed by at least three Program Committee members. The Program Committee also selected the winners of the two Yandex Best Paper Awards.

The Best Paper Award: Jayakrishnan Madathil, Saket Saurabh, and Meirav Zehavi, "Max-Cut Above Spanning Tree Is Fixed Parameter Tractable"

The Best Student Paper Award: Alexander Kozachinskiy, "Recognizing Read-Once Functions from Depth-Three Formulas"

Many people and organizations contributed to the smooth running and the success of CSR 2016. In particular our thanks go to:

- All authors who submitted their current research to CSR
- All invited speakers who agreed to give a talk at the conference
- Our reviewers and additional reviewers, whose expertise flowed into the decision process
- The members of the Program Committee, who graciously gave their time and energy
- The members of the local Organizing Committee, who made the conference possible
- The EasyChair conference management system for hosting the evaluation process
- The Higher School of Economics for hosting the conference
- Yandex for supporting the conference and providing Best Paper Awards
- The European Association for Theoretical Computer Science (EATCS) for the scientific support of the conference

March 2018

Fedor V. Fomin
Vladimir Podolskii

Organization

Program Committee

Maxim Babenko	Moscow State University, Russia
Mikołaj Bojńaczyk	Warsaw University, Poland
Holger Dell	Saarland University and Cluster of Excellence, MMCI, Germany
Edith Elkind	University of Oxford, UK
Fedor Fomin	University of Bergen, Norway
Fabrizio Grandoni	IDSIA, University of Lugano, Switzerland
Dmitry Itsykson	Steklov Institute of Mathematics at St. Petersburg, Russia
Mikko Koivisto	University of Helsinki, Finland
Antonina Kolokolova	Memorial University of Newfoundland, Canada
Alexander Kulikov	Steklov Mathematical Institute at St. Petersburg, Russia
Jakob Nordström	KTH Royal Institute of Technology, Sweden
Alexander Okhotin	St. Petersburg State University, Russia
Vladimir Podolskii	Steklov Mathematical Institute, Russia
Ilya Razenshteyn	Microsoft Research Redmond, USA
Saket Saurabh	The Institute of Mathematical Sciences, Chennai, India
Alexander Shen	LIRMM CNRS and University Montpellier 2, France, on leave from IITP RAS, Moscow
Arseny Shur	Ural Federal University, Russia
Dirk Oliver Theis	University of Tartu, Estonia
René van Bevern	Novosibirsk State University, Russia
Meirav Zehavi	Ben-Gurion University, Israel

Additional Reviewers

Aganezov, Sergey
Aghighi, Meysam
Akhmedov, Maxim
Bliznets, Ivan
Boiret, Adrien
De Oliveira Oliveira, Mateus
de Rezende, Susanna F.
Dey, Palash
Gerasimov, Alexander
Glinskih, Ludmila
Golovach, Petr

Golovnev, Alexander
Guillon, Bruno
Gálvez, Waldo
Göös, Mika
Hermelin, Danny
Iskhakov, Timur
Johannsen, Jan
Kalenkova, Anna
Karpov, Nikolai
Karthik, C. S.
Kaski, Petteri
Knop, Alexander

Kolesnichenko, Ignat
Korhonen, Janne H.
Kosolobov, Dmitry
Kuske, Dietrich
Kutrib, Martin
Kuznetsov, Stepan
Lasota, Sławomir
M. S., Ramanujan
Mukhopadhyay, Sagnik
Nasre, Meghana
Nichterlein, André
Nikulin, Yury

Oparin, Vsevolod
Pettie, Seth
Pyatkin, Artem
Radhakrishnan, Jaikumar
Rajan, Varun
Raskin, Mikhail
Reinhardt, Klaus
Risse, Kilian
Robere, Robert

Romashchenko, Andrei
Roth, Marc
Salomaa, Kai
Savchenko, Ruslan
Schaeffer, Luke
Scharf, Nadja
Scquizzato, Michele
Segev, Danny
Seidl, Helmut

Shapoval, Alexander
Sokolov, Dmitry
Speranski, Stanislav O.
Stephan, Frank
Suchy, Ondrej
Swernofsky, Joseph
Törmä, Ilkka
Vyalyi, Mikhail

Abstract of Invited Talks

Constructive and Non-constructive Combinatorics

Noga Alon[1,2]

[1] Princeton University, Princeton, NJ 08544, USA
[2] Tel Aviv University, Tel Aviv, 69978, Israel
nalon@math.princeton.edu

Purely combinatorial proofs of combinatorial statements usually provide efficient procedures for solving the corresponding algorithmic problems, even if they deal with NP-hard invariants. One representative classical example out of many is Dirac's Theorem that asserts that any simple graph with $n \geq 3$ vertices and minimum degree at least $n/2$ contains a Hamilton cycle. The proof supplies a simple polynomial time algorithm for finding a Hamilton cycle in any such graph, although the problem of deciding whether or not a given input graph contains a Hamilton cycle is NP-hard. Similar examples include Turán's Theorem, Vizing's Theorem or the Four Color Theorem.

Modern combinatorics often applies more sophisticated, non-combinatorial tools, including probabilistic, topological or algebraic techniques. Probabilistic proofs usually supply efficient (deterministic or randomized) algorithms, and in many cases the randomized algorithms can be derandomized and converted into deterministic ones. See [5, 8, 10] for some prominent examples obtained during the last decade. In contrast, proofs based on topological and algebraic reasoning are often non-constructive and provide no efficient solutions for the corresponding algorithmic problems. Finding such solutions is an intriguing challenge. In the lecture I will describe several old and new examples of this type. A representative example is the following, known as the Cycle and Triangles question. While it may look somewhat special, I have chosen it as it can be solved either by applying topological techniques or by using algebraic tools, and yet there is neither a known combinatorial solution nor a known algorithmic one.

Du, Hsu and Wang [6] conjectured that if a graph on $3n$ vertices is the edge disjoint union of a Hamilton cycle of length $3n$ and n vertex disjoint triangles then its independence number is n. Erdős conjectured that in fact any such graph is 3 colorable. Using the algebraic approach in [4], Fleischner and Stiebitz [7] proved this conjecture in a stronger form - any such graph is in fact 3-list-colorable, namely, for every assignment of a list of 3 colors for each vertex, there is a proper coloring assigning to each vertex a color from its list.

As proved in a recent paper [1] the original conjecture, in a slightly stronger form, can be derived quickly from a result of Schrijver about critical subgraphs of the Kneser graph [11]. Strengthening Lovász theorem he proved, using the Borsuk-Ulam

Research supported in part by an ISF grant and by a GIF grant.

Theorem, that in any coloring of the independent sets of vertices of size k in a cycle of length n by less than $n - 2k + 2$ colors there are two disjoint independent sets having the same color. This supplies a short proof of the following statement: Let $C_{3n} = (V, E)$ be cycle of length $3n$ and let $V = A_1 \cup A_2 \cup \ldots \cup A_n$ be a partition of its vertex set into n pairwise disjoint sets, each of size 3. Then there exist two disjoint independent sets in the cycle, each containing one point from each A_i.

Here is a proof. Define a coloring of the independent sets of size n in C_{3n} as follows. If S is such an independent set and there is an index i so that $|S \cap A_i| \geq 2$, color S by the smallest such i. Otherwise, color S by the color $n + 1$. By the above result of Schrijver there are two disjoint independent sets S_1, S_2 with the same color. This color cannot be any $i \leq n$, since if this is the case then

$$|(S_1 \cup S_2) \cap A_i| = |S_1 \cap A_i| + |S_2 \cap A_i| \geq 2 + 2 = 4 > 3 = |A_i|,$$

which is impossible. Thus S_1 and S_2 are both colored $n + 1$, meaning that each of them contains exactly one element of each A_i. ☐

Several extensions are proved in [1]. The Fleischner-Stiebitz theorem implies that the representing set in the Du-Hsu-Huang conjecture can be required to contain any given vertex. This can also be deduced from the topological version of Hall's Theorem of Aharoni and Haxell, as shown in [2]. None of the above proofs supplies an efficient algorithm for finding the desired independent set.

In the lecture I will describe several additional non-constructive proofs of combinatorial statements proved by applying the Borsuk-Ulam theorem, as well as additional algebraic proofs that supply no efficient algorithms. Certain results of this type can be found in [3, 9], and I will mention a few more recent examples. I will also discuss the reasons that suggest that the derivation of constructive proofs may be hard.

References

1. Aharoni, R., et al.: Fair representation by independent sets. In: Loebl, M., Nešetřil, J., Thomas, R. (eds.) A Journey Through Discrete Mathematics, pp. 31–58. Springer, Cham (2017)
2. Aharoni, R., Holzman, R., Howard, D., Sprüsell, P.: Cooperative colorings and systems of independent representatives. Electron. J. Combin. 22(2), 14 (2015). Paper 2.27
3. Alon, N.: Discrete mathematics: methods and challenges. In: Proceedings of the International Congress of Mathematicians (ICM), Beijing 2002, China, pp. 119–135. Higher Education Press (2003)
4. Alon, N., Tarsi, M.: Chromatic numbers and orientations of graphs. Combinatorica **12**, 125–134 (1992)
5. Bansal, N.: Constructive algorithms for discrepancy minimization. In: Proceedings of 2010 IEEE FOCS, pp. 3–10 (2010)
6. Du, D.Z., Hsu, D.F., Hwang, F.K.: The Hamiltonian property of consecutive-d digraphs, in graph-theoretic models in computer science, II (Las Cruces, NM, 1988. 1990). Math. Comput. Model. **17**, 61–63 (1993)
7. Fleischner, H., Stiebitz, M.: A solution to a colouring problem of P. Erdős. Discrete Math. **101**, 39–48 (1992)

8. Lovett, S., Meka, R.: Constructive discrepancy minimization by walking on the edges. In: Proceedings of 2012 IEEE FOCS, pp. 61–67 (2012)
9. Matoušek, J.: Using the Borsuk-Ulam theorem. In: Lectures on Topological Methods in Combinatorics and Geometry, Springer, Heidelberg (2003)
10. Moser, R.A., Tardos, G.: A constructive proof of the general Lovász local lemma. J. ACM **57**, **15** (2010). Art. 11
11. Schrijver, A.: Vertex-critical subgraphs of Kneser graphs. Nieuw Arch. Wisk. **26**, 454–461 (1978)

Physarum Solves Non-negative Undirected Linear Programs

Ruben Becker[1], Vincenzo Bonifaci[2], Andreas Karrenbauer[1],
Pavel Kolev[1], and Kurt Mehlhorn[1]

[1] Max Planck Institute for Informatics, Saarland Informatics Campus,
Saarbrücken, Germany
[2] Institute for the Analysis of Systems and Informatics,
National Research Council of Italy (IASI-CNR), Rome, Italy

Physarum polycephalum is a slime mold that apparently is able to solve shortest path problems. Nakagaki, Yamada, and Tóth [1] report about the following experiment; see Fig. 1. They built a maze, covered it by pieces of Physarum (the slime can be cut into pieces which will reunite if brought into vicinity), and then fed the slime with oatmeal at two locations. After a few hours the slime retracted to a path that follows the shortest path in the maze connecting the food sources. The authors report that they repeated the experiment with different mazes; in all experiments, Physarum retracted to the shortest path.

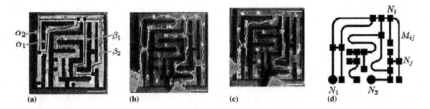

Fig. 1. The experiment in [1] (reprinted from there): (a) shows the maze uniformly covered by Physarum; yellow color indicates presence of Physarum. Food (oatmeal) is provided at the locations labeled AG. After a while the mold retracts to the shortest path connecting the food sources as shown in (b) and (c). (d) shows the underlying abstract graph. The video [2] shows the experiment.

The paper [3] proposes a mathematical model for the behavior of the slime and argues extensively that the model is adequate. Physarum is modeled as an electrical network with time varying resistors. We have a simple *undirected* graph $G = (N, E)$ with distinguished nodes s_0 and s_1 modeling the food sources. Each edge $e \in E$ has a positive length c_e and a positive capacity $x_e(t)$; c_e is fixed, but $x_e(t)$ is a function of time. The resistance $r_e(t)$ of e is $r_e(t) = c_e/x_e(t)$. In the electrical network defined by these resistances, a current of value 1 is forced from s_0 to s_1. For an (arbitrarily oriented) edge $e = (u, v)$, let $q_e(t)$ be the resulting current over e. Then, the capacity of e evolves according to the differential equation

$$\dot{x}_e(t) = |q_e(t)| - x_e(t), \tag{1}$$

where $\dot{x}_e$ is the derivative of x_e with respect to time. In equilibrium ($\dot{x}_e = 0$ for all e), the flow through any edge is equal to its capacity. In non-equilibrium, the capacity grows (shrinks) if the absolute value of the flow is larger (smaller) than the capacity. In the sequel, we will mostly drop the argument t as is customary in the treatment of dynamical systems. It is well-known that the electrical flow q is the feasible flow minimizing energy dissipation $\sum_e r_e q_e^2$ (Thomson's principle).

Miyaji and Ohnishi were the first to analyze convergence for special graphs (parallel links and planar graphs with source and sink on the same face) in [4]. In [5] convergence was proven for *all* graphs. We state the result from [5] for the special case that the shortest path is unique.

Theorem 1 *([5]). Assume $c > 0$ and that the undirected shortest path P^* from s_0 to s_1 w.r.t. the cost vector c is unique. Assume $x(0) > 0$. Then x(t) in (1) converges to P^*. Namely, $x_e(t) \to 1$ for $e \in P^*$ and $x_e \to 0$ for $e \notin P^*$ as $t \to \infty$.*

[5] also proves an analogous result for the undirected transportation problem; [6] simplified the argument under additional assumptions. The paper [7] studies a more general dynamics and proves convergence for parallel links.

In this paper[1], we extend this result to *non-negative undirected linear programs*

$$\min\{c^T x : Af = b, \ |f| \le x\}, \tag{2}$$

where $A \in \mathrm{IR}^{n \times m}$, $b \in \mathrm{IR}^n$, $x \in \mathrm{IR}^m$, $c \in \mathrm{IR}^m_{\ge 0}$, and the absolute values are taken componentwise. Undirected LPs can model a wide range of problems, e.g. optimization problems such as shortest path and min-cost flow in undirected graphs, and the Basis Pursuit problem in signal processing [9].

We use n for the number of rows of A and m for the number of columns, since this notation is appropriate when A is the node-edge-incidence matrix of a graph. A vector f is *feasible* if $Af = b$. We assume that the system $Af = b$ has a feasible solution and that there is no non-zero f in the kernel of A with $c_e f_e = 0$ for all e. A vector f lies in the kernel of A if $Af = 0$. The vector q in (1) is now the *minimum energy feasible solution*

$$q(t) = \mathrm{argmin}_{f \in \mathrm{IR}^m} \left\{ \sum_{e : x_e \ne 0} \frac{c_e}{x_e(t)} f_e^2 : Af = b \wedge f_e = 0 \text{ whenever } x_e = 0 \right\}. \tag{3}$$

We remark that q is unique. If A is the incidence matrix of a graph (the column corresponding to an edge e has one entry $+1$, one entry -1 and all other entries are equal to zero), (2) is a transshipment problem with flow sources and sinks encoded by a demand vector b. The condition that there is no solution in the kernel of A with $c_e f_e = 0$ for all e states that every cycle contains at least one edge of positive cost. In that setting,

---

[1] The full paper [8] is available on the arxiv.

$q(t)$ as defined by (3) coincides with the electrical flow induced by resistors of value $c_e/x_e(t)$.

Theorem 2. *Let* $c \geq 0$ *satisfy* $c^T |f| > 0$ *for every nonzero* f *in the kernel of* A. *Let* x^* *be an optimum solution of (2) and let* $X_{\star}$ *be the set of optimum solutions. Assume* $x(0) > 0$. *The following holds for the dynamics (1) with* q *as in (3):*

(i)   *The solution* $x(t)$ *exists for all* $t \geq 0$.
(ii) *The cost* $c^T x(t)$ *converges to* $c^T x^*$ *as* t *goes to infinity.*
(iii) *The vector* $x(t)$ *converges to* $X_{\star}$.
(iv) *For all* e *with* $c_e > 0$, $x_e(t) - |q_e(t)|$ *converges to zero as* t *goes to infinity.[2] If* x^* *is unique,* $x(t)$ *and* $q(t)$ *converge to* x^* *as* t *goes to infinity.*

Item (i) was previously shown in [10] for the case of a strictly positive cost vector. The result in [10] is stated for the cost vector $c = \mathbf{1}$. The case of a general positive cost vector reduces to this special case by rescaling the solution vector x. We stress that the dynamics (1) is biologically-grounded. It was proposed to model a biological system and not as an optimization method. Nevertheless, it can solve a large class of non-negative LPs. Table 1 summarizes our first main result and puts it into context.

Table 1. Convergence results for the continuous undirected Physarum dynamics (1).

Reference	Problem	Existence of solution	Convergence to OPT	Comments
[4]	Undirected shortest path	Yes	Yes	Parallel edges, planar graphs
[5]	Undirected shortest path	Yes	Yes	All graphs
[10]	Undirected positive LP	Yes	No	$c \geq 0$
Our Result	Undirected Nonnegative LP	Yes	Yes	(1) $c \geq 0$ (2) $\forall v \in \ker(A) : c^T |v| > 0$

References

1. Nakagaki, T., Yamada, H., Tóth, A.: Maze-solving by an amoeboid organism. Nature **407**, 470 (2000)
2. http://people.mpi-inf.mpg.de/∼mehlhorn/ftp/SlimeAusschnitt.webm
3. Tero, A., Kobayashi, R., Nakagaki, T.: A mathematical model for adaptive transport network in path finding by true slime mold. J. Theor. Biol. 553–564 (2007)
4. Miyaji, T., Ohnishi, I.: Physarum can solve the shortest path problem on Riemannian surface mathematically rigorously. Int. J. Pure Appl. Math. **47**, 353–369 (2008)
5. Bonifaci, V., Mehlhorn, K., Varma, G.: Physarum can compute shortest paths. J. Theor. Biol. **309**, 121–133 (2012)

---

[2] We conjecture that this also holds for the edges e with $c_e = 0$.

6. Bonifaci, V.: Physarum can compute shortest paths: a short proof. Inf. Process. Lett. **113**, 4–7 (2013)
7. Bonifaci, V.: A revised model of network transport optimization in Physarum Polycephalum (2015)
8. Becker, R., Bonifaci, V., Karrenbauer, A., Kolev, P., Mehlhorn, K.: Two results on slime mold computations. Technical report (2017). https://arxiv.org/abs/1707.06631
9. Chen, S.S., Donoho, D.L., Saunders, M.A.: Atomic decomposition by basis pursuit. SIAM J. Sci. Comput. **20**, 33–61 (1998)
10. Straszak, D., Vishnoi, N.K.: IRLS and slime mold: equivalence and convergence (2016). CoRR, abs/1601.02712

Limitations of Algebraic Lower Bound Proofs

Markus Bläser

Saarland University
mblaeser@cs.uni-saarland.de

Abstract. Algebraic natural proofs were recently introduced by Forbes et al. [FSV17] and independently by Grochow et al. [GKSS17]. Assume we are given some polynomial of which we think that it is hard to compute, say the permanent $\text{per}_n \in K[X_{1,1}, \ldots, X_{n,n}]$ of size $n \times n$. per_n is a multilinear polynomial of degree n in n^2 variables. The space of all such polynomials has dimension $\binom{n^2}{n}$. For such a polynomial f, let c_f denote its coefficient vector, which has length $\binom{n^2}{n}$. An algebraic proof that the permanent is hard is a polynomial P that vanishes on all polynomials of low complexity but not on the permanent, that is, $P(c_f) = 0$ for all f as above of low complexity but $P(c_{\text{per}_n}) \neq 0$. What is the complexity of such a P? If such a P has high complexity, then this means that proving a circuit lower bound for per_n is hard.

For Boolean circuit complexity, Razborov and Rudich [RR97] introduced so-called natural proofs. The objects they consider are truth tables of Boolean functions (instead of coefficient vectors). A natural proof P is a predicate on the set of truth tables that has two properties: The first one is *largeness*, that is, P is true for a sufficiently large fraction of all Boolean functions. The second one is *constructivity*, that is, P is computable by small circuits. Razborov and Rudichs famous barrier result states that natural proofs can only yield superpolynomial bounds if certain pseudorandom generators do not exist.

In the algebraic setting, largeness comes for free, the zero set of any nonzero polynomial is small in the sense that almost all inputs do not lie in the zero set. We can now ask the question whether there is some sort of barrier in the algebraic world, too, or could it be that there is a polynomial P that is easy to compute, vanishes on all coefficient vectors of polynomials of low complexity, but $P(c_{\text{per}_n}) \neq 0$

Unfortunately, there is no known analogous theory of pseudorandomness in the algebraic setting. Therefore, Forbes et al. use a concept called succinct hitting sets instead. This assumption is related to polynomial identity testing, but it is currently not clear how plausible this assumption is. Forbes et al. are only able to construct succinct hitting sets against rather weak models of arithmetic circuits.

Let $S \subseteq K[X_1, \ldots, X_n]$ be a set of polynomials. $H \subseteq K^n$ is called a *hitting set* for S, if for all $p \in S$, there is an $x \in H$ such that $p(x) \neq 0$. If P is a polynomial that vanishes on all cofficient vectors c_f of polynomials f of low complexity and P has itself low complexity, then this can be interpreted as follows: Polynomials of low complexity do not have simple hitting sets. This is the main idea behind the concept of succinct hitting sets.

There is one further complication: If a polynomial vanishes on a particular set, it also vanishes on the Zariski closure of this set. So an algebraic proof against some class S will vanish on polynomials f that are not contained in S, but are contained in the closure $\overline{S}$. Polynomials in the border $\overline{S} \setminus S$ have higher complexity than polynomials in S (otherwise, they would be in S), yet they cannot be distinguished by an algebraic proof from polynomials in S, independently of any barrier. Therefore, to study algebraic proofs properly, one needs to look at Zariski closed classes of polynomials.

In the setting above, the proof polynomial P has much more variables than the polynomials f. There are also settings were these numbers are polynomially related, tensors are such an example. One can think of a tensor as a "three-dimensional matrix" $t = (t_{h,i,j}) \in K^{\ell \times m \times n} := K^\ell \otimes K^m \otimes K^n$. A rank-one tensor is a tensor of the form $u \otimes v \otimes w$ with $u \in K^\ell$, $v \in K^m$ and $w \in K^n$. The rank $R(t)$ of t is the smallest number r of rank-one tensors $s_1, \ldots, s_r$ such that $t = s_1 + s_2 + \ldots + s_r$. Let $S_r = \{s \in K^\ell \otimes K^m \otimes K^n \mid R(s) \leq r\}$ be the set of all tensors of rank at most r. An algebraic proof that $R(t) > r$ is a polynomial P in $\ell m n$ variables such that P vanishes on S_r and $P(t) \neq 0$. However, the set S_r is not Zariski-closed. That is, it is not the vanishing set of a set of polynomials. So we look at the Zariski closure X_r of S_r instead. These tensors are called the tensors of border rank $\leq r$. As seen above, the appropriate quantity to study when considering algebraic proofs is the border rank.

Given a tensor t and a bound b, it is NP-hard to decide whether $R(T) \leq b$ as shown by Håstad [Hås90], however, it is not known whether this holds for the border rank. We define a similar quantity, which we call *(border) completion rank*, and proof that completion rank and even border completion rank are NP-hard to compute. Next we construct a small family of tensors (small means that they come from a closed, even low dimensional set) such that not all of these tensors can have algebraic proofs of polynomial size against the set of all tensors of completion rank $\leq r$ for some appropriately chosen r. This means that there is a tensor t such that any polynomial P with $P(t) \neq 0$ that vanishes on all tensor of completion rank r has superpolynomial circuit complexity. This result if of course conditional, but it is based on the widely believed assumption that coNP $\not\subseteq \exists$BPP. One can view this as a meta-result: Proving lower bounds via algebraic proofs is difficult. At least, if we want to represent the proof by an algebraic circuit.

Even the geometric complexity approach initiated by Mulmuley and Sohoni [MS01] eventually produces an algebraic proof. However, it is produced from some intermediate representation, which can be more compact. We provide one such example: We show, of course conditionally, that matrices with nonzero permanent cannot have small algebraic proofs. However, geometric complexity theory provides very short proofs for them.

(Parts of the presentation are joint work with Christian Ikenmeyer, Gorav Jindal, and Vladimir Lysikov)

References

[FSV17]   Forbes, M.A., Shpilka, A., Volk, B.L.: Succinct hitting sets and barriers to proving algebraic circuits lower bounds. In: Hatami, H., McKenzie, P., King, H. (eds.) Proceedings of the 49th Annual ACM SIGACT Symposium on Theory of Computing, STOC 2017, Montreal, QC, Canada, 19–23 June 2017, pp. 653–664. ACM (2017)

[GKSS17]   Grochow, J.A., Kumar, M. Saks, M.E., Saraf, S.: Towards an algebraic natural proofs barrier via polynomial identity testing (2017). CoRR, abs/1701.01717

[Hås90]   Håstad, J.: Tensor rank is NP-complete. J. Algorithms 11(4), 644–654 (1990)

[MS01]   Mulmuley, K., Sohoni, M.A..: Geometric complexity theory I: an approach to the P vs. NP and related problems. SIAM J. Comput. 31(2), 496–526 (2001)

[RR97]   Razborov, A.A., Rudich, S.: Natural proofs. J. Comput. Syst. Sci. 55(1), 24–35 (1997)

Complexity of Generation

Vladimir Gurvich[1,2]

[1] RUTCOR & RBS, Rutgers University, 100 Rockafeller Road,
Piscataway NJ 08854, USA
[2] National Research Institute Higher School of Economics (HSE),
Moscow, Russia
vladimir.gurvich@gmail.com

Abstract. In this talk I summarize the results obtained in 1999–2008 by Leonid Khachiyan, Endre Boros, Konrad Borys, Khaled Elbassiony, Kazuhisa Makino, and myself, on complexity of generation algorithms. These algorithms can be partitioned into three groups: supergraph, flash-light (backtrack), and dual-bounded generation. We will call a problem *tractable* if it can be solved by a polynomial (n^{const}) or quasi-polynomial ($n^{polylog(n)}$) time algorithm. More generally, for any positive non-decreasing function $g = g(n)$, generating can be performed in total or incremental time g, or with g-delay. Most of the polynomial delay algorithms are provided by the flash-light (backtrack) method. As for the incremental algorithms, generating the next object is equivalent with just verifying its existence, which is a standard decision problem. Thus, incremental generation, in contrast to the delay one, may be NP-hard or NP-complete. For example, we show that generating all vertices of a polyhedron, given by its facets, is NP-complete (while the complexity status is still open in case of the polytopes, that is, bounded polyhedra). This problem is reduced to generating all negative cycles of a weighted digraph, which is NP-complete (for graphs, too). Generating all minimal transversals to a hypergraph, so-called dualization, plays an important role. For this problem an incremental quasi-polynomial algorithm (but no polynomial one) is known. We outline several wide classes of generation problems that can be reduced to dualization and, thus, solved in incremental quasi-polynomial time. We survey algorithms and complexity bounds for the above and many other generation problems.

This work was supported in part by the Russian Academic Excellence Project '5-100'.

Lower Bounds for Unrestricted Boolean Circuits: Open Problems

Alexander S. Kulikov

St. Petersburg Department of Steklov Institute of Mathematics,
St. Petersburg, Russia

Abstract. To prove that P $\neq$ NP, it suffices to prove a superpolynomial lower bound on Boolean circuit complexity of a function from NP. Currently, we are not even close to achieving this goal: we do not know how to prove a $4n$ lower bound. What is more depressing is that there are almost no techniques for proving circuit lower bounds.

In this note, we briefly review various approaches that could potentially lead to stronger linear or superlinear lower bounds for unrestricted Boolean circuits (i.e., circuits with no restriction on depth, fan-out, or basis).

Online Labeling: Algorithms, Lower Bounds and Open Questions

Michael Saks

Department of Mathematics, Rutgers University, New Brunswick,
NJ 08854, USA
saks@math.rutgers.edu

Abstract. The online labeling problem (also known as the file maintenance problem), is a natural algorithmic problem that has arisen as a buidling block for data structures. A stream of distinct integer items is to be assigned labels online from a label set $\{1, \ldots, m\}$ so that the order of the labels respects the natural order of the items. Maintaining order on the labels may require relabeling items. The algorithm pays 1 each time an item is labeled or relabeled and the goal of the algorithm is to minimize the total cost.

We survey upper and lower bounds and open problems in both the deterministic and randomized setting.

Reading MCSP Through SAT

Rahul Santhanam

Department of Computer Science, University of Oxford, UK
rahul.santhanam@cs.ox.ac.uk

Abstract. The Minimum Circuit Size Problem (MCSP) and the Circuit Satisfiability Problem (SAT) are fundamental problems in theoretical computer science. These problems are in a sense dual to each other - while MCSP asks if a Boolean function given by its truth table has small circuits, SAT and its variants ask about properties of the Boolean function corresponding to a given circuit. SAT has featured in some of the most important and influential results in complexity theory over the past few decades, including the Cook-Levin theorem, the PCP theorem and Williams' connection between SAT algorithms and circuit lower bounds. MCSP, however, remains mysterious. This lecture will describe a research program aiming for a deeper understanding of MCSP guided by a comparative analysis with SAT, emphasizing the following phenomena:

1. Explicit constructions: While it is easy to compute explicitly positive and negative instances of SAT of any given length, the corresponding question for MCSP is intimately tied to long sought-after circuit lower bounds.
2. NP-hardness: While the classical Cook-Levin theorem establishes the NP-hardness of SAT, the NP-hardness of MCSP remains open. There are some NP-hardness results for variants of MCSP where the circuit class is strongly restricted. On the other hand, there is a recent line of work ruling out NP-hardness under restricted reductions or deriving hard-to-prove complexity consequences from NP-hardness under standard reductions. In sum, there is no strong evidence for or against NP-completeness.
3. Search to decision reductions: The complexity of deciding SAT is polynomial-time equivalent to the complexity of finding a satisfying assignment for a satisfiable circuit, using the downward self-reducibility of SAT. MCSP does not appear to have a similar downward self-reducibility property. Recent work gives an approximate search-to-decision reduction for MCSP using an intriguing connection with learning theory.
4. Average-case complexity: SAT for k-CNFs appears hard empirically for high enough clause density under a natural distribution where clauses are picked uniformly and independently at random. This motivates Feige's hypothesis, which is known to imply strong hardness of approximation results for various natural NP problems. The complexity of MCSP under the uniform distribution on inputs has been studied under the guise of the "natural proofs" of Razborov and Rudich, which have strong connections to cryptography and proof complexity. Though SAT is not known to reduce to MCSP in the worst case, a recent result shows that the MKTP problem, a close cousin of MCSP, is hard on average under the uniform distribution if Feige's hypothesis holds.

Keywords: Minimum circuit size problem · Satisfiability · NP-hardness Learning · Natural proofs

References

1. Carmosino, M., Impagliazzo, R., Kabanets, V., Kolokolova, A.: Learning algorithms from natural proofs. In: Proceedings of 31st Conference on Computational Complexity (CCC), pp. 10:1–10:24 (2016)
2. Feige, U.: Relations between average case complexity and approximation complexity. In: Proceedings of 34th Annual ACM Symposium on Theory of Computing (STOC), pp. 534–543 (2002)
3. Hirahara, S., Oliveira, I. C., Santhanam, R.: NP-hardness of minimum circuit size problem for OR-AND-MOD circuits. Electronic colloquium on computational complexity (ECCC), pp. 25–30 (2018)
4. Hirahara, S., Santhanam, R.: On the average-case complexity of MCSP and its variants. In: Proceedings of 32nd Computational Complexity Conference (CCC), pp. 7:1–7:20 (2017)
5. Kabanets, V., Cai, J.-Y.: Circuit minimization problem. In: Proceedings of 32nd Annual ACM Symposium on Theory of Computing (STOC), pp. 73–79 (2000)
6. Razborov, A., Rudich, S.: Natural proofs. J. Compt. Syst. Sci. **55**(1), 24–35 (1997)
7. Trakhtenbrot, B.: A survey of Russian approaches to perebor algorithms. IEEE Ann. Hist. Comput. **6**(4), 384–400 (1984)
8. Williams, R.: Improving exhaustive search implies superpolynomial lower bounds. SIAM J. Comput. **42**(3), 1218–1244 (2013)

Recent Developments on the Asymmetric Traveling Salesman Problem

László A. Végh

London School of Economics and Political Science, London, WC2A 2AE, UK
l.vegh@lse.ac.uk
http://personal.lse.ac.uk/veghl

Abstract. The talk presents an overview of recent developments on the approximability of the Asymmetric Traveling Salesman Problem

Keywords: Traveling salesman problem · Approximation algorithms

In the traveling salesman problem (TSP), the input is given by n cities and their pairwise distances, and the goal is to find a shortest tour visiting every city and returning to the starting city. It is one of the best known optimization problems, with variants studied since the 19th century. The problem is NP-complete; if no assumptions are made on the distance function, it cannot be approximated within any constant factor. It is therefore common to assume that the distances satisfy the triangle inequality, or equivalently, that the traveler is allowed to visit some cities more than once.

In this talk, we focus on approximation algorithms for TSP. A classical algorithm by Christofides from 1976 [5] gives a $\frac{3}{2}$-approximation algorithm if the distance function is assumed to be *symmetric*. Forty years on, this simple algorithm is still the best known approximation algorithm for general symmetric costs. Improved guarantees have been given in recent work for the special setting of unweighted shortest path metrics, [9, 13–15].

In contrast to the tight guarantees for the symmetric case, our understanding of the more general asymmetric traveling salesman problem (ATSP) is far from complete. A classical LP relaxation was obtained by Held and Karp in the 1970s [10, 11]. This has been used as the lower bound in all approximation algorithms (both symmetric and asymmetric). Whereas the best lower bound on the integrality gap is 2 in the asymmetric case [4], even finding a constant factor approximation guarantee has remained open until very recently.

An elegant approximation algorithm for ATSP was given by Frieze, Galbiati and Maffioli [7], with approximation guarantee of $\log_2(n)$. This algorithm constructs a sequence of cycle covers, gradually growing the connected components of the solution. A series of papers [3, 6, 12] have improved the approximation factor by constant factors, but finding an $o(\log_n)$ approximation guarantee remained open for a longer time period.

This was first achieved in the breakthrough result Asadpour et al. [2], who obtained an $O(\log n / \log \log n)$-approximation factor for ATSP. They introduced a new and influential approach, making a connection between the approximability of ATSP and the existence of *thin trees*, a problem studied in the context of graph theory. In particular, they showed that finding a tree of constant "thinness" would imply a constant factor approximation for ATSP. Following this approach, Oveis Gharan and Saberi gave a constant factor approximation for graphs with bounded genus [8]. More recently, Anari and Oveis Gharan have obtained an upper bound $O(\text{poly} \log \log n)$ [1] on the integrality gap of the Held-Karp relaxation. This, however, does not provide an efficient approximation algorithm of the same guarantee, since their arguments are non-constructive.

An ATSP solution must satisfy three properties simultaneously: *connectivity, Eulerian degree constraints*, and *integrality*. For any two among these three requirements, a minimum cost solution can be found efficiently. Indeed, connected and Eulerian (but not necessarily integral) vectors are exactly the feasible solutions to the Held-Karp relaxation; connected and integral (but not necessary Eulerian) edge sets are the spanning trees; and Eulerian and integral (but not necessary connected) edge sets are the cycle covers. At a very high level, all ATSP approximation algorithms start by either relaxing the Eulerian constraints, or relaxing connectivity.

The thin tree approach start by relaxing the Eulerian constraint, starting with a special spanning tree. The thinness property can be then used to fix the violations of the Eulerian constraints. In contrast, the Frieze, Galbiati and Maffioli [7] algorithm start by relaxing connectivity: they construct a sequence of cycle covers whose union becomes connected.

Svensson introduced a new approach via relaxing connectivity, by giving a reduction from ATSP to a problem called *Local-Connectivity ATSP* [16]. An "α-light" solution to this problem implies an $O(\alpha)$ approximation to ATSP. The paper [16] used this approach for giving a $(27 + \varepsilon)$-approximation algorithm for the special case of node-weighted metrics, where Local-Connectivity ATSP is relatively easy to implement. We have generalized this to graphs with at most two different edge weights in subsequent work [17], however, this already required substantial technical effort.

In our recent paper [18], we have built upon and generalized both of these results to obtain the first constant-factor approximation algorithm for ATSP for arbitrary metrics. The constant factor in the paper is 5500; however, this is not optimized for the sake of simplicity in the presentation.

In contrast to the two edge weights result [17], we do not try to tackle Local-Connectivity ATSP directly in arbitrary weighted graphs. Instead, we introduce a series of natural reduction steps to reduce the problem of approximating ATSP in general to that of approximating ATSP on special, structured instances called *vertebrate pairs*. These instances enjoy properties that make them amenable for Local-Connectivity ATSP.

All these reductions build on classical techniques from mathematical optimization and from graph theory. We start by applying the uncrossing technique for the optimal solution to the dual of the Held-Karp LP to reduce general ATSP instances to *laminarly-weighted ATSP instances*. These enjoy a special weight structure defined by a laminar family of vertex subsets.

The subsequent steps define a natural contraction operation and, using a recursive algorithm, reduce the problem to *irreducible instances*. The very same property that makes these instances difficult for the reduction in turn enables us to construct a *backbone*, a special subtour that visits most vertices in the instance. In the final step, we take advantage of the backbone for implementing Local-Connectivity ATSP.

We believe that, by further optimizing these techniques, the integrality gap of the LP relaxation can be upper-bounded by the hundreds. In order to achieve an upper bound closer to the current lower bound 2, even to say 50 is likely to require some substantial new ideas.

References

1. Anari, N., Gharan, S.O.: Effective-resistance-reducing flows, spectrally thin trees, and asymmetric TSP. In: IEEE 56th Annual Symposium on Foundations of Computer Science (FOCS), pp. 20–39 (2015)
2. Asadpour, A., Goemans, M.X., Madry, A., Gharan, S.O., Saberi, A.: An O(log n/ log log n)-approximation algorithm for the asymmetric traveling salesman problem. In: Proceedings of the Twenty-First Annual ACM-SIAM Symposium on Discrete Algorithms, SODA 2010, pp. 379–389 (2010)
3. Bläser, M.: A new approximation algorithm for the asymmetric TSP with triangle inequality. ACM Trans. Algorithms 4(4) (2008)
4. Charikar, M., Goemans, M.X., Karloff, H.J.: On the integrality ratio for the asymmetric traveling salesman problem. Math. Oper. Res. 31(2), 245–252 (2006)
5. Christofides, N.: Worst-case analysis of a new heuristic for the travelling salesman problem. Technical report, Graduate School of Industrial Administration, CMU (1976)
6. Feige, U., Singh, M.: Improved approximation ratios for traveling salesperson tours and paths in directed graphs. In: Charikar, M., Jansen, K., Reingold, O., Rolim, J.D.P. (eds.) APPROX/RANDOM -2007. LNCS, vol. 4627, pp. 104–118. Springer, Heidelberg (2007). https://doi.org/10.1007/978-3-540-74208-1_8
7. Frieze, A.M., Galbiati, G., Maffioli, F.: On the worst-case performance of some algorithms for the asymmetric traveling salesman problem. Networks 12(1), 23–39 (1982)
8. Gharan, S.O., Saberi, A.: The asymmetric traveling salesman problem on graphs with bounded genus. In: Proceedings of the Twenty-Second Annual ACM-SIAM Symposium on Discrete Algorithms, SODA 2011, pp. 967–975 (2011)
9. Gharan, S.O., Saberi, A., Singh, M.: A randomized rounding approach to the traveling salesman problem. In: IEEE 52nd Annual Symposium on Foundations of Computer Science, FOCS 2011, pp. 550–559 (2011)
10 Held, M., Karp, R.M.: The traveling-salesman problem and minimum spanning trees. Oper. Res. 18(6), 1138–1162 (1970)
11. Held, M., Karp, R.M.: The traveling-salesman problem and minimum spanning trees: Part II. Math. Program. 1(1), 6–25 (1971)
12. Kaplan, H., Lewenstein, M., Shafrir, N., Sviridenko, M.: Approximation algorithms for asymmetric TSP by decomposing directed regular multigraphs. J. ACM 52(4), 602–626 (2005)
13. Mömke, T., Svensson, O.: Removing and adding edges for the traveling salesman problem. J. ACM 63(1), 2:1–2:28 (2016)

14. Mucha, M.: 13/9-approximation for graphic TSP. In: 29th International Symposium on Theoretical Aspects of Computer Science, STACS 2012, pp. 30–41 (2012)
15. Sebő, A., Vygen, J.: Shorter tours by nicer ears: 7/5-approximation for the graph-TSP, 3/2 for the path version, and 4/3 for two-edge-connected subgraphs. Combinatorica **34**(5), 597–629 (2014)
16. Svensson, O.: Approximating ATSP by relaxing connectivity. In: FOCS 2015: Proceedings of the 56th Annual IEEE Symposium on Foundations of Computer Science (2015). http://arxiv.org/abs/1502.02051
17. Svensson, O., Tarnawski, J., Végh, L.A.: Constant factor approximation for ATSP with two edge weights. In: Louveaux, Q., Skutella, M. (eds.) Integer Programming and Combinatorial Optimization. IPCO 2016. LNCS, vol. 9682, pp. 226–237. Springer, Cham (2016)
18. Svensson, O., Tarnawski, J., Végh, L.A.: A constant-factor approximation algorithm for the asymmetric traveling salesman problem. In: STOC (2018, forthcoming)

Contents

Complexity of Generation

Vladimir Gurvich[1,2]($\boxtimes$)

[1] RUTCOR and RBS, Rutgers University, 100 Rockafeller Road,
Piscataway, NJ 08854, USA
vladimir.gurvich@gmail.com

[2] National Research Institute, Higher School of Economics (HSE),
Moscow, Russia

Abstract. In this talk I summarize the results obtained in 1999–2008 by Leonid Khachiyan, Endre Boros, Konrad Borys, Khaled Elbassiony, Kazuhisa Makino, and myself, on complexity of generation algorithms. These algorithms can be partitioned into three groups: supergraph, flash-light (backtrack), and dual-bounded generation. We will call a problem *tractable* if it can be solved by a polynomial (n^{const}) or quasi-polynomial ($n^{polylog(n)}$) time algorithm. More generally, for any positive non-decreasing function $g = g(n)$, generating can be performed in total or incremental time g, or with g-delay. Most of the polynomial delay algorithms are provided by the flash-light (backtrack) method. As for the incremental algorithms, generating the next object is equivalent with just verifying its existence, which is a standard decision problem. Thus, incremental generation, in contrast to the delay one, may be NP-hard or NP-complete. For example, we show that generating all vertices of a polyhedron, given by its facets, is NP-complete (while the complexity status is still open in case of the polytopes, that is, bounded polyhedra). This problem is reduced to generating all negative cycles of a weighted digraph, which is NP-complete (for graphs, too). Generating all minimal transversals to a hypergraph, so-called dualization, plays an important role. For this problem an incremental quasi-polynomial algorithm (but no polynomial one) is known. We outline several wide classes of generation problems that can be reduced to dualization and, thus, solved in incremental quasi-polynomial time. We survey algorithms and complexity bounds for the above and many other generation problems.

Keywords: Generation algorithm · Polynomial · Quasi-polynomial
Total · Incremental · Dualization · Dual or transversal hypergraph
Supergraph · Backtrack · Flash-light · Dual-bounded
Independent set · Matroid · Submodular function
Polymatroid function

This work was supported in part by the Russian Academic Excellence Project '5-100'.

F. V. Fomin and V. V. Podolskii (Eds.): CSR 2018, LNCS 10846, pp. 1–14, 2018.
https://doi.org/10.1007/978-3-319-90530-3_1

1   How They Measure Complexity of Generation

Generating all vertices of the n-cube $\{-1 \leq x_i \leq 1 \mid i = 1, \ldots, n\}$, obviously, requires exponential in n time, just because there are 2^n vertices. Yet, it would be naive to treat this problem as a hard one.

To introduce the "right scale of complexity", it is conventional to measure the generation time in both initial input and output size, $T = T(|II|, |O|)$. We say that a generation problem is solved in *total* or *output* polynomial (P) or quasi-polynomial (QP) time, whenever T is a P or QP function of its two variables.

However, when the output is large, it may be a bad good idea to wait (for a very long time, maybe) and to obtain all required objects only at the very end. We would prefer to output the objects one by one in some order, say, $O = \{o_1, \ldots, o_k, \ldots, o_N\}$. Denote by $O_k = \{o_1, \ldots, o_k\}$ the first $k - 2$ objects and by $T(o_k)$ the time between outputting o_{k-1} and o_k. We assume that $O_0 = \emptyset$ and $T(O_0) = 0$. A generation algorithm is called

- (I) *incremental* P or QP, if $T(o_k)$ is P or QP in k and in $|II|$;
- (D) *P or QP delay*, if for each k, time $T(o_k)$ is P or QP in $|II|$.

Such an approach was suggested in [45]; see also [3,9,15,26–29,37,42,44,47,48], cited in the chronological order. We have to add several remarks:

- Obviously, (I) is weaker than (D), just because for (I) the generation time is allowed to increase with k.
- One can equivalently reformulate (D) requiring that $T(O_k)$ is at most k times P or QP in $|II|$, which assumes a trick. It is not necessary to output all generated objects immediately, instead one can delay their output, which may help with keeping the required inequality.
- In case of (I), generating the next object is equivalent [37,45] with just verifying its existence, which is a standard decision problem. If it is NP-complete or NP-hard, we say this about the generation problem itself.
- In contrast, in case of (D) we are not aware of any indicator showing that it is NP-hard to solve a problem with P or QP delay.
- Of course, QP is growing faster than P, but it is much closer to P than to an exponent: $n^c = 2^{c \log n} < n^{\log n} = 2^{\log^2 n} \ll 2^n$ or, more generally, $n^{c_1} < n^{poly \log n} \ll c_2^n$, where c_1, c_2, and c are positive real constants, $c_2 > 1$; where > (resp., $\gg$) means asymptotically larger (resp., "much larger").
- More generally, for any positive non-decreasing function $g = g(n)$, generation may be performed in total or incremental time g, or with g-delay.
- For some problems it may be NP-hard already to verify the existence of required objects. Presume, for example, that we have to output all Hamiltonian cycles of a graph. Then, it is hard even to start generation.
- In contrast, if we have already output exponentially many, in $|II|$, objects of O, then "we are safe", because allowed to spend even exponential time generating the next object.

- Thus, solving a typical hard generation problem, we successfully output poly-nomially many, in $|II|$, objects and then "suddenly get stuck"; see the next section for examples.
- At first glance, it seems that hardness of a generation problem may depend on the order in which we output the objects. But in fact it cannot. Indeed, suppose that generating o_k is NP-hard. Then, it is hard to decide if the list O_{k-1} is already complete, or it can be extended. Assume for contradiction that there exists another algorithm (outputting the same objects in some other order) which is efficient, say, QP incremental. But then, we could solve the above hard problem efficiently as follows. Let us try to output k objects by the new algorithm. Either it will output exactly O_{k-1} and stop, thus, showing that there exists no o_k; or on some step, which number is at most k, it will output an object $o \notin O_{k-1}$, thus, showing that O_{k-1} is not a complete list and its extending is not a hard problem.
- The same arguments prove that no total QP algorithm can exist either. Assume the opposite. Then, we just let such an algorithm run for a time $T > QP(k)$ to obtain a solution of the above NP-hard problem. As before, we either output O_{k-1} and stop, or get an $o \notin O_{k-1}$.
- Standardly, we always assume that $P \neq NP$ and also that $QP \neq NP$.

2   Hard Generation Problems

2.1   Examples

For a multi-hypergraph H on the ground-set V and positive integer k generate all k-*unions*, that is, all minimal subsets of V that contain at least k hyperedges of H, counted with their multiplicity. Standardly, the terms *minimal* and *maximal* applied to a subset mean *inclusion-minimal* and *inclusion-maximal*.

Consider the following decision problem: given a list of k-unions, whether there are more of them, or the list is complete? Obviously, the problem is in NP; we will show that it is NP-complete reducing to it the following classical NP-complete problem: given a simple graph $G = (V, E)$ whether it contains an independent (edgeless) set $V' \subseteq V$ of size k. To each G we assign an H as follows: treat every edge (resp., vertex) of G as a hyperedge of H of multiplicity $k - 2$ (resp., 1). Then, $\{v', v''\} \subseteq V$ is a k-union of H whenever it is an edge $e = (v', v'') \in E$. Thus, it is easy to generate $m = |E|$ (linearly many in $|II|$) k-unions. Yet, one more exists if and only if G has an independent set of size k, and this decision problem is NP-complete. The problem remains NP-complete if we replace the multi-hypergraphs by standard ones; see [39, 40].

NP-completeness in the above example is based on the presence of an integer parameter k. An example of a "purely combinatorial" NP-complete generation problem, which was found first, is generating all minimal directed cuts in a (strongly connected) digraph; see [35] and Sect. 5.3 below.

Another important example is "Generating all vertices of a polyhedron given by its facets"; see more details in [15, 31], and in the rest of this sections.

2.2 Circulation Cones and Polyhedra of Digrahs

Given a digraph $G = (V, E)$ in which loops are not allowed, standardly we define a function $sign : V \times E \to \{0, \pm 1\}$: for each $v \in V$ and $e \in E$ we set $sign(v, e) = 1$ if e is going from v, $sign(v, e) = -1$ if e is going into v, and $sign(v, e) = 0$ if v and e are not incident.

Also we define two more mappings $x : E \to \mathbb{R}_+$ and $w : E \to \mathbb{R}_+$ that will be interpreted as a flow and weight functions. Then, the *circulation cone* is defined by the system of equalities

$$\sum_{e \in E} x(e) sign(v, e) = 0 \ \forall v \in V.$$

(Recall that $x_e \geq 0$.) A *circulation polyhedron* we obtain by adding to the above linear constraints one more hyperplane

$$\sum_{e \in E} w(e) x(e) = -1.$$

It is not difficult to see [15,35] that its vertices are in one-to-one correspondence with the directed cycles of the negative total weight, so-called *negative cycles*.

Theorem 1 ([15,35]). *Generating all negative (directed) cycles of a (directed) graph is NP-complete.*

Corollary 1 ([15,35]). *Generating all vertices of a polyhedron is NP-complete.*

The proof is immediate from the above arguments and Theorem 1. The proof of the latter will be briefly sketched in Sects. 2.3–2.5.

In [15,35] the extreme rays are also characterized, and it appears that they exist in all non-trivial circulation polyhedra, implying that they are unbounded. Corollary 1 does not extend this case and, hence, for the bounded polyhedra (that is, for polytopes) vertex-generation may be tractable.

2.3 Verifying Boolean Equalities and Inequalities

Let C and D be a monotone CNF, resp., DNF of variables $x_1, y_1, \ldots, x_k, y_k$ and

$$C_0 = (x_1 \vee y_1) \wedge \ldots \wedge (x_k \vee y_k), \ D_0 = x_1 y_1 \vee \ldots \vee x_k y_k.$$

Theorem 2 ([15,25]). *The following 12 claims are polynomially equivalent:*

C *is not satisfiable*, $C \Rightarrow D_0$, $D_0 \geq C$, $D_0 \vee C = D_0$, $D_0 \wedge C = C$, $C\overline{D}_0 = 0$;

D *is not a tautology*, $D \Leftarrow C_0$, $C_0 \leq D$, $C_0 \wedge D = C_0$, $C_0 \vee D = D$, $\overline{C} \vee D_0 = 1$.

Hence, all 12 are co-NP-complete. In contrast, verifying $\overline{C}D_0 = 0$, $C \vee \overline{D}_0 = 1$, and $D \leq C_0$ are polynomial (in fact, very simple) problems, while verifying the equality $C = D$ is exactly dualization, see Sect. 4.2.

2.4   Sausage Lemma

The properties of Theorem 2 can be reformulated in many different ways. First time, some of them were stated in terms of graphs and digraphs in [25]; see also [30]. The following weak version will already imply Theorem 1.

Let $G_S = (V_S, E_S)$ and $G_T = (V_T, E_T)$ be a pair of (non-directed) graphs that have one vertex in common, $V_S \cap V_T = \{v\}$, and the same number of edges. Fix two poles $s \in V_S \setminus \{v\}$ and $t \in V_T \setminus \{v\}$ and a one-to-one correspondence $\phi : E_S \to E_T$.

Lemma 1 ([25]). *It is co-NP-complete to decide if there exists a path between s and t that contains no pair of edges identified by ϕ. The problem remains co-NP-complete when (G_S, s, v) and (G_T, v, t) are series-parallel two-pole networks of depth 3. If depth is bounded by 2, the problem is polynomially equivalent with dualization; see Sect. 4.2.*

2.5   A Sketch of the Proof of Theorem 1 for Digraphs

It is well-known that in a non-directed series-parallel two-pole network (G, s, t), all s-t pathes pass every edge in the same direction. Let us direct each edges accordingly, from s to t, subdivide it by three inner vertices, and assign weights $(+1, -1, -1, +1)$ to the obtained four edges.

For any two edges $e' \in E_S$ and $e'' \in E_T$ such that $\phi(e') = e''$ let us merge the first inner vertex (resp., the third one) of e' and the third inner vertex (resp., the first one) of e'', thus, getting a negative directed cycle $(-1, -1, -1, -1)$; see Fig. 1. Since we are doing such merging for $m = |E_S| = |E_T|$ edges, we obtain a "pretty messy" graph with m negative cycles.

Consider G_S and G_T, with all merging of their vertices considered above; add one more edge of weight (-1) directed from t to s, and denote the obtained digraph by G. Let us try to generate all negative directed cycles in G. We already have m such cycles, yet, it is co-NP-complete to decide if there are more. Indeed, it is not difficult to show [15,35] that the answer is positive if and only if there exists an s-t directed path in G that contains no pair of edges identified by ϕ, and this is hard to decide, by Lemma 1.

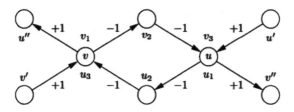

Fig. 1. Edges e' and e'' such that $\phi(e') = e''$ are subdivided by vertices u_1, u_2, u_3 and v_1, v_2, v_3, respectively; furthermore, $v_1 = u_3$ and $v_3 = u_1$.

3   Flash-Light and Supergraph

Flash-light. Consider, for example, generating all s-t paths in a graph $G = (V, E)$. We order edges of E and then start to generate all paths from s in the lexicographical order. It may happen that such a path cannot already be extended to an s-t path and we will spend in vane exponential in $n = |V|$ time trying to find one. Yet, this complication can be easily avoided, since the corresponding decision problem "whether a given path from s can be extended to an s-t path" is polynomial. Solving it after each extension of the current path (which is somewhat similar to using a flash-light in a dark place) we generate all s-t paths with polynomial delay. In [45] this method was applied to generating (directed) pathes, cycles, and spanning trees in (directed) graphs. More general problems were considered in [32, 33, 36].

Most polynomial delay generation algorithms are obtained by the flash-light method. However, many generation problems can be solved efficiently, while the corresponding flash-light decision problems are NP-hard. For example, generating minimal cuts is incremental P, while recognition of an edge-subset of such cut is NP-complete [32, 33, 36]; generating minimal transversals of a hypergraph (dualization, see Sect. 4.2) is incremental QP, while recognition of a subtransversal is NP-complete already for graphs [7]. It is NP-complete to decide whether a set of vertices contains a maximal clique.

Supergraph. This method assigns a vertex $o \in O$ to each object that we have to generate and an edge $e = (o', o'')$ to each pair of objects that are neighbors in a sense; for example, O consists of all bases of a matroid and E is determined by the exchange property. Then, we generate all objects traversing the obtained "supergraph". We have either to prove its connectivity or to traverse every connected component; see, for example, [5, 32, 33].

4   Generating Dual-Bounded Hypergraphs

4.1   Monotone Generation

Given a ground-set V and a property of its subsets defined by a set-function $f : 2^V \to \{0, 1\}$. The property are called *monotone* if f is monotone non-decreasing. Then the (monotone generation) problem is to output all minimal subsets of $V' \subseteq V$ such that $f(V') = 1$. In this section we restrict ourselves by such problems.

4.2   Dualization

Given a hypergraph $H = (V, E)$ its dual $H^d = (V', E')$ is defined on the same ground-set V as the family of all minimal transversals to E, that is, $e \cap e' \neq \emptyset$ for all $e \in E, e' \in E'$, and for every e'' such that $e \cap e'' \neq \emptyset$ for all $e \in E$ there exists a $e' \in E'$ such that $e' \subseteq e''$. By definition, $V' \subseteq V$ and H^d is a Sperner hypergraph, that is, $e'_1 \subseteq e'_2$ for no $e'_1, e'_2 \in E'$. If H is Sperner too then $V = V'$ and $H^{dd} = H$.

Given H and a family of its transversals H', verify the equality $H^d = H'$ assuming that both H and H' are Sperner. This is the famous *dualization* problem. Solving it successively several times one can, given H, generate H^d.

Note that the complement in V to a (minimal) transversal of H is a (maximal) independent set (MIS) of H. Indeed, by definition, it contains no edge of H. Thus, dualization and generating all MIS(H) are trivially equivalent.

Given a set and a family of its subsets, generate all minimal set-covers. This is another equivalent reformulation.

No P time algorithm for dualization is known, yet, a QP one was constructed in [20]. Later, this result was generalized in several directions. In particular, the Boolean cube was replaced by the product of discrete intervals [13,14], of posets, and of forests [21–24].

For the hypergraphs of bounded dimension, $dim(H) = \max_{e \in E} |e|$ (resp., degree, $deg(H) = \max_{v \in V}(|E'| \mid v \in e' \,\forall\, e' \in E' \subseteq E)$) dualization can be solved in polynomial (of degree $dim(H) + 1$ (resp., $deg(H) + 1$)) time. Moreover, in case of bounded dimension, dualization can be efficiently solved in parallel: the problem is in NC for $dim(H) \leq 3$ and in RNC for $dim(H) = 4, 5, \ldots$; see [8].

Dualization is also tractable for the hypergraphs of bounded edge-intersection [11]. Let $A(k, r)$ denote the class of hypergraphs in which the intersection of any k distinct edges is of size at most r. It is shown in [11] that for the hypergraphs of bounded dimension, $A(1, c)$, as well as for the hypergraphs of bounded degree, $A(c, 0)$, dualization can be solved in incremental polynomial time and in space polynomial only in the size of H. This result is extended for $A(k, r)$ with bounded $k + r$. For this class dualization is NC-reducible to generating a single minimal transversal for a partial subhypergraph of $H \in A(k, r)$. Somewhat surprisingly, the latter problem seems difficult to solve in parallel and the above observation results only in an efficient parallel algorithms for the incremental (sic!) dualization in classes $A(1, c)$, $A(c, 0)$, and $A(2, 1)$.

Thus, we can always generate H^d by a QP incremental algorithm. In many cases QP can be relaced by P; in particular, in all above cases, see also [19].

Given a hypergraph $H = (V, E)$, denote by $H^c = (V, E')$ the edge-complement hypergraph, $E' = \{V \setminus e \mid e \in E\}$. Clearly $H^{cc} = H$ for any H. Let B be the bases of a matroid M. Then, by definition, B^c consists of the bases of the dual matroid M^d. Furthermore, B^d are the transversals to bases; B^{cd} are the minimal sets not contained in a base of M, or in other words, the circuits C of M. Hence, $C^{dc} = B^{cddc} = B$.

All hypergraphs listed above can be generated in incremental P time [10,46], due to the exchange axioms. Note that very long sequences $cdcdcd, \ldots$ and $dcdcdc, \ldots$ may consist of pairwise distinct hypergraphs.

If we have P or QP reduction of a generation problem S to dualization, we say that S is *tractable*, since dualization is QP and, hence, it is very unlikely that S is NP-hard. On the other hand, when dualization is reduced to S, we say that S is *difficult*, since no P time algorithm for dualization is known. This is an example of the policy of double standards.

4.3   Joint Generation [6, 9, 25]

Let a hypergraph H be given by a P or QP oracle and we want to output both H and H^d, assuming that they are both Sperner. This can be done by solving dualization successively. Indeed, given partial hypergraphs $H' \subseteq H$ and $H'' \subseteq H^d$, we can verify their duality in QP time. If they are dual, we output $H = H', H^d = H''$ and stop; if they are not, dualization and the oracle allow us to extend either H' or H'' still keeping both containments $H' \subseteq H$ and $H'' \subseteq H^d$ [6, 25]. However, we do not control, which one will be extended.

Thus, joined generation of H and H^d simultaneously can be realized by an incremental QP time algorithm; in contrast, generating H, or H^d, or even both, but separately, can be NP-hard; [25]. This looks like a paradox but can be simply explained. Suppose, we are interested in H, while H^d is just garbage. Yet, joint generation may output edges of H separated by long sequences of edges of H^d which lengths are growing exponentially. Still, joint generation is effective under certain extra assumptions.

4.4   Generating (Uniformly) Dual-Bounded Hypergraphs

Let $|H|$ denote the size (that is, the number of edges) of H. As we saw, joint generation of H and H^d may be not efficient for H when $|H^d|$ is exponential in $|H|$. A class $\mathcal{H}$ of hypergraphs is called P or QP dual-bounded (DB) if in this class $|H^d|$ is bounded by a P or QP in $|H|$ and in the size $|II|$ of the initial input. Obviously, generating P or QP DB hypergraphs can be performed in the *total* P or QP time, respectively. Moreover, we can even obtain an *incremental* P or QP algorithm, rather than a total one, requiring a stronger inequality.

A class of hypergraphs $\mathcal{H}$ is called P or QP *uniformly DB* if $|H'^d|$ is bounded by a P or QP in $|H'|$ for all partial hypergraphs $H' \subseteq H \in \mathcal{H}$ simultaneously.

All hypergraphs from a P or QP uniformly DB class can be generated in incremental P or QP time.

By definition, class $\mathcal{H}$ is uniformly DB whenever it is DB and hereditary (that is, $H' \in \mathcal{H}$ for any $H' \subseteq H \in \mathcal{H}$). In addition, many DB but non-hereditary classes are uniformly DB, as well. Moreover, all "natural" DB classes that we know are uniformly DB, although "artificial" exceptions exist; see more details in [9, 14, 16, 18]. This is somewhat surprising, because uniformity looks like an essential addition to DB.

5   More Examples of Generation Problems

5.1   Monotone Integer Programming [12–14, 36]

Consider the following system of linear inequalities

(*) $Ax \geq b, \ \bar{0} \leq x \leq c$, where $A \in \mathbb{R}^{m \times n}$, $b \in \mathbb{R}^m_+$, $c \in \overline{\mathbb{R}}^n_+$ (that is, coordinates of c can take value $+\infty$), $x \in \mathbb{Z}^n_+$, and $\bar{0} \in \mathbb{R}^n_+$ is the origin.

System (*) is called *monotone* if $Ax \geq b \Rightarrow Ax' \geq b \ \forall x, x' \mid \bar{0} \leq x \leq x' \leq c$.

Non-negativity of the matrix, $A \in \mathbb{R}_+^{m \times n}$, is sufficient for monotonicity, but not necessary. Note that without monotonicity, already verification of solvability of (*) is NP-complete, while for monotone systems it is trivial. In this case we will consider monotone generation.

Problem (K): generate all minimal integer solutions $Z(A, b, c)$ to (*).

If A is a $(0, 1)$ matrix, x is a $(0, 1)$ n-vector, and $b = e$ is the vector of m ones, then (K) is the dualization of the hypergraph H defined by the rows of A; then, a (minimal) solution x to (*) is a (minimal) transversal to H.

We generalize, replacing a $(0, 1)$ vector x by a non-negative integer one, $b = e$ by an arbitrary $b \in \mathbb{R}_+^m$, and finally, $H(A, b, c)$ by $Z(A, b, c)$, thus, replacing the Boolean cube by the direct product of discrete intervals (a box). These replacements keeps all basic properties [9].

Now, let us Consider the dual problem.

Problem (L): generate $MIS(Z(A, b, c)) = Z^{dc}(A, b, c)$, in other words, all maximal $x \in \mathbb{Z}_+^n$ not feasible to the monotone system (*), that is, $x \notin Z(A, b, c)$.

Problem (K) is uniformly DB, satisfying the inequality inequality

$$|Z^d(A, b, c)| \leq mn|Z(A, b, c)|,$$

which is "almost" precise. Namely, there exists an example with

$$|Z^d(A, b, c)| \geq \frac{mn}{2log^2m}|Z(A, b, c)|.$$

An incremental QP generation algorithm for $Z(A, b, c)$ is provided by the method of joint generation, while generating $Z^d(A, b, c)$ is NP-complete [12].

5.2 Maximal Frequent and Minimal Infrequent Sets in Data Mining [1, 17, 24]

Given a binary $m \times n$ matrix A and positive integer t, a set C of columns is called t-frequent if there exist at least t rows such that in the obtained $t \times |C|$ submatrix all entries are 1s; otherwise C is called t-infrequent. Let α and β denote the numbers of the maximal t-frequent and minimal t-infrequent sets, respectively. The DB inequality $\alpha \leq (m - t + 1)\beta$, implies that generating all minimal t-infrequent sets is incremental QP. In contrast, generating all maximal t-frequent sets is NP-complete. See [17] for more details.

5.3 Minimal Strongly Connected Subgraphs and Dicuts [35]

Given a (strongly connected) digraph $G = (V, E)$, generate all

- (K) minimal edge-sets $E' \subseteq E$ such that the digraph $G' = (V, E')$ is strongly connected;
- (L) minimal dicuts, that is, minimal edge-sets $E'' \subseteq E$ such that the digraph $G' = (V, E \setminus E'')$ is not strongly connected.

These dual problems are not DB; the size of each set may be exponential in the size of the other. Problem (K) is solved in incremental P time by the supergraph method, while (L) is NP-complete [35]. The proof of the latter claim is based on the "sausage lemma" (Sect. 2.4 and [25]) but requires extra tricks [35].

5.4   Generating Generalized Paths, Cuts, and Spanning Sets in Graphs [32, 33, 36, 45]

Given a (non-directed connected) graph $G = (V, E)$ and k edge-subsets $E_1, \ldots,$ $E_k \subseteq E$, for every $I \subseteq [k]$ define the graph $G_I = (V, \cup_{i \in I} E_i)$ and consider the following two pairs of problems. Generate all

– (K1) generalized spanning sets, that is, all minimal $I \subseteq [k] = \{1, \ldots, k\}$ such that graph G_I is connected;
– (L1) generalized complementary cuts, that is, all maximal $I \subseteq [k]$ such that graph G_I is not connected.

Given also two poles $s, t \in V$, generate all

– (K2) generalized s-t-paths, that is, all minimal $I \subseteq [k]$ such that s and t belong to one connected component of G_I.
– (L2) generalized complementary s-t-cuts, that is, all maximal $I \subseteq [k]$ such that s and t are in distinct connected component of G_I.

Problem (K1) is QP uniformly DB and, hence, can be solved in incremental QP time. Three other problems are NP-complete; all three proofs are based on the sausage lemma.

5.5   Spanning Linear Spaces by Linear Subspaces [10, 34]

Let $\mathcal{L} = \{L_1, \ldots, L_k\}$ be a set of linear subspaces in a space $L = F^d$ of dimension d over a field F and $t \leq d$ be a positive integer threshold, for every $I \subseteq [k]$ set $L_I = \cup_{i \in I} L_i$, or in other words, L_I is spanned by $L_i, i \in I$;

Consider the following pair of dual problems. Generate all

– (K) minimal subsets $I \subseteq [k]$ such that $dim(L_I) \geq t$;
– (L) maximal subsets $I \subseteq [k]$ such that $dim(L_I) < t$.

If $t = d$ then $L_I = L$, that is, $L_i, i \in I$, span the whole space. Problem (K) is QP uniformly DB and, hence, is incremental QP, while (L) is NP-complete.

Let us fix a basis B in L and assume that subspaces $L_i, i \in I$, are defined by subsets of B. In this case problem (K) is reduced to generating all minimal covers of B by these subsets. This problem is equivalent to dualization.

5.6   Polymatroid Functions and Systems of Polymatroid Inequalities [10, 34, 38, 41]

An integer non-negative set-function $f : 2^V \to \mathbb{Z}_+$ is called *polymatroid* if it is submodular, $f(I' \cup I'') + f(I' \cap I'') \leq f(I') + f(I'')$ for any $I', I'' \in V$, monotone non-decreasing, and $f(\emptyset) = 0$.

For example, $f(I) = dim(L_I)$ from the previous section is a polymatroid function. A special case of this example is the number of vertices minus the number of connected components of a graph. In other words, given a graph $G = (V, E)$ and a family I of its edge-sets $\{E_i \subseteq E \mid i \in I\}$, we set $f(I) = |V| - C(G(V, E_I))$, where $E_I = \cup_{i \in I} E_i$ and $C(G)$ is the number of connected components of G. Another example is the number of degrees of freedom of a mechanical system. See [38] for more examples and details.

Given a system of polymatroid inequalities

(*) $f_j(I) \geq t_j$, $j \in [r] = \{1, \ldots, r\}$, where functions f_j are polymatroid and thresholds t_j are QP in $n = |V|$, generate all

(K)   minimal $I \subseteq V$ satisfying (*); (L) maximal $I \subseteq V$ not satisfying (*).

Problem (K) is QP uniformly DB, and, hence, it is incremental QP, while (L) is NP-complete.

5.7   Uniformly DB Inequalities for Polymatroid Systems [10, 34]

Let $H(f, t)$ be the family of all minimal subsets of V satisfying (*).

Standardly, $MIS(H(f, t))$ denote the family of all maximal subsets of V not satisfying (*). Let $\beta = |H(f, t)|$ and $\alpha = |MIS(H(f, t))|$ denote the numbers of sets in these families. Then,

$$\alpha \leq \beta^{\log t / c(n, \beta)} \text{ for } \beta \geq 2 \text{ and } \alpha \leq n \text{ when } \beta = 1,$$

where $c = c(n, \beta)$ is the (unique) root of the equation

$$2^c (n^{c/\log \beta} - 1) = 1.$$

This is a DB inequality for (K). It is QP whenever t is QP in $n = |V|$. This DB inequality can be replaced by a much simpler, although slightly weaker, one:

$$\alpha \leq (n\beta)^{\log t}.$$

This bound can be viewed as a generalization, from graphs to hypergraphs, of an upper bound for the number of MIS. For graphs we have

$$2^p \leq |MIS(G)| \leq \delta^p + 1,$$

where $p = p(G)$ is the maximum size of the induced matchings in G and $\delta = \delta(G)$ is the maximum number of pairs of vertices at distance 2; in particular, $\delta \leq \binom{n-1}{2}$, where the equality is achieved only on stars. This bound was proven in [4]; slightly weaker results were obtained in [2, 43].

For the systems of r polymatroid inequalities, we generalize as follows:

$$\alpha \leq r \max(n, \beta^{\log t / c(n, \beta)}), \text{ where } t = \max(t_1, \ldots, t_r).$$

Interestingly, the coefficient $1/c = 1/c(n, \beta)$ in the exponent is accurate and cannot be reduced.

References

1. Agrawal, R., Mannila, H., Srikant, R., Toivonen, H., Verkamo, A.I.: Fast discovery of association rules. In: Advances in Knowledge Discovery and Data Mining, pp. 307–328 (1996)
2. Alekseev, V.E.: On the number of dead-end independent sets in graphs from hereditary classes. Comb.-Algebraic Methods Discret. Optim. **6**, 5–8 (1991). (Nizhni Novgorod, in Russian)
3. Bertolazzi, P., Sassano, A.: An $O(mn)$ time algorithm for regular set-covering problems. Theor. Comput. Sci. **54**, 237–247 (1987)
4. Balas, E., Yu, C.S.: On graphs with polynomially solvable maximal-weight clique problem. Networks **19**, 247–253 (1989)
5. Bixby, R., Cunningham, W.: Matroid optimization and algorithms. In: Handbook of Combinatorics, pp. 550–609 (1995)
6. Bioch, J.C., Ibaraki, T.: Complexity of identification and dualization of positive Boolean functions. Inf. Comput. **123**, 50–63 (1995)
7. Boros, E., Gurvich, V., Hammer, P.L.: Dual subimplicants of positive Boolean functions. Optim. Methods Softw. **10**, 147–156 (1998). RUTCOR Research report 11-1993, Rutgers University
8. Boros, E., Elbassioni, K., Gurvich, V., Khachiyan, L.: An efficient incremental algorithm for generating all maximal independent sets for hypergraphs of bounded dimension. Parallel Proc. Lett. **10**(4), 253–266 (2000)
9. Boros, E., Elbassioni, K., Gurvich, V., Khachiyan, L.: Generating dual-bounded hypergraphs. Optim. Methods Softw. **17**(5), 749–781 (2002)
10. Boros, E., Elbassioni, K., Gurvich, V., Khachiyan, L.: An inequality for polymatroid functions and its applications. Discret. Appl. Math. **131**(2), 255–281 (2003). Special Issue on Submodularity
11. Boros, E., Elbassioni, K., Gurvich, V., Khachiyan, L.: On dualization of the hypergraphs with bounded edge-intersections and other related classes of hypergraphs. Theor. Comput. Sci. **382**, 139–150 (2008). RUTCOR Research report RRR-37-2003, Rutgers University
12. Khachiyan, L., Boros, E., Elbassioni, K., Gurvich, V.: Generating all minimal integral solutions to monotone $\land, \lor$-systems of linear, transversal and polymatroid inequalities. In: Jędrzejowicz, J., Szepietowski, A. (eds.) MFCS 2005. LNCS, vol. 3618, pp. 556–567. Springer, Heidelberg (2005). https://doi.org/10.1007/11549345_48
13. Boros, E., Elbassioni, K., Gurvich, V., Khachiyan, L., Makino, K.: On generating all minimal integer solutions for a monotone system of linear inequalities. In: Orejas, F., Spirakis, P.G., van Leeuwen, J. (eds.) ICALP 2001. LNCS, vol. 2076, pp. 92–103. Springer, Heidelberg (2001). https://doi.org/10.1007/3-540-48224-5_8
14. Boros, E., Gurvich, V., Elbassioni, K., Khachiyan, L., Makino, K.: Dual-bounded generating problems: all minimal integer solutions for a monotone system of linear inequalities. SIAM J. Comput. **31**(5), 1624–1643 (2002)
15. Boros, E., Elbassioni, K., Gurvich, V., Makino, K.: Generating vertices of polyhedra and related monotone generation problems. CRM Proc. Lect. Notes Polyhedral Comput. **48**, 15–44 (2009). Special volume dedicated to Leonid Khachiyan and Victor Klee
16. Boros, E., Gurvich, V., Khachiyan, L., Makino, K.: Dual-bounded generation: partial and multiple transversals of a hypergraph. SIAM J. Comput. **30**(6), 2036–2050 (2001)

17. Boros, E., Gurvich, V., Khachiyan, L., Makino, K.: On maximal frequent and minimal infrequent sets in binary matrices. Ann. Math. Artif. Intell. **39**, 211–221 (2003)

18. Boros, E., Gurvich, V., Khachiyan, L., Makino, K.: Dual-bounded generating problems: weighted transversals of a hypergraph. Discret. Appl. Math. **142**(1–3), 1–15 (2004)

19. Crama, Y.: Dualization of regular Boolean functions. Discret. Appl. Math. **16**, 79–85 (1987)

20. Fredman, M.L., Khachiyan, L.: On the complexity of dualization of monotone disjunctive normal forms. J. Algorithms **21**, 618–628 (1996)

21. Elbassioni, K.M.: An algorithm for dualization in products of lattices and its applications. In: Möhring, R., Raman, R. (eds.) ESA 2002. LNCS, vol. 2461, pp. 424–435. Springer, Heidelberg (2002). https://doi.org/10.1007/3-540-45749-6_39

22. Elbassioni, K.M.: On dualization in products of forests. In: Alt, H., Ferreira, A. (eds.) STACS 2002. LNCS, vol. 2285, pp. 142–153. Springer, Heidelberg (2002). https://doi.org/10.1007/3-540-45841-7_11

23. Elbassioni, K.M.: Incremental algorithms for enumerating extremal solutions of monotone systems of submodular inequalities and their applications. Ph.D. thesis, Chap. 4, Dualization in Products of Chains, Semi-lattices, and forests, Rutgers (2002)

24. Elbassioni, K.M.: Finding all minimal infrequent multi-dimensional intervals. In: Correa, J.R., Hevia, A., Kiwi, M. (eds.) LATIN 2006. LNCS, vol. 3887, pp. 423–434. Springer, Heidelberg (2006). https://doi.org/10.1007/11682462_40

25. Gurvich, V., Khachiyan, L.: On generating the irredundant conjunctive and disjunctive normal forms of monotone Boolean functions. Discret. Appl. Math. **96–97**, 363–373 (1999). Rutcor Research report, RRR-35-1995, Rutgers University

26. Jerrum, M.R., Valiant, L.G., Vazirani, V.V.: Random generation of combinatorial structures from a uniform distribution. Theor. Comput. Sci. **43**, 169–188 (1986)

27. Johnson, D.S., Yannakakis, M., Papadimitriou, C.H.: On generating all maximal independent sets. Inf. Process. Lett. **27**, 119–123 (1988)

28. Karp, R., Luby, M.: Monte-Carlo algorithms for enumeration and reliability problems. In: FOCS-24, pp. 56–64 (1983). https://doi.org/10.1109/SFCS.1983.35

29. Karp, R., Upfal, E., Wigderson, A.: The complexity of parallel search. J. Comput. Syst. Sci. **38**, 225–253 (1988)

30. Khachiyan, L.: Transversal hypergraphs and families of polyhedral cones. In: Pardalos, P. (ed.) Advances in Convex Analysis and Global Optimization, Proceedings of the International Conference on Dedicated to the Memory of K. Carathéodory, Samos, Greece, June 2000. Kluwer (2001)

31. Khachiyan, L., Boros, E., Borys, K., Elbassioni, K., Gurvich, V.: Generating all vertices of a polyhedron is hard. In: SODA-17, pp. 758–765 (2006). Discret. Comput. Geom. **39**(1–3), 174–190 (2008)

32. Khachiyan, L., Boros, E., Borys, K., Elbassioni, K., Gurvich, V., Makino, K.: Enumerating spanning and connected subsets in graphs and matroids. J. Oper. Res. Soc. Jpn. **50**(4), 325–338 (2007)

33. Khachiyan, L., Boros, E., Elbassioni, K., Gurvich, V.: Enumerating disjunctions and conjunctions of paths and cuts in reliability theory. Discret. Appl. Math. **155**(2), 137–149 (2007)

34. Khachiyan, L., Boros, E., Elbassioni, K., Gurvich, V.: Generating all minimal integral solutions to AND-OR systems of monotone polymatroid inequalities: conjunctions are simpler than disjunctions. RUTCOR Research report 41-2004, Rutgers University (2004). Discret. Appl. Math. **156**(11), 2020–2034 (2008). Special Volume in Memory of Leonid Khachiyan (1952–2005)

35. Boros, E., Elbassioni, K., Gurvich, V., Khachiyan, L.: Enumerating minimal dicuts and strongly connected subgraphs and related geometric problems. In: Bienstock, D., Nemhauser, G. (eds.) IPCO 2004. LNCS, vol. 3064, pp. 152–162. Springer, Heidelberg (2004). https://doi.org/10.1007/978-3-540-25960-2_12. Algorithmica **50**(1), 159–172 (2008)

36. Khachiyan, L., Boros, E., Elbassioni, K., Gurvich, V., Makino, K.: Enumerating cut conjunctions in graphs and related problems. Algorithmica **51**(3), 239–263 (2008)

37. Lawler, E., Lenstra, J.K., Rinnooy Kan, A.H.G.: Generating all maximal independent sets: NP-hardness and polynomial-time algorithms. SIAM J. Comput. **9**, 558–565 (1980)

38. Lovász, L.: Submodular functions and convexity. In: Bachem, A., Korte, B., Grötschel, M. (eds.) Mathematical Programming: The State of the Art, pp. 235–257. Springer, Berlin (1983). https://doi.org/10.1007/978-3-642-68874-4_10

39. Makino, K., Ibaraki, T.: Interor and exterior functions of Boolean functions. Discret. Appl. Math. **69**, 209–231 (1996)

40. Makino, K., Ibaraki, T.: Inner-core and outer-core functions of partially defined Boolean functions. Discret. Appl. Math. **96–97**(1–3), 307–326 (1999). Rutcor Research report, RRR-45-95, Rutgers University

41. McDiarmid, C.G.H.: Rado's theorem for polymatroids. Math. Proc. Camb. Phil. Soc. **78**, 263–281 (1975)

42. Morris, B., Sinclair, A.: Random walks on truncated cubes and sampling 0-1 knapsack problem, FOCS-40, pp. 230–240 (1999)

43. Prisner, E.: Graphs with few cliques. Graph Theory Comb. Algorithms **1–2**, 945–956 (1995)

44. Provan, J.S.: Efficient enumeration of the vertices of polyhedra associated with network LP's. Math. Program. **64**, 47–64 (1994)

45. Read, R.C., Tarjan, R.E.: Bounds on backtrack algorithms for listing cycles, paths, and spanning trees. Networks **5**, 237–252 (1975)

46. Seymour, P.D.: A note on hyperplane generation. J. Comb. Theory Ser. B **61**, 88–91 (1994)

47. Stockmeyer, L.J., Vazirani, V.: NP-completeness of some generalizations of the maximum matching problem. Inf. Process. Lett. **15**(1), 14–19 (1982)

48. Tsukiyama, S., Ide, M., Ariyoshi, H., Shirakawa, I.: A new algorithm for generating all maximal independent sets. SIAM J. Comput. **6**, 505–517 (1977)

Lower Bounds for Unrestricted Boolean Circuits: Open Problems

Alexander S. Kulikov$^{(\boxtimes)}$ (iD)

St. Petersburg Department of Steklov Institute of Mathematics,
St. Petersburg, Russia
kulikov@logic.pdmi.ras.ru

Abstract. To prove that P $\neq$ NP, it suffices to prove a superpolynomial lower bound on Boolean circuit complexity of a function from NP. Currently, we are not even close to achieving this goal: we do not know how to prove a $4n$ lower bound. What is more depressing is that there are almost no techniques for proving circuit lower bounds.

In this note, we briefly review various approaches that could potentially lead to stronger linear or superlinear lower bounds for unrestricted Boolean circuits (i.e., circuits with no restriction on depth, fan-out, or basis).

1 Computational Model: Boolean Circuits

A *straight-line program* is a simple and natural program for computing a Boolean function $f \colon \{0,1\}^n \to \{0,1\}$. The input to such a program is variables $x_1, \ldots, x_n$ and each line of the program computes the value of a new Boolean variable by applying a binary Boolean operation to some of two previous variables. A *circuit* is a convenient way of representing a straight-line program as a directed acyclic graph. Below we show an example of a program and the corresponding circuit of size four for the majority function on three input bits x_1, x_2, x_3 (that outputs 1 iff $x_1 + x_2 + x_3 \geq 2$).

$$x_4 = x_1 \wedge x_3$$
$$x_5 = x_1 \oplus x_2$$
$$x_6 = x_5 \wedge x_3$$
$$x_7 = x_4 \oplus x_6$$

To prove that P $\neq$ NP, it suffices to find a Boolean function from NP that cannot be computed by polynomial size circuits (more precisely, a family of functions $\{f_n\}_{n=1}^{\infty}$ such that f_n has n inputs, $\bigcup_{n=1}^{\infty} f_n^{-1}(1) \in$ NP, and circuit size of f_n grows superpolynomially in n). This problem turned out to be extremely difficult: we do not know how to prove $4n$ lower bound, not to mention superlinear or superpolynomial lower bounds. Most of the known lower bounds are proved using the so-called gate elimination method which is difficult to use to

© Springer International Publishing AG, part of Springer Nature 2018
F. V. Fomin and V. V. Podolskii (Eds.): CSR 2018, LNCS 10846, pp. 15–22, 2018.
https://doi.org/10.1007/978-3-319-90530-3_2

beat the $4n$ barrier. In the rest of this note, we briefly review various approaches that could potentially lead to stronger linear or superlinear lower bounds. The focus of this note is unrestricted circuit model: we do not pose any restriction on the depth of a circuit, the fan-out of its gates, or the basis of allowed operations computed at gates. For various restricted circuit classes, such as monotone circuits (circuits using $\wedge$ and $\vee$ operations only), constant depth circuits, and formulas (circuit of fan-out 1), much stronger lower bounds are known. What is more important, various beautiful techniques for proving such lower bounds have been developed. An exposition of these bounds and techniques can be found in an excellent recent book by Jukna [1].

2   Lower Bounds: Approaches and Open Problems

Notation

We use $B_{n,m}$ to denote the set of all Boolean functions with n inputs and m outputs. By default, we will assume that $m = 1$, that is, we consider Boolean *predicates*. We use B_n as a shortcut for $B_{n,1}$. By a function f we mean (unless stated otherwise) a family of functions: $f = \{f_n \colon f_n \in B_n\}_{n=1}^{\infty}$.

2.1   Known Lower Bounds and Gate Elimination Method

How to prove, say, a $3n - o(n)$ lower bound for a Boolean function f? One way to do this is by induction: first show that f is resistant to $n - o(n)$ substitutions of some type (say, $x_i \leftarrow c$, where $c \in \{0,1\}$, or $x_i \leftarrow \bigoplus_{j \in J} x_j \oplus c$); then show that for any circuit computing f one can find a substitution eliminating at least three gates. This type of argument is known as *gate elimination* and it is used in most of the known lower bounds proofs, in particular, in the proof of the currently strongest lower bound $(3 + 1/86)n - o(n)$ for affine dispersers by Find et al. [2]. A gate elimination proof usually consists of many cases depending on how the top part of a circuit looks like. The stronger is the lower bound the larger is the number of cases: if one wants to prove, say, $4n$ lower bound, one needs to carefully check that no two of the four gates eliminated at each iteration coincide. This makes gate elimination proofs quite tedious. Moreover, it was recently shown by Golovnev et al. [3] that certain formalizations of the gate elimination method are not able to prove stronger than cn lower bounds for a small constant c. For example, they constructed a simple function f such that no substitution of the form $x_i \leftarrow g$, where g is an *arbitrary* function on all the remaining variables, can reduce the circuit size of f by more than 5. *Can one at least prove a $4n$ lower bound using gate elimination?*

Further reading. An exposition of the proofs based on the gate elimination method is given by Wegener [4, Chap. 5]. A more recent survey is given by Golovnev et al. [3, Sect. 2].

2.2 Multi-output Functions

Can one prove stronger lower bounds for functions with multiple outputs? In this case, we assume that for each output of such a function, a circuit contains a gate computing this output. Computing several functions simultaneously is definitely not easier than computing any one of them. However, currently, we do not know how to exploit this fact in lower bounds proofs: the strongest lower bound for functions with $o(n)$ outputs is the same as for functions with a single output (up to additive $o(n)$ terms). When the number of outputs becomes linear, one can use the following observation by Lamagna and Savage [5]: the circuit complexity of computing k different functions $f_1, \ldots, f_k \in B_n$ simultaneously is at least $(\min_i \text{gates}(f_i) - 1) + k$. This is just because none of the topologically first $\min_i \text{gates}(f_i) - 1$ gates can compute any of the outputs and one needs at least k gates to compute all outputs. This allows one to prove $(c + 1)n - O(1)$ lower bounds for functions from $B_{n,n}$ from cn lower bounds for functions from $B_{n,1}$: given $f \in B_n$, consider $g = (g_1, \ldots, g_n) \in B_{n,n}$ where $g_i(x) = f(x) \oplus x_i$; then, $\text{gates}(g_i) \geq \text{gates}(f) - 1$ and hence $\text{gates}(g) \geq \text{gates}(f) + n - 2$. *How to prove a $5n$ lower bound for a function from $B_{n,n}$?*

Further reading. A survey of lower bounds for multi-output functions is given by Hiltgen [6, Chap. 4].

2.3 Non-gate-Elimination Lower Bounds

Are there approaches other than gate elimination for proving lower bounds for unrestricted circuits. There are a few lower bounds that are *not* based on gate elimination techniques. Alas, none of them is currently known to give a stronger than $2n$ lower bound. Blum and Seysen [7] proved that *any* optimal circuit that computes simultaneously AND and NOR of n input bits consists of two formulas (that is, each output is computed by a tree) and hence has size $2n - 2$. Note that the gate elimination method with bit-fixing substitutions cannot be used for this particular function: assigning a constant to an input variable immediately trivializes one of the two output functions (and one loses a possibility to proceed by induction). Melanich [8] came up with a similar, but simpler argument. She considered the following multi-output function from $B_{n,o(n)}$: there are $n = \binom{k}{2}$ inputs $x_{\{i,j\}}$, where $1 \leq i \neq j \leq k$, and $k = o(n)$ outputs; the i-th output computes the AND of variables $\{x_{\{i,j\}}\}_{j \neq i}$. Each input contributes to two outputs and hence the function can be computed by a circuit of size $2n - o(n)$. Melanich proves that this straightforward circuit is optimal by showing that in any circuit (computing this function) one can reduce the number of gates shared between several outputs without increasing the size of the circuit. Chashkin [9] proved a $2n - o(n)$ for a function $f \in B_{n,\log_2 n}$ that has the form $f(x) = Ax$ where the matrix $A \in \{0,1\}^{\log_2 n \times n}$ has n pairwise distinct columns. He showed that any circuit computing this function has at least $n - o(n)$ branching gates (i.e., gates of out-degree at least 2). The lower bound then follows by counting the number of edges. *Can any of these non-gate-elimination methods be extended to get stronger than $2n$ lower bounds?*

2.4   Symmetric Functions

Can one prove a superlinear lower bound for a symmetric function (i.e., a function whose output depends on the sum of input bits only)? In fact, one cannot: while basic symmetric functions like parity, MOD_3, and majority are used to prove superpolynomial lower bounds in, e.g., constant depth circuit model, any symmetric function can be computed by a circuit of size $4.5n + o(n)$ as shown by Demenkov et al. [10]. The strongest known lower $2.5n - O(1)$ is proved by Stockmeyer [11]. Hence, it is not excluded that there exist symmetric functions of circuit size, say, $4n$. Note that the multi-output function $SUM_n \in B_{n, \lceil \log_2(n+1) \rceil}$ that outputs the binary encoding of the sum of n input bits is not easier than any symmetric function $f \in B_n$: f can be computed by a circuit of size gates(SUM_n) + $o(n)$. *What is the circuit size of* SUM_n?

Further reading. Known lower and upper bounds on complexity of symmetric functions in various models are summarized in Jukna's book [1, end of Chap. 1].

2.5   Satisfiability Algorithms

Given a circuit with n inputs, how hard is it to find an assignment making this circuit to output 1? Williams [12] recently developed a general framework of getting circuit lower bounds from faster than brute force search satisfiability algorithms. Extending Williams' results, Jahanjou et al. [13] proved that one can prove a $2cn$ lower bound (for a function from $B_{n,2}$) by designing an $O(2^n/n^{\omega(1)})$-time algorithm for checking satisfiability of circuits of size $2cn$. In a sense, results like this show that designing fast satisfiability algorithms is not easier than proving circuit lower bounds. This also reflects the state-of-the-art on satisfiability algorithms: we only know how to beat the brute force search for circuits of size at most $2.99n$ [14]. Hence, the known satisfiability algorithms for small size (unrestricted) circuits currently do not give improved lower bounds. *Can one improve the brute force search for the satisfiability problem on circuits of size $4n$? Do non-trivial satisfiability algorithms for circuits of size cn imply cn lower bounds?*

Further reading. A good starting point is a recent survey by Williams [15].

2.6   Mass Production

Can one take a sufficiently hard function with constant number of inputs and cook out of it a family of functions of high circuit complexity? About 70 years ago, Shannon [16] showed that almost all functions from B_n have circuit complexity $\Omega(2^n/n)$ (by showing that the total number 2^{2^n} of functions is greater than the number of circuits of size $o(2^n/n)$). This implies that for any constant c, one can find a function $f_k \in B_k$, where $k = k(c)$, of circuit size at least ck just by enumerating functions one by one. A natural attempt to cook a family of functions out of f_k would be to define a function $f_n \in B_{n, \frac{n}{k}}$ as follows: split n input bits into $\frac{n}{k}$ blocks of k bits and apply f_k to each of the blocks. In other words, we compute f_k on $\frac{n}{k}$ independent blocks of size k. The function f_n can be computed by a circuit of size $\frac{n}{k} \cdot \text{gates}(f_k)$. If this naive way of computing f_k was close to optimal, one would get a close to cn lower bound on the circuit size of f_n. We, however, do not know how to prove this. Still, this is what is known.

For a positive integer r and a function $f \in B_n$, by $r \times f$ we denote a function from $B_{rn,r}$ that applies f to r independent blocks of size n. We say that a *mass production* effect occurs for f when $\text{gates}(r \times f)$ is (much) smaller than $r \cdot \text{gates}(f)$. For *very simple* functions like $f = x_1 \oplus \cdots \oplus x_n$ (or any other function whose optimal circuit is a read-once formula) there is no mass production effect: $\text{gates}(r \times f) = r \cdot \text{gates}(f)$. This can be shown just by counting wires: f depends essentially on all its variables, hence there is at least one outgoing wire for every input; since each internal (non-output) gate reduces the number of outgoing wires at most by one, we conclude that $\text{gates}(f) = n - 1$ and $\text{gates}(r \times f) = rn - r = r \cdot (n - 1)$. Hiltgen [6] also shows that mass poduction effect occurs for many functions of circuit size about $2n$. On the other hand, for *very hard* function f one can show that $\text{gates}(r \times f)$ is almost the same as $\text{gates}(f)$ even if r is superpolynomial in n. More precisely, Ulig [17] showed that $\text{gates}(r \times f) \leq 2^n/n + o(2^n/n)$ for any $f \in B_n$ and $r = 2^{o(n/\log n)}$. *What are the functions avoiding mass production effect?*

Further reading. More on mass production can be found in Wegener's book [4, Sect. 10.2] and Hiltgen's PhD thesis [6, Sect. 4.4].

2.7 Logarithmic Depth Circuits

Can we at least prove superlinear lower bounds on circuits of logarithmic (i.e., $O(\log n)$) depth? Alas, currently, it is not known. However, if we further restrict the depth to be constant (in this case, one needs to allow arbitrary fan-in and to specify the operations allowed at gates), then one can prove even superpolynomial lower bounds! Moreover, Valiant [18] showed the following connection between these two models: if a function can be computed by a circuits of logarithmic depth and linear size, then it can also be computed by a subexponential depth 3 circuit, more precisely by an OR of CNF's of total size $2^{O(n/\log\log n)}$ (here, the constant inside $O(\cdot)$ depends on constants a, b where the size and depth of the original circuit is an and $b \log n$). *Currently, the strongest lower bounds known for such depth 3 circuits are of the form $2^{\Omega(n^{1/2})}$, though exponential lower bounds are known if we further restrict the length of clauses in CNF's to be constant.*

Further reading. An exposition of Valiant's reduction is given in the book by Viola [19, Chap. 2], while known results on constant depth circuits are summarized in the book by Jukna [1, Chaps. 11–12].

2.8   Linear Circuits and Matrix Rigidity

Can we at least prove superlinear lower bounds for circuits consisting of parity gates only? This question makes sense for multi-output functions. Specifically, let us focus on functions of the form $f(x) = Ax$ where $A \in \{0,1\}^{n \times n}$. Non-constructively, one can show that for almost all matrices A, the size of the smallest linear circuit computing Ax is $\Omega(n^2/\log n)$ (and there is a matching upper bound by Lupanov [20]). Alas, we do not have superlinear lower bounds even for this restricted model, even when we additionally restrict the depth to be $O(\log n)$. Interestingly, Valiant's depth reduction mentioned in Sec. 2.7 can be used to relate the circuit size to the notion of matrix rigidity introduced by Grigoriev [21] and Valiant [22]. Roughly speaking, for a parameter r, the rigidity of A, $R_A(r)$, is the Hamming distance from A to the set of matrices of rank (over $\mathbb{F}$) at most r. Valiant shows that if $R_A(\epsilon n) \geq n^{1+\delta}$ for positive constants ϵ, δ, then the function Ax cannot be computed by linear circuits of logarithmic depth of size $O(n)$. *So far, we have no such examples of explicit matrices.*

Further reading. More on circuit complexity and matrix rigidity can be found in the book by Lokam [23, Chap. 2]. Lower bounds for *constant depth linear circuits* (where superlinear lower bounds are known!) are summarized in the recent book by Jukna and Sergeev [24].

2.9   Multiplicative Complexity

What if some gates are given for free? Basically, each gate in a binary Boolean circuit is either an XOR-type gate, i.e., computes a binary operation of the form $x \oplus y \oplus a$ where $a \in \{0,1\}$, or an AND-type gate, i.e., computes $(x \oplus y) \wedge (y \oplus b) \oplus c$ where $a, b, c \in \{0,1\}$. It is well known that XOR-type gates are avoidable: any function can be computed by a circuit in the basis $U_2 = B_2 \setminus \{\oplus, \equiv\}$. On the other hand, AND-type gates are unavoidable and it was shown by Nechiporuk [25] that almost all Boolean functions require about $2^{n/2}$ such gates. The minimum number of AND-type gates required to compute f is known as *multiplicative complexity* of f, $\mathrm{mc}(f)$. Of course, $\mathrm{mc}(f) \leq \mathrm{gates}(f)$ and the known lower bounds on multiplicative complexity are even weaker than those on circuit complexity. At the same time, one can prove lower bounds on mc without analyzing the structure of a circuit: as shown by Schnorr [26], a circuit with k AND-type gates computes a function of degree at most $k + 1$. Here, the degree of a function is the degree of its polynomial over $\mathbb{F}_2$. This immediately gives a lower bound $n-1$ on multiplicative complexity of functions of full degree: e.g., $\mathrm{mc}(\mathrm{AND}) = n - 1$. *Strangely enough, this is the strongest known lower bound: we do not know how to prove* $\mathrm{mc}(f) \geq n$, *let alone proving* $\mathrm{mc}(f) \geq (1 + \epsilon)n$.

Acknowledgments. The research is supported by Russian Science Foundation (project 16-11-10123). The author is thankful to Alexander Golovnev and Edward A. Hirsch for fruitful discussions and many useful comments.

References

1. Jukna, S.: Boolean Function Complexity – Advances and Frontiers. Algorithms and Combinatorics. Springer, Heidelberg (2012). https://doi.org/10.1007/978-3-642-24508-4
2. Find, M.G., Golovnev, A., Hirsch, E.A., Kulikov, A.S.: A better-than-3n lower bound for the circuit complexity of an explicit function. In: Dinur, I. (ed.) IEEE 57th Annual Symposium on Foundations of Computer Science, FOCS 2016, Hyatt Regency, New Brunswick, NJ, USA, 9–11 October 2016, pp. 89–98. IEEE Computer Society (2016)
3. Golovnev, A., Hirsch, E.A., Knop, A., Kulikov, A.S.: On the limits of gate elimination. [27], pp. 46:1–46:13
4. Wegener, I.: The Complexity of Boolean Functions. Wiley-Teubner, Hoboken (1987)
5. Lamagna, E.A., Savage, J.E.: On the logical complexity of symmetric switching functions in monotone and complete bases. Technical report, Brown University (1973)
6. Hiltgen, A.P.: Cryptographically relevant contributions to combinational complexity theory. Ph.D. thesis, ETH Zurich, Zürich, Switzerland (1994)
7. Blum, N., Seysen, M.: Characterization of all optimal networks for a simultaneous computation of AND and NOR. Acta Inf. **21**, 171–181 (1984)
8. Melanich, O.: Technical report (2012)
9. Chashkin, A.V.: On complexity of Boolean matrices, graphs and corresponding Boolean matrices. Diskretnaya matematika **6**(2), 43–73 (1994). (in Russian)
10. Demenkov, E., Kojevnikov, A., Kulikov, A.S., Yaroslavtsev, G.: New upper bounds on the Boolean circuit complexity of symmetric functions. Inf. Process. Lett. **110**(7), 264–267 (2010)
11. Stockmeyer, L.J.: On the combinational complexity of certain symmetric Boolean functions. Math. Syst. Theory **10**, 323–336 (1977)
12. Williams, R.: Improving exhaustive search implies superpolynomial lower bounds. SIAM J. Comput. **42**(3), 1218–1244 (2013)
13. Jahanjou, H., Miles, E., Viola, E.: Local reductions. In: Halldórsson, M.M., Iwama, K., Kobayashi, N., Speckmann, B. (eds.) ICALP 2015. LNCS, vol. 9134, pp. 749–760. Springer, Heidelberg (2015). https://doi.org/10.1007/978-3-662-47672-7_61
14. Golovnev, A., Kulikov, A.S., Smal, A.V., Tamaki, S.: Circuit size lower bounds and #SAT upper bounds through a general framework. [27], pp. 45:1–45:16
15. Williams, R.R.: Some ways of thinking algorithmically about impossibility. SIGLOG News **4**(3), 28–40 (2017)
16. Shannon, C.E.: The synthesis of two-terminal switching circuits. Bell Syst. Tech. J. **28**(1), 59–98 (1949)
17. Ulig, D.: On the synthesis of self-correcting schemes from functional elements with a small number of reliable elements. Math. Notes Acad. Sci. USSR **15**(6), 558–562 (1974)
18. Valiant, L.G.: Exponential lower bounds for restricted monotone circuits. In: Johnson, D.S., Fagin, R., Fredman, M.L., Harel, D., Karp, R.M., Lynch, N.A., Papadimitriou, C.H., Rivest, R.L., Ruzzo, W.L., Seiferas, J.I. (eds.) Proceedings of the 15th Annual ACM Symposium on Theory of Computing, Boston, Massachusetts, USA, 25–27 April 1983, pp. 110–117. ACM (1983)
19. Viola, E.: On the power of small-depth computation. Found. Trends Theor. Comput. Sci. **5**(1), 1–72 (2009)

20. Lupanov, O.B.: On rectifier and switching-and-rectifier schemes. Dokl. Akad. Nauk SSSR **111**, 1171–1174 (1956)
21. Grigoriev, D.: An application of separability and independence notions for proving lower bounds of circuit complexity. Notes Sci. Semin. LOMI **60**, 38–48 (1976)
22. Valiant, L.G.: Graph-theoretic arguments in low-level complexity. In: Gruska, J. (ed.) MFCS 1977. LNCS, vol. 53, pp. 162–176. Springer, Heidelberg (1977). https://doi.org/10.1007/3-540-08353-7_135
23. Lokam, S.V.: Complexity lower bounds using linear algebra. Found. Trends Theor. Comput. Sci. **4**(1–2), 1–155 (2009)
24. Jukna, S., Sergeev, I.: Complexity of linear Boolean operators. Found. Trends Theor. Comput. Sci. **9**(1), 1–123 (2013)
25. Nechiporuk, E.I.: Complexity of schemes in certain bases containing nontrivial elements with zero weights. Dokl. Akad. Nauk SSSR **139**(6), 1302–1303 (1961)
26. Schnorr, C.P.: The multiplicative complexity of Boolean functions. In: Mora, T. (ed.) AAECC 1988. LNCS, vol. 357, pp. 45–58. Springer, Heidelberg (1989). https://doi.org/10.1007/3-540-51083-4_47
27. Faliszewski, P., Muscholl, A., Niedermeier, R. (eds.): 41st International Symposium on Mathematical Foundations of Computer Science, MFCS 2016. LIPIcs, vol. 58, Kraków, Poland, 22–26 August 2016. Schloss Dagstuhl - Leibniz-Zentrum fuer Informatik (2016)

Online Labeling: Algorithms, Lower Bounds and Open Questions

Michael Saks[(⊠)]

Department of Mathematics, Rutgers University, New Brunswick, NJ 08854, USA
saks@math.rutgers.edu

Abstract. The online labeling problem (also known as the file maintenance problem), is a natural algorithmic problem that has arisen as a
buidling block for data structures. A stream of distinct integer items is
to be assigned labels online from a label set $\{1, \ldots, m\}$ so that the order
of the labels respects the natural order of the items. Maintaining order
on the labels may require relabeling items. The algorithm pays 1 each
time an item is labeled or relabeled and the goal of the algorithm is to
minimize the total cost.

We survey upper and lower bounds and open problems in both the
deterministic and randomized setting.

Keywords: Online labeling · Data structures

1 The File Maintenance Problem

The *online labeling problem* (also known as the *file maintenance problem)* is
a simple and appealing algorithmic problem that was introduced in the early
1980's by Itai et al. [17], involving the online labeling of items in a stream of
distinct elements from a totally ordered set. The problem appears in connection
with various data structure problems. In this extended abstract we survey known
results, both upper and lower bounds, for this problem, as well as open questions.

The setting of the problem is a stream of distinct data items arriving online
that are assigned labels from a label set $\{1, \ldots, m\}$. The items come from a
totally ordered set, which we take to be the set $\{1, \ldots, r\}$. The items arrive
one at a time, and each item must be labeled when it arrives. At all times, the
labeling of the items must respect the ordering of the items; if items y and z
satisfy $y < z$ then the label of y must be less than the label of z. Maintaining this
order requirement may require relabeling items when new items arrive. We are
assessed a charge of 1 when an item is initially labeled and each time an item is
relabeled. The goal is to minimize the total cost of all labelings and relabelings.

An alternative formulation for this problem is the *file maintenance problem*.
In this formulation items are not assigned labels from the set $\{1, \ldots, m\}$ but
instead are stored in an array of length m. The items may be located anywhere in
the array (and there are in general many empty locations) but we must maintain

F. V. Fomin and V. V. Podolskii (Eds.): CSR 2018, LNCS 10846, pp. 23–28, 2018.
https://doi.org/10.1007/978-3-319-90530-3_3

the condition that the left to right order of items in the array respects the intrinsic ordering on the items. We pay 1 for storing an item, and 1 each time we move an item. Viewing the index of the array location where containing an item as its label, it's easy to see that file maintenance and online labeling are equivalent. In this paper we'll use the online labeling formulation.

This problem was introduced in [17] in the context of an application to priority queue implementation. Other applications in algorithm design include design of cache-oblivious B-trees (Bender et al. [4] and Brodal et al. [8]), and cache-oblivious dynamic dictionaries (Bender et al. [6]). It was also shown (Emek and Korman [16]) that lower bounds on the cost of file maintenance can be used to obtain lower bounds on a problem in distributed resource allocation called the Distributed Controller problem (introduced in [1]).

The online labeling problem is parameterized by the *label range* m, the *universe size* r and the *list size* n. If the universe size r is less than or equal to the label range m, one can simply label each arriving item by itself. This algorithm requires no relabelings, and so the cost is simply n. Note also that the problem is obviously infeasible if $n > m$. Thus the interesting range of parameters is $n \leq m < r$.

To formalize the problem: A *labeling function* for the subset $Y \subseteq [1, r]$ of items is a map $f : Y \longrightarrow [1, m]$ that is strictly order preserving, i.e., for $x, y \in Y$ if $x < y$ then $f(x) < f(y)$. In particular f is one-to-one, so $|Y| \leq m$. Labels that are in the image of f are *occupied* and the others are *unoccupied*. A *configuration* is a pair (Y, f) where Y is a set of items and f is a labeling function for Y.

The online labeling problem with parameters m, r, n as above, can be described as a two player game $G^n(m, r)$, called the *online labeling game* between the *algorithm* and the *adversary*. The game is played in a sequence of n rounds. In each round t the *adversary* selects an item y^t that was not previously selected, i.e, from the set $\{1, \ldots, r\} - \{y^1, \ldots, y^{t-1}\}$. The algorithm then chooses a labeling function f^t for the set $Y^t = \{y^1, \ldots, y^t\}$. We say that item y^t is *arrives* at round t. (Y^t, f^t) is called the *configuration at the end of round* t and also *the configuration at the beginning of round* $t + 1$.

Formally, an adversary strategy is a sequence $(y^1, \ldots, y^n)$ of distinct items and an algorithm is any map that associates a labeling function to every sequence $(y^1, \ldots, y^t)$ (where $t \leq n$).

We say that item y^t is labeled in round t, and that an item $y \in Y^{t-1}$ is *relabeled in round* t if $f^t(y) \neq f^{t-1}(y)$. The *cost up to round* t is $\chi^t = \sum_{i=1}^{t} |L^i|$, where L^i is the set of items labeled or relabeled in round i. The objective of the algorithm is to minimize χ^n and the objective of the adversary is to maximize χ^n. We also define the *amortized cost per item* $\alpha^n = \chi^n / n$.

The cost of an algorithm is the maximum over all adversary sequences of the cost of the algorithm on the sequence. We write $\chi^n(m, r)$ for the value of the game, which is the minimum cost of an algorithm. We define $\alpha^n(m, r)$ to be $\chi^n(m, r)$: this is the optimal *amortized cost per item*.

One may consider both the case where the algorithm is deterministic, and the case that the algorithm is randomized. In the language of games, a randomized algorithm is a mixed strategy, i.e. a probability distribution over deterministic strategies. The cost of a randomized algorithm on an adversary sequence is the expected cost (over the random choices of the algorithm), and the (worst case expected) cost of the algorithm is the maximum over all adversary sequences of the expected cost of the algorithm on that sequence. The minimum cost of any randomized algorithm is denoted $\tilde{\chi}^n(m, r)$, and the amortized cost is $\tilde{a}^n(m, r) = \tilde{\chi}^n(m, r)/n$.

2   Deterministic Algorithms with Arbitrary Universe

The universe size r plays a secondary role in results on the problem: the main known upper bounds are independent of r, and hold for any (even infinite) totally ordered universe (e.g., the real numbers). For now we focus on this case, and mention the role of the universe size later.

The label range m must (of course) be at least the number n of items to be labeled. It is also easy to show that if $m \geq 2^n$ then there is an algorithm that never relabels any items, so its cost is exactly n. So we assume $n \leq m < 2^n$. In the case of arbitrary universe, there are upper and lower bounds known that are matching up to a constant or weakly superconstant factor for all m.

There are two natural range of parameters that have received the most attention. In the case of *linear label range* we have $m = (1 + C)n$ for some $C > 0$, and in the case of *polynomial label range* we have $m = \theta(n^{1+C})$ for some constant $C > 0$.

All known upper bounds are provided by analysis of explicit algorithms. The published algorithms all work for arbitrary (even infinite) universes, and r plays no role in their analysis. Itai et al. [17] gave an algorithm for the case of linear label range having worst case amortized cost $O((\log n)^2)$ per item. Improvements and simplifications were given by Willard [20] and Bender et al. [4]. In the special case that $m = n$, algorithms with amortized cost $O((\log n)^3)$ per item were given [7, 21]. It is also well known that the algorithm of Itai et al. can be adapted to give amortized cost $O(\log n)$ per item in the case of polynomial label range. An algorithm with $O(\log n)$ worst case cost per item for polynomial label range was given by Kopelowitz [19]. Bulánek et al. (see [3]) gave an algorithm that has amortized cost $O(\log n / \log \log m)$, for m large enough (bigger than $2^{\log n^3}$ but at most 2^n).

As mentioned, the known lower bound results for deterministic algorithms come close to matching these upper bounds.

For the case of polynomial label range, Dietz et al. [13] (also in [21]) proved an amortized lower bound $\Omega(\log n)$ which matches the upper bound. This lower bound was extended, and the proof simplified, by Babka et al. [2, 3] who gave an amortized lower bound of $\Omega(\frac{\log n}{1 + \log \log m - \log \log n})$ that is valid for m between n and 2^n. This bound includes the lower bound of [13] for polynomial label range, and also matches the above-mentioned upper bound of $O(\log n / \log \log m)$ when m is at least $2^{\log n^3}$.

For the case of linear label range, Bulánek et al. [9] proved an amortized lower bound of $\Omega(\log^2 n)$ matching the upper bound in the initial paper of [17]. This lower bound built on earlier work of Dietz and Zhang ([12,15], also available in Zhang's Ph.D. thesis [21]) that proved the same lower bound for a restricted class of algorithms, called smooth algorithms. In [9] an amortized lower bound of $\Omega(\log^3 n)$ was proved in the case $m = n$, matching the upper bound of [7,21].

Table 1, adapted from [3], summarizes the known results for deterministic online labeling with arbitrary universe.

Table 1. Known bounds for amortized cost of deterministic online labeling.

Array size (m)	Asymptotic bound	Lower bound	Upper bound
$m = n$	$\Theta\big((\log n)^3\big)$	[9]	[21]
$m = cn$, constant $c > 1$	$\Theta\big((\log n)^2\big)$	[9]	[17]
$m = n^C$, constant $C > 1$	$\Theta(\log n)$	[13]	[17]
$m = n^{\omega(1)}$	$\Omega\left(\frac{\log n}{1+\log\log m - \log\log n}\right)$	[2]	
$m \geq 2^{1+\log^3 n}$	$\Theta\left(\frac{\log n}{\log\log m}\right)$	[2]	[3,9]

3   The Role of the Universe Size

Thus far, we've ignored the possible role of the universe size r, and allowed r to be infinite. All of the algorithmic results described above hold for arbitrary universe size, and it is unclear the extent to which restricting the universe size can be exploited by algorithms. As noted earlier, if the universe size r is restricted below the label range m, then we can use the identity labeling, which has a cost of only n. What if the universe size is not much bigger than m, for example, is bounded by some constant multiple of m. Surprisingly, in the case of linear label range, the tight amortized $\Omega(\log^2 n)$ lower bound shown in [9] for linear array size holds even if the universe size r is bounded by a (sufficiently large) constant multiple of m. However, the known proofs of the lower bounds for polynomial (or larger) label range seem to require r to be exponential in n. This leaves open the intriguing question of whether one can improve on the best algorithms when r is bounded by a not too large function of m. For example, in the case of polynomal label range, $m = n^{1+C}$ for $C > 0$, can one can improve on the $O(\log n)$ amortized cost algorithm if the range r is bounded above by a polynomial in m, or by $O(m \log m)$?

4   Randomized Online Labeling

As discussed earlier, a randomized online labeling algorithm can be viewed as a probability distribution over deterministic online labeling algorithms. The cost of an input sequence is the expected cost of the input sequence over the randomness of the algorithm.

In the deterministic setting, we can think of the game as happening in rounds, where in round t the adversary selects y^t knowing the current labeling function of Y^{t-1}. In the randomized setting, the adversary does not know the current labeling function of Y^{t-1}; he only knows the probability distribution induced by the algorithm over labelings of Y^{t-1}.

As usual, randomized algorithms are at least as powerful as deterministic algorithms, and the natural question is: does randomness help for online labeling?

In the case of polynomial label range, Bulánek et al. [10] proved that the $\Omega(\log n)$ amortized lower bound that was proved for deterministic algorithms extends to the randomized case. However, in other ranges of m it is not known whether randomized algorithms can improve on deterministic algorithms. An especially interesting open question is whether the tight $\Omega(\log^2 n)$ lower bound for the case $m = (1 + C)n$ (linearly many labels) extends to randomized algorithm, or is there a randomized algorithm whose cost beats the $O(\log^2 n)$ deterministic algorithm of [17]?

Acknowledgements. Thanks to my collaborators Michal Koucky, Jan Bulánek, Martin Bobka and Václav Cunat on several of the results mentioned here. I also gratefully acknowledge the support of the Simons Foundation under Project 332622.

References

1. Afek, Y., Awerbuch, B., Plotkin, S.A., Saks, M.E.: Local management of a global resource in a communication network. J. ACM **43**(1), 1–19 (1996)
2. Babka, M., Bulánek, J., Čunát, V., Koucký, M., Saks, M.: On online labeling with polynomially many labels. In: Epstein, L., Ferragina, P. (eds.) ESA 2012. LNCS, vol. 7501, pp. 121–132. Springer, Heidelberg (2012). https://doi.org/10.1007/978-3-642-33090-2_12
3. Babka, M., Bulánek, J., Čunát, V., Koucký, M., Saks, M.: On online labeling with large label set. SIAM J. Disc. Math. (2017, to appear). http://sites.math.rutgers.edu/~saks/PUBS/large-label.pdf
4. Bender, M.A., Cole, R., Demaine, E.D., Farach-Colton, M., Zito, J.: Two simplified algorithms for maintaining order in a list. In: Möhring, R., Raman, R. (eds.) ESA 2002. LNCS, vol. 2461, pp. 152–164. Springer, Heidelberg (2002). https://doi.org/10.1007/3-540-45749-6_17
5. Bender, M.A., Demaine, E.D., Farach-Colton, M.: Cache-oblivious B-trees. SIAM J. Comput. **35**(2), 341–358 (2005). https://doi.org/10.1137/S0097539701389956
6. Bender, M.A., Duan, Z., Iacono, J., Wu, J.: A locality-preserving cache-oblivious dynamic dictionary. J. Algorithms **53**(2), 115–136 (2004). https://doi.org/10.1016/j.jalgor.2004.04.014
7. Bird, R.S., Sadnicki, S.: Bird and Stefan Sadnicki: minimal on-line labelling. Inf. Process. Lett. **101**(1), 41–45 (2007). https://doi.org/10.1016/j.ipl.2006.07.011
8. Brodal, G.S., Fagerberg, R., Jacob, R.: Cache oblivious search trees via binary trees of small height. In: Proceedings of the Thirteenth Annual ACM-SIAM Symposium on Discrete Algorithms, SODA 2002, pp. 39–48. Society for Industrial and Applied Mathematics (2002) http://dl.acm.org/citation.cfm?id=545381.545386

9. Bulánek, J., Koucký, M., Saks, M.: Tight lower bounds for the online labeling problem. SIAM J. Comput. **44**(6), 1765–1797 (2015). https://doi.org/10.1145/2213977.
   2214083. (Preliminary version: Proceedings of the Forty-Fourth Annual ACM Symposium on Theory of Computing, STOC 2012, pp. 1185–1198. ACM, New York
   (2012))
10. Bulánek, J., Koucký, M., Saks, M.: On randomized online labeling with polynomi-
    ally many labels. In: Fomin, F.V., Freivalds, R., Kwiatkowska, M., Peleg, D. (eds.)
    ICALP 2013. LNCS, vol. 7965, pp. 291–302. Springer, Heidelberg (2013). https://
    doi.org/10.1007/978-3-642-39206-1_25
11. Dietz, P.F.: Maintaining order in a linked list. In: Proceedings of the 14th Annual
    ACM Symposium on Theory of Computing, STOC, pp. 122–127 (1982)
12. Dietz, P.F., Seiferas, J.I., Zhang, J.: Lower bounds for smooth list labeling.
    Manuscript (2005). (Listed in the references of [13])
13. Dietz, P.F., Seiferas, J.I., Zhang, J.: A tight lower bound for online monotonic list
    labeling. SIAM J. Discret. Math. **18**, 626–637 (2005)
14. Dietz, P., Sleator, D.: Two algorithms for maintaining order in a list. In: Proceed-
    ings of the Nineteenth Annual ACM Symposium on Theory of Computing, STOC
    1987, pp. 365–372, ACM, New York (1987). https://doi.org/10.1145/28395.28434
15. Dietz, P.F., Zhang, J.: Lower bounds for monotonic list labeling. In: Gilbert, J.R.,
    Karlsson, R. (eds.) SWAT 1990. LNCS, vol. 447, pp. 173–180. Springer, Heidelberg
    (1990). https://doi.org/10.1007/3-540-52846-6_87
16. Emek, Y., Korman, A.: New bounds for the controller problem. Distrib. Comput.
    24(3–4), 177–186 (2011). https://doi.org/10.1007/s00446-010-0119-z
17. Itai, A., Konheim, A.G., Rodeh, M.: A sparse table implementation of priority
    queues. In: Even, S., Kariv, O. (eds.) ICALP 1981. LNCS, vol. 115, pp. 417–431.
    Springer, Heidelberg (1981). https://doi.org/10.1007/3-540-10843-2_34
18. Korman, A., Kutten, S.: Controller and estimator for dynamic networks. In: PODC,
    pp. 175–184 (2007)
19. Kopelowitz, T.: On-line indexing for general alphabets via predecessor queries on
    subsets of an ordered list. In: Proceedings of the 2012 IEEE 53rd Annual Sympo-
    sium on Foundations of Computer Science, FOCS 2012, Washington, DC, USA,
    pp. 283–292. IEEE Computer Society (2012). https://doi.org/10.1109/FOCS.2012.
    79
20. Willard, D.E.: A density control algorithm for doing insertions and deletions in a
    sequentially ordered file in a good worst-case time. Inf. Comput. **97**(2), 150–204
    (1992). https://doi.org/10.1016/0890-5401(92)90034-D
21. Zhang, J.: Density control and on-line labeling problems. Ph.D. thesis, University
    of Rochester, Rochester (1993)

Maintaining Chordal Graphs Dynamically: Improved Upper and Lower Bounds

Niranka Banerjee[1($\boxtimes$)], Venkatesh Raman[1], and Srinivasa Rao Satti[2]

[1] The Institute of Mathematical Sciences, HBNI, CIT Campus,
Taramani, Chennai 600 113, India
{nirankab,vraman}@imsc.res.in
[2] Seoul National University, 1 Gwanak-ro, Gwanak-gu, Seoul 151-744, Korea
ssrao@cse.snu.ac.kr

Abstract. We study upper and lower bounds for the problem of maintaining a chordal graph G under edge insertions and deletions. Let G be a chordal graph on n vertices and m edges and let (u, v) be the edge to be deleted or inserted.

- Let k be the size of the maximum clique in G. Our first result is an improved analysis of an earlier approach due to Ibarra [12] to support edge deletions. We can construct a data structure in $O(nk^2)$ time such that we can report in $O(1)$ time if $G\backslash(u, v)$ is chordal and if it is, we can update the structure in $O(n+k^2)$ time. We then show using a charging argument that the update time can be improved to $O(n^2/\Delta + k^2)$ amortized time over a sequence of Δ deletions.
- We develop a data structure to maintain a *perfect elimination ordering* (PEO) of chordal graphs where we can detect whether $G\backslash(u, v)$ is chordal in $O(\min\{degree(u), degree(v)\})$ time, and if it is chordal, we can update the structure in $O(degree(u) + degree(v))$ time. In graphs of bounded degree, our query and update bounds are a constant.
- Finally, we show that we can obtain a PEO of the graph from a clique-tree in $O(n)$ time after an edge insertion or deletion (against a naive $O(m + n)$ time). This answers a question posed by Ibarra [12].

Regarding lower bounds, we show that any dynamic structure to maintain a chordal graph requires $\Omega(\log n)$ amortized time per edge addition or deletion or per query to detect chordality, in the cell probe model with word size $\log n$.

1 Introduction

A graph is chordal if every cycle of size four or more in the graph has a chord. The study of chordal graphs has quite a rich history and the class of graphs has found use in a wide range of areas such as in biology, artificial intelligence, database systems and facility location problems [4,7,18]. There are $O(m + n)$ time algorithms to detect whether a graph on n vertices and m edges is chordal by computing what is called a *perfect elimination ordering* or *a clique tree decomposition* of the graph [17,18]. A perfect elimination ordering in a graph is an ordering of the vertices of the graph such that for each vertex x, x and the neighbors of x that occur after x in the ordering form a clique. A graph is chordal if and only if it has a perfect elimination ordering (PEO) of its vertices. Another characterization of chordal graphs is in the form of clique trees. A *clique tree* of a graph is a tree decomposition of the graph, where the bags in each node of the decomposition induce a maximal clique. A graph is chordal if and only if it has a clique tree [3,19].

We consider the problem of maintaining chordal graphs under edge insertions and deletions. As in the well-studied area of dynamic graph algorithms, we want our algorithm to report and update faster than what would require in the static algorithm to test chordality of the resulting graph from scratch, i.e. better than $O(m + n)$ time. Sometimes it is convenient to restrict the update operations on the graph. If we are allowed only insertions on the graph, then a structure supporting such an operation is said to work in the incremental setting. Similarly, if we are allowed only delete operations, such a structure is said to work in the decremental setting. A structure that allows both insertions and deletions is called a fully dynamic structure.

Our structures (as in the case of previous ones for the problem) always maintain a chordal graph in that whenever the addition or deletion of an edge makes the graph non-chordal, the algorithm reports that it is non-chordal and does not perform the update. We call the operation that detects chordality and returns a yes or no answer as a *query*, and the operation to update the resulting graph (if the resulting graph is chordal) as the *update* operation.

1.1   Previous Work

Ibarra [12] developed two fully dynamic algorithms for maintaining chordality. First one has a query and update time of $O(n)$ with a preprocessing time of $O(m + n)$, while the other has a query time of $O(\sqrt{m})$ and the update time of $O(m + n)$ with a preprocessing time of $O(mn + n^2)$. The latter is particularly useful for sparse graphs. Mezzini [13] developed a fully dynamic algorithm with $O(1)$ query and $O(n^2)$ update time, for both addition and deletion of edges. Berry et al. [1] gave an algorithm which takes amortized $O(n)$ time for insertion, deletion and deletion queries and an amortized $O(1)$ time for insertion queries.

Tarjan and Yannakakis [18] give an algorithm to convert what are called acyclic hypergraphs (clique trees are acyclic hypergraphs) to PEO in $O(m + n)$ time. In this paper, we give a data structure to augment clique trees and an algorithm to obtain a PEO from that, to support maintenance of PEO under insertions and deletions of edges in $O(n)$ time.

Logarithmic time lower bounds for update and query times for graph problems were first developed by Fredman et al. [10], who gave an $\Omega(\log n / \log \log n)$ amortized query and update times for fully dynamic connectivity in the cell probe model with word size $O(\log n)$. The cell probe model [20] is a useful model for proving lower bounds of algorithms; computation in this model is framed as querying a set of memory cells. Patrascu et al. [16] improved the bounds to $\Omega(\log n)$ amortized. For maintaining special classes of graphs, Hell et al. [9] gave an $\Omega(\log n/\log \log n)$ amortized lower bound per update and query operation for fully dynamic recognition of proper interval graphs with word size of $O(\log n)$. It is not difficult to improve the lower bound to $\Omega(\log n)$ by applying the result of Patrascu et al. [16]. Hell et al. [9] also state without proof that a similar technique may be applied to get lower bounds for dynamic recognition of chordal graphs. In this paper, we give a formal proof of the $\Omega(\log n)$ lower bound that also applies for a few other subclasses of chordal graphs.

Amortization bounds in all these structures report the time per update/query required by the algorithm over a long sequence of query and update operations. In particular, the algorithm will not perform an update if the graph property is not satisfied on the resulting graph. For example if at some update, an edge is deleted and the resulting graph is not chordal then we insert the edge back again and continue to the next update.

1.2 Our Results

Our first structure follows Ibarra's approach in maintaining chordality by maintaining a clique tree decomposition of the chordal graph. However, we design and analyze our structures based on the maximum size k of a bag in the clique tree. Specifically we show that we can construct a data structure in $O(nk^2)$ time such that given an edge e to be deleted from G, we can report in $O(1)$ time if $G \backslash e$ is chordal and if it is, we can update the structure in $O(n + k^2)$ time. For example, for planar graphs where the maximum size of a clique is a constant, our structure supports an $O(1)$ query and $O(n)$ update time in the worst case. Using a careful charging argument, we show that the update time is actually $O(n^2/\Delta + k^2)$ amortized over Δ edge deletions. Hence, in particular if Δ is at least n^2/k^2, the amortized bound becomes $O(k^2)$. For example, if $k = \Theta(n^{1/4})$ and the initial chordal graph has $\Theta(n^{3/2}) = \Theta(n^2/k^2)$ edges, over all these edge deletions, our amortized bound for an update is $O(k^2)$ which is $O(\sqrt{n})$ while the query is still supported in constant time.

Our next results uses a perfect elimination ordering of chordal graphs.

– We show that a PEO can be represented by a dynamic list [6] so that given a query edge (u, v) to be deleted, we can detect chordality in $O(\min\{degree(u), degree(v)\})$ and update the resulting chordal graph in $O(degree(u) + degree(v))$ time. Our bounds match the bounds in Ibarra's result for the decremental setting in the worst case, but give better results when u and v have low degree. In particular, for chordal graphs with bounded degree, our method takes a constant time to update a PEO under edge deletions.
– We then give a method to augment a clique-tree decomposition of a chordal graph with simple data structures and show that we can obtain a PEO of the graph from a clique-tree in $O(n)$ time after an edge insertion or deletion(against a naive $O(m + n)$ time algorithm). This answers a question posed

by Ibarra [12]. The only non-trivial algorithms for problems such as minimum coloring, maximum independent set, minimum clique cover on chordal graphs are known via PEO [8]. Thus, our conversion from clique tree to PEO means that all of Ibarra's results which took $O(n)$ query and update time or more can now be directly translated to PEO and hence all these problems on chordal graphs can also be solved more efficiently.

Finally, we give the first non-trivial lower bound for the fully dynamic maintenance of chordal graphs. By giving a reduction from the problem of dynamically maintaining a forest under edge insertions and deletions, we show that any structure to maintain a chordal graph requires $\Omega(\log n)$ amortized time for a query or an update.

1.3   Organization of the Paper

In Sect. 2, we develop a structure using clique trees if only deletion of edges are allowed. We analyze this structure in two different ways to give a worst case and an amortized bound. Section 3 gives a worst case algorithm, on deletion of edges using a PEO ordering of the graph.

We then give an algorithm to obtain a PEO from a clique tree efficiently. Section 4 gives the lower bound for our problem.

We conclude in Sect. 5 with open problems and further directions of research.

2   Decremental Algorithms Using Clique Tree

We assume that the vertices in the graph are labelled $1, 2, ..., n$. The neighborhood of a vertex u refers to the adjacent vertices of u in the graph and $degree(u)$ refers to the number of neighbors of a vertex u. We start with the definiton of a clique tree decomposition of a graph. Given a graph $G = (V, E)$ with $|V| = n$ and $|E| = m$, a tree decomposition of G is a pair $(T, \{X_i\}_{i \in V(T)})$, where T is a tree, $V(T)$ is its vertex set, and there is a set $X_i \subseteq V$ associated with each node i of the tree, with the following properties [5]: (1) The union of all sets X_i equals V. (2) For every edge (u, v) in the graph, there is a subset X_i that contains both u and v. (3) For each vertex v of the graph, all the nodes X_i that v belongs to form a subtree of T.

A clique tree of a graph G is a tree decomposition where the subsets X_i in each node induce a maximal clique. To distinguish between vertices of the graph and its associated tree decomposition, we call the vertices of the tree as nodes. We will use bags and nodes interchangeably to denote the sets X_i when there is no confusion. We define the neighbors of a node in the clique tree to be its parent and all its children.

2.1   Structure with a Worst Case Update Time

We provide structures here that use the clique-tree decomposition of chordal graphs. They mainly use the following characterizations of chordal graphs.

Theorem 1 [3,19]. *A graph G is chordal if and only if G has a clique tree.*

Lemma 1 [12]. *Given a chordal graph G, and an edge $e = (u, v)$, $G\backslash e$ is chordal if and only if u and v are together present in exactly one maximal clique, and hence in only one bag of the clique tree.*

Using this, we prove the following:

Theorem 2. *Let G be a chordal graph. Let k be the maximum size of a clique in G. We can construct a data structure in $O(nk^2)$ time such that given an edge (u, v) to be deleted from G, we can report in $O(1)$ time if $G\backslash(u, v)$ is chordal and if it is, we can update the structure in $O(n + k^2)$ time.*

Proof. We first give a high level description of Ibarra's algorithm in maintaining a clique tree of the chordal graph under edge deletions. We then explain the data structures to implement it to support the operations in the claimed bounds.

Algorithm

1. Check if the given edge (u, v) is present in only one bag, if not report a negative answer, and if yes, then we need to update the clique tree.
2. If Y is the unique bag containing the edge (u, v), the node corresponding to Y is split into two nodes, Y_1 and Y_2. Y_1 now contains $Y\backslash u$ and Y_2 contains $Y\backslash v$. Y_2 becomes the parent of Y_1. From the children of Y_1 remove all nodes which contain u and make them children of Y_2. The other children remain as children of Y_1.
3. Check whether the bags of any neighbor of these newly formed nodes is a superset of the node. If yes, we "absorb" these nodes into the corresponding neighbor.

   To check whether one node is a superset of the other, Ibarra maintains the intersection size of two adjacent nodes X and Y. We denote this to be the *int* value between nodes X and Y. Let Y be the node which has been split and let ℓ be the size of Y before splitting. If $|X \cap Y| = \ell - 1$ then X absorbs the new Y. Check [12] for details.

Now we give details of the structures used to implement the algorithm. First, we build a clique tree from the given graph G. The clique tree can be represented by a pointer representation where each node points to its parent in the tree. Furthermore, we maintain the following structures.

- For each edge in the graph G we store,
  - a counter indicating the number of nodes of the clique tree to which the edge belongs, and
  - two way pointers from/to each edge to/from all the nodes it belongs to.

  We can store this structure as an adjacency matrix, with each position (u, v) in the matrix having the counter and the list of pointers. Accessing the information corresponding to edge (u, v) can be done in $O(1)$ time.
- Similarly for each vertex of the graph G we maintain a counter and a list of two way pointers to all the nodes it belongs to.

– For each node X in the clique tree, we store
  - the list of vertices sorted according to their labels,
  - For each node Y in the clique tree which is a neighbor of X, we store $|X \cap Y|$ in non-increasing order of values in an array associated with the bag X with a pointer from each cell in the array to the node it corresponds to.

Now we explain how to support the delete operation. Given a query (u, v), we first look at the counter value of (u, v) in the adjacency matrix. If it is more than one, we report that $G \backslash (u, v)$ is not chordal. Otherwise, we need to update the clique tree.

Let the node which is pointed to by the cell (u, v) in the adjacency matrix be Y and its parent be Y_{par}. Also let $|Y| = \ell$. We create two new nodes Y_1 which contains $Y \backslash u$ and Y_2 that contains $Y \backslash v$. Y_2 becomes the parent of Y_1. Y_{par} becomes Y_2's parent. For all the children of Y, all nodes which contain u become children of Y_2 and all nodes which contain v become children of Y_1. The connected subtree property ensures that v does not appear in Y_2 or any of its ancestors. So the int values between the nodes Y_{par}, Y_1, Y_2 and all its children remain the same as it was between Y and these children. The new nodes formed may now be subsets of any of its adjacent nodes. To maintain the clique tree, we now consider four distinct cases:

Case 1. If none of the intersections between Y_1, Y_2 and their neighborhood is $\ell - 1$ (we can find this from the sorted lists associated with Y_1 and Y_2) then neither Y_1 nor Y_2 is absorbed. In this case, update the int value of Y_2 and its parent by creating a sorted list for Y_2 and inserting the int value of $Y_2 \cap Y_{par}$ in the array of Y_{par}. Add pointers of all edges and vertices which are part of Y_2 to point at the node and change the counters.

Case 2. If Y_1 gets absorbed into one of its neighbors (check int of Y_1 with its neighbors and see which one is $\ell - 1$, in case of a tie choose any one), delete the node Y_1, adjust the parent pointer of its neighbor to now point at Y_2, and adjust all the parent pointers of all the other neighbors of Y_1, to point at this new node. Merge the two sorted int lists of Y_1 and its neighbor together.

Case 3. Our algorithm is similar to Case 2 if Y_2 gets absorbed into one of its neighbors.

Case 4. If both Y_1 and Y_2 are absorbed into one of their neighbors, the parent pointer of the neighbor which absorbs Y_1, now points to the neighbor which absorbed Y_2. The int value between these two nodes becomes $\ell - 2$.

Update the sorted lists in each of the new nodes formed in the clique tree.

Now, we analyze the runtime for construction of our structures. Building a clique tree requires $O(m + n)$ time. Let k be the maximum size of a node in the clique tree. From the property of chordal graphs, we know that there are a maximum of n nodes in the tree. Then there are a total of $m = O(nk^2)$ edges of the graph in the clique tree. Thus, building a clique tree takes $O(nk^2)$ time giving a total preprocessing time of $O(nk^2)$. Storing a counter and the pointers for each edge and vertex of the clique tree takes a total of $O(nk^2)$ time.

For deletion query, we look at the concerned cell of the adjacency matrix and check if the counter value is 1 in $O(1)$ time.

For update, for each of the cases above it takes $O(k^2)$ time to update the counters and the pointers for all edges (there are at most k^2 edges in a node) and $O(k)$ time to update for all the vertices. In $O(1)$ time we can, from the sorted *int* lists find if a node will be absorbed or not. In Case 1, updating the parent pointer information takes $O(1)$ time. Updating the sorted *int* list takes $O(\log n)$ time. In Cases 2, 3 and 4, where the nodes Y_1 and/or Y get absorbed, we need to update the pointers of all neighbors of Y and also merge two sorted lists. This takes $O(n)$ time. Updating the vertex information of each node takes $O(k)$ time. Thus the total time taken to update is $O(n + k^2)$. $\square$

2.2   Amortized Analysis

We now give a better amortized runtime bound for the above algorithm by analyzing it differently. We show

Theorem 3. *Let G be a chordal graph. We can, in $O(nk^2)$ time, construct a data structure such that given a sequence of Δ edge deletions, we can support deletion query in $O(1)$ time and deletion update in $O(n^2/\Delta + k^2)$ amortized time.*

Proof. Updation of the structures involve the time to split a node in the clique tree and also to absorb the node into one of its neighbors and updating the clique tree. We deal with the total time taken to perform the split and absorb operations seperately.

First, we look at the total time spent for the split operations for each node. Let Y be a node in the clique tree before any deletion operation and let d be the degree of the node Y. Over the course of edge deletions, Y gets split into multiple nodes. Let us denote these set of nodes to be Y_{split}. Whenever a node from Y_{split} splits the node size decreases by one and the total cost incurred is the degree of that node. To analyze the runtime we can imagine a binary tree whose root node is Y with a node size of k. Y has two children each (because of a split) with each node of size $k - 1$. They have four children each of which correspond to a node of size $k - 2$ and so on. The total cost incurred at each level is d. The maximum height of this tree is k. So the total time spent by Y is $O(kd)$. Now, $k \sum d$ is at most $k(n - 1)$ and hence we have the total time taken by the algorithm for splitting nodes is $O(kn)$.

Now, let us analyze the total runtime for absorption of nodes in the algorithm. Let $Y's$ neighbor where it gets absorbed be Y_{nbr}. Let d be the degree of the node Y, and d_{nbr} be the degree of the node Y_{nbr} before absorption. The cost of absorption to update the pointers of Y_{nbr} is equal to d.

We associate a charge with every node to account for part of the work done during the absorption. Eventually the sum of the charges in the (existing) nodes account for the total work done for absorptions. Let Y be a node which is absorbed into Y_{nbr} at some point in the sequence of deletions. The amount of work done for this absorption is the number of children of Y (which now become the children of Y_{nbr}) to update the child pointers of Y_{nbr}. We account for this by adding a charge of d to the node Y_{nbr}. In addition, we pass the charge accumulated in Y to Y_{nbr}. So the new charge at Y_{nbr} is the old charge in that node plus the charge at Y plus d.

We first claim that the charge accumulated at any node with degree d is at most d^2. If this was true before, then the new charge at Y_{nbr} is at most its old charge plus $d + d^2$ (as the charge at Y was at most d^2 by induction hypothesis and its degree is at most d). The old charge at Y_{nbr} by induction hypothesis is at most d_{nbr}^2, and its new degree is $d + d_{nbr}$. The new charge is at most $d_{nbr}^2 + d^2 + d \leq (d + d_{nbr})^2$ which proves the claim.

Hence the total charge on the existing nodes at any point of time is at most $4n^2$ as the sum of the degrees is at most $2n$. We spend another $O(k^2)$ time for each update to update the nodes corresponding to every pair of vertices in the bag that got split. Thus, the amortized time for Δ edge deletions is $O(n^2/\Delta + k^2)$. $\qquad\square$

We can, in $O(nk^2)$ time, construct a data structure such that given a sequence of $\Omega(n^2/k^2)$ edge deletions, we can support deletion query in $O(1)$ time and deletion update in $O(k^2)$ total time. In particular if the graph has at least $m = \Omega(n^{3/2})$ edges and the size of the maximum clique is $O(n^{1/4})$, then we have an $O(1)$ query and $O(n^{1/2})$ update time.

3   Dynamic Maintenance of Perfect Elimination Ordering

3.1   Decremental Algorithm

We now give a decremental algorithm using perfect elimination ordering (PEO). Towards that we first state the following characterization.

Lemma 2 [14]. *Let G be a chordal graph, and let $e = (u, v)$ be an edge. $G\backslash(u, v)$ is chordal if and only if all the common neighbors of u and v are adjacent to each other, i.e., they form a clique.*

Using the above characterization and the well-known dynamic list to represent the PEO, we obtain the following result.

Theorem 4. $\star$[1] *Let G be a chordal graph represented by its adjacency list and adjacency matrix. We can, in $O(m + n)$ time, construct a PEO of G, such that whenever an edge (u, v) is deleted, we can determine if $G\backslash(u, v)$ is chordal in $O(\min\{degree(u), degree(v)\})$ time, and update the structures if it is the case, in $O(degree(u) + degree(v))$ time.*

If the chordal graph has bounded degree, we get the following.

Corollary 1. *Let G be a chordal graph with bounded degree given by its adjacency matrix and adjacency list. We can in $O(m + n)$ time construct a PEO of the vertices of G such that whenever an edge (u, v) is deleted, we can in $O(1)$ time determine if $G\backslash(u, v)$ is chordal and if yes, we can update the structure in $O(1)$ time.*

---

[1] Proof deferred to the full version.

3.2   Fully Dynamic Maintenance of PEO

We show how to convert a clique tree decomposition of a chordal graph to its PEO ordering in $O(n)$ time even under edge insertions and deletions, thus answering a question posed by Ibarra [12]. In $O(m + n)$ time we can obtain a clique tree from a graph G as well as store the intersection values between every pair of adjacent nodes in a clique tree [17,18]. In addition, we show that for the lists associated with two adjacent nodes, defined as A and B, we can also store the values $A\backslash B$ and $B\backslash A$ and update them on edge addition and deletion efficiently. This added information helps to convert from a clique tree to PEO efficiently in $O(n)$ time. Using the help of the following lemma (proof in appendix) we show how to maintain this structure for edge additions/deletions.

Lemma 3. $\star^2$ *Let G be a chordal graph given with its clique-tree decomposition. We can construct a data structure in $O(nk \log k)$ time (where k refers to the maximum node size in the clique tree) and store the vertices differing between two adjacent nodes in the clique tree i.e. for the lists A and B associated with the two adjacent nodes, we store the values $A\backslash B$ and $B\backslash A$ on the edge connecting the nodes. We can update this structure and the clique-tree in $O(n)$ time on addition or deletion of an edge from the graph G.*

Using this lemma, we look at converting a clique tree to a PEO ordering efficiently. The proof of the following theorem gives details of implementation in $O(n)$ time.

Theorem 5. *Let G be a chordal graph given with its clique-tree decomposition. We can augment it in $O(nk \log k)$ time (where k refers to the maximum node size in the clique tree) such that we can convert the clique tree to a PEO in $O(n)$ time on addition or deletion of an edge provided the modified graph remains chordal.*

Proof. Let A and B denote the lists associated with two adjacent nodes. Define $P = A\backslash B$ and $Q = B\backslash A$. Using Lemma 3 we construct and store the clique tree data structure augmented with the sets P and Q on each edge. Arbitrarily root the clique tree and do a depth first search traversal ordered by the start times of the nodes. We take all the vertices in that node and push it into a stack. We continue our DFS traversal and whenever we arrive at a node A, we push the vertices $A\backslash B$ into the stack, where B is its parent. At the end of the traversal, pop the vertices from the stack. The order in which they are popped is the order of the PEO.

For correctness we need to show that this traversal maintains the PEO at any time instant. Initially when we consider the root node, they form a maximal clique, so pushing them in any order in the stack does not violate the PEO property. Let us take a node A at some intermediate step of the traversal and push $A\backslash B$ into the stack. For a vertex $a \in A\backslash B$ to violate the PEO ordering, a has to be a neighbor of b and c, two vertices already in the stack below a but

---

^{2} Proof deferred to the full version.

b and c are not adjacent to each other. But this cannot happen. As b and c are already in the stack they have been visited earlier in the traversal. We show b and c are both in node A. If not, they cannot appear after A in the traversal as it violates the connected subtree property. A is the first node in the traversal which contains a and as (a,b) and (a,c) are neighbors they have to appear in some node by definition of clique trees. Therefore, the vertices b and c are both in node A as well. As each node is a maximal clique, vertices b and c are also neighbors. So we see that the algorithm does not violate the PEO ordering.

During the traversal we push each vertex into the stack only once and pop them out once. Hence, the total time spent is $O(n)$.   □

4   Lower Bound

We first observe that the reduction [16] from the Query-Sum problem to dynamic connectivity also holds for fully dynamic connectivity on forests to show the following.

Theorem 6 [16]. *Consider any dynamic data structure that performs a sequence of n edge insertions and deletions that maintain the forest structure starting from an edgeless graph. Suppose the structure also supports queries of the form whether a pair of vertices are in the same connected component. Then such a structure requires $\Omega(\log n)$ amortized time per query and update to support a sequence of n query and update operations in the cell probe model of word size $\log n$.*

We use this observation to give a reduction to our problem to prove a similar lower bound.

Theorem 7. *Any dynamic structure that maintains a chordal graph under edge insertions and deletions requires $\Omega(\log n)$ amortized time per update or query in the cell probe model of word size $\log n$.*

Proof. The main idea is to ensure that when a query for a pair (u,v) comes, we add a new path of length three between u and v and check whether the resulting graph is chordal. If the pair of vertices are in different components, then the new additions don't add any cycle, and if they are in the same component, then new additions create a chordless cycle of length greater than three. Hence we can test the reachability question using the chordality query. We give the details below.

Given an instance I of the fully dynamic connectivity problem on forests with n vertices, we create a graph on $n+2$ vertices where the first n vertices correspond to the original vertices, and there are two new vertices s and t with an edge between s and t. Whenever an edge $\{u,v\}$ is added to I, we call the addition of edge $\{u,v\}$ to I'. Whenever an edge $\{u,v\}$ is deleted from I, we delete the same edge from I'. The forest maintenance property of the instance I ensures that these addition or deletion of edges always ensures a forest is maintained in I' as well.

When a query between a pair of vertices u and v comes, we simply add the edges $\{u, s\}$ and $\{t, v\}$ and ask whether the resulting graph is chordal. If it is, then we declare that u and v are in different components of the forest, and otherwise they are in the same component. We then delete the edges $\{u, s\}$ and $\{t, v\}$ from the graph. If u and v are in the same component, then the path in the component between u and v along with edges $\{u, s\}, \{s, t\}$ and $\{t, v\}$ form a chordless cycle. This proves the correctness of the reduction.

Thus every connectivity query in I is implemented by two edge additions, a chordality query and two edge deletions in I'. Furthermore, every update in I is implemented by the same update in I'. Thus from Theorem 6, the theorem follows.

We observe that the only property of chordal graphs we used in the above reduction is that trees are chordal and any induced cycle of length greater than three is not chordal. Hence the same reduction works for any subclass of chordal graphs that contains the class of trees. Thus we have

Corollary 2. *Any dynamic structure that maintains a Ptolemaic graph or a k-tree or a strongly chordal graph (for definitions refer [2, 11, 15]) under edge insertions and deletions requires $\Omega(\log n)$ amortized time per update or query.*

5   Conclusions

We have presented improved upper and lower bounds for maintaining chordal graphs under edge deletions and insertions. graphs. We also showed that we can shift between different decompositions of chordal graphs in $O(n)$ time which helps to solve applications that require different decompositions. An interesting open problem is to prove a super logarithmic lower bound for the query and update operations for maintenance of chordal graphs. We have given a structure to maintain a PEO under edge insertions and deletions in $O(n)$ time by augmenting the clique tree decomposition. It would be an interesting problem to see if the optimization problems (like maximum clique and independent set) that use PEO can be updated in $O(n)$ time under edge insertions and deletions.

Acknowledgement. The first author would like to thank Keerti Choudhary for useful discussions leading to Theorem 5.

References

1. Berry, A., Sigayret, A., Spinrad, J.: Faster dynamic algorithms for chordal graphs, and an application to phylogeny. In: Kratsch, D. (ed.) WG 2005. LNCS, vol. 3787, pp. 445–455. Springer, Heidelberg (2005). https://doi.org/10.1007/11604686_39
2. Brandstädt, A., Dragan, F.F., Chepoi, V., Voloshin, V.I.: Dually chordal graphs. SIAM J. Discrete Math. **11**(3), 437–455 (1998)
3. Buneman, P.: A characterisation of rigid circuit graphs. Discrete Math. **9**(3), 205–212 (1974)

4. Deshpande, A., Garofalakis, M.N., Jordan, M.I.: Efficient stepwise selection in decomposable models. In: UAI 2001: Proceedings of the 17th Conference in Uncertainty in Artificial Intelligence, University of Washington, Seattle, Washington, USA, 2–5 Aug 2001, pp. 128–135 (2001)
5. Diestel, R.: Graph Theory. GTM, vol. 173, 4th edn. Springer, Heidelberg (2012)
6. Dietz, P.F., Sleator, D.D.: Two algorithms for maintaining order in a list. In: Proceedings of the 19th Annual ACM Symposium on Theory of Computing 1987, New York, NY, USA, pp. 365–372 (1987)
7. Fagin, R.: Degrees of acyclicity for hypergraphs and relational database schemes. J. ACM **30**(3), 514–550 (1983)
8. Gavril, F.: Algorithms for minimum coloring, maximum clique, minimum covering by cliques, and maximum independent set of a chordal graph. SIAM J. Comput. **1**(2), 180–187 (1972)
9. Hell, P., Shamir, R., Sharan, R.: A fully dynamic algorithm for recognizing and representing proper interval graphs. SIAM J. Comput. **31**(1), 289–305 (2001)
10. Henzinger, M.R., Fredman, M.L.: Lower bounds for fully dynamic connectivity problems in graphs. Algorithmica **22**(3), 351–362 (1998)
11. Howorka, E.: A characterization of ptolemaic graphs. J. Graph Theory **5**(3), 323–331 (1981)
12. Ibarra, L.: Fully dynamic algorithms for chordal graphs and split graphs. ACM Trans. Algorithms **4**(4), 40:1–40:20 (2008)
13. Mezzini, M.: Fully dynamic algorithm for chordal graphs with $O(1)$ query-time and $O(n^2)$ update-time. Theor. Comput. Sci. **445**, 82–92 (2012)
14. Mezzini, M., Moscarini, M.: Simple algorithms for minimal triangulation of a graph and backward selection of a decomposable markov network. Theor. Comput. Sci. **411**(7–9), 958–966 (2010)
15. Nesetril, J.: Structural properties of sparse graphs. Electron. Notes Discrete Math. **31**, 247–251 (2008)
16. Patrascu, M., Demaine, E.D.: Logarithmic lower bounds in the cell-probe model. SIAM J. Comput. **35**(4), 932–963 (2006)
17. Rose, D.J., Tarjan, R.E., Lueker, G.S.: Algorithmic aspects of vertex elimination on graphs. SIAM J. Comput. **5**(2), 266–283 (1976)
18. Tarjan, R.E., Yannakakis, M.: Simple linear-time algorithms to test chordality of graphs, test acyclicity of hypergraphs, and selectively reduce acyclic hypergraphs. SIAM J. Comput. **13**(3), 566–579 (1984)
19. Walter, J.R.: Representations of chordal graphs as subtrees of a tree. J. Graph Theory **2**(3), 265–267 (1978)
20. Yao, A.C.-C.: Should tables be sorted? J. ACM **28**(3), 615–628 (1981)

Distributed Symmetry-Breaking Algorithms for Congested Cliques

Leonid Barenboim$^{(\boxtimes)}$ and Victor Khazanov

Open University of Israel, Ra'anana, Israel
leonidb@openu.ac.il, viktorkh@gmail.com

Abstract. The *Congested Clique* is a distributed-computing model for single-hop networks with restricted bandwidth that has been very intensively studied recently. It models a network by an n-vertex graph in which any pair of vertices can communicate one with another by transmitting $O(\log n)$ bits in each round. Various problems have been studied in this setting, but for some of them the best-known results are those for general networks. For other problems, the results for Congested Cliques are better than on general networks, but still incur significant dependency on the number of vertices n. Hence the performance of these algorithms may become poor on large cliques, even though their diameter is just 1. In this paper we devise significantly improved algorithms for various symmetry-breaking problems, such as forests-decompositions, vertex-colorings, and maximal independent set.

We analyze the running time of our algorithms as a function of the arboricity a of a clique subgraph that is given as input. The arboricity is always smaller than the number of vertices n in the subgraph, and for many families of graphs it is significantly smaller. In particular, trees, planar graphs, graphs with constant genus, and many other graphs have bounded arboricity, but unbounded size. We obtain $O(a)$-forest-decomposition algorithm with $O(\log a)$ time that improves the previously-known $O(\log n)$ time, $O(a^{2+\epsilon})$-coloring in $O(\log^* n)$ time that improves upon an $O(\log n)$-time algorithm, $O(a)$-coloring in $O(a^\epsilon)$-time that improves upon several previous algorithms, and a maximal independent set algorithm with $O(\sqrt{a})$ time that improves at least quadratically upon the state-of-the-art for small and moderate values of a.

Those results are achieved using several techniques. First, we produce a forest decomposition with a helpful structure called *H-partition* within $O(\log a)$ rounds. In general graphs this structure requires $\Theta(\log n)$ time, but in Congested Cliques we are able to compute it faster. We employ this structure in conjunction with partitioning techniques that allow us to solve various symmetry-breaking problems efficiently.

This research has been supported by ISF grant 724/15 and Open University of Israel research fund. Full version of this paper is available online: https://arxiv.org/pdf/1802.07209.pdf.

© Springer International Publishing AG, part of Springer Nature 2018
F. V. Fomin and V. V. Podolskii (Eds.): CSR 2018, LNCS 10846, pp. 41–52, 2018.
https://doi.org/10.1007/978-3-319-90530-3_5

1   Introduction

1.1   The Congested Clique Model and Problems

In the message-passing $LOCAL$ model of distributed computing a network is represented by an n-vertex graph $G = (V, E)$. Each vertex has its own processing unit and memory of unrestricted size. In addition, each vertex has a unique identity number (ID) of size $O(\log n)$. Computation proceeds in synchronous rounds. In each round vertices perform local computations and send messages to their neighbors. The running time in this model is the number of rounds required to complete a task. Local computation is not counted towards running time. Message size is not restricted. Therefore, this model is less suitable for networks that are constrained in message size as a result of limited channel bandwidth. To handle such networks, a more realistic model has been studied. This is the $CONGEST$ model that is similar to the LOCAL model, except that each edge is only allowed to transmit $O(\log n)$ bits per round. An important type of CONGEST networks that has been intensively studied recently is the *Congested Clique* model. It represents single-hop networks with limited bandwidth. Although the diameter of such networks is 1, which would make any problem on such graphs trivial in the LOCAL model, in the Congested Cliques various tasks become very challenging. Note that the Congested Clique is equivalent to a general n-vertex graph in which any pair of vertices (not necessarily neighbors) can exchange messages of size $O(\log n)$ in each round. Such a general graph corresponds to a subgraph of an n-clique. The subgraph constitutes the input, while the clique constitutes the communication infrastructure.

The study of the problem of Minimum Spanning Tree (henceforth, MST) was initiated in the Congested Clique model by Lotker et al. [16]. They devised a deterministic $O(\log \log n)$-rounds algorithm that improved a straight-forward $O(\log n)$ solution. In the sequel, randomized $O(\log \log \log n)$-rounds- [10,17], $O(\log^* n)$-rounds[1]- [8], and $O(1)$-rounds [12] algorithms for MST in Congested Cliques were devised. These algorithms, however, may fail with certain probabilities. Thus obtaining deterministic algorithms that never fail seems to be a more challenging task in this setting. Since the publication of the result of [16] many additional problems have been studied in the Congested Clique setting [4–7,11]. In particular, several *symmetry-breaking* problems were investigated. Solving such problems is very useful in networks in order to allocate resources, schedule tasks, perform load-balancing, and so on. Hegeman and Pemmaraju [11] obtained a randomized $O(\Delta)$-coloring algorithm with $O(1)$ rounds if the maximum degree Δ is at least $\Omega(\log^4 n)$, and $O(\log \log n)$-time otherwise. We note that although in a clique it holds that $\Delta = n - 1$, and an $O(\Delta)$-coloring algorithm is trivial (by choosing unique vertex identifiers as colors), the problem is defined in a more general way. Specifically, we are given a clique $Q = (V, E)$, and a subgraph $G' = (V, E'), E' \subseteq E$. The goal is computing a solution for G' as a function of $\Delta = \Delta(G')$, rather then $\Delta(Q)$. In this

---

[1] $\log^* n$ is the number of times the $\log_2$ function has to be applied iteratively until we arrive at a number smaller than 2. That is, $\log^* 2 = 1$, and for $n > 2$, $\log^* n = 1 + \log^*(\log n)$.

case the $O(\Delta)$-coloring problem becomes non-trivial at all. We are not aware of previously-known deterministic algorithms for coloring in the Congested Clique that outperform algorithms for general graphs. (Except an algorithm of [5] that is not applicable in general, but rather if $\Delta = O(n^{1/3})$. In this case its running time is $O(\log \Delta)$.)

Another symmetry-breaking problem that was studied in the Congested Clique is Maximal Independent Set (henceforth, MIS). The goal of this problem is to compute a subset of non-adjacent vertices that cannot be extended. Again, this problem is interesting in subgrahs of the Congested Clique, rather than the Congested Clique as a whole. A deterministic algorithm for this problem with running time $O(\log \Delta \log n)$ was devised in [5]. If $\Delta = O(n^{1/3})$ then the running time of the algorithm of [5] improves to $O(\log \Delta)$. Ghaffari [7] devised a randomized MIS algorithm for the Congested Clique that requires $\tilde{O}(\log \Delta/\sqrt{\log n}+1) \le \tilde{O}(\sqrt{\log \Delta})$ rounds. Interestingly, when Δ is not restricted, all above-mentioned deterministic algorithms and most randomized ones have significant dependency on the clique size n. Obtaining a deterministic algorithm for these problems that does not depend on n is an important objective, since very large clique subgraphs may have some bounded parameters (e.g., bounded arboricity) that can be utilized in order to improve running time.

1.2   Our Results and Techniques

We devise improved *deterministic* symmetry-breaking algorithms for the Congested Clique that have very loose dependency on n, or not at all. Specifically, for clique subgraphs with *arboricity*[2] a we obtain $O(a)$-coloring in $O(a^\epsilon)$ time (for an arbitrarily small constant $\epsilon > 0$), $O(a^{1+\epsilon})$-coloring in $O(\log^2 a)$ time, $O(a^{(2+\epsilon)})$-coloring in $O(\log^* n)$ time and Maximal Independent Set in $O(\sqrt{a})$ time. The best previously-known algorithms for these coloring problems are those for general graphs, and incur a multiplicative factor of $\log n$. See table below. Moreover, in general graphs, the $\log n$ factor is unavoidable when solving the coloring problems in which the number of colors is a function of a [1]. Our results demonstrate that in Congested Cliques much better solutions are possible. Our MIS algorithm outperforms the results of [4] when there is a large gap between a and Δ or between a and n. For example, trees, planar graphs, graphs of constant genus, and graphs that exclude any fixed minor, all have arboricity $a = O(1)$. On the other hand, their maximum degree Δ and size n are unbounded.

Our main technical tool is an $O(a)$-forests-decomposition algorithm that requires $O(\log a)$ rounds in the Congested Clique. This is in contrast to general graphs where $O(a)$-forests-decomposition requires $\Theta(\log n)$ rounds [1]. Once we compute such a forests decomposition, each vertex knows its $O(a)$ parents in the $O(a)$ forests of the decomposition. We orient edges towards parents. The union of all edges that point towards parents constitute the edge set E' of the

---

[2] The arboricity is the minimum number of forests that graph edges can be partitioned into. It always holds that $a(G') \le \Delta(G')$, and often the arboricity of a graph is significantly smaller than its maximum degree.

Our results (deterministic)		Previous results (deterministic and randomized)	
	Running time		Running time
Forest-decomposition	$O(\log a)$	Forest-decomposition [1]	$O(\log n)$
$O(a^{2+\varepsilon})$-coloring	$O(\log^* n)$	$O(a^{2+\varepsilon})$-coloring [1]	$O(\log n)$
$O(a^2)$-coloring	$O(\log a) + \log^* n$	$O(a^2)$-coloring [1]	$O(\log n)$
$O(a^{1+\varepsilon})$-coloring	$O(\log^2 a)$	$O(a^{1+\varepsilon})$-coloring [2]	$O(\log a \log n)$
$O(a)$-coloring	$O(a^\varepsilon)$	$O(a)$-coloring [2]	$O(\min(a^\varepsilon \log n, a^\varepsilon + \log^{1+\varepsilon} n))$
MIS	$O(\sqrt{a})$	MIS [1]	$O(a + \log n)$
		MIS [5]	$O(\log \Delta \log n)$
		MIS (rand.) [7]	$\tilde{O}(\sqrt{\log \Delta})$
$O(\Delta)$-coloring	$O(a^\varepsilon)$	$O(\Delta)$-coloring (rand.) [11]	$O(\log \log n)$

input. This is because for each edge, one of its endpoint is oriented outwards, and is considered in the union. Note also that the out degree of each vertex is $O(a)$. Then, within $O(a)$ rounds each vertex can broadcast the information about all its outgoing edges to all other vertices in the graph. Indeed, each outgoing edge can be represented by $O(\log n)$ bits using IDs of endpoints. Then, in round $i \in O(a)$, each vertex broadcasts to all vertices the information of its ith outgoing edges. After $O(a)$ rounds all vertices know all edge of E' and are able to construct locally (in their internal memory) the input graph $G' = (V, E')$.

Once vertices know the input graph they can solve any computable problem (for unweighted graphs or graphs with weights consisting of $O(\log n)$ bits) locally. The vertices run the same deterministic algorithm locally, and obtain a consistent solution (the same in all vertices). Then each vertex deduces its part from the solution of the entire graph. This does not require communication whatsoever, and so the additional (distributed) running time for this computation is 0. Thus our results demonstrate that any computable problem can be solved in the Congested Clique in $O(a)$ rounds deterministically. This is an alternative way of showing what follows from Lenzen's [14] routing scheme, since a graph with arboricity a has $O(n \cdot a)$ edges that can be announced within $O(a)$ rounds of Lenzen's algorithm. But the additional structure of forests-decomposition that we obtain is useful for speeding up certain computations, as we discuss below. We note that although in this model it is allowed to make unrestricted local computation, in this paper we do not abuse this ability, and devise algorithms whose local computations are reasonable (i.e., polynomial).

Since any computable problem can be solved in $O(a)$ rounds, our next goal is obtaining algorithms with a better running time. We do so by partitioning the input into subgraphs of smaller arboricity. We note that vertex disjoint subgraphs are Congested Cliques by themselves that can be processed in parallel. For example, partitioning the input graph into $O(a^{1-\varepsilon})$-subgraphs of arboricity $O(a^\varepsilon)$, and coloring subgraphs in parallel using disjoint palettes, makes it possible to color the entire input graph with $O(a)$ colors in $O(a^\varepsilon)$ time rather than $O(a)$. Partitioning also works for MIS, although this problem is more difficult to parallelize. (In the general CONGEST model the best algorithm in terms of a has running time $O(a + \log^* n)$.) Nevertheless, using our new partitioning techniques we obtain an MIS with $O(\sqrt{a})$ time in the Congested Clique. We believe that this technique is of independent interest, and may be applicable more broadly. Specifically, by quickly partitioning the input into subgraphs of small arboricity, we can solve any computable problem in these subgraphs in $O(a^\varepsilon)$ time, rather than $O(a)$. Given a method that efficiently combines these solutions, it would be possible to obtain a solution for the entire input significantly faster than $O(a)$.

1.3   Related Work

Lenzen [14] devised a communication scheme for the Congested Clique. Specifically, if each vertex is required to send $O(n)$ messages of $O(\log n)$ bits each, and if each vertex needs to receive at most $O(n)$ messages, then this communication can be performed within $O(1)$ rounds in the Congested Clique. Algebraic methods for the Congested Clique were studied in [4,6]. Symmetry-breaking problems were very intensively studied in general graphs. Many of these results apply to the Congested Clique. In particular, Goldberg et al. [9] devised a $(\Delta+1)$-coloring algorithm with running time $O(\Delta \log n)$. Linial [15] devised an $O(\Delta^2)$-coloring algorithm with running time $O(\log^* n)$. Kuhn and Wattenhofer [13] obtained a $(\Delta + 1)$-coloring algorithm with running time $O(\Delta \log \Delta + \log^* n)$. Barenboim and Elkin [2] devised an $O(\min(a^\varepsilon \log n, a^\varepsilon + \log^{1+\varepsilon} n))$ time algorithm for $O(a)$-coloring, and $O(\log a \log n)$ time algorithm for $O(a^{1+\varepsilon})$-coloring.

2   Preliminaries

In this section we provide some basic definition. In the full version of this paper [3] we also survey several known procedures that are used in our algorithms that we describe in the next sections.

2.1   Definitions

Given a graph $G = (V, E)$, the *k-vertex-coloring* problem goal is finding a proper coloring $\varphi : V \to 1, 2, \ldots, k$ that satisfies $\varphi(v) \neq \varphi(u), \forall (u, v) \in E$. The *out-degree* of a vertex v in a directed graph is the number of edges incident to v that are oriented out of v. An *orientation* μ of (the edge set of) a graph is an assignment of direction to each edge $(u, v) \in E$ either towards u or towards v. In our work we use a concept of *partial orientations*, which was employed by Barenboim and Elkin [2]. A partial orientation is allowed not to orient some edges of the graph. By this definition, a partial orientation σ has deficit at most $d \geq 0$, if for every vertex v in the graph the number of edges incident to v that σ does not orient is no greater than d. Another important parameter of a partial orientation is its length l. This is the length of the longest path P in which all edges are oriented consistently by σ. (That is, each vertex in the path has out-degree and in-degree at most 1 in the path.) An *H-partition* $(H_1, H_2, \ldots, H_\ell)$ of $G = (V, E)$ with degree A, for some parameter A, is a partition of V, such that for any vertex in a set H_i, $i \in [\ell]$, the number of its neighbors in $H_i \cup H_{i+1} \cup \ldots \cup H_\ell$ is at most A.

3   Forest-Decomposition-CC

In this section we describe our Forest-Decomposition algorithm for the Congested Clique. Our Forest-Decomposition algorithm starts with computing an H-partition. This computation is performed faster in Congested Cliques than in general graphs thanks to the following observation. Once the first $O(\log a)$ H-sets are computed (within $O(\log a)$ time), the subgraph induced by the remaining active vertices has at most $O(n)$ edges. (We prove this in Lemma 1 below.) Consequently, all these vertices can learn this entire subgraph using Lenzen's algorithms within $O(1)$ rounds. Then each vertex can locally compute the H-set it belongs to. This is in contrast to the algorithm for general graphs where the running time is $\Theta(\log n)$, even for graphs with $O(n)$ edges.

First we provide a procedure which computes an H-partition within $O(1)$ rounds, on graphs with edge set of size at most $O(n)$. This procedure is based on Lenzen's routing scheme. The main idea of the procedure is that each vertex can transmit all edges adjacent on it to all other vertices in the graph. This is because the overall number of messages each vertex receives in this case is $O(n)$. Indeed, each edge can be encoded as a message of size $O(\log n)$ that contains the IDs of the edge endpoints, and the number of messages is bounded by the number of edges in the graph. Since the number of sent messages of each vertex is also bounded by $O(n)$, Lenzen's scheme allows all vertices to transmit all their edges to all other vertices within constant number of rounds, as long as the number of edges is $O(n)$. Once a vertex receives all the edges of the graph, it constructs the graph in its local memory. All vertices construct the same graph, and perform a local computation of the H-partition. This does not require any communication whatsoever, but since all vertices hold the same graph, the resulting H-partition is consistent in all vertices. This completes the description of the procedure.

Next, we provide a general procedure to compute an H-partition in graphs with any number of edges in the Congested Clique model. The procedure is called *Procedure H-Partition-CC*. The computation is done by first reducing the number of edges to $O(n)$ within $O(\log a)$ rounds, and then invoking Procedure

---

Algorithm 1. H-partition of an input graph G with arboricity a and $O(n)$ edges

---

1: **procedure** SPARSE-PARTITION(G, a, ε)
2:      Each node u in G broadcasts its degree to every other node v in G
3:      Using Lenzen's scheme, send all information about all edges to all vertices of G
4:      Each vertex $v \in V$ perfomrs locally the following operations:
5:      Initially, all vertices of G are marked as active.
6:      $i = \left\lceil \frac{2}{\varepsilon} \log a + 1 \right\rceil$
7:      **while** $i \leq \frac{2}{\varepsilon} \log n$ **do**
8:          **if** v is active and has at most $(2 + \varepsilon) \cdot a$ active neighbors **then**
9:              make v inactive
10:             add v to H_i
11:         $i = i + 1$

Sparse-Partition on the remaining subgraph. The reduction phase (lines 3–13 of the algorithm below) operates similarly to Procedure Sparse-Partition, but the partition into H-sets is performed in a distributed manner, rather than locally, and the number of iterations is just $O(\log a)$, rather than $O(\log n)$. In the next lemmas we show that this is sufficient to reduce the number of edges to $O(n)$.

Algorithm 2. Computing an H-partitions of a general graph G with arboricity a in the Congested Clique model

---

1: **procedure** H-PARTITION-CC(a, ε)
2:    *An algorithm for each vertex v V :*
3:    $i = 1$
4:    **while** $i \leq \lceil \frac{2}{\varepsilon} \cdot \log a \rceil$ **do**
5:       **if** v is active and has at most $(2 + \varepsilon) \cdot a$ active neighbors **then**
6:          make v inactive
7:          add v to H_i
8:          send the messages "inactive" and "v joined H_i" to all the neighbors
9:       **for** each received "inactive" message **do**
10:             mark the sender neighbor as inactive
11:       **end for**
12:       $i = i + 1$
13:    **end while**
14:    $H_i, H_{i+1} \ldots, H_{O(\log n)}$ = invoke Procedure Sparse-Partition on the subgraph induced by remaining active vertices

---

Lemma 1. *After $\lceil \frac{2}{\varepsilon} \log a \rceil$ rounds (lines 4–13 in Algorithm 2), the number of edges whose both endpoints are incident to nodes that are still active is $O(n)$.*

Proof. Consider the ith iteration. By [1], the graph G_i induced by the remaining active vertices in the round i has $(\frac{2}{2+\varepsilon})^i \cdot |V|$ vertices. Recall that a graph with arboricity a has no more than $n \cdot a$ edges. The number of edges in the graph G_i is at most: $(\frac{2}{2+\varepsilon})^i \cdot n \cdot a$. Then in the round $i = \lceil \frac{2}{\varepsilon} \log a \rceil$, the graph G_i has $(\frac{2}{2+\varepsilon})^{\lceil \frac{2}{\varepsilon} \log a \rceil} \cdot n \cdot a = O(n)$ edges. $\qquad\square$

The next lemma states the correctness of Algorithm 2, and analyzes its running time.

Lemma 2. *Algorithm 2 computes an H-partion in $O(\log a)$ rounds.*

Proof. The correctness of Algorithm 1 follows from the correctness of H-partition of [1] in conjunction with Lenzen's routing scheme. Specifically, within $O(\log a)$ rounds the algorithm properly computes the H-sets $H_1, H_2, \ldots, H_{O(\log a)}$, and within an additional round the remaining subgraph is learnt by all vertices using Lenzen's scheme, and all H-sets of this subgraph, up to $H_{O(\log n)}$, are computed locally by each vertex. Thus, each vertex can deduce the index of its H-set within $O(\log a)$ rounds from the beginning of the algorithm. $\qquad\square$

We summarize the properties of Procedure H-Partition-CC below.

Theorem 1. *Procedure H-Partition-CC invoked on a graph G with arboricity $a(G)$ and a parameter ε, $0 < \varepsilon \leq 2$ computes an H-partition of size $l = O(\log n)$ with degree at most $O(a)$. The running time of the procedure is $O(\log a)$.*

We next devise a forest-decomposition algorithm for the Congested Clique model, called *Procedure Forest-Decomposition-CC*. It accepts as input the parameters a and ε. In the first step, it computes an *H-Partition-CC*, with degree at most $(2+\varepsilon) \cdot a$. In the next step, it invokes a procedure called Procedure Orientation [2] as follows.

Procedure Orientation: For each edge $e = (u,v)$, if the endpoints u, v are in different sets $H_i, H_j, i \neq j$, then the edge is oriented towards the vertex in the set with a greater index. Otherwise, if $i = j$, the edge e is oriented towards the vertex with a greater *ID* among the two vertices u and v. The orientation μ produced by this step is acyclic. Each vertex has out-degree at most $(2 + \varepsilon) \cdot a$. The correctness of the procedure follows from the correctness of Procedure Orientation from [1].

The last step of the algorithm is partitioning the edge set of the graph into forests as follows: each vertex is in charge of its outgoing edges, and it assigns each outgoing edge a distinct label from the set $\{1, 2, \ldots, (2 + \varepsilon) \cdot a\}$. This completes the description of the algorithm. Its pseudocode and analysis are provided below.

Algorithm 3. Partitioning of the edge set of G into ($\lfloor (2+\varepsilon) \cdot a \rfloor$) forests in the Congested-Clique model

---

1: **procedure** FORESTS-DECOMPOSITION-CC(a, ε)
2:     invoke Procedure H-Partition-CC(a, ε)
3:     $\mu =$ Orientation()
4:     assign a distinct label to each μ-outgoing edge of v from the set $[\lfloor (2 + \varepsilon) \cdot a \rfloor]$

Lemma 3. *The time complexity of Procedure Forests-Decomposition-CC is $O(\log a)$.*

Proof. Procedure H-Partition-CC takes $O(\log a)$ time, and steps (2) and (3) of Forests-Decomposition-CC require O(1) rounds each. Therefore, the overall time of Procedure Forests-Decomposition-CC is $O(\log a)$.   □

Theorem 2. *For a graph G with arboricity $a = a(G)$, and a parameter $\varepsilon, 0 < \varepsilon \leq 2$, in Congested Clique, Procedure Forests-Decomposition-CC (a, ε) partitions the edge set of G into ($\lfloor (2 + \varepsilon) \cdot a \rfloor$) forests in $O(\log a)$ rounds. Moreover, as a result of its execution each vertex v knows the label and the orientation of every edge (v, u) adjacent to v.*

4 A General Solution with $O(a)$ Time in Congested Clique

In this section we describe how to solve any computable problem in $O(a)$ time in the Congested Clique. We note that since any graph with arboricity a has $O(a \cdot n)$ edges, this is possible to achieve by directly applying $O(a)$ rounds of Lenzen's scheme [14]. However, in this section we present an alternative solution that employs forest-decompositions. Given a forest-decomposition in which the number of parents (i.e. outgoing edges) of each vertex is bounded by $O(a)$, we can solve any computable problem within this number of rounds. Specifically, once Procedure Forests-Decomposition-CC is invoked, it partitions the edge set of G into ($\lfloor (2 + \varepsilon) \cdot a \rfloor$) forests in $O(\log a)$ rounds. As a result of its execution, each vertex v knows the label and the orientation of every edge (v, u) adjacent to u. An outgoing edge from a vertex v to a vertex u labeled with a label i means that u is the parent of v in a tree of the ith forest F_i. Therefore, by transmitting the information of a distinct parent in a round, each vertex can inform all other vertices of the graph about all its parents. This will require an overall of $O(a)$ rounds - one round per parent. Then, each vertex knows all parents of all vertices in the graph G. But this information is sufficient to construct the graph G locally. Indeed, for each edge e of the graph G, one of its endpoints is a parent of the other in some forest i, and thus this edge is announced to all vertices in round i. Within $O(a)$ rounds, all edges are announced, and so the entire graph is known to all vertices. Therefore, we can solve any computable problem on G locally (without any additional communication), by executing the same deterministic algorithm on the same graph that is known to all. This guarantees a consistent solution in all vertices. Thus, we obtain a general solution with $O(a)$ time to any computable problem in the Congested Clique. (Note that this is true either if the input graph G is unweighted or if G has weights on edges that require $O(\log n)$ bits per edge. In the latter case, the information about weights can be transmitted together with the information about parents without affecting the running time bound $O(a)$. Recall, however, that all our algorithms in this paper are for unweighted graphs.) Therefore, it would be more interesting to find faster than $\Theta(a)$ algorithms for various problems. We obtain such algorithms in the next sections.

5 $O(a^2)$-Coloring in $O(\log a + \log^* n)$ Time

Note that in the synchronous message-passing model of distributed computing a proper $O(a^2)$-coloring requires $\Theta(\log n)$ time [15]. However, in Congested Clique we can improve the running time and reach even better result of $O(\log a)$ + $\log^* n$.

In this section we employ Procedure Forests-Decomposition-CC to provide an efficient algorithm that colors the input graph G of arboricity $a = a(G)$ in $O(a^2)$ colors. The running time of the algorithm is $O(\log a) + \log^* n$. For computing an $O(a^2)$-coloring we will use Procedure Arb-Linial described in [1].

Procedure Arb-Linial accepts a graph G with arboricity $a(G)$. Given an $O(a)$-forests-decomposition of G, the procedure computes a proper coloring φ of the graph using $O(a^2)$ colors in $O(\log^* n)$ running time. During the execution of this procedure, each vertex transmits at most $O(\log n)$ bits over each edge in each round.

Procedure Forest-Decomposition-CC has better running time than the respective procedure on general graphs, which allows us to compute a proper $O(a^2)$-coloring of the graph very quickly. We devise a procedure called *Procedure Arb-Coloring-CC* that works in the following way. The procedure starts by executing Procedure Forest-Decomposition-CC with the input parameter $a = a(G)$. This invocation returns an H-partition of G of size $l \leq \lceil \frac{2}{\varepsilon} \log n \rceil$, and degree at most $A = (2 + \varepsilon) \cdot a$. Then, we invoke Procedure Arb-Linial on the forest-decomposition. Since the procedure requires each vertex to send only its current color to its neighbors (which is of size $O(\log n)$), Procedure Arb-Linial can be invoked as-is in the congested clique. In our case we execute Procedure Arb-Linial with an input parameter $A = (2 + \varepsilon) \cdot a$. In Procedure Arb-Linial each vertex considers only the colors of its parents in forests $F_1, F_2, \ldots, F_A$. By [1,15], the algorithm computes $O(((2 + \varepsilon) \cdot a)^2) = O(a^2)$-coloring. This completes the description of Procedure Arb-Coloring-CC. Its pseudocode and running time analysis are provided below.

Algorithm 4. $O(a^2)$-coloring in the Congested Clique

1: **procedure** ARB-COLORING-CC(a, ε)
2:     $H = (H_1, H_2, \ldots, H_l)$ = invoke Procedure Forest-Decomposition-CC
3:     invoke Procedure Arb-Linial $(H, A = (2 + \varepsilon) \cdot a)$

Theorem 3. *Procedure Arb-Coloring-CC computes a proper $O(a^2)$-coloring in the Congested Clique in $O(\log a + \log^* n)$ rounds.*

Proof. The correctness of the procedure follows from the above discussion. The running time of step (1) is $O(\log a)$ rounds, by Lemma 3. Step (2) requires $O(\log^* n)$ rounds, by [1,15]. Thus, the overall running time of the procedure is $O(\log a) + \log^* n$.   □

6 $O(a^{2+\varepsilon})$-Coloring in $O(\log^* n)$ Time

In this section we show that the factor of $\log a$ can be eliminated from the running time of Theorem 3 in the expense of slightly increasing the number of colors to $O(a^{2+\varepsilon})$, for an arbitrarily small positive constant ε. To this end, we invoke Procedure H-Partition-CC with second parameter set as a^ε, rather than ε. (And the number of iterations of line 4 is now set to a sufficiently large constant, instead of $\lceil \frac{2}{\varepsilon} \log a \rceil$.) We show below that this way the running time of forests-decompositions becomes just $O(1)$. However, the number of forests produced is now $O(a^{(1+\varepsilon)})$, rather than $O(a)$. Moreover, once Procedure Forest-Decomposition-CC terminates, we invoke Arb-Linial-CC algorithm on the result of the forest decomposition to compute $O((a^{(1+\varepsilon)})^2)$-Coloring.

Lemma 4. *Invoking Procedure H-Partition-CC with the second parameter set as $q = a^{\varepsilon}$ requires $O(1)$ rounds.*

Proof. In each round the number of active vertices is reduced by a factor of $\Theta(a^{\varepsilon})$. For $i = 1, 2, \ldots$, the number of edges in the subgraph induced by active vertices in round i is at most $O(\frac{(a \cdot n)}{(a^{\varepsilon})^i})$. Thus, after $i = O(\frac{1}{\varepsilon})$ rounds, the number of remaining edges will be $O(n)$. Then we can employ Lenzen's scheme, broadcast these edges to all vertices within $O(1)$ rounds, and compute the remaining H-sets locally. Therefore, the overall running time is $O(\frac{1}{\varepsilon}) = O(1)$.   □

Lemma 5. *For graphs G with $a(G) = a$, and a parameter, $q = a^{\varepsilon}$, for an arbitrarily small positive constant ε, Procedure Forest-Decomposition-CC partitions the edge set of G into $A = O(a^{1+\varepsilon})$ oriented forests in $O(1)$ rounds in Congested Clique.*

Proof. By Lemma 4, Procedure H-Partitions-CC executes in $O(1)$ rounds, the second stage is an orientation that is computed in $O(1)$ rounds, and assigning labels to outgoing edges is computed in $O(1)$ rounds as well. Therefore, the overall time of is $O(1)$.   □

The next theorem follows directly from Lemmas 4–5.

Theorem 4. *For graphs G with $a(G) = a$ and with a parameter $q = a^{\varepsilon}$, for a positive constant ε, Procedure Arb-Coloring-CC computes $O(a^{(2+\varepsilon)})$-coloring within $O(\log^* n)$ time in Congested Clique.*

7 $O(a^{1+\varepsilon})$-Coloring in $O(\log^2 a)$ Time, $O(a)$-Coloring in $O(a^{\varepsilon})$ Time and MIS in $O(\sqrt{a})$ Time

Due to space limitations, we provide here the statements of our results for $O(a^{1+\varepsilon})$-coloring, $O(a)$-coloring and MIS. The coloring and MIS algorithms and their proofs appear in the full version of this paper [3].

Theorem 5. *The running time of the Procedure Proper-Coloring-CC on a graphs G with arboricity $a(G) = a$ is $O(\log^2 a)$. The procedure colors an input graph G with $O(a^{1+\varepsilon})$ colors, for an arbitrarily small positive constant ε.*

Theorem 6. *Invoking Procedure Proper-Coloring-CC on a graph G with arboricity a with the parameter $p = \lceil a^{\varepsilon/3} \rceil$, produces a proper $O(a)$-coloring of G within $O(a^{\varepsilon})$ time.*

Theorem 7. *Procedure MIS-CC computes a proper MIS of the input graph G. The running time of Procedure MIS-CC is $O(\sqrt{a})$.*

References

1. Barenboim, L., Elkin, M.: Sublogarithmic distributed MIS algorithm for sparse graphs using Nash-Williams decomposition. In: Proceedings of the 27th ACM Symposium on Principles of Distributed Computing, pp. 25–34 (2008)
2. Barenboim, L., Elkin, M.: Deterministic distributed vertex coloring in polylogarithmic time. J. ACM **58**(5), 23 (2011)
3. Barenboim, L., Khazanov, V.: Distributed symmetry-breaking in congested cliques. https://arxiv.org/pdf/1802.07209.pdf
4. Censor-Hillel, K., Kaski, P., Korhonenz, J., Lenzen, C., Paz, A., Suomela, J.: Algebraic methods in the congested clique. In: Proceedings of the 34th ACM Symposium on Principles of Distributed Computing, pp. 143–152 (2015)
5. Censor-Hillel, K., Parter, M., Schwartzman, G.: Derandomizing local distributed algorithms under bandwidth restrictions. In: Proceedings of the 31st International Symposium on Distributed Computing (2016)
6. Le Gall, F.: Further algebraic algorithms in the congested clique model and applications to graph-theoretic problems. In: Gavoille, C., Ilcinkas, D. (eds.) DISC 2016. LNCS, vol. 9888, pp. 57–70. Springer, Heidelberg (2016). https://doi.org/10.1007/978-3-662-53426-7_5
7. Ghaffari, M.: Distributed MIS via All-to-All communication. In: Proceedings of the 36th ACM Symposium on Principles of Distributed Computing, pp. 141–149 (2017)
8. Ghaffari, M., Parter, M.: MST in log-star rounds of congested clique. In: 35th ACM Symposium on Principles of Distributed Computing (PODC), pp. 19–28 (2016)
9. Goldberg, A., Plotkin, S., Shannon, G.: Parallel symmetry-breaking in sparse graphs. SIAM J. Discrete Math. **1**(4), 434–446 (1988)
10. Hegeman, J., Pandurangan, G., Pemmaraju, S., Sardeshmukh, V., Scquizzato, M., Toward optimal bounds in the congested clique: graph connectivity and MST. In: Proceedings 34th ACM Symposium on Principles of Distributed Computing, pp. 91–100 (2015)
11. Hegeman, J., Pemmaraju, S.: Lessons from the congested clique applied to mapreduce. Theor. Comput. Sci. **608**, 268–281 (2015)
12. Jurdzinski, T., Nowicki, K.: MST in $O(1)$ rounds of the congested clique. In: Proceedings of 29th ACM-SIAM Symposium on Discrete Algorithms, pp. 2620–2632 (2018)
13. Kuhn, F., Wattenhofer, R.: On the complexity of distributed graph coloring. In Proceedings of 25th ACM Symposium on Principles of Distributed Computing, pp. 7–15 (2006)
14. Lenzen, C.: Optimal deterministic routing and sorting on the congested clique. In: Proceedings 32nd ACM Symposium on Principles of Distributed Computing, pp. 42–50 (2013)
15. Linial, N.: Locality in distributed graph algorithms. SICOMP **21**(1), 193–201 (1992)
16. Lotker, Z., Pavlov, E., Patt-Shamir, B., Peleg, D.: MST construction in $O(\log \log n)$ communication rounds. In: the Proceedings of the Symposium on Parallel Algorithms and Architectures, pp. 94–100. ACM (2003)
17. Pemmaraju, S., Sardeshmukh, V.: Minimum-weight spanning tree construction in $O(\log \log \log n)$ rounds on the congested clique. http://arxiv.org/abs/1412.2333

The Clever Shopper Problem

Laurent Bulteau[1], Danny Hermelin[2], Anthony Labarre[1(✉)],
and Stéphane Vialette[1]

[1] Université Paris-Est, LIGM (UMR 8049), CNRS, ENPC, ESIEE Paris, UPEM,
77454 Marne-la-Vallée, France
{laurent.bulteau,anthony.labarre,stephane.vialette}@u-pem.fr
[2] Department of Industrial Engineering and Management,
Ben-Gurion University of the Negev, Beersheba, Israel
hermelin@bgu.ac.il

Abstract. We investigate a variant of the so-called INTERNET SHOPPING problem introduced by Blazewicz et al. (2010), where a customer wants to buy a list of products at the lowest possible total cost from shops which offer discounts when purchases exceed a certain threshold. Although the problem is NP-hard, we provide exact algorithms for several cases, e.g. when each shop sells only two items, and an FPT algorithm for the number of items, or for the number of shops when all prices are equal. We complement each result with hardness proofs in order to draw a tight boundary between tractable and intractable cases. Finally, we give an approximation algorithm and hardness results for the problem of maximising the sum of discounts.

1 Introduction

Blazewicz et al. [3] introduced and described the INTERNET SHOPPING problem as follows: given a set of shops offering products at various prices and the delivery costs for each set of items bought from each shop, find where to buy each product from a shopping list at a minimum total cost. The problem is known to be NP-hard in the strong sense even when all products are free and all delivery costs are equal to one, and admits no polynomial $(c \ln n)$-approximation algorithm (for any $0 < c < 1$) unless $\mathsf{P} = \mathsf{NP}$.

A more realistic variant takes into account discounts offered by shops in some cases. These could be offered, for instance, when the shopper's purchases exceed a certain amount, or in the case of special promotions where buying several items together costs less than buying them separately. Blazewicz et al. [4] investigated such a variant, which features a concave increasing discount function on the products' prices. They showed that the problem is NP-complete in the strong sense even if each product appears in at most three shops and each shop sells exactly three products, as well as in the case where each product is available at three different prices and each shop has all products but sells exactly three of them at the same value. A variant where two separate discount functions

© Springer International Publishing AG, part of Springer Nature 2018
F. V. Fomin and V. V. Podolskii (Eds.): CSR 2018, LNCS 10846, pp. 53–64, 2018.
https://doi.org/10.1007/978-3-319-90530-3_6

are taken into account (one for the deliveries, the other for the prices) was also recently introduced and studied by Blazewicz et al. [5].

In this work, we investigate the case where a shopper aims to buy n books from m shops with free shipping; additionally, each shop offers a discount when purchases exceed a certain threshold (discounts and thresholds are specific to each shop). We show that the associated decision problem, which we call the CLEVER SHOPPER problem, is already NP-complete when only two shops are available, or when all books are available from two shops and each shop sells exactly three books. We also obtain parameterised hardness results: namely, that CLEVER SHOPPER is W[1]-hard when the parameter is m or the number of shops in a solution, and that it admits no polynomial-size kernel. On the positive side, we give a polynomial-time algorithm for the case where every shop sells at most two books, an XP algorithm for the case where few shops sell books at small prices, an FPT algorithm with parameter n, and another FPT algorithm with parameter m.

Let us now formally define CLEVER SHOPPER. For $n \in \mathbb{N}$, let $[n] = \{1, 2, \ldots, n\}$. Let B be a set of books to buy, S be a set of shops; $E \subseteq B \times S$ encodes the availability of the books in the shops, and $w : E \to \mathbb{N}$ encodes the prices. A subset $E' \subseteq E$ describes from which shop each book should be bought; each book is *covered exactly once* (i.e., any $b \in B$ has degree 1 in E'). A *discount* $d_s \in \mathbb{R}^+$ is associated to each shop s and offered when a *threshold* $t_s \in \mathbb{R}^+$ is reached, which is formally defined using the following *threshold function*:

$$\delta(s, E', d_s, t_s) = \begin{cases} d_s & \text{if } \sum_{(b,s) \in E'} w(e) \geq t_s, \\ 0 & \text{otherwise.} \end{cases}$$

We refer to the function $\mathscr{D}$ that maps each shop s to the pair (d_s, t_s) as the *discount function*. The problem we study is formally stated below, and generalises well-studied problems such as BIN COVERING [1] and H-INDEX MANIPULATION [12].

CLEVER SHOPPER

Input: an edge-weighted bipartite graph $G = (B \cup S, E, w)$; a discount function $\mathscr{D}$; a bound $K \in \mathbb{N}$.

Question: is there a subset $E' \subseteq E$ that covers each element of B exactly once and such that $\sum_{e \in E'} w(e) - \sum_{s \in S} \delta(s, E', d_s, t_s) \leq K$?

2   Hardness Results

We prove in this section several hardness results under various restrictions, both with regards to classical complexity theory and parameterised complexity theory. We show that CLEVER SHOPPER is NP-complete even if there are only two shops to choose from. For this first hardness result, we need book prices to be encoded in binary (i.e. they can be exponentially high compared to the input size).

Proposition 1. CLEVER SHOPPER *is NP-complete in the weak sense (i.e., prices are encoded in binary), even when* $|S| = 2$.

Proof (reduction from PARTITION*).* Recall the well-known NP-complete PARTITION problem [11]: given a finite set A and a size $\omega(a) \in \mathbb{N}$ for each element in A, decide whether there exists a subset $A' \subseteq A$ such that $\sum_{a \in A'} \omega(a) = \sum_{a \in A \setminus A'} \omega(a)$.

Let $\mathcal{I} = (A, \omega)$ be an instance of PARTITION, and $T = \sum_{a \in A} \omega(a)$. We obtain an instance $\mathcal{I}'$ of CLEVER SHOPPER as follows: introduce two shops s_1 and s_2 with $(d_{s_1}, t_{s_1}) = (d_{s_2}, t_{s_2}) = (1, T/2)$. Each item $a \in A$ is a book that shops s_1 and s_2 sell at the same price — namely, $\omega(a)$. It is now clear that there exists a subset $A' \subseteq A$ such that $\sum_{a \in A'} \omega(a) = \sum_{a \in A \setminus A'} \omega(a)$ if and only if all books can be purchased for a total cost of $T - 2$. □

This NP-hardness result allows arbitrarily high prices (the reduction from PARTITION requires prices of the order of $2^{|B|}$). In a more realistic setting, we might assume a polynomial bound on prices, i.e., they can be encoded in unary. As we show below, the problem remains hard for a few shops in the sense of W[1]-hardness. We complement this result with an XP algorithm in Proposition 7.

Proposition 2. CLEVER SHOPPER *is* W*[1]-hard for* $m = |S|$ *in the strong sense* (*i.e., even when prices are encoded in unary*).

Proof (reduction from BIN PACKING*).* Recall the well-known BIN PACKING problem: given n items with weights $w_1, w_2, \ldots, w_n$ and m bins with the same given capacity W, decide whether each item can be assigned to a bin so that the total weight of the items in any bin does not exceed W. BIN PACKING is NP-complete in the strong sense and W[1]-hard for parameter m, even when $\sum_{i=1}^{n} w_i = mW$ and all weights are encoded in unary [10].

We build an instance $\mathcal{I}$ of CLEVER SHOPPER from an instance of BIN PACKING with the aforementioned restrictions as follows. Create m identical shops, each with $t_s = W$ and $d_s = 1$. Create n books, where book i is available in every shop at price w_i. The budget is $m(W - 1)$. In other words, any solution requires to obtain the discount from every shop, which is only possible if purchases amount to a total of exactly W per shop before discount. Therefore, the solutions to $\mathcal{I}$ correspond exactly to the solutions of the original instance of BIN PACKING. □

We can obtain another hardness result under the assumption that all books are sold at a unit price. Here we cannot bound the total number of shops (we give an FPT algorithm for parameter m in Proposition 8 in that setting), but only the number of *chosen* shops (i.e., shops where at least one book is purchased).

Proposition 3. CLEVER SHOPPER *with unit prices is* W*[1]-hard for the parameter "number of chosen shops".*

Proof (reduction from PERFECT CODE*).* Given a graph $G = (V, E)$ and a positive integer k, PERFECT CODE asks for a size-k subset $V' \subseteq V$ such that for each vertex $u \in V$ there is precisely one vertex in $N[v] \cap V'$ (where $N[v]$ is the *closed* neighbourhood of v, i.e., v and its adjacent vertices, as opposed to the

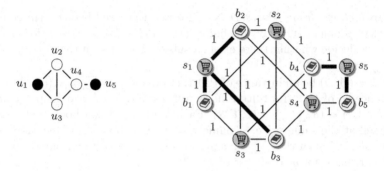

Fig. 1. Reducing PERFECT CODE to CLEVER SHOPPER. Left: the input graph with a size-2 perfect code (bold). Right: the corresponding bipartite graph and a solution with total cost $5 - 2 = 3$ (bold).

open neighbourhood $N(v) = N[v] \setminus \{v\}$). This problem is known to be W[1]-hard for parameter k [7].

Let $\mathcal{I} = (G = (V, E), k)$ be an instance of PERFECT CODE. Write $V = \{u_1, u_2, \ldots, u_n\}$. We obtain an instance $\mathcal{I}'$ of CLEVER SHOPPER as follows. Let us first define a bipartite graph $G' = (B \cup S, E')$ where $B = \{b_i : u_i \in V\}$, $S = \{s_i : u_i \in V\}$ and $E' = \{\{b_j, s_i\} : u_j \in N_G[u_i]\}$. All shops sell books at a unit price. As for the discount function, for each shop $s_i \in S$ we have $\mathscr{D}(s_i) = (1, d_G(u_i) + 1)$ (i.e., a unit discount will be applied, from $d_G(u_i) + 1$ of purchase). Figure 1 illustrates the construction.

We claim that there exists a size-k perfect code for G if and only if all books can be bought for a total cost of $n - k$.

$\boxed{\Rightarrow}$ Let $V' \subseteq V$ be a size-k perfect code in G. For every $u_i \in V$, let $u_{\mathrm{pc}(i)}$ be the unique vertex in $N[v] \cap V'$ (pc is well-defined since V' is a perfect code). Then buying each book $b_i \in B$ at shop $b_{\mathrm{pc}(i)}$ yields a solution for $\mathcal{I}'$, and it is simple to check that its cost is $n - k$.

$\boxed{\Leftarrow}$ Suppose that all books can be bought for a total cost of $n - k$. Since n books must be bought at unit price and shops only offer a unit discount, k shops must be chosen in the solution. Let $S' \subseteq S$ denote these k shops. Since $\mathscr{D}(s_i) = (1, d_G(u_i) + 1)$ for each shop $s_i \in S$, we conclude that for each book $b_i \in B$ there is precisely one shop in $N[b_i] \cap S'$. Then $\{u_i : s_i \in S'\}$ is a size-k perfect code in G.

Note that the number of visited shops corresponds exactly to the total discount received (i.e. to parameter k in the reduction).   □

We now prove[1] the non-existence of polynomial kernels (under standard complexity assumptions) for CLEVER SHOPPER parameterised by the number of books. To this end, we use the OR-COMPOSITION technique [6]: given a problem $\mathscr{P}$ and a parameterised problem $\mathscr{Q}$, an OR-COMPOSITION is a reduction taking t instances $(I_1, \ldots, I_t)$ of $\mathscr{P}$, and building an instance (J, k) of $\mathscr{Q}$, with k bounded

---

[1] Details will appear in the full version.

by a polynomial on $\max_{t' \leq t} |I_{t'}| + \log t$, such that (J, k) is a yes-instance if and only if there exists $t' \leq t$ such that $I_{t'}$ is a yes-instance. If $\mathscr{P}$ is NP-hard, then $\mathscr{Q}$ does not admit a polynomial kernel unless NP $\subseteq$ coNP/poly [6].

Proposition 4. CLEVER SHOPPER *admits no polynomial kernel unless* NP $\subseteq$ coNP/poly.

3   Positive Results

We now give exact algorithms for CLEVER SHOPPER: a polynomial-time algorithm for the case where every shop sells at most two books, and three parameterised algorithms based respectively on the number of books, the number of shops, and a bound on the prices.

We give a polynomial time algorithm for the case where each shop sells at most two books. As we shall see in Sect. 4, this bound is best possible. Its running time is dominated by the time required to find a maximum matching in a graph with $|B \cup S|$ vertices.

Proposition 5. CLEVER SHOPPER *is in* P *if every shop sells at most two books.*

Proof. Let $\mathcal{I}$ be an instance of CLEVER SHOPPER given by an edge-weighted bipartite graph $G = (B \cup S, E, w)$ and a pair (d_s, t_s) for each $s \in S$, where $d_s, t_s \in \mathbb{R}^+$. Vertices in S (resp. in B) have degree at most 2 (resp. at least 1). Note that vertices in S can be made to have degree exactly 2, by adding dummy edges with arbitrarily high costs, with no impact on the solution. For $b \in B$, let $p(b)$ be the cheapest available price for book b (discount excluded), i.e., $p(b) = \min\{w(\{b, s\}) \mid s \in S\}$.

Construct a new (non-bipartite) graph $G' = (B \cup S, E', w')$, as follows: for every shop $s \in S$, let $\{b_1, b_2\} = N_G(s)$ (i.e., the two books available at shop s).

- For each $i \in \{1, 2\}$, if $w(\{b_i, s\}) \geq t_s$, then add an edge $\{b_i, s\}$ to E' with weight $w'(\{b_i, s\}) = d_s + p(b_i) - w(\{b_i, s\})$.
- If $w(\{b_1, s\}) + w(\{b_2, s\}) \geq t_s$, add an edge $\{b_1, b_2\}$ to E' with weight $w'(\{b_1, b_2\}) = d_s + p(b_1) - w(\{b_1, s\}) + p(b_2) - w(\{b_2, s\})$. If edge $\{b_1, b_2\}$ existed already, keep only the one with maximum weight.

Note that edges with negative weights may remain: they may be safely ignored, but we keep them to avoid case distinctions in the rest of this proof. Figure 2 illustrates the construction. Since a maximum weight matching for G' can be found in polynomial time [8], it is now enough to prove the following claim: G' admits a matching of weight at least W if and only if instance $\mathcal{I}$ of CLEVER SHOPPER admits a solution of total cost at most $\sum_{b \in B} p(b) - W$.

⇐ Assume that instance $\mathcal{I}$ admits a solution $E^* \subseteq E$ of total cost $\sum_{b \in B} p(b) - W$. Note that $W \geq 0$ (the sum of the minimum prices of the books is an upper bound of the optimal solution). We build a matching M of G' as follows. Let $s \in S$ be any *discount shop*, i.e., a shop whose discount is claimed, and let b_1 and b_2 be its neighbours. Then at least one of them has to be bought from s to get the discount.

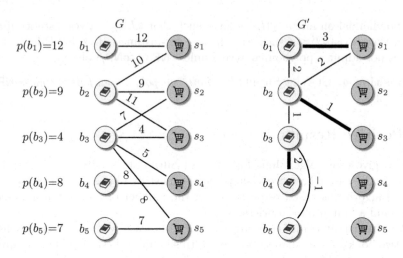

Fig. 2. Each shop offers a discount of 3 on a purchase of value ≥ 10. Bold edges indicate how to obtain optimal discounts: buy book b_1 from shop s_1, book b_2 from shop s_3, and books b_3 and b_4 from shop s_4. The remaining books are bought at their cheapest available price (so here we buy b_5 from s_5). Our clever customer used the discounts to buy all books for 6 less than if she had bought each book at its lowest price: 3 for b_1, 1 for b_2, 2 for b_3 and b_4 together.

- If $\{b_1, s\} \in E^*$ and $\{b_2, s\} \notin E^*$, add $\{b_1, s\}$ to M. The amount spent at this shop is $w(\{b_1, s\}) - d_s = p(b_1) - w'(\{b_1, s\})$.
- Similarly, if $\{b_2, s\} \in E^*$ and $\{b_1, s\} \notin E^*$, add $\{b_2, s\}$ to M. The amount spent at this shop is $w(\{b_2, s\}) - d_s = p(b_2) - w'(\{b_2, s\})$.
- Finally, if $\{b_1, s\} \in E^*$ and $\{b_2, s\} \in E^*$, then add $\{b_1, b_2\}$ to M. The amount spent at this shop is $w(\{b_1, s\}) + w(\{b_2, s\}) - d_s \geq p(b_1) + p(b_2) - w'(\{b_1, b_2\})$.

Note that edges added to M are indeed present in E', since in order to obtain the discount from s, the book prices must satisfy the same condition as for creating the corresponding edges. Note also that M is a matching, since each book can be bought from at most one shop. Let B^* be the set of books bought from discount shops. Summing over all these shops, the total price paid for the books in B^* is at least $\sum_{b \in B^*} p(b) - \sum_{e \in M} w'(e)$.

The books in $B \setminus B^*$ do not yield any discount, so the total price paid for them is at least $\sum_{b \in B \setminus B^*} p(b)$. Overall, the cost of the books is at least $\sum_{b \in B} p_b - \sum_{e \in M} w'(e)$, therefore $\sum_{e \in M} w'(e) \geq W$.

$\boxed{\Rightarrow}$ Let M be a maximum weight matching of G' of weight W. For each edge $e \in M$, let s_e be the shop for which e was introduced. For an edge $e = \{b, s_e\} \in M$, buy book b from shop s_e. The price is high enough to reach the threshold for the discount, so we pay $w(\{b, s_e\}) - d_e = p(b) - w'(e)$. For an edge $e = \{b_1, b_2\} \in M$, buy books b_1 and b_2 together from shop s_e. We again get the discount, and pay $w(\{b_1, s_e\}) + w(\{b_2, s_e\}) - d_e = p(b_1) + p(b_2) - w'(e)$. Note that for $e \neq f \in M$, $s_e \neq s_f$, so we never count the same discount twice. For every other book, buy them at the cheapest possible price $p(b)$, without expecting to get any discount. The total price paid is at most $\sum_{b \in B} p(b) - \sum_{e \in M} w'(e) = \sum_{b \in B} p(b) - W$. □

We now give a dynamic programming FPT algorithm with the number of books as parameter.

Proposition 6. CLEVER SHOPPER *admits an* FPT *algorithm for parameter n with running time $O(m3^n)$.*

Proof. Given $j \in [m]$ and $B' \subseteq B$, let $p_j(B')$ be the price for buying all books in B' together from shop s_j (discount included), and $p_{\leq j}(B')$ be the lowest price that can be obtained when purchasing all books in B' from a subset of $\{s_1, \ldots, s_j\}$. Our goal is to compute $p_{\leq m}(B)$.

For $j = 1$, clearly $p_{\leq 1}(B') = p_1(B')$ for every B'. For any other j, consider an optimal way of buying the books in B' from shops $s_1, \ldots, s_j$. This way the customer buys some (possibly empty) subset B'' of books in s_j, and the rest, i.e., $B' \setminus B''$, at the lowest price from shops $s_1, \ldots, s_{j-1}$. Therefore:

$$p_{\leq j}(B') = \begin{cases} p_j(B') & \text{if } j = 1, \\ \min_{B'' \subseteq B'} \{p_j(B'') + p_{\leq j-1}(B' \setminus B'')\} & \text{otherwise.} \end{cases}$$

The values of $p_j(B')$ for all j and B' can be computed in $O(m2^n)$ time. Then the dynamic programming table requires to enumerate, for all j, all subsets B' and B'' such that $B'' \subseteq B' \subseteq B$. Any such pair B'', B' can be interpreted as a vector $v \in \{0, 1, 2\}^n$, where $i \in B'' \Leftrightarrow v_i = 2$ and $i \in B' \Leftrightarrow v_i \geq 1$. Therefore, filling the dynamic table takes $m3^n$ steps, each requiring constant time.   □

As usual with dynamic programming, this algorithm yields the optimal price that can be obtained. One gets the actual solution (i.e., where to buy each book) with classic backtracing techniques.

The NP-hardness of CLEVER SHOPPER for two shops (using large prices, encoded in binary) and its W[1]-hardness when the parameter is the number of shops leave a very small opening for positive results: we can only consider small prices (encoded in unary) for a constant number of shops. The following result proves the tractability of this case.

Proposition 7. CLEVER SHOPPER *admits an* XP *algorithm running in time $O(nm\mathcal{W}^m)$, where $\mathcal{W}$ is the sum of all the prices of the instance, n is the number of books, and m is the number of shops.*

Proof. We propose the following dynamic programming algorithm, which generalises the classical pseudo-polynomial algorithm for PARTITION. Let $i \in [n]$ and $p_s \in [\mathcal{W}]$ for $s \in S$. Define $T[i, p_{s_1}, \ldots, p_{s_m}]$ as 1 if it is possible to buy books 1 to i by paying exactly p_s (discount excluded) in shop s; and 0 otherwise. For $i = 0$, $T[0, p_{s_1}, \ldots, p_{s_m}] = 1$ if and only if $p_s = 0$ for all $s \in S$. The following formula allows to fill the table recursively for $i \geq 1$:

$$T[i, p_{s_1}, \ldots, p_{s_m}] = \max_{e \in E, i \in e} T[i-1, p'_{s_1}, \ldots, p'_{s_m}] \text{ where } p'_s = \begin{cases} p_s - w(e) & \text{if } s \in e, \\ p_s & \text{otherwise.} \end{cases}$$

It remains to be checked whether the table contains a valid solution, which requires us to take the discounts into account. Clearly, an entry $T[n, p_{s_1}, \ldots, p_{s_m}] = 1$ leads to a solution if the following holds:

$$\sum_{s \in S} p_s - \sum_{s \in S, p_s \geq t_s} d_s \leq K.$$

The running time corresponds exactly to the time needed to fill the table: any of the nW^m cells requires at most m look-ups, which yields the claimed running time.   □

Proposition 8. CLEVER SHOPPER *admits an* FPT *algorithm for parameter* m *when all prices are equal.*

Proof. We assume without loss of generality that all prices are equal to 1. Let $S' \subseteq S$. We write $f_{S'} : B \cup S \to \mathbb{N}$ for the following function:

$$f_{S'}(b) = 1 \text{ for } b \in B,$$
$$f_{S'}(s) = t_s \text{ for } s \in S',$$
$$f_{S'}(s) = 0 \text{ for } s \notin S'.$$

We write $d_{S'} = \sum_{s \in S'} d_s$ and $t_{S'} = \sum_{s \in S'} t_s$. An f-*star subgraph* of $G = (B \cup S, E)$ is a subgraph G' such that the degree of each vertex $u \in B \cup S$ is at most $f(u)$ in G', and every connected component of G' is isomorphic to $K_{1,p}$ for some integer p.

Let $\mathcal{I} = (B \cup S, E, w, \mathscr{D}, K)$ be an instance of CLEVER SHOPPER with $w(e) = 1$ for all $e \in E$. We show that $\mathcal{I}$ is a yes-instance if and only if there exists $S' \subseteq S$ with $|B| - d_{S'} \leq K$ such that $(B \cup S, E)$ admits an $f_{S'}$-star subgraph with $t_{S'}$ edges. An FPT algorithm follows easily from this characterisation: enumerate all subsets S' of S in time $2^{|S|}$, and for each subset, compute a maximum $f_{S'}$-star subgraph in time $O(|E| \log |B \cup S|)$ [9].

⇒ Let $E' \subseteq E$ be a solution and S' be the set of shops whose threshold t_s is reached. Since the total price is $|B| - d_{S'}$, we have $|B| - d_{S'} \leq K$. Since every weight equals 1, all vertices of S' have degree at most t_s in E'. Let $E'' \subseteq E'$ be a subset obtained by keeping exactly t_s edges incident to each $s \in S'$ and no edge incident to $s \notin S'$. Then E'' is an $f_{S'}$-star subgraph of size $t_{S'}$.

⇐ Let $G' = (B \cup S, E')$ be an $f_{S'}$-star factor of G of size $t_{S'}$ with $S' \subseteq S$, and $|B| - d_{S'} \leq K$. The degree and size constraints force all vertices in S' to have degree exactly t_s in G'. We build a solution as follows: for each book $b \in B$, if E' contains an edge (b, s) incident to b, then buy b from shop s, otherwise buy b from any other shop. Overall, at least t_s books are purchased from a shop $s \in S'$, so the total price is at most $|B| - d_{S'}$.   □

4   Approximations

Since variants of CLEVER SHOPPER are, by and large, hard to solve exactly, it is natural to look for approximation algorithms. However, our hardness proofs can be modified to imply the NP-hardness of deciding whether the total price (including discounts) is 0 or more. For instance, in Proposition 1, we can set the discounts to $T/2$ instead of 1, so the PARTITION instance reduces to checking whether the optimal solution has cost 0. Therefore, we start with the following bad news:

Corollary 1. CLEVER SHOPPER *admits no approximation unless* P = NP.

Since this result seems resilient to most natural restrictions on the input structure (bounded prices, bounded degree, etc.), our proposed angle is to maximise the total discount rather than minimise the total cost. However, maximising the total discount is only relevant when the base price of the books is the same in all solutions (otherwise the optimal solution might not be the one with maximum discount), i.e., each book b has a fixed price w_b, and $w(\{b, s\}) = w_b$ for every $\{b, s\} \in E$. We call this variant MAX-DISCOUNT CLEVER SHOPPER. This "fixed price" constraint is not strong (all reductions from Sect. 2 satisfy it). In this setting, Proposition 1 shows that it is NP-hard to decide whether the optimal discount is 1 or 2. This yields the following corollary:

Corollary 2. MAX-DISCOUNT CLEVER SHOPPER *is* APX-*hard: it does not admit a* $(2 - \epsilon)$-*approximation unless* P = NP.

Whether or not MAX-DISCOUNT CLEVER SHOPPER admits a fixed-ratio approximation remains open.

Proposition 9. MAX-DISCOUNT CLEVER SHOPPER *is* APX-*hard even when each shop sells at most 3 books, and each book is available in at most 2 shops.*

Proof. We reduce from MAX 3-SAT (the problem of satisfying the maximum number of clauses in a 3-SAT instance), known to be APX-hard when each literal occurs exactly twice [2]. Let $\varphi = C_1 \wedge C_2 \wedge \cdots \wedge C_m$ be such a 3-CNF formula over a set $X = \{x_1, x_2, \ldots, x_n\}$ of boolean variables. For every $1 \leq i \leq m$ and $1 \leq j \leq 3$, let $\ell_{i,j}$ be the j-th literal of clause C_i. We obtain an instance $\mathcal{I}$ of MAX-DISCOUNT CLEVER SHOPPER by first building a bipartite graph $G = (B \cup S, E)$ as follows (for ease of presentation, C_i, x_i and $\ell_{i,j}$ will be used both to denote respectively clauses, variables and literals in 3-CNF formula context, and the corresponding vertices in G):

$$B = \{\ell_{i,j} : 1 \leq i \leq m \text{ and } 1 \leq j \leq 3\} \cup \{x_i : 1 \leq i \leq n\}$$
$$S = \{C_i : 1 \leq i \leq m\} \cup \{t_i, f_i : 1 \leq i \leq n\}$$
$$E = E_1 \cup E_{2,p} \cup E_{2,n} \cup E_3$$

where

$$E_1 = \{\{\ell_{i,j}, C_i\} : 1 \leq i \leq m \text{ and } 1 \leq j \leq 3\}$$
$$E_{2,p} = \{\{\ell_{i,j}, t_i\} : 1 \leq i \leq m \text{ and } \ell_{i,j} \text{ is the positive literal } x_i\}$$
$$E_{2,n} = \{\{\ell_{i,j}, f_i\} : 1 \leq i \leq m \text{ and } \ell_{i,j} \text{ is the negative literal } \overline{x_i}\}$$
$$E_3 = \{\{x_i, t_i\}, \{x_i, f_i\} : 1 \leq i \leq n\}.$$

Observe that each shop sells exactly 3 books and that each book is sold in exactly 2 shops. We now turn to defining the prices, the thresholds and the discounts. All shops sell books at a unit price. For the shops C_i, $1 \leq i \leq m$, a purchase of value 1 yields a discount of 1. For the shops t_i and f_i, $1 \leq i \leq n$, a purchase of value 3 yields a discount of 2. This discount policy implies that, for every $1 \leq i \leq n$, a customer cannot obtain a 2 discount both in shop t_i and in shop f_i (this follows from the fact that the book x_i is sold by both shops t_i and f_i).

First, it is easy to see that the largest discount that can be obtained is $2n + m$ (the upper bound is achieved by obtaining a discount in every shop C_i for $1 \leq i \leq m$, and in either the shop t_i or the shop f_i for $1 \leq i \leq n$). On the other side, for any truth assignment τ for φ satisfying k clauses, a $2n + k$ discount can be obtained as follows.

- For any variable x_i, $1 \leq i \leq n$, if $\tau(x_i) = $ false, then buy 3 books from shop t_i, and if $\tau(x_i) = $ true then buy 3 books from shop f_i. Intuitively, if a variable is true, then all negative literals are "removed" by f_i, and all positive literals remain available for the corresponding clauses.
- For any clause $C_i = \ell_{i,1} \vee \ell_{i,2} \vee \ell_{i,3}$ satisfied by the truth assignment τ, buy book $\ell_{i,j}$ from shop C_i, where $\ell_{i,j}$ is a literal satisfying the clause C_i.

Then it follows that

$$\text{opt}(I) = 2n + \text{opt}(\varphi) = 3m/2 + \text{opt}(\varphi) \qquad \text{(since } 4n = 3m)$$
$$\leq 3\,\text{opt}(\varphi) + \text{opt}(\varphi) \qquad \text{(since } 2\,\text{opt}(\varphi) \geq m)$$
$$\leq 4\,\text{opt}(\varphi).$$

Suppose now that we buy all books in B for a total discount of k'. First, we may clearly assume that $k' \geq 2n$ since a total $2n$ discount can always be achieved by buying 3 books either from shop t_i or from shop f_i, for every $1 \leq i \leq n$. Second, we may also assume that, for every $1 \leq i \leq n$, we buy either exactly 3 books from shop t_i or exactly 3 books from shop f_i. Indeed, if there exists an index $1 \leq i \leq n$ for which this is false, then buying either exactly 3 books from shop t_i or exactly 3 books from shop f_i instead results in a total k'' discount with $k'' \geq k'$ (this follows from the fact that we can get a 2 discount from t_i or f_i but only a 1 discount from any shop C_j, $1 \leq j \leq m$). We now obtain a truth assignment τ for φ as follows: for any variable x_i, $1 \leq i \leq n$, set $\tau(x_i) = $ false if we buy 3 books from shop t_i, and set $\tau(x_i) = $ true if we buy 3 books from shop f_i (the truth assignment τ is well-defined since, for $1 \leq i \leq n$, we cannot simultaneously buy 3 books from shop t_i and 3 books from shop f_i because of book x_i). Therefore, a clause C_i is satisfied by τ if and only if the corresponding shop C_i contains at least one book $l_{i,j}$ which is not bought from some other shop t_i or f_i. If we let k stand for the number of clauses satisfied by τ, then we obtain $k \geq k' - 2n$. It then follows that

$$\text{opt}(\varphi) - k = \text{opt}(I) - 2n - k \leq \text{opt}(I) - 2n - k' + 2n = \text{opt}(I) - k'.$$

Therefore, our reduction is an L-reduction (i.e., $\text{opt}(\mathcal{I}) \leq \alpha_1 \text{opt}(\varphi)$ and $\text{opt}(\varphi) - k \leq \alpha_2 (\text{opt}(\mathcal{I}) - k')$) with $\alpha_1 = 4$ and $\alpha_2 = 1$. $\qquad\square$

Proposition 10. MAX-DISCOUNT CLEVER SHOPPER *where each shop sells at most k books admits a k-approximation.*

Proof. Let B_s be the set of books sold by shop s. Our approximation algorithm proceeds as follows: start with a set of selected shops $S' = \emptyset$, a set of available books $B' = B$ and sort the shops by decreasing value of d_s. Then for each shop s, let $B'_s = B_s \cap B'$. If the books in B'_s are enough to get the discount ($\sum_{b \in B'_s} \geq t_s$), then assign all books of B'_s to shop s, add s to S' and set $B' = B' \setminus B'_s$. Finally, assign the remaining books to arbitrary shops that sell them.

We now prove the approximation ratio. For any $b \in B$, if $b \in B'_s$ for some $s \in S'$ then let $\delta(b) = d_s$, and $\delta(b) = 0$ otherwise. Thus, for any shop $s \in S'$, $d_s = \frac{1}{|B'_s|} \sum_{b \in B'_s} \delta(b) \geq \frac{1}{k} \sum_{b \in B'_s} \delta(b)$ due to the degree-k constraint. Note that for each shop of S', the amount spent at s is at least t_s, so the total discount obtained with this algorithm is $D \geq \sum_{s \in S'} d_s \geq \frac{1}{k} \sum_{b \in B} \delta(b)$.

We now compare the result of the algorithm with any optimal solution. For such a solution, let D^* be its total discount, S^* be the set of shops where purchases reach the threshold, and, for any $s \in S^*$, let B_s^* be the (non-empty) set of books purchased in shop s. Note that $D^* = \sum_{s \in S^*} d_s$.

Consider a shop $s \in S^*$. We show that there exists a book $b^*(s) \in B_s^*$ with $\delta(b^*(s)) \geq d_s$. If $s \in S^* \cap S'$, then we take $b^*(s)$ to be any book in B_s^*. Either $b^*(s) \in B'_s$, in which case $\delta(b^*(s)) = d_s$, or $b^*(s) \notin B'_s$, in which case $b^*(s)$ was assigned by the algorithm to a shop with a larger discount, i.e., $\delta(b^*(s)) \geq d_s$. If $s \in S^* \setminus S'$, since $s \notin S'$, at least one book in B_s^* is not available at the time the algorithm considers shop s; let $b^*(s)$ be such a book. Since it is not available, it has been selected as part of $B'_{s'}$ for some earlier shop s' (i.e., $d_s \leq d_{s'}$). Therefore, $b^*(s) \in B_s^* \cap B'_{s'}$ and $\delta(b^*(s)) = d_{s'} \geq d_s$. Since the sets B_s^* are pairwise disjoint for $s \in S^*$, we have $\sum_{s \in S^*} \delta(b^*(s)) \leq \sum_{b \in B} \delta(b)$. Putting it all together, we obtain:

$$D^* = \sum_{s \in S^*} d_s \leq \sum_{s \in S^*} \delta(b^*(s)) \leq \sum_{b \in B} \delta(b) \leq kD.$$

$\square$

5   Conclusion

We introduced the CLEVER SHOPPER problem, a variant of INTERNET SHOPPING with free deliveries and shop-specific discounts based on shop-specific thresholds. We proved a number of hardness results, both in the classical complexity setting and from a parameterised complexity point of view. We also gave efficient algorithms for particular cases where restrictions apply to the number of books, the number of shops, or the nature of prices.

An interesting angle for future work is that of designing efficient exact algorithms for the general cases in which our FPT algorithms are not sufficient. Furthermore, it would be of interest to determine whether the CLEVER SHOPPER problem is FPT for parameter *maximum price + number of shops*.

References

1. Assmann, S., Johnson, D., Kleitman, D., Leung, J.-T.: On a dual version of the one-dimensional bin packing problem. J. Algorithms **5**, 502–525 (1984)
2. Berman, P., Karpinski, M., Scott, A.D.: Approximation hardness of short symmetric instances of MAX-3SAT. In: Electronic Colloquium on Computational Complexity (ECCC) (2003)
3. Blazewicz, J., Kovalyov, M.Y., Musial, J., Urbanski, A.P., Wojciechowski, A.: Internet shopping optimization problem. Appl. Math. Comput. Sci. **20**, 385–390 (2010)
4. Blazewicz, J., Bouvry, P., Kovalyov, M.Y., Musial, J.: Internet shopping with price sensitive discounts. 4OR **12**, 35–48 (2014)
5. Blazewicz, J., Cheriere, N., Dutot, P.-F., Musial, J., Trystram, D.: Novel dual discounting functions for the internet shopping optimization problem: new algorithms. J. Sched. **19**, 245–255 (2016)
6. Bodlaender, H.L., Jansen, B.M.P., Kratsch, S.: Kernelization lower bounds by cross-composition. SIAM J. Discret. Math. **28**, 277–305 (2014)
7. Cesati, M.: Perfect code is W[1]-complete. Inf. Process. Lett. **81**, 163–168 (2002)
8. Edmonds, J.: Paths, trees and flowers. Canad. J. Math. 449–467 (1965)
9. Gabow, H.N.: A note on degree-constrained star subgraphs of bipartite graphs. Inf. Process. Lett. **5**, 165–167 (1976)
10. Jansen, K., Kratsch, S., Marx, D., Schlotter, I.: Bin packing with fixed number of bins revisited. J. Comput. Syst. Sci. **79**, 39–49 (2013)
11. Karp, R.M.: Reducibility among combinatorial problems. In: Miller, R.E., Thatcher, J.W. (eds.) Proceedings of a Symposium on the Complexity of Computer Computations. The IBM Research Symposia Series, pp. 85–103. Plenum Press, Yorktown Heights (1972)
12. van Bevern, R., Komusiewicz, C., Niedermeier, R., Sorge, M., Walsh, T.: H-index manipulation by merging articles: models, theory, and experiments. Artif. Intell. **240**, 19–35 (2016)

A Tight Lower Bound for Steiner Orientation

Rajesh Chitnis[1][(✉)] and Andreas Emil Feldmann[2]

[1] Department of Computer Science, University of Warwick, Coventry, UK
rajeshchitnis@gmail.com
[2] Charles University in Prague, Prague, Czechia
feldmann.a.e@gmail.com

Abstract. In the STEINER ORIENTATION problem, the input is a mixed graph G (it has both directed and undirected edges) and a set of k terminal pairs $\mathscr{T}$. The question is whether we can orient the undirected edges in a way such that there is a directed $s \rightsquigarrow t$ path for each terminal pair $(s, t) \in \mathscr{T}$. Arkin and Hassin [DAM'02] showed that the STEINER ORIENTATION problem is NP-complete. They also gave a polynomial time algorithm for the special case when $k = 2$.

From the viewpoint of exact algorithms, Cygan, Kortsarz and Nutov [ESA'12, SIDMA'13] designed an XP algorithm running in $n^{O(k)}$ time for all $k \geq 1$. Pilipczuk and Wahlström [SODA '16] showed that the STEINER ORIENTATION problem is W[1]-hard parameterized by k. As a byproduct of their reduction, they were able to show that under the Exponential Time Hypothesis (ETH) of Impagliazzo, Paturi and Zane [JCSS'01] the STEINER ORIENTATION problem does not admit an $f(k) \cdot n^{o(k/\log k)}$ algorithm for any computable function f. That is, the $n^{O(k)}$ algorithm of Cygan et al. is almost optimal.

In this paper, we give a short and easy proof that the $n^{O(k)}$ algorithm of Cygan et al. is asymptotically optimal, even if the input graph has genus 1. Formally, we show that the STEINER ORIENTATION problem is W[1]-hard parameterized by the number k of terminal pairs, and, under ETH, cannot be solved in $f(k) \cdot n^{o(k)}$ time for any function f even if the underlying undirected graph has genus 1. We give a reduction from the GRID TILING problem which has turned out to be very useful in proving W[1]-hardness of several problems on planar graphs. As a result of our work, the main remaining open question is whether STEINER ORIENTATION admits the "square-root phenomenon" on planar graphs (graphs with genus 0): can one obtain an algorithm running in time $f(k) \cdot n^{O(\sqrt{k})}$ for PLANAR STEINER ORIENTATION, or does the lower bound of $f(k) \cdot n^{o(k)}$ also translate to planar graphs?

R. Chitnis—Supported by ERC grant CoG 647557 "Small Summaries for Big Data". Part of this work was done when the author was at the Weizmann Institute of Science (and supported by Israel Science Foundation grant #897/13), and visiting Charles University in Prague, Czechia.
A. E. Feldmann—Supported by project CE-ITI (GAČR no. P202/12/G061) of the Czech Science Foundation.

F. V. Fomin and V. V. Podolskii (Eds.): CSR 2018, LNCS 10846, pp. 65–77, 2018.
https://doi.org/10.1007/978-3-319-90530-3_7

1   Introduction

In the STEINER ORIENTATION problem, the input is a mixed graph $G = (V, E)$ (it has both directed and undirected edges) and a set of terminal pairs $\mathscr{T} \subseteq V \times V$. The question is whether we can orient the undirected edges in a way such that there is a directed $s \leadsto t$ path for each terminal pair $(s, t) \in \mathscr{T}$.

STEINER ORIENTATION
Input: A mixed graph G, and a set $\mathscr{T}$ of k terminal pairs
Question: Is there an orientation of the undirected egdes of G such that the resulting graph has an $s \leadsto t$ path for each $(s, t) \in \mathscr{T}$
Parameter: k

Hassin and Megiddo [8] showed that STEINER ORIENTATION is polynomial time solvable if the input graph G is completely undirected, i.e., has no directed edges. If the input graph G is actually mixed then Arkin and Hassin [1] showed that STEINER ORIENTATION is NP-complete. They also gave a polynomial time algorithm for the special case when $k = 2$. Cygan et al. [7] generalized this by giving an $n^{O(k)}$ algorithm for all $k \geq 1$, i.e., STEINER ORIENTATION is in XP parameterized by k. Although the algorithm of Cygan et al. is polynomial time for fixed k, the degree of the polynomial changes as k changes. This left open the question of whether one could design an FPT algorithm for STEINER ORIENTATION parameterized by k, i.e., an algorithm which runs in time $f(k) \cdot n^{O(1)}$ for some computable function f independent of n.

Pilipczuk and Wahlström [17] answered this question negatively by showing that STEINER ORIENTATION is W[1]-hard parameterized by k. As a byproduct of their reduction, they were able to show that under the Exponential Time Hypothesis (ETH) of Impagliazzo and Paturi [9,10] the STEINER ORIENTATION problem does not admit a $f(k) \cdot n^{o(k/\log k)}$ time algorithm for any computable function f. That is, the $n^{O(k)}$ algorithm of Cygan et al. is almost asymptotically optimal. This left open the following two questions:

- Can we close the gap between the $n^{O(k)}$ algorithm and the $f(k) \cdot n^{o(k/\log k)}$ hardness for STEINER ORIENTATION on general graphs?
- Is STEINER ORIENTATION FPT on planar graphs, or can we obtain an improved runtime such as $f(k) \cdot n^{O(\sqrt{k})}$?

In this paper, we answer the first question completely and make partial progress towards the second question. Formally, we show that:

Theorem 1. *The STEINER ORIENTATION problem is W[1]-hard parameterized by the number k of terminal pairs, even if the underlying undirected graph of the input graph has genus 1. Moreover, under ETH, STEINER ORIENTATION (on graphs of genus 1) cannot be solved in $f(k) \cdot n^{o(k)}$ time for any function f.*

Note that Theorem 1 only leaves open the case of graphs with genus 0, i.e., planar graphs. The open question is whether STEINER ORIENTATION admits the "square-root phenomenon" on planar graphs, i.e., can one obtain a $f(k) \cdot n^{O(\sqrt{k})}$ time algorithm[1] for PLANAR STEINER ORIENTATION, or does the lower bound of $f(k) \cdot n^{o(k)}$ also translate to planar graphs? To the best of our knowledge, even the NP-hardness of PLANAR STEINER ORIENTATION is not known.

Our reduction uses some ideas given by Pilipczuk and Wahlström [17], who obtained a lower bound of $f(k) \cdot n^{o(\sqrt{k})}$ via a rather involved reduction from MULTICOLORED CLIQUE. This was later [16] improved to $f(k) \cdot n^{o(k/\log k)}$ via the standard trick of reducing from the COLORED SUBGRAPH ISOMORPHISM problem [12] instead. To obtain our tight lower bound in Theorem 1 for genus 1 graphs, we use some of the gadgets provided by Pilipczuk and Wahlström [17], but instead give a reduction from the GRID TILING problem introduced by Marx [11]. This way we obtain a cleaner and arguably simpler proof than the one given in [17]. The GRID TILING problem is defined as follows, where we use the standard notation $[n] = \{1, 2, \ldots, n\}$.

$k \times k$ **GRID TILING**

Input : Integers k, n, and k^2 non-empty sets $S_{i,j} \subseteq [n] \times [n]$ where $i, j \in [k]$.
Question: For each $1 \leq i, j \leq k$ does there exist a value $s_{i,j} \in S_{i,j}$ such that

- if $s_{i,j} = (x, y)$ and $s_{i,j+1} = (x', y')$ then $x = x'$, and
- if $s_{i,j} = (x, y)$ and $s_{i+1,j} = (x', y')$ then $y = y'$.

We denote an instance of GRID TILING by $(k, n, \{S_{i,j}\}_{1 \leq i,j \leq k})$. The GRID TILING problem has turned out to be a convenient starting point for parameterized reductions for problems on planar graphs, and has been used recently in several W[1]-hardness proofs [3–5,13–15]. Under ETH, it was shown by Chen et al. [2] that k-CLIQUE[2] does not admit an algorithm running in time $f(k) \cdot n^{o(k)}$ for any function f. There is a simple reduction (see Theorem 14.28 from [6]) from k-CLIQUE to $k \times k$ GRID TILING implying the same runtime lower bound for the latter problem.

2   The Reduction

We begin with describing the reduction from an instance of $k \times k$ GRID TILING to an instance of STEINER ORIENTATION with $O(k)$ terminal pairs. We will then prove that a solution to the GRID TILING instance implies a solution to STEINER ORIENTATION in the constructed instance. To finalize the proof of Theorem 1 we then prove the reverse implication as well.

2.1   Construction

Consider an instance $I = (k, n, \{S_{i,j}\}_{1 \leq i,j \leq k})$ of GRID TILING. We now build an instance $(G, \mathscr{T})$ of STEINER ORIENTATION as follows (refer to Fig. 1).

---

[1] Or even an FPT algorithm.
[2] The k-CLIQUE problem asks whether there is a clique of size $\geq k$.

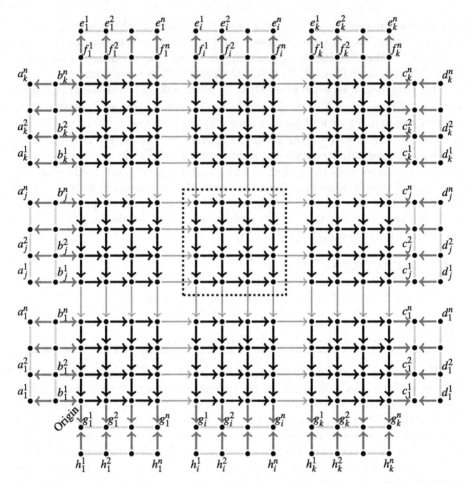

Fig. 1. The instance of STEINER ORIENTATION created from an instance of GRID TILING (before the splitting operation). At this point, the only undirected edges are the green edges. For clarity, we do not show the (directed) perfect matching (which we denote by yellow edges) given by $d_i^j \to a_i^j$ and $h_i^j \to e_i^j$ for each $i \in [k], j \in [n]$. The gadget $G_{i,j}$ is highlighted by a dotted rectangle. (Color figure online)

- We first fix the Origin as marked in black[3].
- The "horizontal right" direction is viewed as the positive X axis and the "vertical upward" is viewed as the positive Y axis.
- **Black Grid Edges:** For each $1 \leq i, j \leq k$ we introduce the $n \times n$ grid $G_{i,j}$ to correspond to the set $S_{i,j}$ of the GRID TILING instance. In Fig. 1 we highlight the gadget $G_{i,j}$ by a dotted rectangle.
  - The bottom left vertex of gadget $G_{i,j}$ is denoted by $v_{i,j}^{1,1}$.

---

[3] This is the unique vertex which has incoming edge from b_1^1 and an outgoing edge to g_1^1.

- Each row of $G_{i,j}$ is horizontal and the number of the row increases as we go vertically upwards. Similarly, each column of $G_{i,j}$ is vertical and the number of the column increases as we go horizontally rightwards. For each $1 \leq h, \ell \leq n$ the unique vertex which is the intersection of the hth column and ℓ^{th} row is denoted by $v_{i,j}^{h,\ell}$.

Orient each horizontal edge of the grid $G_{i,j}$ to the right, and each vertical edge to the bottom.

- We now define four special sets of vertices for the gadget $G_{i,j}$ given by
  - $\texttt{Left}(G_{i,j}) = \{v_{i,j}^{1,\ell} : \ell \in [n]\}$
  - $\texttt{Right}(G_{i,j}) = \{v_{i,j}^{n,\ell} : \ell \in [n]\}$
  - $\texttt{Top}(G_{i,j}) = \{v_{i,j}^{\ell,n} : \ell \in [n]\}$
  - $\texttt{Bottom}(G_{i,j}) = \{v_{i,j}^{\ell,1} : \ell \in [n]\}$
- **Horizontal** Orange **Inter-Grid Edges:** For each $1 \leq i \leq k-1, 1 \leq j \leq k$
  - Add the directed perfect matching from vertices of $\texttt{Right}(G_{i,j})$ to $\texttt{Left}(G_{i+1,j})$ given by the set of edges $\{v_{i,j}^{n,\ell} \to v_{i+1,j}^{1,\ell} : \ell \in [n]\}$.
- **Vertical** Orange **Inter-Grid Edges:** For each $2 \leq j \leq k, 1 \leq i \leq k$
  - Add the directed perfect matching from vertices of $\texttt{Bottom}(G_{i,j})$ to $\texttt{Top}(G_{i,j-1})$ given by the set of edges $\{v_{i,j}^{\ell,1} \to v_{i,j-1}^{\ell,n} : \ell \in [n]\}$.
- We introduce $8k \cdot n$ red vertices given by
  - $A := \{a_i^j \mid i \in [k], j \in [n]\}$
  - $B := \{b_i^j \mid i \in [k], j \in [n]\}$
  - $C := \{c_i^j \mid i \in [k], j \in [n]\}$
  - $D := \{d_i^j \mid i \in [k], j \in [n]\}$
  - $E := \{e_i^j \mid i \in [k], j \in [n]\}$
  - $F := \{f_i^j \mid i \in [k], j \in [n]\}$
  - $G := \{g_i^j \mid i \in [k], j \in [n]\}$
  - $H := \{h_i^j \mid i \in [k], j \in [n]\}$
- **Blue Edges:**
  - For each $i \in [k], j \in [n]$
    * add the directed edge $h_i^j \to g_i^j$,
    * add the directed edge $v_{i,1}^{j,1} \to g_i^j$,
    * add the directed edge $f_i^j \to e_i^j$,
    * add the directed edge $f_i^j \to v_{i,k}^{j,n}$.
  - For each $i \in [k], j \in [n]$
    * add the directed edge $d_i^j \to c_i^j$,
    * add the directed edge $v_{k,i}^{n,j} \to c_i^j$,
    * add the directed edge $b_i^j \to a_i^j$,
    * add the directed edge $b_i^j \to v_{1,i}^{1,j}$.
- Yellow **Edges** (these are left out in Fig. 1): For each $i \in [k], j \in [n]$
  - <u>Category I</u>: add the directed edge $d_i^j \to a_i^j$,
  - <u>Category II</u>: add the directed edge $h_i^j \to e_i^j$.
- Green **Edges:** For each $i \in [k]$

- add the undirected path $a_i^1 - a_i^2 - a_i^3 - \ldots\ldots - a_i^{n-1} - a_i^n$, and denote this path[4] by A_i,
- add the undirected path $b_i^1 - b_i^2 - b_i^3 - \ldots\ldots - b_i^{n-1} - b_i^n$, and denote this path by B_i,
- add the undirected path $c_i^1 - c_i^2 - c_i^3 - \ldots\ldots - c_i^{n-1} - c_i^n$, and denote this path by C_i,
- add the undirected path $d_i^1 - d_i^2 - d_i^3 - \ldots\ldots - d_i^{n-1} - d_i^n$, and denote this path by D_i,
- add the undirected path $e_i^1 - e_i^2 - e_i^3 - \ldots\ldots - e_i^{n-1} - e_i^n$, and denote this path by E_i,
- add the undirected path $f_i^1 - f_i^2 - f_i^3 - \ldots\ldots - f_i^{n-1} - f_i^n$, and denote this path by F_i,
- add the undirected path $g_i^1 - g_i^2 - g_i^3 - \ldots\ldots - g_i^{n-1} - g_i^n$, and denote this path by G_i,
- add the undirected path $h_i^1 - h_i^2 - h_i^3 - \ldots\ldots - h_i^{n-1} - h_i^n$, and denote this path by H_i.

- For each $1 \le i, j \le k$ and each $1 \le x, y \le n$ we perform the following operation on the vertex $v_{i,j}^{x,y}$:
  - If $(x, y) \in S_{i,j}$ then we keep the vertex $v_{i,j}^{x,y}$ as is.
  - Otherwise we **split** the vertex $v_{i,j}^{x,y}$ into two vertices $v_{i,j,\mathrm{LB}}^{x,y}$ and $v_{i,j,\mathrm{TR}}^{x,y}$. Note that $v_{i,j}^{x,y}$ had 4 incident edges: two incoming (one each from the left and the top) and two outgoing (one each to the right and the bottom). We change the edges as follows (see Fig. 2):
    * Make the left incoming edge and bottom outgoing edge incident on $v_{i,j,\mathrm{LB}}^{s,t}$ (denoted by red color in Fig. 2).
    * Make the top incoming edge and right outgoing edge incident on $v_{i,j,\mathrm{TR}}^{s,t}$ (denoted by blue color in Fig. 2).
    * Add an **undirected** edge between $v_{i,j,\mathrm{LB}}^{s,t}$ and $v_{i,j,\mathrm{TR}}^{s,t}$ (denoted by the dotted edge in Fig. 2).

- The set $\mathscr{T}$ of terminal pairs are given by
  - **Type I:** $(b_j^n, a_j^1), (b_j^1, a_j^n), (d_j^n, c_j^1)$ and (d_j^1, c_j^n) for each $j \in [k]$
  - **Type II:** $(f_j^n, e_j^1), (f_j^1, e_j^n), (h_j^n, g_j^1)$ and (h_j^1, g_j^n) for each $j \in [k]$
  - **Type III:** (d_j^n, a_j^1) and (d_j^1, a_j^n) for each $j \in [k]$
  - **Type IV:** (h_j^n, e_j^1) and (h_j^1, e_j^n) for each $j \in [k]$
  - **Type V:** (b_j^1, c_j^n) and (b_j^n, c_j^1) for each $j \in [k]$
  - **Type VI:** (f_j^1, g_j^n) and (f_j^n, g_j^1) for each $j \in [k]$

Note that the total number of terminal pairs is $16k$.

Remark 1. Note that the graph G constructed above can be drawn on the surface of a torus without any two edges crossing: removing the yellow edges, the graph is clearly planar (as depicted in Fig. 1) and can be drawn on a square polygon. Identifying the right edge R and left edge L of the square such that the lower left corner equals the lower right corner, and also the square's top T

---

[4] Sometimes we also abuse notation slightly and use A_i to denote this set of vertices.

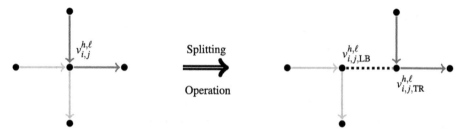

Fig. 2. The splitting operation for vertex $v_{i,j}^{h,\ell}$ when $(h,\ell) \notin S_{i,j}$. The idea behind this splitting is that no matter which way we orient the undirected dotted edge we cannot go **both** from left to right and from top to bottom. However, if we just want to go from left to right (top to bottom) then it is possible by orienting the dotted edge to the right (left), respectively. (Color figure online)

and bottom B edges such that the upper left corner equals the lower left corner, gives an orientable surface of genus 1 (i.e. a torus). The horizontal yellow edges of Category I can connect through $L = R$, and the vertical yellow edges of Category II can connect through $T = B$, without any edges crossing.

Remark 2. For simplicity, we add two "dummy" indices 1 and n which do not belong to any of the sets in the GRID TILING instance. Hence no vertices on the boundary of the grids $G_{i,j}$ (for any $1 \leq i, j \leq k$) are split.

Before proving the correctness of the reduction, we first introduce some notation concerning orientations of the green edges (i.e. potential solutions) of the instance.

Definition 1. *For any $i \in [n]$, a path on n vertices $a_1 - a_2 - \ldots\ldots - a_n$ is said to be oriented towards (away from) i if every edge $a_{j-1} - a_j$ is oriented towards (away from) a_j for every $j \leq i$ and every edge $a_j - a_{j+1}$ is oriented towards (away from) a_j for every $j \geq i$, respectively.*

2.2   GRID TILING has a Solution $\Rightarrow$ STEINER ORIENTATION has a Solution

Suppose that the instance $I = (k, n, \{S_{i,j}\}_{1 \leq i,j \leq k})$ of GRID TILING has a solution, i.e., for each $1 \leq i, j \leq k$ there exists an element $s_{i,j} \in S_{i,j}$ such that

- if $s_{i,j} = (x_{i,j}, y_{i,j})$ and $s_{i,j+1} = (x_{i,j+1}, y_{i,j+1})$ then $x_{i,j} = x_{i,j+1}$,
- if $s_{i,j} = (x_{i,j}, y_{i,j})$ and $s_{i+1,j} = (x_{i+1,j}, y_{i+1,j})$ then $y_{i,j} = y_{i+1,j}$.

That is, there exist elements $\alpha_1, \alpha_2, \ldots, \alpha_k$ and $\beta_1, \beta_2, \ldots, \beta_k$ such that for each $1 \leq i, j \leq k$ we have $(\alpha_i, \beta_j) = s_{i,j} \in S_{i,j}$. We now show that the instance $(G, \mathscr{T})$ of STEINER ORIENTATION has a solution as well. Orient the undirected green edges as follows (note that α_i and β_i are elements from $[n]$ and therefore represent row and column indices in the gadget $G_{i,j}$). For each $i \in [k]$

- orient A_i and C_i away from β_i,
- orient B_i and D_i towards β_i,
- orient E_i and G_i away from α_i,
- orient F_i and H_i towards α_i.

It is easy to see that the above orientations ensure that all terminal pairs of **Types I-IV** are satisfied. We now show that terminal pairs of **Type V** and **Type VI** are also satisfied. First we need some definitions:

Definition 2 (horizontal canonical paths). *Fix $j \in [k]$. For $1 \le \ell \le n$, we denote by Q_j^ℓ the unique (horizontal) directed $b_j^\ell \to c_j^\ell$ path whose second vertex is $v_{1,j}^{1,\ell}$ and second-last vertex is $v_{k,j}^{n,\ell}$. This path starts with the blue edge $(b_j^\ell, v_{1,j}^{1,\ell})$ and ends with the blue edge $(v_{k,j}^{n,\ell}, c_j^\ell)$. The intermediate edges are obtained by selecting the paths of black edges given by the ℓth rows of each gadget $G_{i,j}$ for $i \in [k]$, and connecting these small paths by horizontal orange edges.*

However, we need to address what to do when we encounter a split vertex on this path. Consider the vertex $v_{i,j}^{r,\ell}$ for some $i \in [k]$ and $r \in [n]$. If $v_{i,j}^{r,\ell}$ is not split, then we don't have to do anything. Otherwise, if $v_{i,j}^{r,\ell}$ is split then we add the edge $v_{i,j,LB}^{r,\ell} \to v_{i,j,TR}^{r,\ell}$ to Q_j^ℓ.

Note that the orientation of G which orients all dotted edges rightwards, i.e., $LB \to TR$, contains each of the horizontal canonical paths defined above.

Definition 3 (vertical canonical paths). *Fix $i \in [k]$. For $1 \le \ell \le n$, we denote by P_i^ℓ the unique (vertical) $f_i^\ell \to g_i^\ell$ path whose second vertex is $v_{i,k}^{\ell,n}$ and second-last vertex is $v_{i,1}^{\ell,1}$. This path starts with the blue edge $(f_i^\ell, v_{i,k}^{\ell,n})$ and ends with the blue edge $(v_{i,1}^{\ell,1}, g_i^\ell)$. The intermediate edges are obtained by selecting the paths of black edges given by the ℓth columns of each gadget $G_{i,j}$ for $j \in [k]$, and connecting these small paths by vertical orange edges.*

However, we need to address what to do when we encounter a split vertex on this path. Consider the vertex $v_{i,j}^{\ell,r}$ for some $j \in [k]$ and $r \in [n]$. If $v_{i,j}^{\ell,r}$ is not split, then we don't have to do anything. Otherwise, if $v_{i,j}^{\ell,r}$ is split then we add the edge $v_{i,j,LB}^{\ell,r} \leftarrow v_{i,j,TR}^{\ell,r}$ to P_i^ℓ.

Note that the orientation of G which orients all dotted edges leftwards, i.e., $LB \leftarrow TR$, contains each of the vertical canonical paths defined above. Observe that both the horizontal canonical paths and vertical canonical paths assign orientations to the dotted edges arising from splitting vertices. Hence, one needs to be careful because the splitting operation (see Fig. 2) is designed to ensure that the existence of a horizontal canonical path implies that some vertical canonical path cannot exist (recall that we are allowed to orient each undirected edge in exactly one direction).

Definition 4 (realizable set of paths). *A set of directed paths $\mathscr{P}$ in a mixed graph G is realizable if there is a orientation G^* of G such that each path $P \in \mathscr{P}$ appears in G^*.*

Lemma 1. *The set of vertical canonical paths* $\{P_i^{\alpha_i} : i \in [k]\}$ *together with the set of horizontal canonical paths* $\{Q_j^{\beta_j} : j \in [k]\}$ *are realizable in G.*

Proof. Suppose to the contrary that this set of directed paths is not realizable in G. The only undirected edges which get oriented on horizontal canonical paths or vertical canonical paths are dotted edges which are created from the splitting operation. This implies that there is an undirected dotted edge, say $v_{i,j,LB}^{\alpha_i,\beta_j} - v_{i,j,TR}^{\alpha_i,\beta_j}$ which gets different orientations by the vertical canonical path $P_i^{\alpha_i}$ and the horizontal canonical path $Q_j^{\beta_j}$, respectively. This means that the black vertex $v_{i,j}^{\alpha_i,\beta_j}$ was split. However, by the property of the GRID TILING solution, we have that $(\alpha_i, \beta_j) \in S_{i,j}$ which contradicts the fact that $v_{i,j}^{\alpha_i,\beta_j}$ was split. □

Observe that for each $j \in [k]$, the horizontal path $Q_j^{\beta_j}$ satisfies the two terminal pairs (b_j^1, c_j^n) and (b_j^n, c_j^1) for each $j \in [k]$ of Type V. Similarly, for each $i \in [k]$, the path $P_i^{\alpha_j}$ satisfies the two terminal pairs (f_j^1, g_j^n) and (f_j^n, g_j^1) of Type VI. Lemma 1 guarantees that these families of canonical vertical and horizontal paths can be realized by some orientation (note that the canonical paths only orient black edges, and not green edges whose orientation was already fixed at the start of this subsection) of G. This implies that the instance $(G, \mathscr{T})$ of STEINER ORIENTATION answers YES, and concludes this direction of the proof.

2.3 STEINER ORIENTATION has a Solution $\Rightarrow$ GRID TILING has a Solution

Since the instance $(G, \mathscr{T})$ of STEINER ORIENTATION has a solution, let G^* be the orientation which satisfies all pairs from $\mathscr{T}$. Note that the set of vertices $B \cup D \cup F \cup H$ has no incoming edges. Similarly, the set of vertices $A \cup C \cup E \cup G$ has no outgoing edges.

Lemma 2. *No yellow edge can be on a path in G^* which satisfies any terminal pair of Type I or II.*

Proof. Fix $j \in [k]$. We just prove the lemma for the terminal pair (b_j^n, a_j^1) since the proof for other terminal pairs is similar. Suppose there is a yellow edge on some path P satisfying the terminal pair (b_j^n, a_j^1). This yellow edge cannot be of Category I since D has no incoming edges, and hence we could not have reached D in the first place starting from b_j^n. However, this yellow edge also cannot be of Category II since H has no incoming edges and hence we could not have reached H in the first place starting from b_j^n. □

The next lemma restricts the orientations of the undirected paths of green edges.

Lemma 3. *In the orientation G^*, for each $i \in [k]$ we have that*

- *there exists an integer $\lambda_i \in [n]$ such that the paths A_i, B_i are oriented away from, and towards λ_i, respectively,*
- *there exists an integer $\mu_i \in [n]$ such that the paths C_i, D_i are oriented away from, and towards μ_i, respectively,*
- *there exists an integer $\delta_i \in [n]$ such that the paths E_i, F_i are oriented away from, and towards δ_i, respectively,*
- *there exists an integer $\epsilon_i \in [n]$ such that the paths G_i, H_i are oriented away from, and towards ϵ_i, respectively.*

Proof. Fix $i \in [k]$. We just prove the lemma for the paths A_i, B_i since the proof for other cases is similar. By Lemma 2, we know that the paths satisfying the terminal pairs (b_i^n, a_i^1) and (b_i^1, a_i^n) cannot contain any yellow edges. Since the only non-yellow edges incoming to A are blue edges from B (and B has no incoming edges), it follows that the terminal pairs (b_i^n, a_i^1) and (b_i^1, a_i^n) of Type I are satisfied by edges from the graph $G^*[A_i \cup B_i]$. The path satisfying the terminal pair (b_i^n, a_i^1) has to travel downwards along B_i, use a blue edge and then finally travel downwards along A_i. Similarly, the path satisfying the terminal pair (b_i^1, a_i^n) has to travel upwards along B_i, use a blue edge and then finally travel upwards along A_i. Since we can only orient each green edge in exactly one direction, it follows that both these paths must use the same blue edge, i.e., there exists an integer $\lambda_i \in [n]$ such that the paths A_i, B_i are oriented away from, and towards λ_i, respectively. □

Lemma 4. *For each $i \in [k]$ and integers λ_i, μ_i, δ_i, ϵ_i as given by Lemma 3 we have that*

- $\lambda_i = \mu_i$,
- $\delta_i = \epsilon_i$.

Proof. Fix $i \in [k]$. We just prove that $\lambda_i = \mu_i$ since the proof for the other case is similar. Consider the terminal pairs (d_i^n, a_i^1) and (d_i^1, a_i^n) of Type III. The only outgoing edges from D are to $A \cup C$. However, $A \cup C$ has not outgoing edges. Hence, the aforementioned terminal pairs are satisfied by edges from $G^*[D \cup A]$. By Lemma 3, we know that A_i is oriented away from λ_i and D_i is oriented towards μ_i. Hence, if $\mu_i > \lambda_i$ then the pair (d_i^n, a_i^1) is not satisfied, and if $\mu_i < \lambda_i$ then the pair (d_i^1, a_i^n) is not satisfied. Thus we have $\lambda_i = \mu_i$. □

Lemma 5. *No yellow edge can be on a path satisfying any terminal pair of Type V or VI.*

Proof. Fix $j \in [k]$. We just prove the lemma for the terminal pair (b_j^1, c_j^n) since the proof for other terminal pairs is similar. Suppose there is a yellow edge on some path P satisfying the terminal pair (b_j^1, c_j^n). This yellow edge cannot be of Category I since D has no incoming edges, and hence we could not have reached D in the first place starting from b_j^1. However, this yellow edge also cannot be of Category II since H has no incoming edges and hence we could not have reached H in the first place starting from b_j^1. □

Lemma 6. *For each* $1 \leq i, j \leq k$, *we have that*

- *any path which satisfies the terminal pair* (b_j^1, c_j^n) **must** *contain the horizontal canonical path* $Q_j^{\lambda_j}$,
- *any path which satisfies the terminal pair* (f_i^n, g_i^1) **must** *contain the vertical canonical path* $P_i^{\delta_i}$.

Proof. Fix $j \in [k]$. Consider the terminal pair (b_j^1, c_j^n), and let P be any path satisfying it. By Lemma 5, we know that P cannot have any yellow edges. By Lemmas 3 and 4, we know that B_j, C_j are oriented towards, and away from λ_j, respectively. We claim that the first edge on P which leaves B_j is $b_j^{\lambda_j} \rightarrow v_{1,j}^{1,\lambda_j}$. Clearly, P cannot have any edge from B_j to A_j, since A_j has no outgoing edges. Hence, the path P is of the following type: the vertical upwards path $b_j^1 \rightarrow b_j^2 \rightarrow \ldots b_j^\tau$ followed by the blue edge $b_j^\tau \rightarrow v_{1,j}^{1,\tau}$. Since B_j is oriented towards λ_j it follows that $\lambda_j \geq \tau$. If $\lambda_j > \tau$ then by orientation of the black grid edges and orange edges (note that the splitting doesn't really change the rows/columns level) it follows that P reaches C_j at a vertex c_j^ψ where $\lambda_j > \tau \geq \psi$. However, C_j is oriented away from λ_j which contradicts that P is a path from b_j^1 to c_j^n. Hence, we have that $\lambda_j = \tau = \psi$. Therefore, P contains a subpath which starts at $b_j^{\lambda_j}$ and ends at $c_j^{\lambda_j}$ and all edges of this subpath (except the first and last blue edges) are contained in the graph $G^* \left[\bigcup_{i=1}^k V(G_{i,j}) \right]$, i.e., P contains the canonical horizontal path $Q_j^{\lambda_j}$.

The proof of the second part of the lemma is similar, and we omit the details here. $\qquad\square$

Lemma 7. *The instance* $(k, n, \{S_{i,j}\}_{1 \leq i,j \leq k})$ *of* GRID TILING *has a solution.*

Proof. We show that $(\delta_i, \lambda_j) \in S_{i,j}$ for each $1 \leq i, j \leq k$. This will imply that GRID TILING has a solution.

Fix any $1 \leq i, j \leq k$. By Lemma 6, we know that the orientation G^* must contain the horizontal canonical path $Q_j^{\lambda_j}$ (to satisfy the pair (b_j^1, c_j^n)) and also the vertical canonical path $P_i^{\delta_i}$ (to satisfy the pair (f_i^n, g_i^1)). We now claim that the vertex $v_{i,j}^{\delta_i, \lambda_j}$ cannot be split: suppose to the contrary that it is split. By Definition 3, the path $P_i^{\delta_i}$ orients the edge $v_{i,j,LB}^{\delta_i,\lambda_j} - v_{i,j,TR}^{\delta_i,\lambda_j}$ as $v_{i,j,LB}^{\delta_i,\lambda_j} \leftarrow v_{i,j,TR}^{\delta_i,\lambda_j}$. However, by Definition 2, the path $Q_j^{\lambda_j}$ orients the edge $v_{i,j,LB}^{\delta_i,\lambda_j} - v_{i,j,TR}^{\delta_i,\lambda_j}$ as $v_{i,j,LB}^{\delta_i,\lambda_j} \rightarrow v_{i,j,TR}^{\delta_i,\lambda_j}$, which is a contradiction. Hence, the vertex $v_{i,j}^{\delta_i,\lambda_j}$ is not split, i.e., $(\delta_i, \lambda_j) \in S_{i,j}$ for each $1 \leq i, j \leq k$. $\qquad\square$

2.4   Obtaining the $f(k) \cdot N^{o(k)}$ Lower Bound

It is easy to see that the graph G has $O(n^2 k^2)$ vertices and can be constructed in poly$(n+k)$ time. Combining the two directions from Subsects. 2.2 and 2.3, we get a paramterized reduction from GRID TILING to STEINER ORIENTATION. Hence, the W[1]-hardness of STEINER ORIENTATION follows from the W[1]-hardness of GRID TILING [11]. Chen et al. [2] showed that, for any function f, the existence of an $f(k) \cdot n^{o(k)}$ algorithm for k-CLIQUE violates ETH. There is a simple reduction (see Theorem 14.28 from [6]) from k-CLIQUE to $k \times k$ GRID TILING implying the same runtime lower bound for the latter problem. Our reduction transforms the problem of $k \times k$ GRID TILING into an instance of STEINER ORIENTATION with $O(k)$ demand pairs. Composing the two reductions, we obtain that under ETH there is no $f(k) \cdot n^{o(k)}$ time algorithm for STEINER ORIENTATION. Recall from Remark 1 that the graph G constructed in the STEINER ORIENTATION instance has genus 1, and hence the $f(k) \cdot n^{o(k)}$ lower bound holds for genus 1 graphs too. This concludes the proof of Theorem 1.

References

1. Arkin, E.M., Hassin, R.: A note on orientations of mixed graphs. Discret. Appl. Math. **116**(3), 271–278 (2002)
2. Chen, J., Huang, X., Kanj, I.A., Xia, G.: Strong computational lower bounds via parameterized complexity. J. Comput. Syst. Sci. **72**(8), 1346–1367 (2006)
3. Chitnis, R., Esfandiari, H., Hajiaghayi, M.T., Khandekar, R., Kortsarz, G., Seddighin, S.: A tight algorithm for strongly connected steiner subgraph on two terminals with demands. Algorithmica **77**(4), 1216–1239 (2017)
4. Chitnis, R., Feldmann, A.E., Manurangsi, P.: Parameterized approximation algorithms for directed Steiner network problems. CoRR, abs/1707.06499 (2017)
5. Chitnis, R.H., Hajiaghayi, M., Marx, D.: Tight bounds for planar strongly connected Steiner subgraph with fixed number of terminals (and extensions). In: SODA, pp. 1782–1801 (2014)
6. Cygan, M., Fomin, F.V., Kowalik, Ł., Lokshtanov, D., Marx, D., Pilipczuk, M., Pilipczuk, M., Saurabh, S.: Parameterized Algorithms. Springer, Cham (2015). https://doi.org/10.1007/978-3-319-21275-3
7. Cygan, M., Kortsarz, G., Nutov, Z.: Steiner forest orientation problems. SIAM J. Discret. Math. **27**(3), 1503–1513 (2013)
8. Hassin, R., Megiddo, N.: On orientations and shortest paths. Linear Algebra Appl. **114**, 589–602 (1989). Special Issue Dedicated to Alan J. Hoffman
9. Impagliazzo, R., Paturi, R.: On the complexity of k-SAT. J. Comput. Syst. Sci. **62**(2), 367–375 (2001)
10. Impagliazzo, R., Paturi, R., Zane, F.: Which problems have strongly exponential complexity? J. Comput. Syst. Sci. **63**(4), 512–530 (2001)
11. Marx, D.: On the optimality of planar and geometric approximation schemes. In: FOCS, pp. 338–348 (2007)
12. Marx, D.: Can you beat treewidth? Theory Comput. **6**(1), 85–112 (2010)
13. Marx, D.: A tight lower bound for planar multiway cut with fixed number of terminals. In: Czumaj, A., Mehlhorn, K., Pitts, A., Wattenhofer, R. (eds.) ICALP 2012. LNCS, vol. 7391, pp. 677–688. Springer, Heidelberg (2012). https://doi.org/10.1007/978-3-642-31594-7_57
14. Marx, D., Pilipczuk, M.: Everything you always wanted to know about the parameterized complexity of Subgraph Isomorphism (but were afraid to ask). In: STACS, pp. 542–553 (2014)

15. Marx, D., Pilipczuk, M.: Optimal parameterized algorithms for planar facility location problems using voronoi diagrams. In: Bansal, N., Finocchi, I. (eds.) ESA 2015. LNCS, vol. 9294, pp. 865–877. Springer, Heidelberg (2015). https://doi.org/10.1007/978-3-662-48350-3_72
16. Pilipczuk, M., Wahlström, M.: Directed multicut is W[1]-hard, even for four terminal pairs. CoRR, abs/1507.02178 (2015)
17. Pilipczuk, M., Wahlström, M.: Directed multicut is W[1]-hard, even for four terminal pairs. In: SODA, pp. 1167–1178 (2016)

Can We Create Large k-Cores by Adding Few Edges?

Rajesh Chitnis[1]([⊠]) and Nimrod Talmon[2]

[1] Department of Computer Science, University of Warwick, Coventry, UK
rajeshchitnis@gmail.com
[2] Ben-Gurion University of the Negev, Be'er Sheva, Israel
talmonn@bgu.ac.il

Abstract. The notion of a k-core, defined by Seidman ['83], has turned out to be useful in analyzing network structures. The k-core of a given simple and undirected graph is the maximal induced subgraph such that each vertex in it has degree at least k. Hence, finding a k-core helps to identify a (core) community where each entity is related to at least k other entities. One can find the k-core of a given graph in polynomial time, by iteratively deleting each vertex of degree less than k. Unfortunately, this iterative *dropping out* of vertices can sometimes lead to unraveling of the entire network; e.g., Schelling ['78] considered the extreme example of a path with $k = 2$, where indeed the whole network unravels.

In order to avoid this unraveling, we would like to edit the network in order to maximize the size of its k-core. Formally, we introduce the EDGE k-CORE problem (EKC): *given a graph G, a budget b, and a goal p, can at most b edges be added to G to obtain a k-core containing at least p vertices?* First we show the following dichotomy: EKC is polytime solvable for $k \leq 2$ and NP-hard for $k \geq 3$. Then, we show that EKC is W[1]-hard even when parameterized by $b + k + p$. In searching for an FPT algorithm, we consider the parameter "treewidth", and design an FPT algorithm for EKC which runs in time $(k + \mathbf{tw})^{O(\mathbf{tw}+b)} \cdot \text{poly}(n)$, where $\mathbf{tw}$ is the treewidth of the input graph. Even though an extension of Courcelle's theorem [Arnborg et al., *J. Algorithms* '91] can be used to show FPT for EKC parameterized by $\mathbf{tw} + k + b$, we obtain a much faster running time as compared to Courcelle's theorem (which needs a tower of exponents) by designing a dynamic programming algorithm which needs to take into account the fact that newly added edges might have endpoints in different bags which cross the separator.

R. Chitnis—Supported by ERC grant CoG 647557 "Small Summaries for Big Data". Part of this work was done when the author was at the Weizmann Institute of Science and supported by Israel Science Foundation grant #897/13.

N. Talmon—Part of this work was done when the author was at the Weizmann Institute of Science.

F. V. Fomin and V. V. Podolskii (Eds.): CSR 2018, LNCS 10846, pp. 78–89, 2018.
https://doi.org/10.1007/978-3-319-90530-3_8

1 Introduction

Graphs are very useful for modeling networks which describe relationships and interactions between sets of people or entities, in various disciplines, such as social sciences [17], life sciences [10], medicine [20], etc. Usually, we assign a vertex to each entity, and there is an edge between two entities if they are related or affect each other in some way. Analyzing graph structure has found applications in several important real-world problems such as targeted advertising [24], fraud detection [18], missing link prediction [16], locating functional modules of interacting proteins [14], etc.

An important problem in analyzing the structure of big networks is to detect large communities of vertices that are "related" to one another. This problem has been widely considered in various sub-areas of computer science [12,13,19]. A reasonable and well-studied notion for a vertex to be "related" within a community is to have a *large* number of neighbors within the community. Bhawalkar et al. [4] considered the following model of *user engagement* within a network: there is a single product and each individual has two options of "engaged" or "drop out". We assume that all individuals are initially engaged, and there is some given threshold parameter k such that a person finds it worthwhile to remain engaged if that person has at least k engaged friends.

In this model of *user engagement* all individuals with less than k friends will drop out immediately. Unfortunately, this can propagate and even those individuals who initially had more than k friends in the network may end up dropping out. An extreme example of this was given by Schelling [22, p. 214]: consider a path on n vertices and let $k = 2$. Note that, while $n-2$ vertices initially have degree two in the network, there will be a *cascade of iterated withdrawals* since each endpoint has degree one, thus it drops out and now its neighbor in the path has only one friend in the network and it drops out as well; eventually, the whole network drops out. Indeed, at the end of the iterated withdrawals process the remaining engaged individuals form a unique maximal induced subgraph whose minimum degree is at least k. This subgraph is called the k-*core* and is a well-known concept in the theory of networks; it was introduced by Seidman [23] and also been studied in various social sciences literature [8,9]. The concept of k-core decompositions (where for each vertex v we find the max k such that v belongs to the k-core in G) has been used in the analysis and visualization of large scale networks [1–3].

A Game-Theoretic Model for k-Core: Consider the following game-theoretical model from [4]: each user in a graph $G = (V, E)$ pays a cost of k to remain engaged, and she receives a profit of 1 from every neighbor who is engaged. If an individual is not engaged, then she receives a payoff of zero. Hence, she remains engaged if she has non-negative payoff, i.e., she has at least k neighbors who are engaged. Then the k-core can be viewed as the unique maximal equilibrium in this setting. Assuming that all the players make decisions simultaneously the model can be viewed as a simultaneous-move game where each individual has two strategies viz. remaining engaged or dropping out. For every strategy profile $\delta \in \{0,1\}^{|V|}$ let $S_\delta = \{i \ : \ \delta_i = 1\}$ denote the set of players who remain engaged. We can easily characterize the set of pure Nash equilibria for this game: a strategy profile δ is a Nash equilibrium if and only if the following two conditions hold:

- No engaged player wants to drop out, i.e., minimum degree of the induced graph $G[S_\delta]$ is $\geq k$
- No player who has dropped out wants to become engaged, i.e., no $v \in V(G) \setminus S_\delta$ has $\geq k$ neighbors in S_δ

In general there can be many Nash equilibria. For example, if G itself has minimum degree $\geq k$ then $S_\delta = \emptyset$ and $S_\delta = V(G)$ are two equilibria (and there may be more). Owing to the fact that it is a maximal equilibrium, the k-core has the special property that it is beneficial to both parties: it maximizes the payoff of every user, while also maximizing the size of the network. Chwe [8,9] and Sääskilahti [21] suggest that one can reasonably expect this maximal equilibrium even in real-life implementations of this game.

The Anchored k-Core Problem (AKC): The unraveling described above in Schelling's example of a path might be highly undesirable if the goal is to keep as many people engaged as possible. One possibility of preventing this is by "buying" the two end-point players into being engaged. This ensures that the whole path remains engaged. Correspondingly, Bhawalkar et al. [4] formalized this notion and defined the ANCHORED k-CORE problem (AKC). In AKC, they overcome the issue of unraveling by allowing some "anchors": these are vertices that remain engaged irrespective of their degree. This can be achieved by giving them extra incentives or discounts. The question in AKC is the following: given three integers b, k, and p, can one use at most b *anchors* and ensure that there is an anchored k-core of size at least p?

Besides defining the AKC problem, Bhawalkar et al. [4] showed that AKC is solvable in polynomial time for $k \leq 2$ but NP-hard for $k \geq 3$. Also it is NP-hard to approximate the approximate the size of the optimal k-core to within an $O(n^{1-\epsilon})$ factor[1] for any $\epsilon > 0$. From the viewpoint of parameterized complexity, they showed that for every fixed $k \geq 3$ the p-AKC problem is W[2]-hard with respect to b, and, on the positive side, they gave a polynomial-time algorithm for graphs of bounded treewidth. In a follow-up work, Chitnis et al. [7] showed that it remains NP-hard on planar graphs for all $k \geq 3$, even if the maximum degree of the graph is $k + 2$; that it becomes FPT on planar graphs (unlike on general graphs) parameterized by b for all $k \geq 7$; and, strengthening the intractability result of Bhawalkar et al. [4], they showed W[1]-hardness of it with respect to p (which is always greater than or equal to b).

The Edge k-Core Problem (EKC): In this paper we consider an alternative way to maximize the size of the k-core. For example, in Schelling's example, instead of anchoring the two end-point players, one could also *add an edge* between these two vertices; this again ensures that the whole path remains

---

[1] That is, distinguishing whether the size of the optimal k-core is $O(b)$ or $\Omega(n)$.

engaged. While this changes the structure of the network, it has the desirable property of ensuring a "pure" k-core, i.e., where *each* vertex has degree at least k. Correspondingly, we ask the following question: can we add "few" edges to the given network and obtain a "large" k-core? Formally, the problem we study in this paper is as follows.

The Edge k-Core Problem (EKC)

Input: A simple, undirected graph $G = (V, E)$ and integers b, k, and p.
Question: Is there a set of vertices $H \subseteq V$ of size $\geq p$ such that there is a set $B \subseteq (\binom{V}{2} \setminus E)$ with $|B| \leq b$ and every $v \in H$ satisfies $\deg_{G'[H]}(v) \geq k$, where $G' = (V, E \cup B)$?

EKC arises naturally in several scenarios concerning network resiliency such as:

- **Peer-to-Peer (P2P) networks:** In P2P networks, users share common resources (bandwidth, disk storage, etc.). For any user to benefit from the network they should be connected to at least a certain threshold k number of other users. EKC then tells the parent company which connections shall be added between the users so that a large number of users can successfully use the P2P network.
- **Distributed networks:** Suppose there is an existing network of computers, connected by some topology. Over time the complexity of the tasks to be executed increases, and one may need more computers to perform the task in a distributed fashion. The EDGE k-CORE problem then guides us on how to edit the network (which connections to add between computers) assuming that we know the threshold k number of computers needed for any task.

Our Results. Besides introducing EKC, we first, in Sect. 2, describe a polynomial-time algorithm for $k \leq 2$ and show NP-hardness for $k \geq 3$, thus providing a complexity dichotomy. Then, in Sect. 3, we begin by showing that EKC is W[1]-hard parameterized by $b + p$ even when $k = 3$. This tells us that we need to consider more parameters if we seek fixed-parameter tractability. As a natural network parameter, we consider the treewidth **tw** and design a dynamic program to show that EKC is fixed-parameter tractable with respect to **tw** $+ k + b$. In our view, this is the most technical part of the paper.

Comparing Anchored k-Core and Edge k-Core. While AKC has been studied before [4,6,7], in this paper we introduce and study the EKC problem. Below we briefly show that these two problems are unrelated in the following sense: there are examples of graphs where for the same values of p and k we need very different number of anchored vertices or edge additions to achieve a k-core of size at least p.

- **Edge Additions > Anchored Vertices:** Let G be a disjoint union of two components G_1 and G_2, where $G_1 = K_{z_1}$ (i.e., a clique on z_1 vertices) and G_2 is a z_2-regular graph on n_2 vertices. Choose $z_1 \ll z_2 \ll n_2$. If $k = z_2$ and $p = z_1 + n_2$, then the number of vertices which need to be anchored for an *anchored* k-core of size p is $b_v = z_1$. However, to get a *pure* k-core of size p we need to add $z_2 - (z_1 - 1)$ edges on each vertex of G_1, and hence $b_e \geq \dfrac{z_1 \cdot (z_2 - z_1 + 1)}{2} \gg z_1 = b_v$. This example is asymptotically tight since adding $(k - \deg)$ new edges to a vertex of degree k simulates anchoring the vertex.

– **Edge Additions < Anchored Vertices:** We build a graph G as follows: let $G_1 = K_{2n}$, let G_2 be K_{2n} with a perfect matching removed, and add a matching of size $2n$ between G_1 and G_2. If $k = 2n$ and $p = 4n$, then the number of vertices which need to be anchored to obtain an *anchored* k-core of size p is $b_v = 2n$. However, to get a *pure* k-core of size p it is enough to add the n edges of the perfect matching which were removed from G_2. Hence $b_e = n < 2n = b_v$. This example is strictly tight since anchoring two endpoints of an edge simulates adding the edge.

2   Classical Complexity: Polytime Algorithms and NP-hardness

In this section we first describe a polynomial-time algorithm which solves EDGE k-CORE whenever $k \leq 2$ (Theorem 1). Then, we show that this result is tight with respect to k, by showing that EDGE k-CORE is NP-hard whenever $k \geq 3$ (Theorem 2).

Theorem 1. [⋆][2] EDGE k-CORE *is polynomial-time solvable for $k \leq 2$.*

Theorem 2 (⋆). EDGE k-CORE *is NP-hard for $k \geq 3$.*

3   Parameterized Complexity: W[1]-Hardness and FPT Algorithms

In this section, we analyze EKC via the framework of parameterized complexity. EKC is para-NP-hard parameterized by k since Theorem 2 shows that EKC is NP-hard for $k = 3$. In fact, taking a closer look at the proof of Theorem 2, we observe the following.

Corollary 1. *EKC is NP-hard for $k = 3$, even on planar graphs of max degree 5.*

Now we show that EKC admits a simple XP algorithm parameterized by b.

Observation 3. *The EKC problem admits an XP algorithm parameterized by b.*

---

[2] Proofs of results marked with [⋆] are deferred to the full version of the paper due to lack of space.

Proof. Since any graph on n vertices can have at most $\binom{n}{2}$ edges (recall that we do not allow parallel edges or self-loops), we can try all possible subsets of non-edges (which are not already present) to be added. This gives an $\binom{\binom{n}{2}}{b} = n^{O(b)}$ algorithm.

Consider a yes-instance I of EKC for which $p \leq k$. Since the minimum degree of any subgraph containing at most k vertices is at most $k - 1$, it follows that the size of the k-core created in the solution of I is at least $k + 1$ (recall the definition of EKC, which asks for a k-core containing **at least** p vertices). Thus, we assume henceforth that instances of EKC satisfy $p > k$. Next, we show that Theorem 3 translates to an XP algorithm for EKC parameterized by p.

Proposition 1. *The EKC problem has an XP algorithm parameterized by p.*

Proof. We can convert any set of p vertices (assuming w.l.o.g. that $p > k$) into a k-core by adding at most $\binom{p}{2}$ edges. Hence, if $b \geq \binom{p}{2}$, then we can answer YES; otherwise, i.e., if $b < \binom{p}{2}$, then the $n^{O(b)}$ algorithm from Theorem 3 is also an $n^{O(p^2)}$ algorithm for the EKC problem. □

Next we show that if one wants to design an FPT algorithm for the EKC problem, then even combining the three parameters p, k, and b is not enough.

Theorem 4. *The EDGE k-CORE problem is W[1]-hard parameterized by $p + b$, for $k = 3$.*

Proof. We reduce from the W[1]-hard CLIQUE problem [11] which, given a graph G and an integer ℓ, asks for the existence of ℓ pairwise adjacent vertices in G. Consider an instance $(G = (V, E), \ell)$ of CLIQUE where $V = (v^1, v^2, \ldots, v^n)$ and construct a new graph $G' = (V', E')$ as follows.

For each $1 \leq i \neq j \leq \ell$ make a copy G_{ij} of the vertex set V (do not add any edges). Each of these vertices is *black*. Let the vertex v^r in the copy G_{ij} be labeled v^r_{ij}. Add the following edges to G' (we use the notation $[n] = \{1, 2, \ldots, n\}$):

- For each $1 \leq i \neq j \leq \ell$ and $r, s \in [n]$ we add an edge between v^r_{ij} and v^s_{ji} if and only if $v^r v^s \in E$. Subdivide each such edge twice by adding two new **green** vertices x^{rs}_{ij} and x^{sr}_{ji}. We refer to x^{rs}_{ij} as a **brother** of x^{sr}_{ji} and vice-versa.
- For each $i \in [\ell], r \in [n]$ add the cycle $v^r_{i1} - v^r_{i2} - \ldots v^r_{i,i-1} - v^r_{i,i+1} - \ldots - v^r_{i\ell} - v^r_{i1}$. Let us denote this cycle by C^r_i.

This completes the construction of G'. Let $k = 3$, $b = \binom{\ell}{2}$ and $p = 4b$. In the full version of the paper, we show the correctness of the reduction.

3.1   FPT Algorithm Parameterized by tw + k + b

Next we sketch the proof of our main technical result, namely that EKC is fixed-parameter tractable for **tw**+k+b. Notice that Theorem 4, which shows that EKC is W[1]-hard even for $k + p + b$, indeed motivates studying further parameters.

Let T be a tree and $B : V(T) \to 2^{V(G)}$. The pair (T, B) is called as a *valid tree decomposition* of an undirected graph G, if T is a tree in which every vertex $x \in V(T)$ has an assigned set of vertices $B_x \subseteq V(G)$ (called a bag) such that the following properties are satisfied:

- **(P1):** $\bigcup_{x \in V(T)} B_x = V(G)$.
- **(P2):** For each $u - v \in E(G)$, there exists an $x \in V(T)$ such that $u, v \in B_x$.
- **(P3):** For each $v \in V(G)$, the set of vertices of T whose bags contain v induces a connected subtree of T.

The *width* of the tree decomposition (T, B) is $\max_{x \in V(T)} |B_x| - 1$. The treewidth of a graph G, usually denoted by $\mathbf{tw}(G)$, is the minimum width over all valid tree decompositions of G.

We will use a special type of tree decompositions called *nice* tree decompositions.

Definition 1. *A tree-decomposition (T, B) of G is said to be* nice *if T is a rooted binary tree such that each vertex $t \in T$ is one of the following four types:*

- **Leaf Node:** *t is a leaf in T and $B_t = \{v\}$ for some $v \in G$.*
- **Introduce Node:** *t has exactly one child t' and $B_t = B_{t'} \cup \{v\}$ for some $v \notin B_{t'}$.*
- **Forget Node:** *t has exactly one child t' and $B_t = B_{t'} \setminus \{v\}$ for some vertex $v \in B_{t'}$.*
- **Join Node:** *t has exactly two children t', t'' such that $B_{t'} = B_t = B_{t''}$.*

The advantage of nice tree-decompositions is that when writing a dynamic program we only need to handle four types of nodes. It is known [5,15] that a general tree decomposition (T, B) (of treewidth $\mathbf{tw}$) can be converted, in linear time, into a nice tree decomposition (T', B') of the same width such that $|T'| = O(\mathbf{tw} \cdot n)$.

Our main result in this section, and what we believe is the most technically-involved result in this paper, is a dynamic programming based FPT algorithm for EKC parameterized by $\mathbf{tw} + k + b$. We remark here that the fixed-parameter tractability of EKC parameterized by $\mathbf{tw} + k + b$ can be shown to follow from Courcelle's theorem, albeit with much worse running time (i.e., tower of exponentials); further, we argue that our dynamic programming is interesting because the operation of adding edges is an "inter-bag" operation: given a bag which separates the graph into two parts, a new edge might have one endpoint in each part. Usually, FPT algorithms parameterized by treewidth are for "inter-bag" operations, where each recursive call in the dynamic program is confined to its subtree, while in our case the structure is more involved. In the interest of space, we provide here a description with some intuition for the proof of Theorem 5. The complete formal proof of correctness and the analysis of the running time is deferred to the full version.

Theorem 5. *The* EDGE *k-*CORE *problem can be solved in $(k + \mathbf{tw})^{O(tw+b)} \cdot poly(n))$ time.*

Proof (Sketch). Given a nice tree decomposition of a given graph (notice that finding a tree decomposition, or at least an approximation of it, can be done in FPT time) we design a dynamic program for solving EKC on it. First we fix an (arbitrary) ordering on the vertices of V, say $\phi = (v_1, v_2, \ldots, v_n)$. We will view the vertices in this order when we consider them in a bag of the tree-decomposition; to this end, for a node t in the tree decomposition which is a bag containing x vertices, let us arbitrarily order those x vertices (while fixing this ordering) and denote them by $[v_1, \ldots, v_x]$. For a node $t \in T$ let T_t denote the subtree of T which is rooted at t. Also let V_t denote the union of vertices in all bags of the nodes in the subtree of T rooted at t, i.e., $V_t = \cup_{t \in T_t} B_t$. For $i \in \mathbb{N}$ let $\mathbf{0}^i, \mathbf{1}^i$ denote the multiset which has i zeroes and i ones respectively. For $q \geq s$ let $\mathcal{H}(q, s) = \{\mathbf{z} \in \{0,1\}^q \; : \; \mathbf{z}$ has Hamming weight exactly $s\}$. Similarly, $\mathcal{H}^*(q, s) = \{\mathbf{z} \in \{0, 1, \ldots, k\}^q \; : \; \mathbf{z}$ has at most s non-zero entries$\}$. Finally, for $\mathbf{z} \in \mathcal{H}(q, s)$ we define the set $\mathcal{H}_{\mathbf{z}}^*(q, s) = \{\mathbf{y} \in \mathcal{H}^*(q, s) \; : \; \mathbf{y}[i] \neq 0 \Rightarrow \mathbf{z}[i] \neq 0\}$.

For a node $t \in T$ in the tree decomposition, we define the following boolean quantity

$$\text{BOOL}[t, b_{\text{in}}, b_{\text{out}}, p_{\text{in}}, p_{\text{out}}, \mathbf{y}, \mathbf{z}, q_0, Q]$$

for each choice of

- $b_{\text{in}}, b_{\text{out}}, p_{\text{in}}, p_{\text{out}} \geq 0$
- $b_{\text{in}} + b_{\text{out}} \leq b$
- $p_{\text{in}} \leq |B_t|$ and $p_{\text{out}} \leq |V_t \setminus B_t|$
- $\mathbf{y} \in \mathcal{H}(|B_t|, p_{\text{in}})$ and $\mathbf{z} \in \mathcal{H}_{\mathbf{y}}^*(|B_t|, p_{\text{in}})$
- $0 \leq q_0 \leq p_{\text{out}}$ (actually, $p_{\text{out}} - b \leq q_0 \leq |Q|$)
- $Q = \{q_1, q_2, \ldots, q_{p_{\text{out}} - q_0}\}$ is a (multi)set of size $p_{\text{out}} - q_0$ such that each element in Q is positive and at most k, i.e., $1 \leq q_i \leq k$ for each $i \in [p_{\text{out}} - q_0]$

We set $\text{BOOL}[t, b_{\text{in}}, b_{\text{out}}, p_{\text{in}}, p_{\text{out}}, \mathbf{y}, \mathbf{z}, q_0, Q] = 1$ if and only if there exist sets $H_t \subseteq V_t$ and $E_t \subseteq H_t \times H_t$ such that the following conditions hold:

- $|H_t \cap B_t| = p_{\text{in}}$
- $|H_t \cap (V_t \setminus B_t)| = p_{\text{out}}$
- Number of edges of E_t which have both endpoints in B_t is b_{in}
- Number of edges of E_t which have at most one endpoint in B_t is b_{out}
- $|E_t| \leq b_{in} + b_{out}$
- $\mathbf{y}[v] = 1$ if and only if $v \in B_t \cap H_t$
- If $v \in B_t \cap H_t$ then the degree of v in the graph $G^*[H_t] = G[H_t] \cup E_t$ is $\geq \mathbf{z}[v]$
- There is a bijection $\phi_t : ((V_t \setminus B_t) \cap H_t) \to (Q \cup \mathbf{0}^{q_0})$ such that for every $w \in (V_t \setminus B_t) \cap H_t$ we have $\deg_{G^*[H_t]}(w) + \phi_t(w) \geq k$

We say that (H_t, E_t, ϕ_t) is a **witness** for $\text{BOOL}[t, b_{\text{in}}, b_{\text{out}}, p_{\text{in}}, p_{\text{out}}, \mathbf{y}, \mathbf{z}, q_0, Q]=1$.

Intuition: Instead of solving the EDGE k-CORE problem in recursion, we design a dynamic program which solves a more general problem. Specifically, this general problem is such that (1) we specify more concretely how this given budget is to be used and which structure the k-core shall have; and (2) we allow some "help" from the "outside world". Let us mention that, for (1), we specify how the budget b shall be split into b_{in} (budget to be used solely in the bag vertices) and b_{out} (budget to be used not solely in the bag vertices); how the p vertices of the k-core shall be split between p_{in} (k-core vertices in the bag vertices) and p_{out} (k-core vertices not in the bag vertices); where we even specify, by the 1-entries of the vector $\mathbf{y}$, exactly which vertices of the bag shall be k-core vertices. Now, recall that in the EDGE k-CORE problem, each vertex in the k-core must have degree at least k; for (2), we allow some "help" from the "outside world", by relaxing this "at least k" requirement for some of the vertices; specifically, we specify the needed degrees for the vertices of the k-core in the bag vertices (those p_{in} vertices whose corresponding y values are 1), since we require for them to have only degrees as specified by the $\mathbf{z}$ vector; and, for the vertices of the k-core which are not in the bag vertices, we use two multisets Q and $\mathbf{0}^{q_0}$ (which are together of size p_{out}). We view this as some "degree help" that we allow those vertices to use (in order to make their degree $\geq k$): the exact way by which we specify the amount of "degree help" that the vertices of the k-core could use is defined by the bijection ϕ. The multiset Q corresponds to those vertices which actually need some "degree help" (and we maintain all such numbers), while the number q_0 corresponds to the number of vertices of $H_t \cap (V_t \setminus B_t)$ which do not need any help at all.

> The crucial idea is that although p_{out} can be as large as n, we have that $|Q| \leq b$ since we only have b edges in the budget to help. (In fact, even $\sum_{x \in Q} x \leq b$ holds.)

Let r be the root of T. Next, we will show how to recursively compute the values of the boolean quantity BOOL; for now, let us mention that we will decide that the given instance (G, b, k, p) of EDGE k-CORE is a yes-instance if and only if the following holds:

$$\bigvee_{\substack{b_{in}+b_{out} \leq b \\ 0 \leq p_{in} \leq |B_r| \\ 0 \leq p_{out} \leq |V_r \setminus B_r| \\ p_{in}+p_{out} \geq p \\ \mathbf{y} \in \mathcal{H}(|B_r|, p_{in})}} \text{BOOL}[r, b_{in}, b_{out}, p_{in}, p_{out}, \mathbf{y}, \mathbf{z} = k^{|B_r|}, q_0 = p_{out}, Q = \emptyset] = 1$$

Below we briefly give some intuition on how to recursively compute the values of BOOL for each type of node in the nice tree decomposition. We defer the formal recurrence, proof of correctness, and analysis of the running time to the full version.

3.1.1 Leaf Node

Intuition: For leaf nodes, there are no further recursive calls; thus, it is enough to check the "sanity" of the given values.

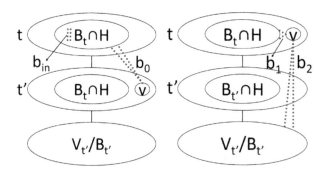

Fig. 1. Recursion on a forget node (left) and on an introduce node (right).

3.1.2 Forget Node

Intuition: Refer to Fig. 1. There are two possibilities for forget node, namely whether the forgotten vertex v is part of the k-core or not. If it is not, then we can call the child with almost exactly the same values; if it is in the k-core, then we guess the exact connections that the forgotten node will have to other vertices in the bag. Given those guesses, we can issue a recursive call almost without worrying about the forgotten node; notice that we guess whether or not v receives non-zero "help" from the outside, since, in the child, v is a bag vertex, and thus does not have a corresponding Q-value or is counted in q_0.

3.1.3 Introduce Node

Intuition: Refer to Fig. 1. Let t' be the child of t such that $B_t = B_{t'} \cup \{v\}$. There are two cases to consider, namely whether v is in the k-core or not. If v is not in the k-core then we can safely issue a recursive call to the child t'. Otherwise, we can fully guess the "new" connections between v to the other vertices in the bag B_t, and then call the child with different z-values, since their degrees will be increased by v. By the definition of a tree-decomposition there cannot be any edges already present between v and any vertex of $V_{t'} \setminus B_{t'}$. However, while staying within FPT time we cannot guess the exact set of vertices from $V_{t'} \setminus B_{t'}$ which get a "help" edge from v. Instead we just guess the number of such edges, and issue the recursive call for t' with the appropriate changes in some Q-values.

3.1.4 Join Node

Intuition: Refer to Fig. 2. First we can guess how the b_{in} edges that will be introduced in the bag of t so we can then call t and t' with $b_{in=0}$. Then we guess the partition of b_{out} into three parts: two parts correspond to the b_{out} edges for t', t'' respectively and the third part corresponds to edges between $V_{t'} \setminus B_{t'}$ and $V_{t''} \setminus B_{t''}$. Note that by the properties of tree decompositions, it follows that there are no edges already present between $V_{t'} \setminus B_{t'}$ and $V_{t''} \setminus B_{t''}$.

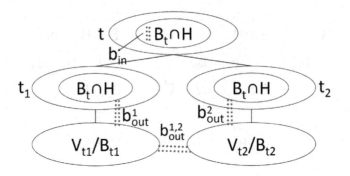

Fig. 2. Recursion on a join node.

4 Conclusions and Future Directions

In this paper, we introduced the EDGE k-CORE problem (EKC), where the goal is to create a "large" k-core by adding only "few" edges, and provided several hardness and algorithmic results. Specifically, we showed that EKC is polynomial-time solvable for $k \leq 2$ but NP-hard for $k \geq 3$; further, we showed that EKC is W[1]-hard for $k+p+b$, but fixed-parameter tractable for $\mathbf{tw}+k+b$.

For future research, one might look at EKC for directed graphs: similar work was done by Chitnis et al. [6] for the AKC problem. Another direction is to study the (in)approximability of EKC. Finally, one can consider a version which combines AKC with EKC: in it, a "large" anchored k-core would be created by anchoring at most b_v vertices and adding at most b_e edges.

Acknowledgments. The authors would like to thank Fedor Fomin, Petr Golovach, and Bart M.P. Jansen for helpful discussions.

References

1. Alvarez-Hamelin, J.I., Dall'Asta, L., Barrat, A., Vespignani, A.: Large scale networks fingerprinting and visualization using the k-core decomposition. In: NIPS 2005, pp. 41–50 (2005)
2. Batagelj, V., Mrvar, A., Zaveršnik, M.: Partitioning approach to visualization of large graphs. In: Kratochvíyl, J. (ed.) GD 1999. LNCS, vol. 1731, pp. 90–97. Springer, Heidelberg (1999). https://doi.org/10.1007/3-540-46648-7_9
3. Baur, M., Brandes, U., Gaertler, M., Wagner, D.: Drawing the AS graph in 2.5 dimensions. In: Pach, J. (ed.) GD 2004. LNCS, vol. 3383, pp. 43–48. Springer, Heidelberg (2005). https://doi.org/10.1007/978-3-540-31843-9_6
4. Bhawalkar, K., Kleinberg, J.M., Lewi, K., Roughgarden, T., Sharma, A.: Preventing unraveling in social networks: the anchored k-core problem. SIAM J. Discret. Math. **29**(3), 1452–1475 (2015)
5. Bodlaender, H.L., Koster, A.: Combinatorial optimization on graphs of bounded treewidth. Comput. J. **51**(3), 255–269 (2008)

6. Chitnis, R., Fomin, F.V., Golovach, P.A.: Parameterized complexity of the anchored k-core problem for directed graphs. Inf. Comput. **247**, 11–22 (2016)
7. Chitnis, R.H., Fomin, F.V., Golovach, P.A.: Preventing unraveling in social networks gets harder. In: Proceedings of AAAI 2013 (2013)
8. Chwe, M.: Structure and strategy in collective action 1. Am. J. Sociol. **105**(1), 128–156 (1999)
9. Chwe, M.: Communication and coordination in social networks. Rev. Econ. Stud. **67**(1), 1–16 (2000)
10. Dezső, Z., Barabási, A.: Halting viruses in scale-free networks. Phys. Rev. E **65**(5), 055103 (2002)
11. Downey, R.G., Fellows, M.R.: Fundamentals of Parameterized Complexity. Texts in Computer Science. Springer, Heidelberg (2013). https://doi.org/10.1007/978-1-4471-5559-1
12. Du, N., Wu, B., Pei, X., Wang, B., Xu, L.: Community detection in large-scale social networks. In: Proceedings of the 9th WebKDD and 1st SNA-KDD 2007 Workshop on Web Mining and Social Network Analysis, pp. 16–25. ACM (2007)
13. Fortunato, S.: Community detection in graphs. Phys. Rep. **486**(3), 75–174 (2010)
14. Gutiérrez-Bunster, T., Stege, U., Thomo, A., Taylor, J.: How do biological networks differ from social networks? (an experimental study). In: ASONAM, pp. 744–751 (2014)
15. Kloks, T. (ed.): Treewidth, Computations and Approximations. LNCS, vol. 842. Springer, Heidelberg (1994). https://doi.org/10.1007/BFb0045375
16. Korovaiko, N., Thomo, A.: Trust prediction from user-item ratings. Soc. Netw. Anal. Min. **3**(3), 749–759 (2013)
17. Mikolajczyk, R.T., Kretzschmar, M.: Collecting social contact data in the context of disease transmission: prospective and retrospective study designs. Soc. Netw. **30**(2), 127–135 (2008)
18. Pandit, S., Chau, D.H., Wang, S., Faloutsos, C.: NetProbe: a fast and scalable system for fraud detection in online auction networks. In: Proceedings of the 16th International Conference on World Wide Web, pp. 201–210. ACM (2007)
19. Papadopoulos, S., Kompatsiaris, Y., Vakali, A., Spyridonos, P.: Community detection in social media. Data Min. Knowl. Discov. **24**(3), 515–554 (2012)
20. Pastor-Satorras, R., Vespignani, A.: Epidemic spreading in scale-free networks. Phys. Rev. Lett. **86**(14), 3200–3203 (2001)
21. Sääskilahti, P.: Monopoly pricing of social goods. Technical report. University Librray of Munich (2007)
22. Schelling, T.: Micromotives and Macrobehavior. W. W. Norton, New York City (1978)
23. Seidman, S.: Network structure and minimum degree. Soc. Netw. **5**(3), 269–287 (1983)
24. Yang, W.-S., Dia, J.-B.: Discovering cohesive subgroups from social networks for targeted advertising. Expert Syst. Appl. **34**(3), 2029–2038 (2008)

Periodicity in Data Streams
with Wildcards

Funda Ergün[1], Elena Grigorescu[2], Erfan Sadeqi Azer[1], and Samson Zhou[2(✉)]

[1] School of Informatics and Computing, Indiana University, Bloomington, IN, USA
{fergun,esadeqia}@indiana.edu
[2] Department of Computer Science, Purdue University, West Lafayette, IN, USA
elena-g@purdue.edu, samsonzhou@gmail.com

Abstract. We investigate the problem of detecting periodic trends within a string S of length n, arriving in the streaming model, containing at most k wildcard characters, where $k = o(n)$. A wildcard character is a special character that can be assigned any other character. We say S has wildcard-period p if there exists an assignment to each of the wildcard characters so that in the resulting stream the length $n - p$ prefix equals the length $n - p$ suffix. We present a two-pass streaming algorithm that computes wildcard-periods of S using $\mathcal{O}\left(k^3 \operatorname{polylog} n\right)$ bits of space, while we also show that this problem cannot be solved in sublinear space in one pass. We then give a one-pass randomized streaming algorithm that computes all wildcard-periods p of S with $p < \frac{n}{2}$ and no wildcard characters appearing in the last p symbols of S, using $\mathcal{O}\left(k^3 \log^9 n\right)$ space.

1 Introduction

We study the problem of detecting repetitive structure in a data stream S containing a small number of *wildcard characters*. Given an alphabet Σ and a special *wildcard* character '$\perp$'[1], let $S \in (\Sigma \cup \{\perp\})^n$ be a stream that contains at most k wildcards. We can assign a value from Σ to each wildcard character in S resulting in many possible values of S. Then we informally say S has *wildcard-period* p if there exists an assignment to each of the wildcard characters in S so that the resulting string consists of the repetition of a block of p characters.

Example 1. The string $S = abcab\perp a\perp c\perp bc$ has wildcard-period 3, since assigning 'c' to the first wildcard character, 'b' to the second wildcard character, and 'a' to the third results in the string 'abcabcabcabc', which consists of repetitions of the substring 'abc' of length 3.

The identification of repetitive structure in data has applications to bioinformatics, natural language processing, and time series data mining. Specifically, finding the smallest period of a string is necessary preprocessing for many algorithms, such as the classic Knuth-Morriss-Pratt [KMP77] algorithm in pattern

---

[1] Although wildcard characters are usually denoted with '?', we use $\perp$ to differentiate from compilation errors - the LaTeX equivalent of wildcard characters.

© Springer International Publishing AG, part of Springer Nature 2018
F. V. Fomin and V. V. Podolskii (Eds.): CSR 2018, LNCS 10846, pp. 90–105, 2018.
https://doi.org/10.1007/978-3-319-90530-3_9

matching, or the basic local alignment search tool (BLAST) [AGM+90] in computational biology.

We consider our problem in the *streaming model*, where we process the input in sequential order and sublinear space. However in practice, some of the data may be erased or corrupted beyond repair, resulting in symbols that we cannot read, '$\bot$'. As a consequence, we attempt to perform pattern matching with optimistic assignments to these values. This motivation has resulted in a number of literature on string algorithms with wildcard characters [MR95, Ind98, CH02, Kal02, CC07, HR14, LNV14, GKP16].

One possible approach to our problem is to generalize the exact periodicity problem, for which [EJS10] give a two-pass streaming algorithm for finding the smallest *exact* period of a string of length n that uses $\mathcal{O}\left(\log^2 n\right)$-space and $\mathcal{O}\left(\log n\right)$ time per arriving symbol. Their results can be easily generalized to an algorithm for finding the wildcard-period of strings using $\mathcal{O}\left(\log^2 n\right)$-space, but at a cost of $\mathcal{O}\left(|\Sigma|^k\right)$ post-processing time, which is often undesirable. More recently, [EGSZ17] study the problem of k-periodicity, where a string is permitted to have up to k permanent changes. The authors give a two-pass streaming algorithm that uses $\mathcal{O}\left(k^4 \log^9 n\right)$ bits of space and runs in $\mathcal{O}\left(k^2 \operatorname{polylog} n\right)$ amortized time per arriving symbol. This algorithm can be modified to recover the wildcard-period. We show how to do this more efficiently in Theorem 6.

1.1   Our Contributions

The challenge of determining periodicity in the presence of wildcard characters can first be approached by working toward an understanding of specific structural properties of strings with wildcard characters. We show in Lemma 2 that the number of possible assignments to the wildcard characters over all periods is "small". This allows us to compress our data into sublinear space. In this paper, given a string S with at most k wildcard characters, we show:

(1) a two-pass randomized streaming algorithm that computes all wildcard-periods of S using $\mathcal{O}\left(k^3 \operatorname{polylog} n\right)$ space, *regardless of period length*, running in $\mathcal{O}\left(k^2 \operatorname{polylog} n\right)$ amortized time per arriving symbol,

(2) a one-pass randomized streaming algorithm that computes all wildcard-periods p of S with $p < \frac{n}{2}$ and no wildcard characters appearing in the last p symbols of S, using $\mathcal{O}\left(k^3 \operatorname{polylog} n\right)$ space, running in $\mathcal{O}\left(k^2 \operatorname{polylog} n\right)$ amortized time per arriving symbol (see the full version of the paper in [EGSZ18]),

(3) a lower bound that any one-pass streaming algorithm that computes all wildcard-periods of S requires $\Omega(n)$ space even when randomization is allowed,

(4) a lower bound that, for $k = o(\sqrt{n})$ with $k > 2$, any one-pass randomized streaming algorithm that computes all wildcard-periods of S with probability at least $1 - \frac{1}{n}$ requires $\Omega(k \log n)$ space, even under the promise that the wildcard-periods are at most $n/2$.

We remark that our algorithm can be easily modified to return the smallest, largest, or any desired wildcard-period of S. Finally, we note in the full version of the paper in [EGSZ18] several results in the related problem of determining distance to p-periodicity. We give an overview of our techniques in Sect. 2.

1.2 Related Work

The study of periodicity in data streams was initiated in [EJS10], in which the authors give an algorithm that detlects the period of a string, using polylog n bits of space. Independently, [BG11] gives a similar result with improved running time. Also, [EAE06] studies mining periodic patterns in streams, and [CM11] studies periodicity via linear sketches, [IKM00] studies periodicity in time-series databases and online data. [EMS10, LN11] study the problem of distinguishing periodic strings from aperiodic ones in the property testing model of sublinear-time computation. Furthermore, [AEL10] studies approximate periodicity in the RAM model under the Hamming and swap distance metrics.

The pattern matching literature is a vast area (see [AG97] for a survey) with many variants. In the data stream model, [PP09, CFP+16] study exact and approximate variants in offline and online settings. We use the sketches from [CFP+16] though there are some other works [AGMP13, CEPR09, RS17, PL07] with different sketches for strings. [CJPS13] also show several lower bounds for online pattern matching problem.

Strings with wildcard characters have been extensively studied in the offline model, usually called "partial words". Blanchet-Sadri [Bla08] presents a number of combinatorial properties on partial words, including a large section devoted to periodicity. Notably, [BMRW12] gives algorithms for determining the period-icity for partial words. Manea et al. [MMT14] improves these results, presenting efficient time offline algorithms for determining periodicity on partial words, minimizing either total time or update time per symbol.

Golan et al. [GKP16] study the pattern matching problem with a small number of wildcards in the streaming model. Prior to this work, several works had studied other aspects of pattern matching under wildcards (See [CH02, CC07, HR14, LNV14]).

Many ideas used in these sublinear algorithms stem from related work in the classical offline model. The well-known KMP algorithm [KMP77] initially used periodic structures to search for patterns within a text. Galil et al. [GS83] later improved the space performance of this pattern matching algorithm. Recently, [Gaw13] also used the properties of periodic strings for pattern matching when the strings are compressed. These interesting properties have allowed several algorithms to satisfy some non-trivial requirements of respective models (see [GKP16, CFP+15] for example).

1.3 Preliminaries

Given an input stream $S[1, \ldots, n]$ of length $|S| = n$ over some alphabet Σ, we denote the ith character of S by $S[i]$, and the substring between locations i and

j (inclusive) $S[i, j]$. We say that two strings $S, T \in \Sigma^n$ have a *mismatch* at index i if $S[i] \neq T[i]$. Then the Hamming distance is the number of such mismatches, denoted $\Delta(S, T) = \left| \{i \mid S[i] \neq T[i]\} \right|$. We denote the concatenation of S and T by $S \circ T$. We denote the greatest common divisor of two integers x and y by $\gcd(x, y)$.

Multiple standard and equivalent definitions of periodicity are often used interchangeably. We say S has period p if $S = B^\ell B'$ where B is a block of length p that appears $\ell \geq 1$ times in a row, and B' is a prefix of B. For instance, *abcdabcdab* has period 4 where $B = abcd$, and $B' = ab$. Equivalently, $S[x] = S[x + p]$ for all $1 \leq x \leq n - p$. Similarly, the following definition is also used for periodicity.

Definition 1. *We say string S has period p if the length $n - p$ prefix of S is identical to its length $n - p$ suffix, $S[1, n - p] = S[p + 1, n]$.*

More generally, we say S has k-period p (i.e., S has period p with k mismatches) if $S[x] = S[x + p]$ for all but at most k (valid) indices x. Equivalently, the following definition is also used for k-periodicity.

Definition 2. *We say string S has k-period p if $\Delta(S[1, n - p], S[p + 1, n]) \leq k$.*

The definition of k-periodicity lends itself to the following observation.

Observation 1. *If p is a k-period of S, then at most k substrings in the sequence of substrings $S[1, p], S[p + 1, 2p], S[2p + 1, 3p], \ldots$ can differ from the preceding substring in the sequence.*

Finally, we use the following definition of wildcard-periodicity:

Definition 3. *We say that a string S has wildcard-period p if there exists an assignment to the wildcard characters, so that $S[1, n - p] = S[p + 1, n]$ (i.e., the resulting string has period p. See Example 1).*

Note that the determinism of the assignments of the characters is very important, as evidenced by Example 2.

Example 2. Consider the string $S = aaa \bot bbb$. To check whether S has wildcard-period 1, we must compare $S[1, n - 1] = aaa \bot bb$ and $S[2, n] = aa \bot bbb$. At first glance, one might think assigning the character 'b' to the wildcard in the prefix $S[1, n - 1]$ and an 'a' in the suffix $S[2, n]$ will make the prefix and the suffix identical. However, this is not a legal move; there is not a single character that the wildcard can be replaced with that makes the above prefix and the suffix the same. Thus, S does not have a wildcard-period of 1.

The following example emphasizes the difference between k-periodicity and wildcard-periodicity:

Example 3. For $k = 1$, the string $S = aaaaabbbbb$ has k-period $p = 1$. However, to obtain wildcard-period $p = 1$, at least five characters in S must be changed to wildcards (for example, all the characters 'a' or 'b').

Therefore, k-periodicity is a good notion for capturing periodicity with respect to long-term, persistent changes, while wildcard-periodicity is a good notion for capturing periodicity against a number of symbols that are errors or erasures.

We shall require data structures and subroutines that allow comparing of strings with mismatches. The below useful fingerprinting algorithm utilizes Karp-Rabin fingerprints [KR87] to obtain general and important properties:

Theorem 2 [KR87]. *Given two strings S and T of length n, there exists a polynomial encoding that uses $\mathcal{O}(\log n)$ bits of space, and outputs whether $S = T$ or $S \neq T$. Moreover, this encoding supports concatenation of strings and can be done in the streaming setting.*

From here, we use the term *fingerprint* to refer to this data structure. We will also need use an algorithm for pattern matching with mismatches, which we call the k-mismatch algorithm.

Theorem 3 [CFP+16]. *Given a string S and an index x, there exists an algorithm which, with probability $1 - \frac{1}{n^2}$, outputs all indices i where $\Delta(S[1, x], S[i + 1, i + x]) \leq k$ using $\mathcal{O}(k^2 \log^8 n)$ bits of space. Moreover, the algorithm runs in $\mathcal{O}(k^2 \operatorname{polylog} n)$ amortized time per arriving symbol.*

Concurrent with our work, Clifford *et al.* [CKP17] provide a nearly-optimal solution to the k-mismatch algorithm, which can potentially be used in the framework of [EGSZ17] to immediately improve over the existing k-periodicity algorithms.

2   Our Approach

To find all the wildcard-periods of S, during our first pass we determine a set $\mathcal{T}$ of *candidate* wildcard-periods, similar to the approach in [EGSZ17], that includes all the true wildcard-periods. We also determine a set $\mathcal{W}$ of positions of the wildcard characters. By a structural result (Lemma 2), we can then use the second pass to verify the candidates and identify the true wildcard-periods.

Pattern matching and periodicity seem to have a symbiotic relationship (for example, exact pattern matching and exact periodicity use each other as subroutines [KMP77,EJS10], as do k-mismatch pattern matching [CFP+16] and k-periodicity [EGSZ17]). It feels tempting and natural to try to apply the algorithm from [GKP16] for pattern matching with wildcards. Unfortunately, there does not seem to be an immediate way of doing this: the [GKP16] algorithm searches for a wildcard-free pattern in text containing up to k wildcards, while we would like to allow wildcards in the pattern *and* the text. We instead choose to use the k-mismatch algorithm from [CFP+16] in the first pass and obtain new structural results about possible assignments to the wildcard characters in the second pass.

In the first pass, we treat wildcards simply as an additional character. We let $\mathcal{T}$ be the set of indices (candidate periods) π that satisfy

$$\Delta(S[1, x], S[\pi + 1, \pi + x]) \leq 2k,$$

for some appropriate value of x that we specify later. Note that each wildcard character can cause up to two mismatches; thus, all true wildcard-periods must satisfy the above inequality. We show that $\mathcal{T}$ can be easily compressed, even though it may contain a linear number of candidates. Specifically, we can succinctly represent $\mathcal{T}$ by adding a few additional "false candidates" into $\mathcal{T}$.

If the correct assignments of the wildcards were known a priori, then the problem would reduce to determining exact periodicity. Unfortunately, we do not know the correct assignments to the wildcard characters prior to the data stream, so most of the difficulty lies in the guessing of assignments, bounding the total number of assignments, and storing these assignments. Thus, the main difference between wildcard-periodicity and both exact periodicity and k-periodicity is the process of verifying candidates. Whereas exact and k-periodicity can be verified by comparing the number of mismatches between the prefix and suffix of length $n - p$, wildcard-periodicity is sensitive to the correct assignments of the wildcards. We address this challenge by noting $\mathcal{W}$, the positions of the wildcard characters in the first pass. Since we also have the list of candidate wildcard-periods following the first pass, we can guess the assignments of the wildcard characters in the second pass by looking at the characters in a few select locations, as in Example 4.

Example 4. The string $S = ababa{\perp}ab$ has wildcard-period $p = 2$. The assignment of the wildcard at position $i = 6$ must be the characters at positions $i \pm p$. Note that $S[i + p] = S[8] = b$ and $S[i - p] = S[4] = b$.

From Example 4, we observe the following:

Observation 4. *If S has wildcard-period p and a wildcard character is known to be at position i, then the assignment of the wildcard must be the character $S[i \pm ap]$, for some integer a, that is not a wildcard.*

We show how to use Observation 4 and the compressed version of $\mathcal{T}$ in the second pass to verify the candidates and output the true wildcard-periods of S.

We note that recent algorithmic improvements to the k-mismatch problem [CKP17] use $\mathcal{O}\left(k \log^2 n\right)$ space. Using this algorithm in place of Theorem 3 as a subroutine in our algorithms improves the space usage to $\mathcal{O}\left(k^3 \log^3 n\right)$ bits in the two-pass algorithm.

3   Two-Pass Algorithm to Compute Wildcard-Periods

In this section, we provide a two-pass, $\mathcal{O}\left(k^3 \log^9 n\right)$-space algorithm to output all wildcard-periods of some string S containing at most k wildcard characters. At a high level, we first identify a list of candidates of the periods of S, detected via the k-mismatch algorithm of [CFP+16] as a black box. Although the number of candidates could be linear, it turns out the string has enough structure that the list of candidates can be succinctly expressed as the union of k arithmetic progressions.

However, this list of candidates is insufficient in identifying the possible assignments to the wildcard characters. To address this issue, we explore the structure of periods with wildcards in order to limit the possible assignments for each wildcard character. Thus, the first pass also records $\mathcal{W}$, the positions of all wildcard characters so that during the second pass, we go over S as well as the compressed data to verify the candidate periods.

We present two algorithms in parallel to find the periods, based on their lengths. The first algorithm identifies all periods p with $p \leq \frac{n}{2}$, while the second algorithm identifies all periods p with $p > \frac{n}{2}$.

3.1   Computing Small Wildcard-Periods

In this section, we describe a two-pass algorithm for finding wildcard-periods of length at most $n/2$. The first pass of the algorithm identifies a set $\mathcal{T}$ of candidate wildcard-periods in terms of indices of S, and maintains its succinct representation $\mathcal{T}^C$, which includes a number of additional indices. It also records $\mathcal{W}$, the positions of all wildcard characters. The second pass of the algorithm recovers each index of $\mathcal{T}$ from $\mathcal{T}^C$ and verifies whether or not the index is a wildcard-period. We can find the assignments of the wildcard characters in the second pass, by looking at the characters in a few locations that we determine via $\mathcal{W}$. We emphasize the following properties of $\mathcal{T}$ and $\mathcal{T}^C$:

(1) All wildcard-periods (possibly as well as additional candidate wildcard-periods that are false positives) are in $\mathcal{T}$.
(2) $\mathcal{T}^C$ can be stored in sublinear space and $\mathcal{T}$ can be fully recovered from $\mathcal{T}^C$.
(3) In the second pass, we can verify and eliminate in sublinear space candidates that are not true periods.

In the first pass, we treat the wildcard characters as a regular, additional alphabet symbol. We observe that if string S with such wildcards has wildcard-period p, there are at most $2k$ indices i such that $S[i] \neq S[i + p]$, caused by the wildcard characters (the converse is not necessarily true). It follows that any wildcard-period p must satisfy

$$\Delta \left(S[1, x], S[p + 1, p + x] \right) \leq 2k$$

for all $x \leq n - p$, and specifically for $x = \frac{n}{2}$. Thus, we set $x = \frac{n}{2}$ and refer to any index p that satisfies $\Delta \left(S[1, x], S[p + 1, p + x] \right) \leq 2k$ as a *candidate wildcard-period*. The set of all candidate wildcard-periods forms the set $\mathcal{T}$. Because $\Delta \left(S[1, x], S[p + 1, p + x] \right) \leq 2k$ is a necessary but not sufficient condition for a wildcard-period p, Property 1 follows.

We give the first pass of the algorithm in full in Algorithm 1.

Algorithm 1. (To determine any wildcard-period p with $p \leq \frac{n}{2}$) First pass

---

Input: A stream S of n symbols $s_i \in \Sigma \cup \{\perp\}$ with at most k wildcard characters $\perp$.
Output: A succinct representation of all candidate wildcard periods and the positions of the wildcard characters.

1: initialize $\pi_j = -1$ for each $0 \leq j < 4k \log n + 2$.
2: initialize $\mathcal{T}^C = \emptyset$.
3: **for** each index i (found using the k-mismatch algorithm) such that

$$\Delta \left(S \left[1, \frac{n}{2} \right], S \left[i+1, \frac{n}{2} + i \right] \right) \leq 2k$$

   do
4:     consider j for which i is in the interval $H_j = \left[\frac{jn}{4(2k \log n + 1)} + 1, \frac{(j+1)n}{4(2k \log n + 1)} \right)$:
5:     **if** there exists no candidate $t \in \mathcal{T}^C$ in the interval H_j **then**
6:         add i to $\mathcal{T}^C$.
7:     **else**
8:         let t be the smallest candidate in $\mathcal{T}^C \cap H_j$ and either $\pi_j = -1$ or $\pi_j > 0$.
9:         **if** $\pi_j = -1$ **then**
10:          set $\pi_j = i - t$.
11:         **else**
12:          set $\pi_j = \gcd (\pi_j, i - t)$.
13: record the positions $\mathcal{W}$ of all wildcard characters.

Here, we show why the remaining properties for $\mathcal{T}$ and $\mathcal{T}^C$ are satisfied. Our algorithm divides the candidates into $\mathcal{O}(k \log n)$ ranges $H_1, H_2, \ldots, H_{\mathcal{O}(k \log n)}$ and stores the candidates in each range $H_j = \left[\frac{jn}{4(2k \log n + 1)} + 1, \frac{(j+1)n}{4(2k \log n + 1)} \right)$ in compressed form as an arithmetic series.

Since we use the k-mismatch algorithm in the first pass, we describe a structural property of the resulting list of candidates:

Theorem 5 [EGSZ17]. *Let p_i be a candidate k-period for a string S, with $p_1 < p_2 < \ldots < p_m$ all contained within H_j. Given the fingerprints of $S[1, n - p_1]$ and $S[p_1 + 1, n]$, we can determine whether or not S has k-period p_i for any $1 \leq i \leq m$ by storing at most $\mathcal{O}\left(k^2 \log n\right)$ additional fingerprints. These fingerprints represent substrings of the form $S[p_1 + a\pi_j, p_1 + (a + 1)\pi_j - 1]$, where $a > 0$ is an integer and $\pi_j = \gcd (p_2 - p_1, p_3 - p_2, \ldots, p_m - p_{m-1})$.*

The structural property can be visualized in Fig. 1. Even though the list of candidates could be linear in size, Theorem 5 enforces a structure upon the list of candidates, so that an arithmetic sequence with first term p_1 and common difference d includes all of $p_1, p_2, \ldots, p_m$. Thus, we can succinctly represent a superset $\mathcal{T}^C$ that contains $\mathcal{T}$ and Property 2 follows.

We now show that any wildcard period p is included among the list of candidates stored by Algorithm 1 during the first pass, and can be recovered from the list.

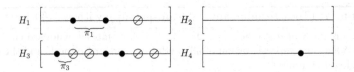

Fig. 1. The dots represent candidate wildcard-periods. For any interval that has more than two dots, it follows that all dots are equally spaced after the first. The black dots represent $\mathcal{T}$ while white dots are artificially inserted to form T, dots that follow an arithmetic sequence.

Lemma 1. If $p < \frac{n}{2}$ is a period and $p \in H_j$, then p can be recovered from $\mathcal{T}^C$ and π_j.

Proof. Suppose $p \in H_j$ is a wildcard period. Then there exists an assignment to the wildcard characters such that $S[1, n-p] = S[p+1, n]$. It follows that for $i = p$,

$$\Delta \left(S\left[1, \frac{n}{2}\right], S\left[i+1, \frac{n}{2}+i\right] \right) \leq 2k,$$

so the index $i = p$ will be reported by the k-mismatch algorithm in the first pass.

If at that time during Pass 1 there is no other index in $\mathcal{T}^C \cap H_j$, then p will be inserted into $\mathcal{T}^C$, so p can clearly be recovered from $\mathcal{T}^C$. If there is another index q in $\mathcal{T}^C \cap H_j$, then π_j will be updated to be a divisor of $p - q$. Hence, $p - q$ is a multiple of π_j. Furthermore, any future update to π_j will result in a value that divides the current value of π_j, due to a greatest common divisor operation. Thus, $p - q$ will remain a multiple of the final value of π_j, and so the set T at the end of the first pass will contain p.

It remains to show that the list of candidate wildcard-periods can be verified in sublinear space in the second pass (Property 3). To do this, we need a combinatorial property for periodicity on strings with wildcard characters.

3.2 Verifying Candidates

Recall that after the first pass, the algorithm maintains $\mathcal{O}(k \log n)$ succinctly represented arithmetic progressions H_j, corresponding to the candidate wildcard periods. The algorithm also maintains $\mathcal{W}$, the list of positions of wildcard characters in S. In the second pass, the algorithm must check, for each $t \in H_j$, $0 \leq j < 2k \log n + 2$, whether $S[1, n-t] = S[t+1, n]$ for an appropriate setting of the wildcard characters. The challenge is computing the fingerprints of both $S[1, n-t]$ and $S[t+1, n]$ in sublinear space, especially if the number of candidates t is linear.

We first set a specific j and note that for the smallest candidate $t \in H_j$, there are at most $\mathcal{O}(k^2 \log n)$ unique substrings $S[t+1, t+\pi_j], S[t+\pi_j+1, t+2\pi_j]$,

$S[t + 2\pi_j + 1, t + 3\pi_j], \ldots$ Since any other candidate $r \in H_j$ satisfies $r = t + a\pi_j$ for some integer $a > 0$, then $S[t + 1, n]$ is the concatenation

$$S[t+1, t+\pi_j] \circ S[t+\pi_j+1, t+2\pi_j] \circ \cdots \circ S[t+(a-1)\pi_j+1, t+a\pi_j] \circ S[r+1, n].$$

Thus, by storing $\mathcal{O}\left(k^2 \log n\right)$ fingerprints and positions, we can recover the fingerprint of the substring $S[r + 1, n]$ for each $r \in H_j$.

The second obstacle is handling wildcard characters in the computation of the fingerprints of $S[1, n - t]$ and $S[t + 1, n]$. To address this challenge, our algorithm delays the calculation of the contribution of wildcard characters to the fingerprints until we know the assignment of the wildcard character with respect to a candidate period. We show that for a specific j, then there are at most $\mathcal{O}\left(k^2 \log n\right)$ possible assignments for the wildcard character $S[w] = S[w \pm t]$ with respect to all candidates $t \in H_j$, across all $w \in \mathcal{W}$, where $\mathcal{W}$ is the positions of all wildcard characters recorded by Algorithm 1. Therefore, we can compute the assignment for each wildcard character with respect to a candidate period in the second pass, and then compute the fingerprint of $S[1, n - t]$ and $S[t + 1, n]$.

Lemma 2. *For a given j, $t \in H_j$ and $w \in \mathcal{W}$, let $\sigma_t(w)$ denote the assignment of $S[w]$. Then $|\{\sigma_t(w)\}| = \mathcal{O}\left(k^2 \log n\right)$.*

Proof. Let t be the smallest candidate in H_j and z be the largest candidate in H_j so that $z = t + a\pi_j$ for some integer $a > 0$. We partition $\mathcal{W}$ into $\mathcal{W}_1$, the set of indices greater than z, and $\mathcal{W}_2$, the set of indices no more than z. We consider the wildcard characters $w_i \in \mathcal{W}_1$, and note that the proof for $\mathcal{W}_2$ is symmetric. Consider the $\mathcal{O}\left(k\right)$ sequences

$$
\begin{array}{cccc}
S[w_1 - t] & S[w_1 - t - \pi_j] & \cdots & S[w_1 - t - a\pi_j] \\
S[w_2 - t] & S[w_2 - t - \pi_j] & \cdots & S[w_2 - t - a\pi_j] \\
\vdots & \vdots & \ddots & \vdots \\
S[w_{|\mathcal{W}_1|} - t] & S[w_{|\mathcal{W}_1|} - t - \pi_j] & \cdots & S[w_{|\mathcal{W}_1|} - t - a\pi_j]
\end{array}
$$

Each term in a sequence that differs from the previous term corresponds to a mismatch between $S[w_i - t - \pi_j + 1, w_i - t]$, $S[w_i - t - 2\pi_j + 1, w_i - t - \pi_j]$, $S[w - t - 3\pi_j + 1, w - t - 2\pi_j], \ldots$ For each j, there are at most $\mathcal{O}\left(k^2 \log n\right)$ unique chains of substrings with length π_j beginning at index $t + 1$. Hence, across all $\mathcal{O}\left(k\right)$ sequences $S[w_i - t]$, $S[w_i - t - \pi_j]$, $S[w_i - t - 2\pi_j], \ldots$, there are at most $\mathcal{O}\left(k^2 \log n\right)$ unique characters. Since the assignment of $S[w_i]$ with respect to any candidate $r \in H_j$ is $S[w_i - r] = S[w_i - t - b\pi_j]$ for some integer $b > 0$, then it follows that there are at most $\mathcal{O}\left(k^2 \log n\right)$ assignments of $S[w]$ across all $w \in \mathcal{W}_1$. As the symmetric proof holds for $\mathcal{W}_2$, then there are at most $\mathcal{O}\left(k^2 \log n\right)$ assignments of $S[w]$ across all $w \in \mathcal{W}$.

Thus, deciding the assignment of $S[w_i]$ with respect to a candidate $t \in H_j$ is simple:

For each j such that $0 \leq j < 4k \log n + 2$:
(1) Let t be the smallest candidate in H_j and z be the largest candidate in H_j so that $z = t + a\pi_j$ for some $a > 0$.
(2) For each $w \in \mathcal{W}$:
   (a) If $w > z$, succinctly record the values of $S[w - t]$, $S[w - t - \pi_j]$, ..., $S[w - t - a\pi_j]$.
   (b) If $w \leq z$, succinctly record the values of $S[w + t]$, $S[w + t + \pi_j]$, ..., $S[w + t + a\pi_j]$.
   Let $r \in H_j$ so that $r = t + b\pi_j$ for some $b > 0$.
(3) The assignment of $S[w]$ with respect to r is any $S[w \pm cr]$ that is not a wildcard character (where c is an integer).

We describe the second pass in Algorithm 2, recalling that at the end of the first pass, the algorithm records $\mathcal{O}(k \log n)$ arithmetic progressions, succinctly represented, as well as the positions of all wildcard characters.

Algorithm 2. (To determine any wildcard-period p with $p \leq \frac{n}{2}$) Second pass

---

Input: A stream S of symbols $s_i \in \Sigma$ with at most k wildcard characters, a succinct representation of all candidate wildcard periods and the position of the wildcard characters.
Output: All wildcard-periods $p \leq \frac{n}{2}$.
1: **for** each t such that $t \in \mathcal{T}^C$ **do**
2:     for each w such that $w \in \mathcal{W}$, implicitly determine the value of $S[w]$ with respect to t.
3:     let j be the integer for which t is in the interval $H_j = \left[\frac{jn}{4(2k \log n+1)} + 1, \frac{(j+1)n}{4(2k \log n+1)}\right)$
4:     **if** $\pi_j > 0$ **then**   ▷ H_j has multiple values in $\mathcal{T}^C$
5:         record up to $128k^2 \log n + 1$ unique fingerprints of length π_j, starting from t.
6:     **else**   ▷ H_j has one value in $\mathcal{T}^C$
7:         record up to $128k^2 \log n + 1$ unique fingerprints of length t, starting from t.
8:     check if $S[1, n - t] = S[t + 1, n]$ and return t if this is true.
9: **for** each t which is in interval $H_j = \left[\frac{jn}{4(2k \log n+1)} + 1, \frac{(j+1)n}{4(2k \log n+1)}\right)$ for some integer j **do**
10:    **if** there exists an index in $\mathcal{T}^C \cap H_j$ whose distance from t is a multiple of π_j **then**
11:        check if $S[1, n - t] = S[t + 1, n]$ and return t if this is true.

For each arithmetic progression, there are $\mathcal{O}(k^2 \log n)$ total possibilities for all of the wildcard characters. Thus, the algorithm maintains the $\mathcal{O}(k^3 \log^2 n)$ characters corresponding to the value of all wildcard characters across all candidate positions.

We now show the ability to construct the fingerprints of $S[1, n - p]$ for any candidate period p.

Lemma 3. *Let p_i be a candidate k-period for a string S, with $p_1 < p_2 < \ldots < p_m$ all contained within H_j. Given the fingerprints of $S[1, n-p_1]$ and $S[p_1+1, n]$, we can determine whether or not S has wildcard-period p_i for any $1 \leq i \leq m$ by storing at most $\mathcal{O}\left(k^2 \log n\right)$ additional fingerprints.*

Proof. Consider a decomposition of S into substrings u_j of length p_i, so that $S = u_1 \circ u_2 \circ u_3 \circ \ldots$. Even though the algorithm does not record a fingerprint for each u_j, each index j for which $u_j \neq u_{j+1}$ corresponds to at least one mismatch. Since the first pass searched for positions that contained at most k mismatches, then it follows from Observation 1 that there are $\mathcal{O}\left(k\right)$ indices j for which $u_j \neq u_{j+1}$. Thus, recording the fingerprints and locations of these indices j suffices to build fingerprints for S, ignoring the wildcard characters. Then we can verify whether or not p_i is a wildcard-period of S if the assignment of the wildcard characters with respect to p_i is also known.

By Theorem 5, the greatest common divisor π_j of the difference between each p_i in H_j is a $\mathcal{O}\left(k^2 \log n\right)$-period. That is, S can be decomposed $S = v \circ v_1 \circ v_2 \circ v_3 \circ \ldots$ so that v has length p_1, and each subsequent substring v_i has length π_j. Then there exist at most $\mathcal{O}\left(k^2 \log n\right)$ indices i for which $v_i \neq v_{i+1}$, by Observation 1. Ignoring wildcard characters, storing the fingerprints and positions of these indices i allows the recovery of the fingerprint of $S[1, n - p_i]$ from the fingerprint of $S[1, n - p_{i-1}]$, since $p_i - p_{i-1}$ is a multiple of π_j. By Lemma 2, we know the values of the wildcard characters with respect to p_i. Therefore, we can confirm whether or not p_i is a wildcard-period. ∎

We now show correctness of the algorithm.

Lemma 4. *For any period $p \leq \frac{n}{2}$, the algorithm outputs p.*

Proof. Since the intervals $\{H_j\}$ cover $\left[1, \frac{n}{2}\right]$, then $p \in H_j$ for some j. It follows from Lemma 1 that after the first pass, p can be recovered from $\mathcal{T}$ and π_j. Thus, the second pass tests whether or not p is a wildcard-period. By Lemma 3, the algorithm outputs p, as desired. ∎

3.3 Computing Large Wildcard-Periods

As in Algorithm 1, we would like to identify candidate periods during the first pass of the algorithm, while treating the wildcard characters as an additional symbol in the alphabet. Unfortunately, if a wildcard-period p is greater than $\frac{n}{2}$, then it no longer satisfies

$$\Delta\left(S\left[1, \frac{n}{2}\right], S\left[p+1, p+\frac{n}{2}\right]\right) \leq 2k,$$

since $p + \frac{n}{2} > n$, and $S\left[p + \frac{n}{2}\right]$ is undefined. However, by treating the wildcard characters as an additional symbol, recall that $\Delta\left(S[1, x], S[p+1, p+x]\right) \leq 2k$

for all $x \leq n - p$. Then we would like to use as large an x as possible while still satisfying $x \leq n - p$ when choosing candidate wildcard periods p. To this effect, the observation in [EJS10] states that we can try exponentially decreasing values of x. Specifically, we run $\log n$ instances of the algorithm in succession, with $x = \frac{n}{2}, \frac{n}{4}, \ldots$. Note that one of these values of x is the largest value as possible while still satisfying $x \leq n - p$. As a result, the corresponding algorithm instance outputs p, while the other instances do not output anything. We detail the first pass in full in the full version of the paper in [EGSZ18].

This partition of $[1, n]$ into the disjoint intervals $\left[1, \frac{n}{2}\right]$, $\left[\frac{n}{2} + 1, \frac{n}{2} + \frac{n}{4}\right]$, ... guarantees that any k-period p is contained in one of these intervals. Moreover, the intervals $\{H_j^{(r)}\}$ partition

$$\left[\frac{n}{2} + \frac{n}{4} + \ldots + \frac{n}{2^{r-1}}, \frac{n}{2} + \ldots + \frac{n}{2^r}\right],$$

and so p can be recovered from $\mathcal{T}_r^C$ and $\{\pi_j^{(r)}\}$. We present the second pass in the full version of the paper in [EGSZ18].

Since correctness follows from the same arguments as the case where $p \leq \frac{n}{2}$, it remains to analyze the space complexity of our algorithm.

Theorem 6. *There exists a two-pass randomized algorithm using $\mathcal{O}\left(k^3 \log^9 n\right)$ bits of space that finds the wildcard-period and runs in $\mathcal{O}\left(k^2 \operatorname{polylog} n\right)$ amortized time per arriving symbol.*

Proof. In the first pass, for each $\mathcal{T}_m$, we maintain a k-mismatch algorithm which requires $\mathcal{O}\left(k^2 \log^8 n\right)$ bits of space, as in Theorem 3. Since $1 \leq m \leq \log n$, we use $\mathcal{O}\left(k^2 \log^9 n\right)$ bits of space in total in the first pass.

In the second pass, we maintain $\mathcal{O}\left(k^2 \log n\right)$ fingerprints for any set of indices in $\mathcal{T}_m$, and there are $\mathcal{O}\left(k \log n\right)$ indices in $\mathcal{T}_m$ for each $1 \leq m \leq \log n$, for a total of $\mathcal{O}\left(k^3 \log^3 n\right)$ bits of space. In addition, we store the $\mathcal{O}\left(k^2 \log n\right)$ assignments for all the wildcard positions in each interval $H_j^{(r)}$, where $1 \leq r \leq \log n$ and $0 \leq j < 2k \log n + 2$. Thus, $\mathcal{O}\left(k^3 \log^9 n\right)$ bits of space suffice for both passes.

The running time of the algorithm is dominated by the time spent for $\log n$ parallel copies of k-mismatch algorithm in the first pass. From Theorem 3, the k-mismatch algorithm runs in $\mathcal{O}\left(k^2 \operatorname{polylog} n\right)$ amortized time per arriving symbol. The rest of the algorithm consists of simple tasks like computing gcd and can be performed very quickly. In the second pass, in total at most $\mathcal{O}\left(k^3 \operatorname{polylog} n\right)$ assignments are determined and stored. Thus, the second pass runs in $\mathcal{O}(1)$ amortized time per arriving symbol.

4   Lower Bounds

We first note that [EJS10] shows computing the period of a string in one-pass requires $\Omega(n)$ space. Since the problem of periodicity for strings containing wildcards is a generalization of exact periodicity, the same lower bound applies.

Theorem 7 (Implied from Theorem 3 from [EJS10] and Theorem 16 from [EGSZ17]). *Given a string S with at most k wildcard characters, any one-pass streaming algorithm that computes the smallest wildcard-period requires $\Omega(n)$ space.*

To show a lower bound that randomized streaming algorithm that computes all wildcard-periods of S with probability at least $1 - \frac{1}{n}$, even under the promise that the wildcard-periods are at most $n/2$, consider the following construction. Define an infinite string $1^1 0^1 1^2 0^2 1^3 0^3 \ldots$, as in [GMSU16], and let ν be the prefix of length $\frac{n}{4}$. Define X to be the set of binary strings of length $\frac{n}{4}$ with Hamming distance $\frac{k}{2}$ from ν. For $x \in X$, let Y_x be the set of binary strings of length $\frac{n}{4}$ with either $\Delta(x,y) = \frac{k}{2}$ or $\Delta(x,y) = \frac{k}{2} + 1$. Pick (x,y) uniformly at random from (X, Y_x). Then Theorem 17 in [EGSZ17] shows a lower bound on the size of the sketches necessary to determine whether $\Delta(x,y) = \frac{k}{2}$ or $\Delta(x,y) = \frac{k}{2} + 1$.

Theorem 8 [EGSZ17]. *Any sketching function S that determines whether $\Delta(x,y) = \frac{k}{2}$ or $\Delta(x,y) > \frac{k}{2}$ from $S(x)$ and $S(y)$, with probability at least $1 - \frac{1}{n}$ for $k = o(\sqrt{n})$, uses $\Omega(k \log n)$ space.*

Suppose Alice has y, along with the locations of the first $\frac{k}{2}$ positions i in which $y[i] \neq x[i]$. Alice replaces these locations with wildcard characters $\perp$, runs the wildcard-period algorithm, and forwards the state of the algorithm to Bob, who has x. Bob then continues running the algorithm on $x \circ x \circ x$ to determine the wildcard-period of the string $S(x,y) = y \circ x \circ x \circ x$. Observe that:

Lemma 5. *If $\Delta(x,y) = \frac{k}{2}$, then the string $S(x,y) = y \circ x \circ x \circ x$ has period $\frac{n}{4}$. On the other hand, if $\Delta(x,y) = \frac{k}{2} + 1$, then $S(x,y)$ has period greater than $\frac{n}{4}$.*

Combining Theorem 8 and Lemma 5:

Theorem 9. *For $k = o(\sqrt{n})$ with $k > 2$, any one-pass randomized streaming algorithm that computes all wildcard-periods of an input string S with probability at least $1 - \frac{1}{n}$ requires $\Omega(k \log n)$ space, even under the promise that the wildcard-periods are at most $\frac{n}{2}$.*

Acknowledgements. We would like to thank the anonymous reviewers for their helpful comments. The work was supported by the National Science Foundation under NSF Awards #1649515 and #1619081.

References

[AEL10] Amir, A., Eisenberg, E., Levy, A.: Approximate periodicity. In: Cheong, O., Chwa, K.-Y., Park, K. (eds.) ISAAC 2010. LNCS, vol. 6506, pp. 25–36. Springer, Heidelberg (2010). https://doi.org/10.1007/978-3-642-17517-6_5

[AG97] Apostolico, A., Galil, Z. (eds.): Pattern Matching Algorithms. Oxford University Press, Oxford (1997)

[AGM+90] Altschul, S.F., Gish, W., Miller, W., Myers, E.W., Lipman, D.J.: Basic local alignment search tool. J. Mol. Biol. **215**(3), 403–410 (1990)

[AGMP13] Andoni, A., Goldberger, A., McGregor, A., Porat, E.: Homomorphic finger-prints under misalignments: sketching edit and shift distances. In: Proceedings of the Forty-Fifth Annual ACM Symposium on Theory of Computing, pp. 931–940 (2013)

[BG11] Breslauer, D., Galil, Z.: Real-time streaming string-matching. In: Giancarlo, R., Manzini, G. (eds.) CPM 2011. LNCS, vol. 6661, pp. 162–172. Springer, Heidelberg (2011). https://doi.org/10.1007/978-3-642-21458-5_15

[Bla08] Blanchet-Sadri, F.: Algorithmic Combinatorics on Partial Words. Discrete Mathematics and its Applications. CRC Press, Boca Raton (2008)

[BMRW12] Blanchet-Sadri, F., Mercas, R., Rashin, A., Willett, E.: Periodicity algorithms and a conjecture on overlaps in partial words. Theor. Comput. Sci. **443**, 35–45 (2012)

[CC07] Clifford, P., Clifford, R.: Simple deterministic wildcard matching. Inf. Process. Lett. **101**(2), 53–54 (2007)

[CEPR09] Clifford, R., Efremenko, K., Porat, E., Rothschild, A.: From coding theory to efficient pattern matching. In: Proceedings of the Twentieth Annual ACM-SIAM Symposium on Discrete Algorithms, pp. 778–784 (2009)

[CFP+15] Clifford, R., Fontaine, A., Porat, E., Sach, B., Starikovskaya, T.: Dictionary matching in a stream. In: Bansal, N., Finocchi, I. (eds.) ESA 2015. LNCS, vol. 9294, pp. 361–372. Springer, Heidelberg (2015). https://doi.org/10.1007/978-3-662-48350-3_31

[CFP+16] Clifford, R., Fontaine, A., Porat, E., Sach, B., Starikovskaya, T.A.: The k-mismatch problem revisited. In: Proceedings of the 27th Annual ACM-SIAM Symposium on Discrete Algorithms, SODA, pp. 2039–2052 (2016)

[CH02] Cole, R., Hariharan, R.: Verifying candidate matches in sparse and wildcard matching. In: Proceedings on 34th Annual ACM Symposium on Theory of Computing (STOC), pp. 592–601 (2002)

[CJPS13] Clifford, R., Jalsenius, M., Porat, E., Sach, B.: Space lower bounds for online pattern matching. Theor. Comput. Sci. **483**, 68–74 (2013)

[CKP17] Clifford, R., Kociumaka, T., Porat, E.: The streaming k-mismatch problem. CoRR, abs/1708.05223 (2017)

[CM11] Crouch, M.S., McGregor, A.: Periodicity and cyclic shifts via linear sketches. In: Goldberg, L.A., Jansen, K., Ravi, R., Rolim, J.D.P. (eds.) APPROX/RANDOM -2011. LNCS, vol. 6845, pp. 158–170. Springer, Heidelberg (2011). https://doi.org/10.1007/978-3-642-22935-0_14

[EAE06] Elfeky, M.G., Aref, W.G., Elmagarmid, A.K.: STAGGER: periodicity mining of data streams using expanding sliding windows. In: Proceedings of the 6th IEEE International Conference on Data Mining (ICDM), pp. 188–199 (2006)

[EGSZ17] Ergün, F., Grigorescu, E., Azer, E.S., Zhou, S.: Streaming periodicity with mismatches. In: Approximation, Randomization, and Combinatorial Optimization. Algorithms and Techniques, APPROX/RANDOM, pp. 42:1–42:21 (2017)

[EGSZ18] Ergün, F., Grigorescu, E., Azer, E.S., Zhou, S.: Periodicity in data streams with wildcards. CoRR, abs/1802.07375 (2018)

[EJS10] Ergün, F., Jowhari, H., Sağlam, M.: Periodicity in streams. In: Serna, M., Shaltiel, R., Jansen, K., Rolim, J. (eds.) APPROX/RANDOM -2010. LNCS, vol. 6302, pp. 545–559. Springer, Heidelberg (2010). https://doi.org/10.1007/978-3-642-15369-3_41

[EMS10] Ergün, F., Muthukrishnan, S., Sahinalp, S.C.: Periodicity testing with sublinear samples and space. ACM Trans. Algorithms **6**(2), 43:1–43:14 (2010)

[Gaw13] Gawrychowski, P.: Optimal pattern matching in LZW compressed strings. ACM Trans. Algorithms (TALG) **9**(3), 25 (2013)

[GKP16] Golan, S., Kopelowitz, T., Porat, E.: Streaming pattern matching with d wildcards. In: 24th Annual European Symposium on Algorithms, pp. 44:1–44:16 (2016)

[GMSU16] Gawrychowski, P., Merkurev, O., Shur, A.M., Uznanski, P.: Tight tradeoffs for real-time approximation of longest palindromes in streams. In: 27th Annual Symposium on Combinatorial Pattern Matching, CPM, pp. 18:1–18:13 (2016)

[GS83] Galil, Z., Seiferas, J.: Time-space-optimal string matching. J. Comput. Syst. Sci. **26**(3), 280–294 (1983)

[HR14] Hermelin, D., Rozenberg, L.: Parameterized complexity analysis for the closest string with wildcards problem. In: Combinatorial Pattern Matching - 25th Annual Symposium, CPM Proceedings, pp. 140–149 (2014)

[IKM00] Indyk, P., Koudas, N., Muthukrishnan, S.: Identifying representative trends in massive time series data sets using sketches. In: VLDB, Proceedings of 26th International Conference on Very Large Data Bases, pp. 363–372 (2000)

[Ind98] Indyk, P.: Faster algorithms for string matching problems: matching the convolution bound. In: 39th Annual Symposium on Foundations of Computer Science, FOCS, pp. 166–173 (1998)

[Kal02] Kalai, A.: Efficient pattern-matching with don't cares. In: Proceedings of the Thirteenth Annual ACM-SIAM Symposium on Discrete Algorithms (SODA), pp. 655–656 (2002)

[KMP77] Knuth, D.E., Morris Jr., J.H., Pratt, V.R.: Fast pattern matching in strings. SIAM J. Comput. **6**(2), 323–350 (1977)

[KR87] Karp, R.M., Rabin, M.O.: Efficient randomized pattern-matching algorithms. IBM J. Res. Dev. **31**(2), 249–260 (1987)

[LN11] Lachish, O., Newman, I.: Testing periodicity. Algorithmica **60**(2), 401–420 (2011)

[LNV14] Lewenstein, M., Nekrich, Y., Vitter, J.S.: Space-efficient string indexing for wildcard pattern matching. In: 31st International Symposium on Theoretical Aspects of Computer Science (STACS), pp. 506–517 (2014)

[MMT14] Manea, F., Mercas, R., Tiseanu, C.: An algorithmic toolbox for periodic partial words. Discret. Appl. Math. **179**, 174–192 (2014)

[MR95] Muthukrishnan, S., Ramesh, H.: String matching under a general matching relation. Inf. Comput. **122**(1), 140–148 (1995)

[PL07] Porat, E., Lipsky, O.: Improved sketching of hamming distance with error correcting. In: Ma, B., Zhang, K. (eds.) CPM 2007. LNCS, vol. 4580, pp. 173–182. Springer, Heidelberg (2007). https://doi.org/10.1007/978-3-540-73437-6_19

[PP09] Porat, B., Porat, E.: Exact and approximate pattern matching in the streaming model. In: 50th Annual IEEE Symposium on Foundations of Computer Science, FOCS, pp. 315–323 (2009)

[RS17] Radoszewski, J., Starikovskaya, T.A.: Streaming k-mismatch with error correcting and applications. In: 2017 Data Compression Conference, DCC, pp. 290–299 (2017)

Maximum Colorful Cycles
in Vertex-Colored Graphs

Giuseppe F. Italiano[2], Yannis Manoussakis[1], Nguyen Kim Thang[3],
and Hong Phong Pham[1(✉)]

[1] LRI, University Paris-Saclay, Orsay, France
phongph.hut@gmail.com
[2] University of Rome Tor Vergata, Rome, Italy
[3] IBISC, University Paris-Saclay, Evry, France

Abstract. In this paper, we study the problem of finding a maximum colorful cycle a vertex-colored graph. Specifically, given a graph with colored vertices, the goal is to find a cycle containing the maximum number of colors. We aim to give a dichotomy overview on the complexity of the problem. We first show that the problem is NP-hard even for simple graphs such as split graphs, biconnected graphs, interval graphs. Then we provide polynomial-time algorithms for classes of vertex-colored threshold graphs and vertex-colored bipartite chain graphs, which are our main contributions.

1 Introduction

In this paper we deal with vertex-colored graphs, which are useful in various situations. For instance, the Web graph may be considered as a vertex-colored graph where the color of a vertex represents the content of the corresponding page (red for mathematics, yellow for physics, etc.) [4]. In a biological population, vertex-colored graphs can be used to represent the connections and interactions between species where different species have different colors. Other applications of vertex-colored graphs can also be found in bioinformatics (Multiple Sequence Alignment Pipeline or for multiple Protein-Protein Interaction networks) [7], or in a number of scheduling problems [15].

Given a vertex-colored graph, a *tropical subgraph* is a subgraph where each color of the initial graph appears at least once. Many graph properties, such as the domination number, the vertex cover number, independent sets, connected components, paths, matchings etc. can be studied in their tropical version. Finding a tropical subgraph in a (biological) population is to look for a subgraph which fully represents the (bio-)diversity of the population. In this paper, we consider a more general question of finding a *maximum colorful* subgraph which is a subgraph with maximum number of colors. Given a vertex-colored graph and some property of a subgraph (for example, paths, cycles, connected components), it could be that the tropical subgraph with the given property does not exist. Hence, one can ask the question of finding a subgraph with the most diverse

© Springer International Publishing AG, part of Springer Nature 2018
F. V. Fomin and V. V. Podolskii (Eds.): CSR 2018, LNCS 10846, pp. 106–117, 2018.
https://doi.org/10.1007/978-3-319-90530-3_10

population. Clearly, a maximum colorful subgraph is tropical if it contains all colors.

This notion of colorful subgraph is close to, but somewhat different from the *colorful* concept considered in [1,12,13], where neighbor vertices must have different colors. It is also related to the concepts of *color patterns* or *colorful* used in bio-informatics [8]. Note that in a *colorful* subgraph considered in our paper, two adjacent vertices may have the same color. In this paper, we study maximum colorful cycles in vertex-colored graphs.

Throughout the paper, we let $G = (V, E)$ denote a simple undirected graph. Given a set of colors $\mathcal{C}$, $G^c = (V, E)$ denotes a vertex-colored graph whose vertices are (not necessarily properly) colored by one of the colors in $\mathcal{C}$. The number of colors of G^c is $|\mathcal{C}|$. Given a subset of vertices $U \subset V$, the set of colors of vertices in U is denoted by $\mathcal{C}(U)$. Moreover, we denote the color of the vertex v by $c(v)$ and denote the number of vertices of U whose colors is c by $v(U, c)$. The set of neighbors of v is denoted by $N(v)$. In this paper, we study the following problem:

Maximum Colorful Cycle Problem (MCCP). Given a vertex-colored graph $G^c = (V, E)$, find a simple cycle with the maximum number of colors of G^c.

Related work. In the special case where each vertex has a distinct color, MCCP reduces to the Hamiltonian cycle problem. The Hamiltonian cycle problem has been widely studied in the literature and it is well known that this problem is NP-complete even for specific classes of graphs such as for undirected planar graphs of maximum degree three [10], for 3-connected 3-regular bipartite graphs [2], etc. However, the Hamiltonian cycle problem can be solved in time $O(m + n)$ for proper interval graphs [3,11].

If a graph G^c is Hamiltonian then it must contain a tropical cycle (which is a maximum colorful cycle) since the set of vertices must contain all colors. The problem of finding a longest cycle has been also studied and this problem can be used to solve the Hamiltonian cycle problem (and thus it is NP-hard). However, for some classes of graphs, there exist polynomial time algorithms for finding the longest cycle in threshold graphs [14], and in bipartite chain graphs [17]. Note that a longest cycle does not necessarily contain the maximum number of colors. However, in our paper, we take advantage of those algorithms to construct a Hamiltonian cycle for a given set of candidate vertices of a maximum colorful cycle.

The tropical subgraph and maximum colorful subgraph problems in vertex-colored graphs have been studied recently. The tropical subgraph problems in vertex-colored graphs such as tropical connected subgraphs, tropical independent sets have been investigated in [9]. Recently, the maximum colorful matching problem [5] and the maximum colorful path problem [6] have been studied, and several hardness results and polynomial-time algorithms were shown for different classes of graphs.

Our contributions. In this paper, we aim to give dichotomy overview on the complexity of MCCP. First, we prove that MCCP is NP-hard even for split graphs, interval graphs and biconnected graphs. Next, we present polynomial-time algorithms for several classes of graphs. First, we show that the MCCP is polynomial for proper interval graphs and split complete graphs. Although those algorithms are not complicated, they provide a sharp separation in term of complexity for interval and split graphs.

Our main contributions are polynomial-time algorithms for threshold graphs and bipartite chain graphs. A graph G is a *threshold* graph if it is constructed from the repetition of two operations: (1) adding an *isolated* vertex to the current graph, or (2) adding a *dominating* vertex to the current graph, i.e.,. one vertex connected to all vertices added earlier. A *bipartite chain* graph $G(X \cup Y, E)$ is a bipartite graph in which vertices in X can be linearly ordered such that $N(x_1) \supseteq N(x_2) \supseteq \ldots \supseteq N(x_{|X|})$. In our approach for both threshold and bipartite chain graphs, we develop connections between maximum colorful cycles and maximum colorful matchings and derive structural properties of maximum colorful cycles. Those properties enable us to identify a small set of candidate vertices for maximum tropical cycles. Subsequently, using longest cycle algorithms [14,17], on these vertices, we can efficiently compute the corresponding maximum tropical cycles for MCCP. The running times of our algorithms are $O(\max\{|\mathcal{C}| \cdot M(m,n)), n(n+m)\})$ and $O(|\mathcal{C}| \cdot \max\{M(m,n), n^3\})$ for threshold graphs and for bipartite graphs respectively, where $|\mathcal{C}|$ is the total number of colors and $M(m,n)$ is the running time for finding a maximum matching in a general graph with m edges and n vertices. (It is known that $M(m,n) = O(\sqrt{n}m)$ [16].) Due to space limit, some results are put in the appendix.

2   Hardness Results for MCCP

Theorem 1. *MCCP is NP-hard for interval graphs and biconnected graphs.*

Proof. We reduce from the SAT problem. Consider a boolean CNF formula B with variables $X = \{x_1, \ldots, x_s\}$ and clauses $B = \{b_1, \ldots, b_t\}$. We construct the following graph. Suppose that $\forall 1 \leq i \leq s$, the variable x_i appears in clauses $b_{i1}, b_{i2}, \ldots, b_{i\alpha_i}$ and $\overline{x_i}$ appears in clauses $b'_{i1}, b'_{i2}, \ldots, b'_{i\beta_i}$ in which $b_{ij} \in B$ and $b'_{ik} \in B$. An intersection model for our graph is constructed as follows. On the real line, we create $(s+1)$ intervals $v_1, v_2, \ldots, v_{s+1}$ such that v_i intersects only v_{i-1} and v_{i+1} $(1 \leq i \leq s-1)$. Next for each variable x_i of X $(1 \leq i \leq s)$, we create α_i same intervals $b_{i1}, b_{i2}, \ldots, b_{i\alpha_i}$ such that these intervals intersect pairwise each other and intersect only v_i and v_{i+1} among other vertices v_j. Similarly, for each $\overline{x_i}$, β_i same intervals $b'_{i1}, b'_{i2}, \ldots, b'_{i\beta_i}$ are drawn such that they intersect pairwise each other and intersect only v_i and v_{i+1}. Additionally, we create one special interval v_0 such that v_0 intersects with all other intervals, except for the intervals b_{1j} and $b'_{1j} (0 \leq j \leq b_{1\alpha_1}, b'_{1\beta_1})$. Note that this graph is both interval and biconnected.

From this intersection model, we obtain the corresponding interval graph and give colors as follows. Every vertex corresponding to the clause b_l has the same color c_l. Observe that vertices $b_{i1}, b_{i2}, \ldots, b_{i\alpha_i}$ make a clique in G^c, similarly for vertices $b'_{i1}, b'_{i2}, \ldots, b'_{i\beta_i}$. For the vertex v_i, we use the color c'_i such that all colors c'_i are distinct and different from the colors c_l. See an illustration in Fig. 1.

Now we claim that there exists a truth assignment to the variables of B satisfying all clauses if and only if G^c contains a cycle with all colors.

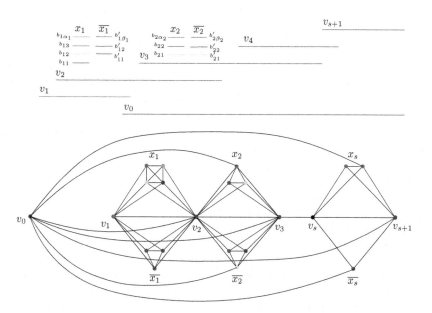

Fig. 1. Reduction of the SAT problem to MCCP for interval graphs. (Color figure online)

Now from a truth assignment B with all satisfied clauses, it is possible to obtain a cycle with all colors as follows. We start with the edge (v_0, v_1). Then, to go from v_i to v_{i+1}, $1 \leq i \leq s$, in the case that the variable x_i is assigned true then we select the sub-path $(v_i \rightarrow b_{i1} \rightarrow b_{i2} \rightarrow \ldots \rightarrow b_{i\alpha_i} \rightarrow v_{i+1})$ into the final path. Otherwise, i.e., x_i is assigned as false, then the sub-path $(v_i \rightarrow b'_{i1} \rightarrow b'_{i2} \rightarrow \ldots \rightarrow b'_{i\beta_i} \rightarrow v_{i+1})$ is selected. Finally, the edge (v_{s+1}, v_0) is added. It is clear that we obtain a cycle with all colors of G^c.

Conversely, from a cycle K containing all colors G^c, we obtain an assignment with all satisfied clauses as follows. Observe first that all vertices v_i must be in K since their colors are distinct. Since v_0, v_1, v_2 are in K and v_0 is not connected to any b_{1j} or $b'_{1j}(0 \leq j \leq b_{1\alpha_1}, b'_{1\beta_1})$, so we must have that the edge (v_0, v_1) must be in K (otherwise, both v_0 and v_1 can not be in K together). Note that in the case that v_0 is directly connected to any v_i or b_{ij} or b'_{ik} $(2 \leq i \leq s)$ then both v_1 and v_{s+1} can not be in K together. Thus v_0 must be connected to v_1 and v_{s+1} in K. Now, for each $1 \leq i \leq s$, K must go from v_i to v_{i+1}. In the case K goes from v_i to v_{i+1} through one path at the side of x_i then we assign the value *true* for x_i, if the side of $\overline{x_i}$ is used then *false* is assigned for x_i, otherwise (the edge (v_i, v_{i+1}) is in K) we assign arbitrarily *true* or *false* for x_i. Clearly this assignment is consistent for each x_i . Since except for all the colors of v_i then all colors of clauses of B must be in K, we obtain that this assignment satisfies all clauses of B. This completes our proof. □

3   An Algorithm for MCCP in Threshold Graphs

Recall that a graph G is a *threshold* graph if it is constructed from the repetition of two operations: (1) adding an *isolated* vertex to the current graph, or (2) adding a *dominating* vertex to the current graph, i.e., one vertex connected to all vertices added earlier. In the following, we denote vertices of type (1) as *isolated vertices* and vertices of type (2) as *dominating vertices*. Let G^c be a vertex-colored threshold graph. Without loss of generality, we can assume that the last added vertex v to G^c is a *dominating* vertex (otherwise v would be an isolated vertex and it would not appear in a maximum colorful cycle, unless we are in the trivial case where the maximum colorful cycle has size one). By this assumption, G^c is connected. It follows from the construction of threshold graphs that any edge must contain at least one dominating vertex and any two dominating vertices must be connected to each other.

We denote by X the set of dominating vertices of G^c, and by Y the set of isolated vertices of G^c. The set of vertices $V(G^c)$ is denoted by $\{v_1, v_2, \ldots, v_m\}$, in the order in which they were added to G^c. We also denote the *number of colors* of any maximum colorful cycle and of any maximum colorful matching in a vertex-colored threshold graph G^c respectively by C_c and by C_m. Recall that C_m can be computed by the algorithm in [5]. In this section we first study the structural properties of maximum colorful cycles and develop connections between maximum colorful cycles and maximum colorful matchings. Next, we will use those properties to design an efficient algorithm for finding a maximum colorful cycle.

Lemma 1. *Let G^c be a vertex-colored threshold graph. Then $C_m - 1 \leq C_c \leq C_m + 1$.*

Proof. We first show that $C_c \geq C_m - 1$. Let M be a maximum colorful matching in G^c with C_m different colors. Note that $|M|$ is at least C_m. Since each edge must contain at least one dominating vertex, choose one dominating vertex from each edge of M: denote those vertices by $x_1, x_2, \ldots, x_{|M|}$, such that x_j was added earlier than x_{j+1}, for $1 \leq j \leq |M| - 1$. Let z_j be the neighbor of x_j in the matching M, for $1 \leq j \leq |M|$. Note that z_j can be an isolated vertex or a dominating vertex. By the order of x-vertices, $N(x_1) \subseteq N(x_2) \subseteq \ldots \subseteq N(x_{|M|})$. Thus, (x_j, z_{j-1}) must be an edge in $E(G^c)$, for $1 \leq j \leq |M|$. As a result, $C' = (x_1, z_1, x_2, z_2, \ldots, x_{|M|-1}, z_{|M|-1}, x_{|M|}, x_1)$ is a cycle containing all vertices in M, except for $z_{|M|}$. The number of colors in the cycle C' is at least $C_m - 1$ since we remove $z_{|M|}$, and thus $C_c \geq C_m - 1$.

Next, suppose by contradiction that $C_c \geq C_m + 2$. Let $K = (v_{i_1}, v_{i_2}, \ldots, v_{i_k}, v_{i_1})$ be a cycle with C_c colors. Let $k = 2t$ if k is even and $k = 2t + 1$ if k is odd. Now, the matching $M = \{(v_{i_1}, v_{i_2}), (v_{i_3}, v_{i_4}), \ldots, (v_{i_{2t-1}}, v_{i_{2t}})\}$ has

$|\mathcal{C}(M)| \geq C_m + 1$ colors, a contradiction. Thus $C_c \leq C_m + 1$ and this completes our proof. □

The following observation allows us to reduce the search space for isolated vertices of maximum colorful cycles:

Lemma 2. *Any maximum colorful cycle can be reduced to another maximum colorful cycle in which any isolated vertex has a color different from the colors of other vertices.*

By Lemma 2, we can restrict our attention only to maximum colorful cycles where each isolated vertex has a distinct color. We now introduce some new terminology. Let $\mathcal{C}_1 := \mathcal{C}(Y)\backslash\mathcal{C}(X) = \{c_{11}, c_{12}, \ldots, c_{1k_1}\}$ be the colors in Y but not in X. Denote $\mathcal{C}_2 := \mathcal{C}(X) = \{c_{21}, c_{22}, \ldots, c_{2k_2}\}$ the set of colors in X. By these definitions, the numbers of colors in $\mathcal{C}_1$ and $\mathcal{C}_2$ are k_1 and k_2, respectively. For each color c_{1i} in $\mathcal{C}_1$, let $\min[c_{1i}]$ be the index of the first vertex in Y with color c_i, i.e., $c(v_{\min[c_{1i}]}) = c_{1i}$ and $c(v_j) \neq c_{1i}$ for every $v_j \in Y$ with $j < \min[c_{1i}]$. Without loss of generality, suppose that $1 < \min[c_{11}] < \min[c_{12}] < \ldots < \min[c_{1k_1}]$. Similarly, for each color c_{2i} in $\mathcal{C}_2$, let $\max[c_{2i}]$ be the index of the last vertex in X such that $c(v_{\max[c_{2i}]}) = c_{2i}$ and $c(v_j) \neq c_{2i}$ for every $v_j \in X$ with $j > \max[c_{2i}]$. Without loss of generality, suppose that $\max[c_{21}] > \max[c_{22}] > \ldots > \max[c_{2k_2}]$. Moreover, for each maximum colorful cycle K, let X_K and Y_K be the sets of dominating vertices and isolated vertices in K, respectively, and let us denote their sets of colors by $\mathcal{C}(X_K)$ and $\mathcal{C}(Y_K)$.

We now consider three different cases, depending on whether $C_c = C_m + 1$, $C_c = C_m$ or $C_c = C_m - 1$.

3.1   Case 1: $C_c = C_m + 1$

Lemma 3. *Suppose that $C_c = C_m + 1$, then any maximum colorful cycle must contain all colors of the given graph G^c.*

Case 1.1: *There exists a maximum colorful cycle K with some edge connecting two dominating vertices.*

Lemma 4. *There exists another maximum colorful cycle K' whose set of vertices $V(K') = V(X) \cup \{v_{\min[c_{11}]}, v_{\min[c_{12}]}, \ldots, v_{\min[c_{1k_1}]}\}$.*

Proof. Let (u, v) be an edge of K such that both u and v are dominating vertices. Recall that any two dominating vertices are connected to each other. Therefore in the case that there exists some dominating vertices which are not in K then we can include them into K by adding into K a path from u to v containing all these dominating vertices and remove the edge (u, v) from K. By doing so it is possible to obtain another maximum colorful cycle K' containing the set $V(X)$. Now since K' contains all colors of the original graph (by Lemma 3) and each isolated vertex of K' (also K) has distinct color (by Lemma 2), we obtain that all colors of $\mathcal{C}_1$ must appears exactly once in K'

and the number of isolated vertices of K' is equal to k_1. Now observe that for any two isolated vertices w and t such that w was added earlier than t in G^c then $N(w) \supseteq N(t)$. This allows to replace all isolated vertices of K' with distinct colors by the set of vertices $\{v_{\min[c_{11}]}, v_{\min[c_{12}]}, \ldots, v_{\min[c_{1k_1}]}\}$. So it is possible to obtain another maximum colorful cycle K' with its set of vertices as $V(X) \cup \{v_{\min[c_{11}]}, v_{\min[c_{12}]}, \ldots, v_{\min[c_{1k_1}]}\}$. □

Case 1.2: *For any maximum colorful cycle, there is no edge of this cycle that connects two dominating vertices.*

For each isolated vertex v, let $X^+(v)$ and $X^-(v)$ be the sets of dominating vertices added to G^c after and before v, respectively. Similarly, let $Y^+(v)$ and $Y^-(v)$ be the sets of isolated vertices added after and before v, respectively. Our maximum colorful cycle in this case will use a special vertex to reduce.

Lemma 5. *There exists exactly one isolated vertex v^* such that*

$$|X^+(v^*)| + |\mathcal{C}(X^+(v^*))| = C_m + 1$$

Moreover, the set of dominating vertices $X_K = X^+(v^*)$ and the number of isolated vertices $|Y_K| = |X^+(v^*)|$.

Proof. If we consider isolated vertices v in the order of the construction of the threshold graph G^c then the value of the sum $|X^+(v)| + |\mathcal{C}(X^+(v))|$ will strictly decrease. Thus there exists at most one isolated vertex v^* satisfying the lemma equality.

In the remainder of the proof, we will show the existence of such vertex v^*. Let v be the *first* isolated vertex in K (in the order of the construction of the threshold graph G^c). We will prove that v is v^*.

We claim that all vertices of $X^+(v)$ are in K and no vertex of $X^-(v)$ is in K.

$X^+(v) \subset K$. Assume that there exists a dominating vertex $u \in X^+(v)$ and u is not in K. As u was added after v, u is connected with v by an edge. Let w be a neighbor of v on K. Since v is an isolated vertex we must have that w is a dominating vertex. Now we remove the edge (v, w) from K and add two edges (v, u) and (u, w) on K then one obtains another maximum colorful cycle in which the edge (u, w) connects two dominating vertices (contradiction to the assumption of Case 1.2). Thus u must be in K.

$X^-(v) \cap K = \emptyset$. Assume that there exists a dominating vertex $t \in X^-(v)$ and t is also in K. Let z be a neighbor of t in K then z must be an isolated vertex. Therefore, z must be added earlier than t (by the construction of threshold graphs). Thus z must be added earlier than v, a contradiction since v is the first isolated vertex in K. So t must not be in K.

Hence, the claim follows.

By the claim, we have $X_K = X^+(v)$. As there is no edge of K connecting two dominating vertices, each edge must have an endpoint as dominating vertex and another endpoint as isolated vertex. So $|X_K| = |Y_K|$. From that the number of isolated vertices of K (i.e. $|Y_K|$) is equal to $|X^+(v)|$.

By Lemma 2, each isolated vertex in this cycle has a distinct color, so the number of colors of all isolated vertices in K is $|X^+(v)|$. Moreover, the number of colors of all dominating vertices of K is $|C(X^+(v))|$. Therefore we obtain that $|X^+(v)| + |C(X^+(v))|$ equals the number of colors in K, which is $C_c = C_m + 1$ by the case assumption. Hence, the lemma equality holds at v, i.e., $|X^+(v)| + |C(X^+(v))| = C_m + 1$. $\qquad\square$

Note that one can detect this vertex v^* efficiently by computing C_m and by checking the above identity over all vertices.

Let $\{c'_{11}, c'_{12}, \ldots, c'_{1k'}\} = C(Y)\backslash C(X^+(v^*))$ be the set of colors in Y but not in $X^+(v^*)$. As we did before, let us define by $v_{\min[c'_{1i}]}$ the first isolated vertex in G^c with color c'_{1i}. Again, without loss of generality, assume that $\min[c'_{11}] < \min[c'_{12}] < \ldots < \min[c'_{1k'}]$. Now we are ready to show the main structural property of a maximum colorful cycle in this case.

Lemma 6. *Let v^* be the unique vertex such that $|X^+(v^*)| + |C(X^+(v^*))| = C_m + 1$ and let K be a maximum colorful cycle. Then, there exists another maximum colorful cycle K' where $V(K') = X^+(v^*) \cup \{v_{\min[c'_{11}]}, v_{\min[c'_{12}]}, \ldots, v_{\min[c'_{1|X^+(v^*)|}]}\}$.*

Proof. By Lemma 5, the number of isolated vertices of K (i.e. $|Y_K|$) equals $|X^+(v^*)|$ and each vertex in Y_K has a distinct color. Observe that $C(Y_K) \subseteq C(Y)\backslash C(X^+(v^*))$. This follows from the fact that if $C(Y_K) \cap C(X^+(v^*)) \neq \emptyset$ then the total number of colors in K is

$$|C(Y_K) \cup C(X_K)| = |C(Y_K) \cup C(X^+(v^*))| < |C(Y_K)| + |C(X^+(v^*))|$$
$$\leq |X^+(v^*)| + |C(X^+(v^*))| = C_m + 1 = C_c$$

where the equalities are due to Lemma 5. This contradicts the fact that K is a maximum colorful cycle. Therefore, the observation holds true and so $|X^+(v^*)| \leq k'$.

Recall that for any two isolated vertices w and t such that w was added earlier than t in G^c then $N(w) \supseteq N(t)$. Hence, by replacing $|X^+(v^*)|$ isolated vertices in K by vertices $\{v_{\min[c'_{11}]}, v_{\min[c'_{12}]}, \ldots, v_{\min[c'_{1|X^+(v^*)|}]}\}$, one gets another cycle K' with the same number of colors as K. This vertex replacing procedure can be done since $|X^+(v^*)| \leq k'$. Note that K' and K may have different sets of colors but their cardinals are the same. Thus, we have a maximum colorful cycle K' where $V(K') = X^+(v^*) \cup \{v_{\min[c'_{11}]}, v_{\min[c'_{12}]}, \ldots, v_{\min[c'_{1|X^+(v^*)|}]}\}$. $\qquad\square$

Since v^* and the sets of vertices $X^+(v^*)$ and $\{v_{\min[c'_{11}]}, v_{\min[c'_{12}]}, \ldots, v_{\min[c'_{1|X^+(v^*)|}]}\}$ can be immediately identified, in the case that there exists a maximum colorful cycle then we can use the algorithm in [14] to construct a Hamiltonian cycle consisting of all these vertices.

3.2   Case 2: $C_c = C_m$

The following lemma helps to limit the search space of the set of colors of dominating vertices of a maximum colorful cycle.

Lemma 7. *For any maximum colorful cycle K, there exists at most one dominating vertex v such that $v \notin K$ and $c(v) \notin \mathcal{C}(K)$.*

By this lemma, the set of colors $\mathcal{C}(X_K)$ is either $\mathcal{C}(X)$ or $\mathcal{C}(X)\backslash c$ for some color c. We distinguish two corresponding sub-cases.

Case 2.1: *There exists a maximum colorful cycle K such that there exists exactly one dominating vertex v^{**} such that $v^{**} \notin K$ and $c(v^{**}) \notin \mathcal{C}(K)$.*

Denote $\{c'_{11}, c'_{12}, \ldots, c'_{1k'}\} = (\mathcal{C}(Y)\backslash\mathcal{C}(X))\cup\{c(v^{**})\}$. Similarly as previously, let $v_{\min[c'_{1i}]}$ be the first isolated vertex with the color c'_{1i} in G^c. Without loss of generality, suppose that $\min[c'_{11}] < \min[c'_{12}] < \ldots < \min[c'_{1k'}]$. Note that, in contrast to the previous case, the vertex v^{**} can not be immediately identified. However, our algorithm will loop over all vertices by considering each as v^{**}. The following lemma helps to replace vertices to obtain another maximum colorful cycle from a maximum colorful cycle in this situation.

Lemma 8. *Let K be a maximum colorful cycle of G^c and v^{**} be a dominating vertex such that $v^{**} \notin K$ and $c(v^{**}) \notin \mathcal{C}(K)$. Then, there exists another colorful cycle K' such that the set of vertices $V(K') = V(X)\backslash\{v^{**}\} \cup \{v_{\min[c'_{11}]}, v_{\min[c'_{12}]}, \ldots, v_{\min[c'_{1\ell}]}\}$ where $\ell = C_m - |\mathcal{C}(X)| + 1$.*

Proof. By the case assumption, $|\mathcal{C}(X_K)| = |\mathcal{C}(X)| - 1$. Since $|\mathcal{C}(K)| = C_m$, we obtain that $|\mathcal{C}(Y_K)| = C_m - |\mathcal{C}(X)| + 1$. Moreover, by Lemma 2, the color of any isolated vertex of K is different to other vertex's color. So $\mathcal{C}(Y_K) \subseteq \mathcal{C}(Y)\backslash\mathcal{C}(X_K)$. Therefore, $C_m - |\mathcal{C}(X)| + 1 \leq |\mathcal{C}(Y)\backslash\mathcal{C}(X)\cup\{c(v^{**})\}| = k'$. Recall the observation that for any two isolated vertices w and t, if w was added earlier than t in G^c then $N(w) \supseteq N(t)$. Denote $\ell = C_m - |\mathcal{C}(X)| + 1$. By replacing ℓ vertices of K by vertices $\{v_{\min[c'_{11}]}, v_{\min[c'_{12}]}, \ldots, v_{\min[c'_{1\ell}]}\}$, one obtains another maximum colorful cycle K' with the same number of colors. Note that $\ell < k'$ so the replacing procedure can always be done. Now the set of vertices $V(K') = V(X)\backslash\{v^{**}\} \cup \{v_{\min[c'_{11}]}, v_{\min[c'_{12}]}, \ldots, v_{\min[c'_{1\ell}]}\}$ as required by the lemma. $\square$

Case 2.2: *For any maximum colorful cycle K, there does not exist any dominating vertex v such that $v \notin K$ and $c(v) \notin \mathcal{C}(K)$.*

Recall that $\mathcal{C}_2 := \{c_{21}, c_{22}, \ldots, c_{2k_2}\}$ is the set of colors in X. Let $\max[c_{2i}]$ be the index of the last vertex in X such that $c(v_{\max[c_{2i}]}) = c_{2i}$ and $c(v_j) \neq c_{1i}$ for every $v_j \in X$ with $j > \max[c_{2i}]$. Without loss of generality, suppose that $\max[c_{21}] > \max[c_{22}] > \ldots > \max[c_{2k_2}]$. Let $X_t(G^c)$ be the set of t last dominating vertices of the set $V(X)\backslash\{v_{\max[c_{21}]}, v_{\max[c_{22}]}, \ldots, v_{\max[c_{2k_2}]}\}$. Now the following lemma helps to reduce a maximum colorful cycle to another maximum colorful cycle which is easier to find.

Lemma 9. *Let K be a maximum colorful cycle of G^c. Then, there exists another colorful cycle K' where $V(K') = X_t(G^c) \cup \{v_{\max[c_{21}]}, v_{\max[c_{22}]}, \ldots, v_{\max[c_{2k_2}]}\} \cup \{v_{\min[c_{11}]}, v_{\min[c_{12}]}, \ldots, v_{\min[c_{1\ell}]}\}$ where $\ell = C_m - |\mathcal{C}(X)|$ and $t = |V(K)| - k_2 - C_m + |\mathcal{C}(X)|$.*

By Lemma 9, given the value $|V(K)|$, all elements in the cycle K' are immediately identified. Therefore in our final algorithm we will vary the value of $|V(K)|$ to find the maximum colorful cycle if this case holds.

3.3 Case 3: $C_c = C_m - 1$

In this case, it is possible to obtain easily a cycle with $C_m - 1$ colors from any colorful matching, based on the first part of the proof of Lemma 1.

3.4 Algorithm for Threshold Graphs

Based on the structural properties of maximum colorful cycles according to different cases, we derive the following algorithm for finding a maximum colorful cycle. The algorithm makes use of the algorithm computing a maximum colorful matching [5] and the algorithm computing a Hamiltonian cycle in threshold graphs [14].

Algorithm 1. Maximum colorful cycle in vertex-colored threshold graphs.

1: $C_m \leftarrow$ the number of colors of a maximum colorful matching (using algorithm [5])
2: $\mathcal{C}_1 := \mathcal{C}(Y)\backslash\mathcal{C}(X) = \{c_{11}, c_{12}, \ldots, c_{1k_1}\}$ and $\mathcal{C}_2 := \mathcal{C}(X) = \{c_{21}, c_{22}, \ldots, c_{2k_2}\}$
3: **if** $\exists$ a Hamiltonian cycle K of $V(X) \cup \{v_{\min[c_{11}]}, \ldots, v_{\min[c_{1k_1}]}\}$ **then** # Case 1.1
4:     **return** K as the maximum colorful cycle   # Lemma 4
5: **else**   # Case 1.2
6:     $v^* \leftarrow$ the unique vertex satisfying $|X^+(v^*)| + |C(X^+(v^*))| = C_m + 1$
7:     $X^+(v^*) \leftarrow$ set of dominating vertices added to the graph after v^*
8:     $\{c'_{11}, c'_{12}, \ldots, c'_{1k'}\} \leftarrow \mathcal{C}(Y)\backslash\mathcal{C}(X^+(v^*))$
9:     **if** $\exists$ a Hamiltonian cycle K of $X^+(v^*) \cup \{v_{\min[c_{11}]}, v_{\min[c_{12}]}, \ldots, v_{\min[c_{1k_1}]}\}$ **then**
10:         **return** K as the maximum colorful cycle   # Lemma 6
11:     **end if**
12: **end if**
13: **for** $v^{**} \in V(G^c)$ **do**   # Case 2.1
14:     $\{c'_{11}, c'_{12}, \ldots, c'_{1k'}\} \leftarrow \mathcal{C}(Y)\backslash\mathcal{C}(X) \cup \{c(v^{**})\}$ and $\ell \leftarrow C_m - |\mathcal{C}(X)| + 1$
15:     **if** $\exists$ a Hamiltonian cycle K of $V(X)\backslash\{v^{**}\} \cup \{v_{\min[c'_{11}]}, v_{\min[c'_{12}]}, \ldots, v_{\min[c'_{1\ell}]}\}$ **then**
16:         **return** K as the maximum colorful cycle   # Lemma 8
17:     **end if**
18: **end for**
19: **for** $0 \leq t \leq |V(X)\backslash\{v_{\max[c_{21}]}, v_{\max[c_{22}]}, \ldots, v_{\max[c_{2k_2}]}\}|$ **do**   # Case 2.2
20:     $X_t(G^c) \leftarrow t$ last vertices in $V(X)\backslash\{v_{\max[c_{21}]}, v_{\max[c_{22}]}, \ldots, v_{\max[c_{2k_2}]}\}$
21:     **if** $\exists$ a Hamiltonian cycle K of $X_t(G^c) \cup \{v_{\max[c_{21}]}, v_{\max[c_{22}]}, \ldots, v_{\max[c_{2k_2}]}\}$
    $\cup \{v_{\min[c_{11}]}, v_{\min[c_{12}]}, \ldots, v_{\min[c_{1\ell}]}\}$ where $\ell = C_m - |\mathcal{C}(X)|$ **then**
22:         **return** K as the maximum colorful cycle   # Lemma 9
23:     **end if**
24: **end for**
25: **return** K as a maximum colorful cycle constructed from any maximum colorful matching based on Lemma 1   # Case 3

Theorem 2. *Algorithm 1 computes a maximum colorful cycle of G^c in time $O(\max\{|\mathcal{C}| \cdot M(m,n), n(n+m)\})$ where $|\mathcal{C}|$ is the number of colors in G^c and $M(m,n)$ is the time for finding a maximum matching in a general graph with m edges and n vertices.*

4   An Algorithm for Bipartite Chain Graphs

A bipartite graph $G = (X, Y, E)$ is said to be a *bipartite chain* graph if its vertices can be linearly ordered such that $N(x_1) \supseteq N(x_2) \supseteq \ldots \supseteq N(x_{|X|})$. As a consequence, we also immediately obtain a linear ordering over Y such that $N(y_1) \supseteq N(y_2) \supseteq \ldots \supseteq N(y_{|Y|})$. It is known that these orderings over X and Y can be computed in $O(n)$ time. Here we will look for a maximum colorful cycle in a vertex-colored bipartite chain graph $G^c = (X, Y, E)$.

Algorithm 2. Maximum colorful cycle in vertex-colored bipartite chain graphs.

1: $C_m \leftarrow$ the number of colors of a maximum colorful matching (using algorithm [5])
2: **for** $C_m \geq C_c \geq C_m - 2$ **do**
3:     **for** $1 \leq m \leq |X|, 1 \leq n \leq |Y|$ **do**
4:         $X_{m,n} \leftarrow \{x_i \in X \,|\, c(x_i) \notin \mathcal{C}(\{x_1, x_2, \ldots, x_m\}) \cup \mathcal{C}(\{y_1, y_2, \ldots, y_n\})\}$
5:         $Y_{m,n} \leftarrow \{y_j \in Y \,|\, c(y_j) \notin \mathcal{C}(\{x_1, x_2, \ldots, x_m\}) \cup \mathcal{C}(\{y_1, y_2, \ldots, y_n\})\}$
6:         Denote $\mathcal{C}(X_{m,n}) := \{c_{11}, c_{12}, \ldots, c_{1k_1}\}$ and $\mathcal{C}(Y_{m,n}) := \{c_{21}, c_{22}, \ldots, c_{2k_2}\}$.
7:         **for** $0 \leq \ell \leq C_c$ **do**
8:             $\ell' \leftarrow \max\{C_c - \ell - |\mathcal{C}(x_1, x_2, \ldots, x_m)| - |\mathcal{C}(y_1, y_2, \ldots, y_n)|, 0\}$
9:             $X_{m,n}^{\ell} \leftarrow \{x_{\min[c_{11}]}, x_{\min[c_{12}]}, \ldots, x_{\min[c_{1\ell}]}\}$ the set of ℓ first vertices (in the ordering of x-vertices) with distinct colors in $X_{m,n}$.
10:             $Y_{m,n}^{\ell'} \leftarrow \{y_{\min[c_{21}]}, y_{\min[c_{22}]}, \ldots, y_{\min[c_{2\ell'}]}\}$ the set of ℓ' first vertices (in the ordering of y-vertices) with distinct colors in $Y_{m,n}$.
11:             **if** $\exists$ a Hamiltonian cycle K of $\{x_1, x_2, \ldots, x_m\} \cup X_{m,n}^{\ell} \cup \{y_1, y_2, \ldots, y_n\} \cup Y_{m,n}^{\ell'}$ **then**
12:                 **return** K as the maximum colorful cycle
13:             **end if**
14:         **end for**
15:     **end for**
16: **end for**

Theorem 3. *Algorithm 2 computes a maximum colorful cycle of G^c in $O(|\mathcal{C}| \cdot \max\{M(m,n), n^3\})$ where $M(m,n)$ is the best known complexity for finding a maximum matching in a general graph with m edges and n vertices.*

References

1. Akbari, S., Liaghat, V., Nikzad, A.: Colorful paths in vertex coloring of graphs. Electron. J. Comb. **18**(1), P17 (2011)
2. Akiyama, T., Nishizeki, T., Saito, N.: Np-completeness of the hamiltonian cycle problem for bipartite graphs. J. Inf. Proc. **3**(2), 73–76 (1979)

3. Bertossi, A.A.: Finding Hamiltonian circuits in proper interval graphs. Inf. Proc. Lett. **17**(2), 97–101 (1983)

4. Bruckner, S., Hüffner, F., Komusiewicz, C., Niedermeier, R.: Evaluation of ILP-based approaches for partitioning into colorful components. In: Bonifaci, V., Demetrescu, C., Marchetti-Spaccamela, A. (eds.) SEA 2013. LNCS, vol. 7933, pp. 176–187. Springer, Heidelberg (2013). https://doi.org/10.1007/978-3-642-38527-8_17

5. Cohen, J., Manoussakis, Y., Pham, H., Tuza, Z.: Tropical matchings in vertex-colored graphs. In: Latin and American Algorithms, Graphs and Optimization Symposium (2017)

6. Cohen, J., Italiano, G.F., Manoussakis, Y., Nguyen, K.T., Pham, H.P.: Tropical paths in vertex-colored graphs. In: Gao, X., Du, H., Han, M. (eds.) COCOA 2017. LNCS, vol. 10628, pp. 291–305. Springer, Cham (2017). https://doi.org/10.1007/978-3-319-71147-8_20

7. Corel, E., Pitschi, F., Morgenstern, B.: A min-cut algorithm for the consistency problem in multiple sequence alignment. Bioinformatics **26**(8), 1015–1021 (2010)

8. Fellows, M.R., Fertin, G., Hermelin, D., Vialette, S.: Upper and lower bounds for finding connected motifs in vertex-colored graphs. J. Comput. Syst. Sci. **77**(4), 799–811 (2011)

9. Foucaud, F., Harutyunyan, A., Hell, P., Legay, S., Manoussakis, Y., Naserasr, R.: Tropical homomorphisms in vertex-coloured graphs. Discrete App. Math. (to appear)

10. Garey, M.R., Johnson, D.S., Stockmeyer, L.: Some simplified NP-complete problems. In: Proceedings 6th Symposium on Theory of Computing, pp. 47–63 (1974)

11. Ibarra, L.: A simple algorithm to find Hamiltonian cycles in proper interval graphs. Inf. Proc. Lett. **109**(18), 1105–1108 (2009b)

12. Li, H.: A generalization of the Gallai-Roy theorem. Graphs Comb. **17**(4), 681–685 (2001)

13. Lin, C.: Simple proofs of results on paths representing all colors in proper vertex-colorings. Graphs Comb. **23**(2), 201–203 (2007)

14. Mahadev, N.V.R., Peled, U.N.: Longest cycles in threshold graphs. Discrete Math. **135**(1–3), 169–176 (1994)

15. Marx, D.: Graph colouring problems and their applications in scheduling. Periodica Polytech. Electr. Eng. **48**(1–2), 11–16 (2004)

16. Micali, S., Vazirani, V.V.: An $O(\sqrt{|V|}|E|)$ algorithm for finding maximum matching in general graphs. In: Proceedings 21st Symposium on Foundations of Computer Science, pp. 17–27 (1980)

17. Uehara, R., Valiente, G.: Linear structure of bipartite permutation graphs and the longest path problem. Inf. Proc. Lett. **103**(2), 71–77 (2007)

Grammar-Based Compression
of Unranked Trees

Adrià Gascón[1], Markus Lohrey[2], Sebastian Maneth[3], Carl Philipp Reh[2(✉)],
and Kurt Sieber[2]

[1] Warwick University and Alan Turing Institute, Warwick, UK
[2] Universität Siegen, Siegen, Germany
reh@eti.uni-siegen.de
[3] Universität Bremen, Bremen, Germany

Abstract. We introduce forest straight-line programs (FSLPs) as a compressed representation of unranked ordered node-labelled trees. FSLPs are based on the operations of forest algebra and generalize tree straight-line programs. We compare the succinctness of FSLPs with two other compression schemes for unranked trees: top dags and tree straight-line programs of first-child/next sibling encodings. Efficient translations between these formalisms are provided. Finally, we show that equality of unranked trees in the setting where certain symbols are associative or commutative can be tested in polynomial time. This generalizes previous results for testing isomorphism of compressed unordered ranked trees.

1 Introduction

Generally speaking, grammar-based compression represents an object succinctly by means of a small context-free grammar. In many grammar-based compression formalisms such a grammar can be exponentially smaller than the object. Henceforth, there is a great interest in problems that can be solved in polynomial time on the grammar, while requiring at least linear time on the original uncompressed object. One of the most well-known and fundamental such problems is testing equality of the strings produced by two context-free string grammars, each producing exactly one string (such grammars are also known as straight-line programs — in this paper we use the term string straight-line program, SSLP for short). Polynomial time solutions to this problem were discovered, in different contexts by different groups of people, see the survey [14] for references.

Grammar-based compression has been generalized from strings to ordered ranked node-labelled trees, by means of linear context-free tree grammars generating exactly one tree [6]. Such grammars are also known as tree straight-line programs, TSLPs for short. Equality of the trees produced by two TSLPs can also be checked in polynomial time: one constructs SSLPs for the pre-order traversals of the trees, and then applies the above mentioned result for SSLPs, see [6]. The tree case becomes more complex when *unordered* ranked trees are considered. Such trees can be represented using TSLPs, by simply ignoring the order of

F. V. Fomin and V. V. Podolskii (Eds.): CSR 2018, LNCS 10846, pp. 118–131, 2018.
https://doi.org/10.1007/978-3-319-90530-3_11

children in the produced tree. Checking isomorphism of unordered ranked trees generated by TSLPs was recently shown to be solvable in polynomial time [16]. The solution transforms the TSLPs so that they generate canonical representations of the original trees and then checks equality of these canonical forms.

The aforementioned result for ranked trees cannot be applied to *unranked* trees (where the number of children of a node is not bounded), which arise for instance in XML document trees. This is unfortunate, because (*i*) grammar-based compression is particularly effective for XML document trees (see [15]), and (*ii*) XML document trees can often be considered unordered (one speaks of "data-centric XML", see e.g. [1,3,5,20]), allowing even stronger grammar-based compressions [17].

In this paper we introduce a generalization of TSLPs and SSLPs that allows to produce ordered unranked node-labelled trees and forests (i.e., ordered sequences of trees) that we call *forest straight-line programs*, FSLPs for short. In contrast to TSLPs, FSLPs can compress very wide and flat trees. For instance, the tree $f(a, a, \ldots, a)$ with n many a's is not compressible with TSLPs but can be produced by an FSLP of size $O(\log n)$. FSLPs are based on the operations of horizontal and vertical forest composition from forest algebras [4]. The main contributions of this paper are the following:

Comparison with Other Formalisms. We compare the succinctness of FSLPs with two other grammar-based formalisms for compressing unranked node-labelled ordered trees: TSLPs for "first-child/next-sibling" (fcns) encodings and top dags. The fcns-encoding is the standard way of transforming an unranked tree into a binary tree. Then the resulting binary tree can be succinctly represented by a TSLP. This approach was used to apply the TreeRePair-compressor from [15] to unranked trees. We prove that FSLPs and TSLPs for fcns-encodings are equally succinct up to constant multiplicative factors and that one can change between both representations in linear time (Propositions 9 and 10).

Top dags are another formalism for compressing unranked trees [2]. Top dags use horizontal and vertical merge operations for tree construction, which are very similar to the horizontal and vertical concatenation operations from FSLPs. Whereas a top dag can be transformed in linear time into an equivalent FSLP with a constant multiplicative blow-up (Proposition 6), the reverse transformation (from an FSLP to a top dag) needs time $O(\sigma \cdot n)$ and involves a multiplicative blow-up of size $O(\sigma)$ where σ is the number of node labels of the tree (Proposition 7). A simple example (Example 8) shows that this σ-factor is unavoidable. The reason for the σ-factor is a technical restriction in the definition of top dags: In contrast to FSLPs, top dags only allow sharing of common subtrees but not of common subforests. Hence, sharing between (large) subtrees which only differ in their root labels may be impossible at all (as illustrated by Example 8), and this leads to the σ-blow-up in comparison to FSLPs. The impossibility of sharing subforests would also complicate the technical details of our main algorithmic results for FSLPs (in particular Proposition 10 and Theorem 13 which is discussed below) for which we make heavy use of a particular normal form for FSLPs that exploits the sharing of proper subforests. We therefore believe that at least for our purposes, FSLPs are a more adequate formalism than top dags.

Testing Equality Modulo Associativity and Commutativity. Our main algorithmic result for FSLPs can be formulated as follows: Fix a set Σ of node labels and take a subset $\mathcal{C} \subseteq \Sigma$ of "commutative" node labels and a subset $\mathcal{A} \subseteq \Sigma$ of "associative" node labels. This means that for all $a \in \mathcal{A}$, $c \in \mathcal{C}$ and all trees $t_1, t_2, \ldots, t_n$ (i) we do not distinguish between the trees $c(t_1, \ldots, t_n)$ and $c(t_{\sigma(1)}, \ldots, t_{\sigma(n)})$, where σ is any permutation (commutativity), and (ii) we do not distinguish the trees $a(t_1, \ldots, t_n)$ and $a(t_1, \ldots, t_{i-1}, a(t_i, \ldots, t_{j-1}), t_j, \ldots, t_n)$ for $1 \le i \le j \le n+1$ (associativity). We then show that for two given FSLPs F_1 and F_2 that produce trees t_1 and t_2 (of possible exponential size), one can check in polynomial time whether t_1 and t_2 are equal modulo commutativity and associativity (Theorem 13). Note that unordered tree isomorphism corresponds to the case $\mathcal{C} = \Sigma$ and $\mathcal{A} = \emptyset$ (in particular we generalize the result from [16] for ranked unordered trees). Theorem 13 also holds if the trees t_1 and t_2 are given by top dags or TSLPs for the fcns-encodings, since these formalisms can be transformed efficiently into FSLPs. Theorem 13 also shows the utility of FSLPs even if one is only interested in say binary trees, which are represented by TSLPs. The law of associativity will yield very wide and flat trees that are no longer compressible with TSLPs but are still compressible with FSLPs.

Missing proofs can be found in the arXiv version of this paper [11].

2   Straight-Line Programs over Algebras

We will produce strings, trees and forests by algebraic expressions over certain algebras. These expressions will be compressed by directed acyclic graphs. In this section, we introduce the general framework, which will be reused several times in this paper.

An algebraic structure is a tuple $\mathcal{A} = (A, f_1, \ldots, f_k)$ where A is the universe and every $f_i \colon A^{n_i} \to A$ is an operation of a certain arity n_i. In this paper, the arity of all operations will be at most two. If $n_i = 0$, then f_i is called a constant. Moreover, it will be convenient to allow partial operations for the f_i. Algebraic expressions over $\mathcal{A}$ are defined in the usual way: if $e_1, \ldots, e_{n_i}$ are algebraic expressions over $\mathcal{A}$, then also $f_i(e_1, \ldots, e_{n_i})$ is an algebraic expressions over $\mathcal{A}$. For an algebraic expression e, $[\![e]\!] \in A$ denotes the element to which e evaluates (it can be undefined).

A *straight-line program* (SLP for short) over $\mathcal{A}$ is a tuple $P = (V, S, \rho)$, where V is a set of *variables*, $S \in V$ is the *start variable*, and ρ maps every variable $A \in V$ to an expression of the form $f_i(A_1, \ldots, A_{n_i})$ (the so called *right-hand side* of A) such that $A_1, \ldots, A_{n_i} \in V$ and the edge relation $E(P) = \{(A, B) \in V \times V | B$ occurs in $\rho(A)\}$ is acyclic. This allows to define for every variable $A \in V$ its value $[\![A]\!]_P$ inductively by $[\![A]\!]_P = f_i([\![A_1]\!]_P, \ldots, [\![A_{n_i}]\!]_P)$ if $\rho(A) = f_i(A_1, \ldots, A_{n_i})$. Since the f_i can be partially defined, the value of a variable can be undefined. The SLP P will be called *valid* if all values $[\![A]\!]_P$ ($A \in V$) are defined. In our concrete setting, validity of an SLP can be tested by a simple syntax check. The value of P is $[\![P]\!] = [\![S]\!]_P$. Usually, we prove properties of SLPs by induction along the partial order $E(P)^*$.

It will be convenient to allow for the right-hand sides $\rho(A)$ algebraic expressions over $\mathcal{A}$, where the variables from V can appear as atomic expressions. By introducing additional variables, we can transform such an SLP into an equivalent SLP of the original form. We define the size $|P|$ of an SLP P as the total number of occurrences of operations $f_1, \ldots, f_k$ in all right-hand sides (which is the number of variables if all right-hand sides have the standard form $f_i(A_1, \ldots, A_{n_i})$).

Sometimes it is useful to view an SLP $P = (V, S, \rho)$ as a directed acyclic graph (dag) $(V, E(P))$, together with the distinguished output node S, and the node labelling that associates the label f_i with the node $A \in V$ if $\rho(A) = f_i(A_1, \ldots, A_{n_i})$. Note that the outgoing edges $(A, A_1), \ldots, (A, A_{n_i})$ have to be ordered since f_i is in general not commutative and that multi-edges have to be allowed. Such dags are also known as algebraic circuits in the literature.

String Straight-Line Programs. A widely studied type of SLPs are SLPs over a free monoid $(\Sigma^*, \cdot, \varepsilon, (a)_{a \in \Sigma})$, where $\cdot$ is the concatenation operator (which, as usual, is not written explicitly in expressions) and the empty string ε and every alphabet symbol $a \in \Sigma$ are added as constants. We use the term *string straight-line programs* (SSLPs for short) for these SLPs. If we want to emphasize the alphabet Σ, we speak of an SSLP over Σ. In many papers, SSLPs are just called straight-line programs; see [14] for a survey. Occasionally we consider SSLPs without a start variable S and then write (V, ρ).

Example 1. Consider the SSLP $G = (\{S, A, B, C\}, S, \rho)$ over the alphabet $\{a, b\}$ with $\rho(S) = AAB$, $\rho(A) = CBB$, $\rho(B) = CaC$, $\rho(C) = b$. We have $[\![B]\!]_G = bab$, $[\![A]\!]_G = bbabbab$, and $[\![G]\!] = bbabbabbbabbbabbab$. The size of G is 8 (six concatenation operators are used in the right-hand sides, and there are two occurrences of constants).

In the next two sections, we introduce two types of algebras for trees and forests.

3   Forest Algebras and Forest Straight-Line Programs

Trees and Forests. Let us fix a finite set Σ of node labels for the rest of the paper. We consider Σ-labelled rooted ordered trees, where "ordered" means that the children of a node are totally ordered. Every node has a label from Σ. Note that we make no rank assumption: the number of children of a node (also called its degree) is not determined by its node label. The set of nodes (resp. edges) of t is denoted by $V(t)$ (resp., $E(t)$). A *forest* is a (possibly empty) sequence of trees. The size $|f|$ of a forest is the total number of nodes in f. The set of all Σ-labelled forests is denoted by $\mathcal{F}_0(\Sigma)$ and the set of all Σ-labelled trees is denoted by $\mathcal{T}_0(\Sigma)$. As usual, we can identify trees with expressions built up from symbols in Σ and parentheses. Formally, $\mathcal{F}_0(\Sigma)$ and $\mathcal{T}_0(\Sigma)$ can be inductively defined as the following sets of strings over the alphabet $\Sigma \cup \{(,)\}$.

- If $t_1, \ldots, t_n$ are Σ-labelled trees with $n \geq 0$, then the string $t_1 t_2 \cdots t_n$ is a Σ-labelled forest (in particular, the empty string ε is a Σ-labelled forest).

- If f is a Σ-labelled forest and $a \in \Sigma$, then $a(f)$ is a Σ-labelled tree (where the singleton tree $a()$ is usually written as a).

Let us fix a distinguished symbol $x \notin \Sigma$ for the rest of the paper (called the parameter). The set of forests $f \in \mathcal{F}_0(\Sigma \cup \{x\})$ such that x has a unique occurrence in f and this occurrence is at a leaf node is denoted by $\mathcal{F}_1(\Sigma)$. Let $\mathcal{T}_1(\Sigma) = \mathcal{F}_1(\Sigma) \cap \mathcal{T}_0(\Sigma \cup \{x\})$. Elements of $\mathcal{T}_1(\Sigma)$ (resp., $\mathcal{F}_1(\Sigma)$) are called tree contexts (resp., forest contexts). We finally define $\mathcal{F}(\Sigma) = \mathcal{F}_0(\Sigma) \cup \mathcal{F}_1(\Sigma)$ and $\mathcal{T}(\Sigma) = \mathcal{T}_0(\Sigma) \cup \mathcal{T}_1(\Sigma)$. Following [4], we define the *forest algebra* $\mathsf{FA}(\Sigma) = (\mathcal{F}(\Sigma), \boxdot, \boxminus, (a)_{a \in \Sigma}, \varepsilon, x)$ as follows:

- $\boxdot$ is the horizontal concatenation operator: for forests $f_1, f_2 \in \mathcal{F}(\Sigma)$, $f_1 \boxdot f_2$ is defined if $f_1 \in \mathcal{F}_0(\Sigma)$ or $f_2 \in \mathcal{F}_0(\Sigma)$ and in this case we set $f_1 \boxdot f_2 = f_1 f_2$ (i.e., we concatenate the corresponding sequences of trees).
- $\boxminus$ is the vertical concatenation operator: for forests $f_1, f_2 \in \mathcal{F}(\Sigma)$, $f_1 \boxminus f_2$ is defined if $f_1 \in \mathcal{F}_1(\Sigma)$ and in this case $f_1 \boxminus f_2$ is obtained by replacing in f_1 the unique occurrence of the parameter x by the forest f_2.
- Every $a \in \Sigma$ is identified with the unary function $a : \mathcal{F}(\Sigma) \to \mathcal{T}(\Sigma)$ that produces $a(f)$ when applied to $f \in \mathcal{F}(\Sigma)$.
- $\varepsilon \in \mathcal{F}_0(\Sigma)$ and $x \in \mathcal{F}_1(\Sigma)$ are constants of the forest algebra.

For better readability, we also write $f\langle g \rangle$ instead of $f \boxminus g$, fg instead of $f \boxdot g$, and a instead of $a(\varepsilon)$. Note that a forest $f \in \mathcal{F}(\Sigma)$ can be also viewed as an algebraic expression over $\mathsf{FA}(\Sigma)$, which evaluates to f itself (analogously to the free term algebra).

First-Child/Next-Sibling Encoding. The first-child/next-sibling encoding transforms a forest over some alphabet Σ into a binary tree over $\Sigma \uplus \{\bot\}$. We define fcns: $\mathcal{F}_0(\Sigma) \to \mathcal{T}_0(\Sigma \uplus \{\bot\})$ inductively by: (i) fcns$(\varepsilon) = \bot$ and (ii) fcns$(a(f)g) = a(\text{fcns}(f)\text{fcns}(g))$ for $f, g \in \mathcal{F}_0(\Sigma)$, $a \in \Sigma$. Thus, the left (resp., right) child of a node in fcns(f) is the first child (resp., right sibling) of the node in f or a $\bot$-labelled leaf if it does not exist.

Example 2. If $f = a(bc)d(e)$ then

$$\text{fcns}(f) = \text{fcns}(a(bc)d(e)) = a(\text{fcns}(bc)\text{fcns}(d(e)))$$
$$= a(b(\bot \text{fcns}(c))d(\text{fcns}(e)\bot)) = a(b(\bot c(\bot\bot))d(e(\bot\bot)\bot)).$$

Forest Straight-Line Programs. A *forest straight-line program* over Σ, FSLP for short, is a valid straight-line program over the algebra $\mathsf{FA}(\Sigma)$ such that $[\![F]\!] \in \mathcal{F}_0(\Sigma)$. Iterated vertical and horizontal concatenations allow to generate forests, whose depth and width is exponential in the FSLP size. For an FSLP $F = (V, S, \rho)$ and $i \in \{0, 1\}$ we define $V_i = \{A \in V \mid [\![A]\!]_F \in \mathcal{F}_i(\Sigma)\}$.

Example 3. Consider the FSLP $F = (\{S, A_0, A_1, \ldots, A_n, B_0, B_1, \ldots, B_n\}, S, \rho)$ over $\{a, b, c\}$ with ρ defined by $\rho(A_0) = a$, $\rho(A_i) = A_{i-1}A_{i-1}$ for $1 \le i \le n$, $\rho(B_0) = b(A_n x A_n)$, $\rho(B_i) = B_{i-1}\langle B_{i-1} \rangle$ for $1 \le i \le n$, and $\rho(S) = B_n\langle c \rangle$. We have $[\![F]\!] = b(a^{2^n} b(a^{2^n} \cdots b(a^{2^n} c a^{2^n}) \cdots a^{2^n})a^{2^n})$, where b occurs 2^n many times. A more involved example can be found in the arXiv version of this paper [11].

FSLPs generalize *tree straight-line programs* (TSLPs for short) that have been used for the compression of ranked trees before, see e.g. [6,15]. We only need TSLPs for binary trees. A TSLP over Σ can then be defined as an FSLP $T = (V, S, \rho)$ such that for every $A \in V$, $\rho(A)$ has the form a, $a(BC)$, $a(xB)$, $a(Bx)$, or $B\langle C\rangle$ with $a \in \Sigma$, $B, C \in V$. TSLPs can be used in order to compress the fcns-encoding of an unranked tree; see also [15]. It is not hard to see that an FSLP F that produces a binary tree can be transformed into a TSLP T such that $[\![F]\!] = [\![T]\!]$ and $|T| \in O(|F|)$. This is an easy corollary of our normal form for FSLPs that we introduce next (see also the proof of Proposition 9).

Normal Form FSLPs. In this paragraph, we introduce a normal form for FSLPs that turns out to be crucial in the rest of the paper. An FSLP $F = (V, S, \rho)$ is in *normal form* if $V_0 = V_0^\top \uplus V_0^\perp$ and all right-hand sides have one of the following forms:

- $\rho(A) = \varepsilon$, where $A \in V_0^\top$,
- $\rho(A) = BC$, where $A \in V_0^\top, B, C \in V_0$,
- $\rho(A) = B\langle C\rangle$, where $B \in V_1$ and either $A, C \in V_0^\perp$ or $A, C \in V_1$,
- $\rho(A) = a(B)$, where $A \in V_0^\perp$, $a \in \Sigma$ and $B \in V_0$,
- $\rho(A) = a(BxC)$, where $A \in V_1$, $a \in \Sigma$ and $B, C \in V_0$.

Note that the partition $V_0 = V_0^\top \uplus V_0^\perp$ is uniquely determined by ρ. Also note that variables from V_1 produce tree contexts and variables from $V_0^\perp$ produce trees, whereas variables from $V_0^\top$ produce forests with arbitrarily many trees.

Let $F = (V, S, \rho)$ be a normal form FSLP. Every variable $A \in V_1$ produces a vertical concatenation of (possibly exponentially many) variables, whose right-hand sides have the form $a(BxC)$. This vertical concatenation is called the spine of A. Formally, we split V_1 into $V_1^\top = \{A \in V_1 \mid \exists B, C \in V_1 : \rho(A) = B\langle C\rangle\}$ and $V_1^\perp = V_1 \setminus V_1^\top$. We then define the *vertical SSLP* $F^\boxdot = (V_1^\top, \rho_1)$ over $V_1^\perp$ with $\rho_1(A) = BC$ whenever $\rho(A) = B\langle C\rangle$. For every $A \in V_1$ the string $[\![A]\!]_{F^\boxdot} \in (V_1^\perp)^*$ is called the *spine* of A (in F), denoted by $\text{spine}_F(A)$ or just $\text{spine}(A)$ if F is clear from the context. We also define the *horizontal SSLP* $F^\square = (V_0^\top, \rho_0)$ over $V_0^\perp$, where ρ_0 is the restriction of ρ to $V_0^\top$. For every $A \in V_0$ we use $\text{hor}(A)$ to denote the string $[\![A]\!]_{F^\square} \in (V_0^\perp)^*$. Note that $\text{spine}(A) = A$ (resp., $\text{hor}(A) = A$) for every $A \in V_1^\perp$ (resp., $A \in V_0^\perp$).

The intuition behind the normal form can be explained as follows: Consider a tree context $t \in \mathcal{T}_1(\Sigma) \setminus \{x\}$. By decomposing t along the nodes on the unique path from the root to the x-labelled leaf, we can write t as a vertical concatenation of tree contexts $a_1(f_1 x g_1), \ldots, a_n(f_n x g_n)$ for forests $f_1, g_1, \ldots, f_n, g_n$ and symbols $a_1, \ldots, a_n$. In a normal form FSLP one would produce t by first deriving a vertical concatenation $A_1\langle\cdots\langle A_n\rangle\cdots\rangle$. Every A_i is then derived to $a_i(B_i x C_i)$, where B_i (resp., C_i) produces the forest f_i (resp., g_i). Computing an FSLP for this decomposition for a tree context that is already given by an FSLP is the main step in the proof of the normal form theorem below. Another insight is that proper forest contexts from $\mathcal{F}_1(\Sigma) \setminus \mathcal{T}_1(\Sigma)$ can be eliminated without significant size blow-up.

Theorem 4. *From a given FSLP F one can construct in linear time an FSLP F' in normal form such that $[\![F']\!] = [\![F]\!]$ and $|F'| \in O(|F|)$.*

4 Cluster Algebras and Top Dags

In this section we introduce top dags [2,12] as an alternative grammar-based formalism for the compression of unranked trees. A *cluster of rank* 0 is a tree $t \in T_0(\Sigma)$ of size at least two. A *cluster of rank* 1 is a tree $t \in T_0(\Sigma)$ of size at least two together with a distinguished leaf node that we call the *bottom boundary node* of t. In both cases, the root of t is called the *top boundary node* of t. Note that in contrast to forest contexts there is no parameter x. Instead, one of the Σ-labelled leaf nodes may be declared as the bottom boundary node. When writing a cluster of rank 1 in term representation, we underline the bottom boundary node. For instance $a(b\,c(\underline{a}\,b))$ is a cluster of rank 1. An *atomic cluster* is of the form $a(b)$ or $a(\underline{b})$ for $a, b \in \Sigma$. Let $C_i(\Sigma)$ be the set of all clusters of rank $i \in \{0, 1\}$ and let $C(\Sigma) = C_0(\Sigma) \cup C_1(\Sigma)$. We write $\text{rank}(s) = i$ if $s \in C_i(\Sigma)$ for $i \in \{0, 1\}$. We define the *cluster algebra* $\mathsf{CA}(\Sigma) = (C(\Sigma), \odot, \oplus, (a(b), a(\underline{b}))_{a,b\in\Sigma})$ as follows:

- $\odot$ is the horizontal merge operator: $s \odot t$ is only defined if $\text{rank}(s) + \text{rank}(t) \leq 1$ and s, t are of the form $s = a(f), t = a(g)$, i.e., the root labels coincide. Then $s \odot t = a(fg)$. Note that at most one symbol in the forest fg is underlined. The rank of $s \odot t$ is $\text{rank}(s) + \text{rank}(t)$. For instance, $a(b\,c(\underline{a}\,b)) \odot a(b\,c) = a(b\,c(\underline{a}\,b)b\,c)$.
- $\oplus$ is the vertical merge operator: $s \oplus t$ is only defined if $s \in C_1(\Sigma)$ and the label of the root of t (say a) is equal to the label of the bottom boundary node of s. We then obtain $s \oplus t$ by replacing the unique occurrence of $\underline{a}$ in s by t. The rank of $s \oplus t$ is $\text{rank}(t)$. For instance, $a(b\,c(\underline{a}\,b)) \oplus a(b\underline{c}) = a(b\,c(a(b\underline{c})\,b))$.
- The atomic clusters $a(b)$ and $a(\underline{b})$ are constants of the cluster algebra.

A *top tree* for a tree $t \in T_0$ is an algebraic expression e over the algebra $\mathsf{CA}(\Sigma)$ such that $[\![e]\!] = t$. A *top dag* over Σ is a straight-line program D over the algebra $\mathsf{CA}(\Sigma)$ such that $[\![D]\!] \in T_0(\Sigma)$. In our terminology, cluster straight-line program would be a more appropriate name, but we prefer to call them top dags.

Example 5. Consider the top dag $D = (\{S, A_0, \dots, A_n, B_0, \dots, B_n\}, S, \rho)$, where $\rho(A_0) = b(a)$, $\rho(A_i) = A_{i-1} \odot A_{i-1}$ for $1 \leq i \leq n$, $\rho(B_0) = A_n \odot b(\underline{b}) \odot A_n$, $\rho(B_i) = B_{i-1} \oplus B_{i-1}$ for $1 \leq i \leq n$, and $\rho(S) = B_n \oplus b(c)$. We have $[\![D]\!] = b(a^{2^n} b(a^{2^n} \cdots b(a^{2^n} b(c)\, a^{2^n}) \cdots a^{2^n})a^{2^n})$, where b occurs $2^n + 1$ many times.

5 Relative Succinctness

We have now three grammar-based formalisms for the compression of unranked trees: FSLPs, top dags, and TSLPs for fcns-encodings. In this section we study their relative succinctness. It turns out that up to multiplicative factors of size $|\Sigma|$ (number of node labels) all three formalisms are equally succinct. Moreover, the transformations between the formalisms can be computed very efficiently. This allows us to transfer algorithmic results for FSLPs to top dags and TSLPs for fcns encodings, and vice versa. We start with top dags:

Proposition 6. *For a given top dag D one can compute in linear time an FSLP F such that $[\![F]\!] = [\![D]\!]$ and $|F| \in O(|D|)$.*

Proposition 7. *For a given FSLP F with $[\![F]\!] \in \mathcal{T}_0(\Sigma)$ and $|[\![F]\!]| \geq 2$ one can compute in time $O(|\Sigma| \cdot |F|)$ a top dag D such that $[\![D]\!] = [\![F]\!]$ and $|D| \in O(|\Sigma| \cdot |F|)$.*

The following example shows that the size bound in Proposition 7 is sharp:

Example 8. Let $\Sigma = \{a, a_1, \ldots, a_\sigma\}$ and for $n \geq 1$ let $t_n = a(a_1(a^m) \cdots a_\sigma(a^m))$ with $m = 2^n$. For every $n > \sigma$ the tree t_n can be produced by an FSLP of size $O(n)$: using $n = \log m$ many variables we can produce the forest a^m and then $O(n)$ many additional variables suffice to produce t_n. On the other hand, every top dag for t_n has size $\Omega(\sigma \cdot n)$: consider a top tree e that evaluates to t_n. Then e must contain a subexpression e_i that evaluates to the subtree $a_i(a^m)$ $(1 \leq i \leq \sigma)$ of t_n. The subexpression e_i has to produce $a_i(a^m)$ using the $\odot$-operation from copies of $a_i(a)$. Hence, the expression for $a_i(a^m)$ has size $n = \log_2 m$ and different e_i contain no identical subexpressions. Therefore every top dag for t_n has size at least $\sigma \cdot n$.

In contrast, FSLPs and TSLPs for fcns-encodings turn out to be equally succinct up to constant factors:

Proposition 9. *Let $f \in \mathcal{F}(\Sigma)$ be a forest and let F be an FSLP (or TSLP) over $\Sigma \uplus \{\bot\}$ with $[\![F]\!] = fcns(f)$. Then we can transform F in linear time into an FSLP F' over Σ with $[\![F']\!] = f$ and $|F'| \in O(|F|)$.*

Proposition 10. *For every FSLP F over Σ, we can construct in linear time a TSLP T over $\Sigma \cup \{\bot\}$ with $[\![T]\!] = fcns([\![F]\!])$ and $|T| \in O(|F|)$.*

Proposition 10 and the construction from [7, Proposition 8.3.2] allow to reduce the evaluation of forest automata on FSLPs (for a definition of forest and tree automata, see [7]) to the evaluation of ordinary tree automata on binary trees. The latter problem can be solved in polynomial time [18], which yields:

Corollary 11. *Given a forest automaton A and an FSLP (or top dag) F we can check in polynomial time whether A accepts $[\![F]\!]$.*

In [2], a linear time algorithm is presented that constructs from a tree of size n with σ many node labels a top dag of size $O(n/\log_\sigma^{0.19} n)$. In [12] this bound was improved to $O(n \log \log n/\log_\sigma n)$ (for the same algorithm as in [2]). In [19] we recently presented an alternative construction that achieves the information-theoretic optimum of $O(n/\log_\sigma n)$ (another optimal construction was presented in [9]). Moreover, as in [2], the constructed top dag satisfies the additional size bound $O(d \cdot \log n)$, where d is the size of the minimal dag of t. With Propositions 6 and 10 we get:

Corollary 12. *Given a tree t of size n with σ many node labels, one can construct in linear time an FSLP for t (or an TSLP for fcns(t)) of size $O(n/\log_\sigma n) \cap O(d \cdot \log n)$, where d is the size of the minimal dag of t.*

6   Testing Equality Modulo Associativity and Commutativity

In this section we will give an algorithmic application which proves the utility of FSLPs (even if we deal with binary trees). We fix two subsets $\mathcal{A} \subseteq \Sigma$ (the set of *associative symbols*) and $\mathcal{C} \subseteq \Sigma$ (the set of *commutative symbols*). This means that we impose the following identities for all $a \in \mathcal{A}$, $c \in \mathcal{C}$, all trees $t_1, \ldots, t_n \in \mathcal{T}_0(\Sigma)$, all permutations $\sigma \colon \{1, \ldots, n\} \to \{1, \ldots, n\}$, and all $1 \leq i \leq j \leq n + 1$:

$$a(t_1 \cdots t_n) = a(t_1 \cdots t_{i-1} a(t_i \cdots t_{j-1}) t_j \cdots t_n) \qquad (1)$$
$$c(t_1 \cdots t_n) = c(t_{\sigma(1)} \cdots t_{\sigma(n)}). \qquad (2)$$

Note that the standard law of associativity for a binary symbol $\circ$ (i.e., $x \circ (y \circ z) = (x \circ y) \circ z$) can be captured by making $\circ$ an (unranked) associative symbol in the sense of (1). Our main result is:

Theorem 13. *For trees s, t we can test in polynomial time whether s and t are equal modulo the identities in (1) and (2), if s and t are given succinctly by one of the following three formalisms: (i) FSLPs, (ii) top dags, (iii) TSLPs for the fcns-encodings of s, t.*

6.1   Associative Symbols

Below, we define the associative normal form $\mathrm{nf}_\mathcal{A}(f)$ of a forest f and show that from an FSLP F we can compute in linear time an FSLP F' with $[\![F']\!] = \mathrm{nf}_\mathcal{A}([\![F]\!])$. For trees $s, t \in \mathcal{T}_0(\Sigma)$ we have that $s = t$ modulo the identities in (1) if and only if $\mathrm{nf}_\mathcal{A}(s) = \mathrm{nf}_\mathcal{A}(t)$. The generalization to forests is needed for the induction, where a slight technical problem arises. Whether the forests $t_1 \cdots t_{i-1} a(t_i \cdots t_{j-1}) t_j \cdots t_n$ and $t_1 \cdots t_n$ are equal modulo the identities in (1) actually depends on the symbol on top of these two forests. If it is an a, and $a \in \mathcal{A}$, then the two forests are equal modulo associativity, otherwise not. To cope with this problem, we use for every associative symbol $a \in \mathcal{A}$ a function $\phi_a \colon \mathcal{F}_0(\Sigma) \to \mathcal{F}_0(\Sigma)$ that pulls up occurrences of a whenever possible.

Let $\bullet \notin \Sigma$ be a new symbol. For every $a \in \Sigma \cup \{\bullet\}$ let $\phi_a \colon \mathcal{F}_0(\Sigma) \to \mathcal{F}_0(\Sigma)$ be defined as follows, where $f \in \mathcal{F}_0(\Sigma)$ and $t_1, \ldots, t_n \in \mathcal{T}_0(\Sigma)$:

$$\phi_a(b(f)) = \begin{cases} \phi_a(f) & \text{if } a \in \mathcal{A} \text{ and } a = b, \\ b(\phi_b(f)) & \text{otherwise,} \end{cases} \qquad \phi_a(t_1 \cdots t_n) = \phi_a(t_1) \cdots \phi_a(t_n).$$

In particular, $\phi_a(\varepsilon) = \varepsilon$. Moreover, define $\mathrm{nf}_\mathcal{A} \colon \mathcal{F}_0(\Sigma) \to \mathcal{F}_0(\Sigma)$ by $\mathrm{nf}_\mathcal{A}(f) = \phi_\bullet(f)$.

Example 14. Let $t = a(a(cd)b(cd)a(e))$ and $\mathcal{A} = \{a\}$. We obtain

$$\phi_a(t) = \phi_a(a(cd)b(cd)a(e)) = \phi_a(a(cd))\phi_a(b(cd))\phi_a(a(e))$$
$$= \phi_a(cd)b(\phi_b(cd))\phi_a(e) = cdb(cd)e,$$
$$\phi_b(t) = a(\phi_a(a(cd)b(cd)a(e))) = a(cdb(cd)e).$$

To show the following simple lemma one considers the terminating and confluent rewriting system obtained by directing the Eq. (1) from right to left.

Lemma 15. *For two forests $f_1, f_2 \in \mathcal{F}_0(\Sigma)$, $nf_\mathcal{A}(f_1) = nf_\mathcal{A}(f_2)$ if and only if f_1 and f_2 are equal modulo the identities in (1) for all $a \in \mathcal{A}$.*

Lemma 16. *From a given FSLP $F = (V, S, \rho)$ over Σ one can construct in time $\mathcal{O}(|F| \cdot |\Sigma|)$ an FSLP F' with $\llbracket F' \rrbracket = nf_\mathcal{A}(\llbracket F \rrbracket)$.*

For the proof of Lemma 16 one introduces new variables A_a for all $a \in \Sigma \cup \{\bullet\}$ and defines the right-hand sides of F' such that $\llbracket A_a \rrbracket_{F'} = \phi_a(\llbracket A \rrbracket_F)$ for all $A \in V_0$ and $\llbracket B_a \langle \phi_b(f) \rangle \rrbracket_{F'} = \phi_a(\llbracket B \langle f \rangle \rrbracket_F)$ for all $B \in V_1$, $f \in \mathcal{F}_0(\Sigma)$, where b is the label of the parent node of the parameter x in $\llbracket B \rrbracket_F$. This parent node exists if we assume the FSLP F to be in normal form.

6.2 Commutative Symbols

To test whether two trees over Σ are equivalent with respect to commutativity, we define a *commutative normal form* $nf_\mathcal{C}(t)$ of a tree $t \in \mathcal{T}_0(\Sigma)$ such that $nf_\mathcal{C}(t_1) = nf_\mathcal{C}(t_2)$ if and only if t_1 and t_2 are equivalent with respect to the identities in (2) for all $c \in \mathcal{C}$.

We start with a general definition: Let Δ be a possibly infinite alphabet together with a total order $<$. Let $\leq$ be the reflexive closure of $<$. Define the function $\mathrm{sort}^<: \Delta^* \to \Delta^*$ by $\mathrm{sort}^<(a_1 \cdots a_n) = a_{i_1} \cdots a_{i_n}$ with $\{i_1, \ldots, i_n\} = \{1, \ldots, n\}$ and $a_{i_1} \leq \cdots \leq a_{i_n}$.

Lemma 17. *Let G be an SSLP over Δ and let $<$ be some total order on Δ. We can construct in time $\mathcal{O}(|\Delta| \cdot |G|)$ an SSLP G' such that $\llbracket G' \rrbracket = \mathrm{sort}^<(\llbracket G \rrbracket)$.*

In order to define the commutative normal form, we need a total order on $\mathcal{F}_0(\Sigma)$. Recall that elements of $\mathcal{F}_0(\Sigma)$ are particular strings over the alphabet $\Gamma := \Sigma \cup \{(,)\}$. Fix an arbitrary total order on Γ and let $<_{\mathrm{llex}}$ be the *length-lexicographic order* on Γ^* induced by $<$: for $x, y \in \Gamma^*$ we have $x <_{\mathrm{llex}} y$ if $|x| < |y|$ or ($|x| = |y|$, $x = uav$, $y = ubv'$, and $a < b$ for $u, v, v' \in \Gamma^*$ and $a, b \in \Gamma$). We now consider the restriction of $<_{\mathrm{llex}}$ to $\mathcal{F}_0(\Sigma) \subseteq \Gamma^*$. For the proof of the following lemma one first constructs SSLPs for the strings $\llbracket F_1 \rrbracket, \llbracket F_2 \rrbracket \in \Gamma^*$ (the construction is similar to the case of TSLPs, see [6]) and then uses [16, Lemma 3] according to which SSLP-encoded strings can be compared in polynomial time with respect to $<_{\mathrm{llex}}$.

Lemma 18. *For two FSLPs F_1 and F_2 we can check in polynomial time whether $\llbracket F_1 \rrbracket = \llbracket F_2 \rrbracket$, $\llbracket F_1 \rrbracket <_{llex} \llbracket F_2 \rrbracket$ or $\llbracket F_2 \rrbracket <_{llex} \llbracket F_1 \rrbracket$.*

From the restriction of $<_{\text{llex}}$ to $T_0(\Sigma) \subseteq \Gamma^*$ we obtain the function $\text{sort}^{<_{\text{llex}}}$ on $T_0(\Sigma)^* = \mathcal{F}_0(\Sigma)$. We define $\text{nf}_{\mathcal{C}} \colon \mathcal{F}_0(\Sigma) \to \mathcal{F}_0(\Sigma)$ by

$$\text{nf}_{\mathcal{C}}(a(f)) = \begin{cases} a(\text{sort}^{<_{\text{llex}}}(\text{nf}_{\mathcal{C}}(f))) & \text{if } a \in \mathcal{C} \\ a(\text{nf}_{\mathcal{C}}(f)) & \text{otherwise,} \end{cases}$$

$$\text{nf}_{\mathcal{C}}(t_1 \cdots t_n) = \text{nf}_{\mathcal{C}}(t_1) \cdots \text{nf}_{\mathcal{C}}(t_n).$$

Obviously, $f_1, f_2 \in \mathcal{F}(\Sigma)$ are equal modulo the identities in (2) for all $c \in \mathcal{C}$ if and only if $\text{nf}_{\mathcal{C}}(f_1) = \text{nf}_{\mathcal{C}}(f_2)$. Using this fact and Lemma 15 it is not hard to show:

Lemma 19. *For $f_1, f_2 \in \mathcal{F}_0(\Sigma)$ we have $\text{nf}_{\mathcal{C}}(\text{nf}_{\mathcal{A}}(f_1)) = \text{nf}_{\mathcal{C}}(\text{nf}_{\mathcal{A}}(f_2))$ if and only if f_1 and f_2 are equal modulo the identities in (1) and (2) for all $a \in \mathcal{A}$, $c \in \mathcal{C}$.*

For our main technical result (Theorem 21) we need a strengthening of our FSLP normal form. Recall the notion of the *spine* from Sect. 3. We say that an FSLP $F = (V, S, \rho)$ is in *strong normal form* if it is in normal form and for every $A \in V_0^\perp$ with $\rho(A) = B\langle C \rangle$ either $B \in V_1^\perp$ or $|\llbracket C \rrbracket_F| \geq |\llbracket D \rrbracket_F| - 1$ for every $D \in V_1^\perp$ which occurs in $\text{spine}(B)$ (note that $|\llbracket D \rrbracket_F| - 1$ is the number of nodes in $\llbracket D \rrbracket_F$ except for the parameter x).

Lemma 20. *From a given FSLP $F = (V, S, \rho)$ in normal form we can construct in polynomial time an FSLP $F' = (V', S, \rho')$ in strong normal form with $\llbracket F \rrbracket = \llbracket F' \rrbracket$.*

For the proof of Lemma 20 we modify the right-hand sides of variables $A \in V_0^\perp$ with $\rho(A) = B\langle C \rangle$ and $|\text{spine}(B)| \geq 2$. Basically, we replace the vertical concatenations $B\langle C \rangle$ by polynomially many vertical concatenations $B_i\langle C_i \rangle$ which satisfy the condition of the strong normal form. We can now prove the main technical result of this section:

Theorem 21. *From a given FSLP F we can construct in polynomial time an FSLP F' with $\llbracket F' \rrbracket = \text{nf}_{\mathcal{C}}(\llbracket F \rrbracket)$.*

Proof. Let $F = (V, S, \rho)$. By Theorem 4 and Lemma 20 we may assume that F is in strong normal form. For every $A \in V_1$ let

$$\text{args}(A) = \{ t \in T_0(\Sigma) \mid |t| \geq |\llbracket D \rrbracket_F| - 1 \text{for each symbol } D \text{ in } \text{spine}(A) \}$$

We want to construct an FSLP $F' = (V', S, \rho')$ with $V_0 \subseteq V_0'$ and $V_1 = V_1'$ such that

(1) $\llbracket A \rrbracket_{F'} = \text{nf}_{\mathcal{C}}(\llbracket A \rrbracket_F)$ for all $A \in V_0$,
(2) $\llbracket A \rrbracket_{F'}\langle \text{nf}_{\mathcal{C}}(t) \rangle = \text{nf}_{\mathcal{C}}(\llbracket A \rrbracket_F\langle t \rangle)$ for all $A \in V_1$, $t \in \text{args}(A)$.

From (1) we obtain $\llbracket F' \rrbracket = \llbracket S \rrbracket_{F'} = \text{nf}_{\mathcal{C}}(\llbracket S \rrbracket_F) = \text{nf}_{\mathcal{C}}(\llbracket F \rrbracket)$ which concludes the proof.

To define ρ', let $V^c = V_0^c \cup V_1^c$ with $V_1^c = \{A \in V_1 \mid \rho(A) = a(BxC) \text{ with } a \in \mathcal{C}\}$ and $V_0^c = \{A \in V_0 \mid \rho(A) = a(B) \text{ with } a \in \mathcal{C} \text{ or } \rho(A) = D\langle C \rangle \text{ with } D \in V_1^c\}$ be the set of *commutative variables*. We set $\rho'(A) = \rho(A)$ for $A \in V \setminus V^c$. For $A \in V^c$ we define $\rho'(A)$ by induction along the partial order of the dag:

1. $\rho(A) = a(B)$: Let M_A be the set of all $C \in V_0^\perp$ which are below A in the dag, and let $w = \mathrm{hor}(B) = [\![B]\!]_{F\square} \in M_A^*$. By induction, ρ' is already defined on M_A, and thus $[\![C]\!]_{F'}$ is defined for every $C \in M_A$. By Lemma 18, we can compute in polynomial time a total order $<$ on M_A such that $C < D$ implies $[\![C]\!]_{F'} \leq_{\mathrm{llex}} [\![D]\!]_{F'}$ for all $C, D \in M_A$. By Lemma 17, we can construct in linear time an SSLP $G_w = (V_w, S_w, \rho_w)$ with $[\![G_w]\!] = \mathrm{sort}^<(w)$, and we may assume that all variables $D \in V_w$ are new. We add these variables to V_0' together with their right hand sides $\rho'(D) = \rho_w(D)$, and we finally set $\rho'(A) = a(S_w)$.

2. $\rho(A) = B\langle C \rangle$: Let $\rho(B) = a(DxE)$. We define $G_w = (V_w, S_w, \rho_w)$ as before, but with $w = [\![DCE]\!]_{F\square}$ instead of $w = [\![B]\!]_{F\square}$, and we set $\rho'(A) = a(S_w)$.

3. $\rho(A) = a(BxC)$: We define $G_w = (V_w, S_w, \rho_w)$ as before, this time with $w = [\![BC]\!]_{F\square}$, and we set $\rho'(B) = a(S_w x)$.

The main idea is that the strong normal form ensures that in right-hand sides of the form $a(DxE)$ with $a \in \mathcal{C}$ one can move the parameter x to the last position (see point 3 above), since only trees that are larger than all trees produced from D and E are substituted for x. $\qquad \square$

Proof of Theorem 13. By Propositions 6 and 9 it suffices to show Theorem 13 for the case that t_1 and t_2 are given by FSLPs F_1 and F_2, respectively. By Lemma 19 and Lemma 18 it suffices to compute in polynomial time FSLPs F_1' and F_2' for $\mathrm{nf}_\mathcal{C}(\mathrm{nf}_\mathcal{A}(t_1))$ and $\mathrm{nf}_\mathcal{C}(\mathrm{nf}_\mathcal{A}(t_2))$. This can be achieved using Lemma 16 and Theorem 21. $\qquad \square$

7 Future Work

We have shown that simple algebraic manipulations (laws of associativity and commutativity) can be carried out efficiently on grammar-compressed trees. In the future, we plan to investigate other algebraic laws. We are optimistic that our approach can be extended by idempotent symbols (meaning that $a(fttg) = a(ftg)$ for forests f, g and a tree t).

Another interesting open problem concerns context unification modulo associative and commutative symbols. The decidability of (plain) context-unification was a long standing open problem that was finally solved by Jeż [13], who showed the existence of a polynomial space algorithm. Jeż's algorithm uses his recompression technique for TSLPs. One might try to extend this technique to FSLPs with the goal of proving decidability of context unification for terms that also contain associative and commutative symbols. For first-order unification and matching [10], context matching [10], and one-context unification [8] there exist algorithms for TSLP-compressed trees that match the complexity of their uncompressed counterparts. One might also try to extend these results to the associative and commutative setting.

Acknowledgements. The first author was supported by the EPSRC grant EP/N510129/1 at the Alan Turing Institute and the EPSRC grant EP/J017728/2 at University of Edinburgh. The second author was supported by the DFG research project LO748/10-1.

References

1. Abiteboul, S., Bourhis, P., Vianu, V.: Highly expressive query languages for unordered data trees. Theor. Comput. Syst. **57**(4), 927–966 (2015)
2. Bille, P., Gørtz, I.L., Landau, G.M., Weimann, O.: Tree compression with top trees. Inf. Comput. **243**, 166–177 (2015)
3. Boiret, A., Hugot, V., Niehren, J., Treinen, R.: Logics for unordered trees with data constraints on siblings. In: Dediu, A.-H., Formenti, E., Martín-Vide, C., Truthe, B. (eds.) LATA 2015. LNCS, vol. 8977, pp. 175–187. Springer, Cham (2015). https://doi.org/10.1007/978-3-319-15579-1_13
4. Bojańczyk, M., Walukiewicz, I.: Forest algebras. In: Proceedings of Logic and Automata: History and Perspectives [in Honor of Wolfgang Thomas], Texts in Logic and Games, vol. 2, pp. 107–132. Amsterdam University Press (2008)
5. Boneva, I., Ciucanu, R., Staworko, S.: Schemas for unordered XML on a DIME. Theory Comput. Syst. **57**(2), 337–376 (2015)
6. Busatto, G., Lohrey, M., Maneth, S.: Efficient memory representation of XML document trees. Inf. Syst. **33**(4–5), 456–474 (2008)
7. Comon, H., Dauchet, M., Gilleron, R., Jacquemard, F., Löding, C., Lugiez, D., Tison, S., Tommasi, M.: Tree automata techniques and applications. http://www.grappa.univ-lille3.fr/tata (2007)
8. Creus, C., Gascón, A., Godoy, G.: One-context unification with STG-compressed terms is in NP. In: Proceedings of RTA 2012, LIPIcs 15, pp. 149–164. Schloss Dagstuhl - Leibniz-Zentrum für Informatik (2012)
9. Dudek, B., Gawrychowski, P.: Slowing down top trees for better worst-case bounds (2018). https://arxiv.org/abs/1801.01059
10. Gascón, A., Godoy, G., Schmidt-Schauß, M.: Unification and matching on compressed terms. ACM Trans. Comput. Logic **12**(4), 26:1–26:37 (2011)
11. Gascón, A., Lohrey, M., Maneth, S., Reh, P., Sieber, K.: Grammar-based compression of unranked trees (2018). https://arxiv.org/abs/1802.05490
12. Hübschle-Schneider, L., Raman, R.: Tree compression with top trees revisited. In: Bampis, E. (ed.) SEA 2015. LNCS, vol. 9125, pp. 15–27. Springer, Cham (2015). https://doi.org/10.1007/978-3-319-20086-6_2
13. Jeż, A.: Context unification is in PSPACE. In: Esparza, J., Fraigniaud, P., Husfeldt, T., Koutsoupias, E. (eds.) ICALP 2014. LNCS, vol. 8573, pp. 244–255. Springer, Heidelberg (2014). https://doi.org/10.1007/978-3-662-43951-7_21
14. Lohrey, M.: Algorithmics on SLP-compressed strings: a survey. Groups Complex. Cryptol. **4**(2), 241–299 (2012)
15. Lohrey, M., Maneth, S., Mennicke, R.: XML tree structure compression using RePair. Inf. Syst. **38**(8), 1150–1167 (2013)
16. Lohrey, M., Maneth, S., Peternek, F.: Compressed tree canonization. In: Halldórsson, M.M., Iwama, K., Kobayashi, N., Speckmann, B. (eds.) ICALP 2015. LNCS, vol. 9135, pp. 337–349. Springer, Heidelberg (2015). https://doi.org/10.1007/978-3-662-47666-6_27
17. Lohrey, M., Maneth, S., Reh, C.P.: Compression of unordered XML trees. In: Proceedings of ICDT 2017, LIPIcs 68, pp. 18:1–18:17. Schloss Dagstuhl - Leibniz-Zentrum für Informatik (2017)
18. Lohrey, M., Maneth, S., Schmidt-Schauß, M.: Parameter reduction and automata evaluation for grammar-compressed trees. J. Comput. Syst. Sci. **78**(5), 1651–1669 (2012)

19. Lohrey, M., Reh, P., Sieber, K.: Optimal top dag construction (2017). https:// arxiv.org/abs/1712.05822
20. Sundaram, S., Madria, S.K.: A change detection system for unordered XML data using a relational model. Data Knowl. Eng. **72**, 257–284 (2012)

Complement for Two-Way Alternating Automata

Viliam Geffert$^{(\boxtimes)}$

Department of Computer Science, P. J. Šafárik University,
Jesenná 5, 04154 Košice, Slovakia
viliam.geffert@upjs.sk

Abstract. We consider the problem of converting a two-way alternating finite automaton (2AFA) with n states to a 2AFA accepting the complement of the original language. Complementing is trivial for *halting* 2AFAs, by inverting the roles of existential and universal decisions and the roles of accepting and rejecting states. However, since 2AFAs do not have resources to detect infinite loops by counting executed steps, the best construction known so far required $\Omega(4^n)$ states. Here we shall show that the cost of complementing is polynomial in n. This complementary simulation does not eliminate infinite loops.

Keywords: Finite automata · Alternation · Descriptional complexity

1 Introduction

Complement is one of the most familiar language operations, both in computational complexity and in theory of formal languages and automata. Once we are given a device A accepting a language $\mathcal{L}$ by the use of some computational resources, it is quite natural to ask what resources are necessary to decide whether the given input is *rejected*. Despite its familiarity, complementing is a difficult operation—the most important open problem of this kind is $\mathsf{NP} \overset{?}{=} \mathsf{co\text{-}NP}$.

Complementing is trivial for a one-way deterministic finite automaton (1DFA) and for complexity[1] classes $\mathsf{DSpace}(s(m))$ with $s(m) \geq \log m$. This can be done by inverting the roles of accepting and rejecting states. The trivial argument does not work for a two-way deterministic finite automaton (2DFA) because such machine may *reject by getting into an infinite loop* and we do not have resources to detect such loops by counting executed steps. For 2DFAs with n states, the known construction for the complement uses $4 \cdot n$ states. This construction first makes the given automaton A halt on every input [8] or, to be precise, it converts A into a reversible 2DFA [13].

V. Geffert—Supported by the Slovak grant contracts VEGA 1/0056/18 and APVV-15-0091.

[1] Throughout the paper, m denotes the length of the input and n the number of states.

© Springer International Publishing AG, part of Springer Nature 2018
F. V. Fomin and V. V. Podolskii (Eds.): CSR 2018, LNCS 10846, pp. 132–144, 2018.
https://doi.org/10.1007/978-3-319-90530-3_12

Complementing a one-way nondeterministic finite automaton (1NFA) requires 2^n states in the worst case [3], that is, it requires to make the machine deterministic. The problem of complementing a two-way nondeterministic finite automaton (2NFA) by the use of a polynomial number of states is open; this is related to the 2NFA-2DFA trade-off and to NSpace($\log m$) versus DSpace($\log m$) problem, of which very little is known.

Consider now a two-way alternating finite automaton (2AFA) [6,9,12,14]. It is trivial to invert the roles of existential and universal decisions and the roles of accepting and rejecting states, which gives a 2AFA for the complement, *if the original machine never gets into an infinite loop*. But, since 2AFAs do not have resources to count executed steps, the trivial construction does not work for 2AFAs with infinite loops. Moreover, we know from [9] that complementing 2AFAs *does cost* something, namely, at least $\Omega(n \cdot \log n)$ states are required in the worst case. So far, it was an open problem whether the cost of complementing a 2AFA is polynomial. The best known construction uses $(2^n-1)^2 + 1 \geq \Omega(4^n)$ states, by transforming the given 2AFA to a 1NFA for the complement [9].

However, quite recently [7], it was shown that sublogarithmic ASpace($s(m)$) is closed under complement. This complementary simulation does not require elimination of infinite loops in the original alternating machine, but the complemented machine itself gets to infinite loops along some computation paths. This raises a natural question, namely, whether 2AFAs cannot also be complemented state-efficiently *without* trying to eliminate infinite loops.

Using the above result as a starting point, we shall show that each 2AFA A with n states can be replaced by a 2AFA A″ accepting the complement of the original language with the number of states polynomial in n. This solves an open problem. We shall proceed as follows. First, we fix some basic definitions. Then, as an intermediate step, we convert A to A′, (a modified version of) a deterministic Turing machine using a worktape of linear size and accepting a carefully chosen homomorphic image of the language $\mathcal{L}(A)$. This machine model is actually far stronger than finite automata, it corresponds to DSpace(m). Nevertheless, the constructed A′ shall have some special properties so that we shall be able to convert it back to a 2AFA A″, this time for the complement of $\mathcal{L}(A)$, state-efficiently. A″ is not loop-free, but the loops in A″ do not correspond to the loops in the original A.

2   Two-Way Alternating Automata, Preliminaries

We assume the reader is familiar with the standard models of finite state automata and regular languages, see [10,11], as well as with the notion of alternation [5]. For more details, see e.g. [2,15–17]. Here we only recall some basic definitions and fix some elementary notation and terminology.

A *two-way alternating finite state automaton* (2AFA, for short) is defined as a sextuplet $A = (Q_\exists, Q_\forall, \Sigma, H, q_0, F)$, in which $Q_\exists$ and $Q_\forall$ are two finite disjoint sets of *existential and universal states*, respectively, Σ is a finite *input alphabet*, $H \subseteq Q \times (\Sigma \cup \{\vdash, \dashv\}) \times D \times Q$ is a set of *transitions*, where $Q = Q_\exists \cup Q_\forall$ is the set of all states, $\vdash, \dashv \notin \Sigma$ are two special symbols, called the *left and right endmarkers*, respectively, $D = \{-1, 0, +1\}$ represents three possible *input head moves* (to the left, no move, to the right), $q_0 \in Q$ is an *initial state*, and $F \subseteq Q$ is a set of *accepting states*. The states in the set $Q \backslash F$ will be called *rejecting*.

The input tape contains $\vdash a_1 \cdots a_m \dashv$, with the respective endmarkers at the positions 0 and $m + 1$. The machine starts with the head at the left endmarker.

A transition from the set H will be presented in the form $\langle q, a \rangle \to \langle d, q' \rangle$, with the following meaning. If A is in the finite control state $q \in Q$ with the input head at a position $k \in \{0, \ldots, m+1\}$ scanning the symbol $a_k = a \in \Sigma \cup \{\vdash, \dashv\}$, it moves its head to the position $k' = k + d$, where $d \in D = \{-1, 0, +1\}$, and then it switches to the state $q' \in Q$. The machine *halts*, if there are no executable transitions for the given pair $\langle q, a \rangle$. Such pair will be called a *halting condition*. Transitions moving to the left of $\vdash$ or to the right of $\dashv$ are not allowed.

A *configuration* is an ordered pair $\langle q, k \rangle \in Q \times \{0, \ldots, m+1\}$. Configurations inherit the status of the finite control states included, so they are partitioned into existential and universal. Depending on the finite control state and the symbol under the head, they are also partitioned into halting and non-halting.

The acceptance is witnessed as follows. For the given input, consider the tree of all computation paths starting in the initial configuration $\langle q_0, 0 \rangle$. In this tree, each son has exactly one parent, but there may exist several (possibly infinitely many) copies of the same configuration. Then:

- The subtree of all computation paths rooted in a non-halting existential configuration is accepting, if at least one subtree rooted in a son of this configuration is accepting.
- The subtree of all computation paths rooted in a non-halting universal configuration is accepting, if all subtrees rooted in the sons of this configuration are accepting.
- The subtree rooted in a halting configuration is accepting, if the finite control state in this configuration is accepting.[2]
- The subtree rooted in a configuration is rejecting, if it cannot be determined as accepting by application of the rules above.

The given input is accepted, if the tree rooted in the initial configuration is accepting. Clearly, rejection may arise not only because of halting in rejecting states, but also due to infinite loops. The language consisting of all input strings that are accepted by A will be denoted by $\mathcal{L}(A)$.

In the subsequent sections, we shall need a fixed linear order on the state set Q. This is obtained by a straightforward enumeration of all states, beginning with the initial state q_0. That is, $Q = \{q_0, q_1, \ldots, q_{n-1}\}$, with $n = \|Q\|$.

Independently of that, we shall also need a fixed linear order on executable transitions, for each $\langle q, a \rangle \in Q \times (\Sigma \cup \{\vdash, \dashv\})$. So, for this condition $\langle q, a \rangle$, let $\{\langle q, a \rangle \to \langle d_1, r_1 \rangle, \langle q, a \rangle \to \langle d_2, r_2 \rangle, \ldots, \langle q, a \rangle \to \langle d_{n_{q,a}}, r_{n_{q,a}} \rangle\}$ be the set of all executable transitions, with $d_1 r_1, d_2 r_2, \ldots, d_{n_{q,a}} r_{n_{q,a}} \in D \cdot Q$ listed in some fixed lexicographically increasing order. It is obvious that $n_{q,a} \leq \|D\| \cdot \|Q\| = 3n$.

---

[2] Such configuration has no sons and the entire subtree degenerates into a single node.

Now, for $i \in \{1, \ldots, 3n + 1\}$, define

$$\delta_{i,q,a} = \begin{cases} d_i r_i, & \text{if } i \leq n_{q,a}, \\ \text{undefined}, & \text{if } i > n_{q,a}. \end{cases}$$

In particular, if $\langle q, a \rangle$ is a halting condition, $\delta_{1,q,a} = \text{undefined}$. Now, let

$$\alpha_{q,a} = \begin{cases} \text{unknown}, & \text{if A does not halt on condition } \langle q, a \rangle, \\ \text{accept}, & \text{if A halts and accepts on } \langle q, a \rangle, \\ \text{reject}, & \text{if A halts and rejects on } \langle q, a \rangle. \end{cases} \tag{1}$$

For any given input $w = a_1 \cdots a_m$, the evaluation of the computation tree is completely determined by the fixed table of values $\delta_{i,q,a}$ and $\alpha_{q,a}$.

From now on, we can assume that $\alpha_{q_0,\vdash} = \text{unknown}$, for, if $\alpha_{q_0,\vdash}$ were different from unknown, that is, if A halts at the very beginning without executing a single step, then either $\mathcal{L}(A) = \Sigma^*$ or $\mathcal{L}(A) = \varnothing$. But then the complement of $\mathcal{L}(A)$ can be accepted by a single-state 2AFA with the empty set of transitions.

3   Deterministic Machines with Linear Space

As an intermediate step before complementing the given 2AFA A, we shall convert it to A′, a special kind of deterministic Turing machine using a worktape of linear size. This new machine model actually corresponds to DSpace(m).

Our new machine model is almost the same as the standard deterministic Turing machine equipped with a finite state control and a single two-way read-write worktape, containing initially the given input. The main difference is that the worktape is composed of two tracks. The first track can be used in a read-only way, the second track in the standard read-write way. Thus, a worktape cell is of the form $\langle L, C \rangle$ and it can be changed to some $\langle L, C' \rangle$, preserving L in the read-only track. Moreover, there are no special worktape endmarkers.

Formally, (a modified version of) a deterministic linear bounded automaton is A′ $= (\mathbb{P}, \mathbb{L}, \mathbb{C}, \mathbb{H}, P_{\text{INI}}, P_{\text{ACC}})$, where $\mathbb{P}$ is a finite set of states, $\mathbb{L}$ and $\mathbb{C}$ are two finite sets of read-only and read-write symbols, called also labels and contents, respectively, $\mathbb{H} \subseteq \mathbb{P} \times \mathbb{L} \times \mathbb{C} \times \mathbb{C} \times D \times \mathbb{P}$ is a set of transitions, where $D = \{-1, 0, +1\}$ represents head moves, and $P_{\text{INI}}, P_{\text{ACC}} \in \mathbb{P}$ are two special states, called initial and accepting, respectively.

A′ starts in P_{INI} with the head at the leftmost worktape cell. A transition in $\mathbb{H}$ is in the form $\langle P, L, C \rangle \to \langle C', d, P' \rangle$, interpreted as follows. If A′ is in the state P with the head scanning a worktape cell labeled by the read-only symbol L and containing the read-write symbol C, the machine replaces C by C', moves its head in the direction $d \in \{-1, 0, +1\}$, and then it switches to the state P'. For each condition $\langle P, L, C \rangle \in \mathbb{P} \times \mathbb{L} \times \mathbb{C}$, there is at most one executable transition in $\mathbb{H}$. The machine halts, if it tries to execute a transition moving the head to the left of the leftmost worktape cell or to the right of the rightmost cell.

A configuration of A′ is an ordered triple $\langle P, \omega, h \rangle$, consisting of $P \in \mathbb{P}$, the current state, $\omega \in (\mathbb{L} \times \mathbb{C})^*$, the current contents of the entire worktape, and $h \in \{1, \ldots, |\omega|\}$, the current position of the worktape head.

Now we are ready to convert the given 2AFA A to A'. To be more precise, the new machine accepts $\mathcal{H}(a_0a_1\cdots a_m a_{m+1})$ if and only if A accepts $a_1\cdots a_m$, where $a_0 = \vdash$, $a_{m+1} = \dashv$, and $\mathcal{H}$ is a homomorphism mapping the input string of A into the worktape string of A'—to be defined below.

The machine A' evaluates the computation tree of A by the use of the classic depth-first search (see, e.g. [1,4]) in the directed graph the nodes of which are configurations and the edges are transitions of A on the given input tape containing $\vdash w \dashv = a_0a_1\cdots a_m a_{m+1}$. The graph does not have multiple copies of the same configuration, but a configuration can have several "parents". A' keeps this graph on the linear worktape as a data structure in which edges are determined implicitly, by the relative positions of the graph nodes along this tape.

More precisely, the worktape consists of blocks corresponding to the $m+2 = |w| + 2$ input tape positions and each block consists of cells corresponding to the $n = \|Q\|$ different configurations at the given position. Finally, a worktape cell corresponding to a configuration $\langle q_j, k \rangle \in Q \times \{0, \ldots, m+1\}$ will contain $\langle q_j a_k, C \rangle$, with the read-only label representing the corresponding pair $\langle q_j, a_k \rangle \in Q \times (\Sigma \cup \{\vdash, \dashv\})$. The read-write component C in the cell will be utilized for storing navigation data in the course of the depth-first search. Initially, C is equal to unknown for non-halting configurations, but to accept and reject for configurations that halt in accepting and rejecting states, respectively. Formally, using (1), the initial value is $C = \alpha_{q_j, a_k}$. During the depth-first search, the read-write contents C in the cell can also save two links $d'r', d''r'' \in D \cdot Q$, pointing to one of the parents and to one of the sons of the configuration $\langle q_j, k \rangle$. For these reasons, we fix the following worktape alphabets:

$$\mathbb{L} = Q \cdot (\Sigma \cup \{\vdash, \dashv\}),$$
$$\mathbb{C} = D \cdot Q \times D \cdot Q \cup \{\text{unknown}, \text{accept}, \text{reject}\}.$$

The initial contents on the worktape for A' is $\mathcal{H}(\vdash w \dashv)$, defined as follows:

$$\mathcal{H}(\vdash w \dashv) = A_0 A_1 \cdots A_k \cdots A_m A_{m+1}, \quad \text{where}$$
$$A_k = B_{k,0} \cdots B_{k,j} \cdots B_{k,n-1}, \tag{2}$$
$$B_{k,j} = \langle q_j a_k, \alpha_{q_j, a_k} \rangle \in \mathbb{L} \times \mathbb{C}.$$

Let us now fix the states. During the depth-first search, A' traverses the computation graph of A, starting from $B_{0,0} = \langle q_0 \vdash, \alpha_{q_0, \vdash} \rangle = \langle q_0 \vdash, \text{unknown} \rangle$ that corresponds to the initial configuration of A. For each explored configuration $\langle q_j, k \rangle$ in the graph, represented by a cell $B_{k,j}$ labeled by $q_j a_k$ on the worktape, A' evaluates whether the subtree of all computation paths rooted in $\langle q_j, k \rangle$ is accepting or rejecting, and saves this information in the corresponding read-write component of the cell. The finite control state depends on from where A' has arrived to $\langle q_j, k \rangle$:

If A' has arrived to $\langle q_j, k \rangle$ from one of its "parents" by following a single-step edge from some $\langle r, k' \rangle$, where $k - k' = d \in D = \{-1, 0, +1\}$ and $r \in Q$, the machine A' is in the state $\langle dr, 0 \rangle$, keeping this way a backup link to the current parent. When required, this link can be used to get back to the original configuration $\langle r, k' \rangle$, by backing up against the direction of the edge connecting $\langle r, k' \rangle$ with the current configuration $\langle q_j, k \rangle$.

Conversely, if A' has arrived to $\langle q_j, k \rangle$ from one of its "sons" by backing up against the direction of an edge connecting $\langle q_j, k \rangle$ with some $\langle r_i, k + d_i \rangle$, the machine A' is in the state $\langle \text{res\_accept}, 0 \rangle$ or $\langle \text{res\_reject}, 0 \rangle$, depending on whether the subtree of all computation paths rooted in $\langle r_i, k + d_i \rangle$ is accepting or rejecting.

The second component of the finite control state will be utilized to implement moves of A' in the graph, i.e., to implement *(i)* moving along an edge that connects two configurations, and *(ii)* backing up against the direction of such edge. We are now ready to define the state set for A':

$$\mathbb{P} = \mathbb{P}_s \times \mathbb{P}_M, \quad \text{where}$$
$$\mathbb{P}_s = D \cdot Q \cup \{\text{res\_accept}, \text{res\_reject}\},$$
$$\mathbb{P}_M = \{-2, -1, +1, +2\} \cdot Q \cup \{0\}.$$

Before passing further, let us show how to navigate the machine A' through the graph. If we want to move the worktape head placed at a cell corresponding to a configuration $\langle q_i, k \rangle$ (labeled by $q_i a$, where $a = a_k$) to a cell that corresponds to some $\langle q_j, k + d \rangle$ (labeled by $q_j a'$, for some a'), and reach the target worktape position in some state $\langle P_s, 0 \rangle$, for any given $q_i, q_j \in Q, d \in \{-1, 0, +1\}$, and $P_s \in \mathbb{P}_s$, we switch the machine A' to the state $\langle P_s, \mu_{d, q_i, q_j} \rangle$, where

$$\mu_{d, q_i, q_j} = \begin{cases} +_2 q_j, & \text{if } d = +1 \text{ and } i \leq j, \\ +_1 q_j, & \text{if } d = +1 \text{ and } i > j, \text{ or } d = 0 \text{ and } i < j, \\ 0, & \text{if } d = 0 \text{ and } i = j, \\ -_1 q_j, & \text{if } d = -1 \text{ and } i < j, \text{ or } d = 0 \text{ and } i > j, \\ -_2 q_j, & \text{if } d = -1 \text{ and } i \geq j. \end{cases}$$

As an example, by switching A' to $\langle P_s, \mu_{+1, q_5, q_7} \rangle = \langle P_s, +_2 q_7 \rangle$ when the worktape head is placed at a cell corresponding to a configuration $\langle q_5, k \rangle$, we activate a routine searching for the *second* occurrence of a worktape cell containing q_7 in its label, starting the search from the current worktape position and moving to the *right*. When such cell has been found—which corresponds to the configuration $\langle q_7, k + 1 \rangle$—the machine A' will switch to $\langle P_s, 0 \rangle$.

Now we are ready to present transitions for the depth-first search. There are the following cases:

(a) A' arrives to a worktape cell corresponding to a configuration $\langle q, k \rangle$ from one of its parents, in a state $\langle dr, 0 \rangle$, keeping this way a backup link to the current parent. Recall that the current cell under the worktape head is labeled by $qa \in \mathbb{L}$, with $a = a_k$, and it contains a read-write symbol $C \in \mathbb{C} = D \cdot Q \times D \cdot Q \cup \{\text{unknown}, \text{accept}, \text{reject}\}$. This gives the following subcases:

    (a.1) $C = \text{unknown}$, that is, $\langle q, k \rangle$ has not been explored yet. In this case, A' replaces $C = \text{unknown}$ by $C' = \langle dr, d_1 r_1 \rangle$, where $d_1 r_1 = \delta_{1,q,a}$. This saves the links to the current parent and to the first son of the configuration $\langle q, k \rangle$ on the worktape. Next, A' switches its state to $\langle d_1 q, \mu_{d_1, q, r_1} \rangle$,

which activates the routine moving along the edge that connects $\langle q, k \rangle$ with its first son $\langle r_1, k + d_1 \rangle$, reaching this son with a new backup link, in the state $\langle d_1 q, 0 \rangle$.

(a.2) $C \in \{\text{accept}, \text{reject}\}$, that is, $\langle q, k \rangle$ has been explored already (we are just visiting $\langle q, k \rangle$ from another parent) or $\langle q, k \rangle$ is a halting configuration. In either case, the subtree of all computation paths rooted in $\langle q, k \rangle$ is accepting or rejecting, in accordance with C. Without making any changes on the worktape, A′ switches to the respective state $\langle \text{res\_accept}, \mu_{-d,q,r} \rangle$ or $\langle \text{res\_reject}, \mu_{-d,q,r} \rangle$. This activates the routine backing up against the direction of the edge that connects the current parent $\langle r, k-d \rangle$ with $\langle q, k \rangle$, reaching this parent with the result of evaluation, in the respective state $\langle \text{res\_accept}, 0 \rangle$ or $\langle \text{res\_reject}, 0 \rangle$.

(a.3) $C = \langle d'r', d''r'' \rangle$, for some $d'r', d''r'' \in D \cdot Q$. This means that, in the course of exploring the subtree of all computation paths rooted in $\langle q, k \rangle$, the machine A′ visits $\langle q, k \rangle$ again, having followed a computation path of A that enters a loop. A′ proceeds therefore in the same way as if, in (a.2), the read-write symbol C were equal to reject. That is, without making any changes on the worktape, A′ switches to $\langle \text{res\_reject}, \mu_{-d,q,r} \rangle$.

(b) A′ arrives to a worktape cell corresponding to $\langle q, k \rangle$ from one of its sons, in the state $\langle \text{res\_accept}, 0 \rangle$. This means that the subtree rooted in the latest visited son is accepting. At this moment, the current cell under the head is labeled by $qa \in \mathbb{L}$, with $a = a_k$, and it contains the read-write symbol $C = \langle dr, d_i r_i \rangle$, representing the backup link for the current parent and the forward link to the latest visited son. This gives the following subcases:

(b.1) If q is universal and $\delta_{i+1,q,a} \neq \text{undefined}$, that is, $\delta_{i+1,q,a} = d_{i+1} r_{i+1} \in D \cdot Q$, the machine A′ has to evaluate the next son. A′ therefore replaces $C = \langle dr, d_i r_i \rangle$ by $C' = \langle dr, d_{i+1} r_{i+1} \rangle$, which saves the link to the next son on the worktape. Next, A′ switches to the state $\langle d_{i+1} q, \mu_{d_{i+1},q,r_{i+1}} \rangle$, which activates the routine moving along the edge that connects $\langle q, k \rangle$ with $\langle r_{i+1}, k+d_{i+1} \rangle$, reaching this son with the proper backup link, in the state $\langle d_{i+1} q, 0 \rangle$.

(b.2) Conversely, if q is existential or $\delta_{i+1,q,a} = \text{undefined}$ (that is, $\langle q, k \rangle$ has no more sons), the subtree rooted in $\langle q, k \rangle$ is accepting. Thus, A′ replaces $C = \langle dr, d_i r_i \rangle$ on the worktape by $C' = \text{accept}$ and, using the backup link saved in C, it switches to the state $\langle \text{res\_accept}, \mu_{-d,q,r} \rangle$. This activates the routine backing up against the direction of the edge that connects the current parent $\langle r, k-d \rangle$ with $\langle q, k \rangle$, reaching $\langle r, k-d \rangle$ in the state $\langle \text{res\_accept}, 0 \rangle$.

(c) A′ arrives to a worktape cell corresponding to $\langle q, k \rangle$ from one of its sons, in the state $\langle \text{res\_reject}, 0 \rangle$. This means that the subtree rooted in the latest visited son is rejecting. This case is very similar to (b): either *(c.1)* q is existential and $\delta_{i+1,q,a} \neq \text{undefined}$, or *(c.2)* q is universal or $\delta_{i+1,q,a} = \text{undefined}$. Depending on this, A′ either evaluates the next son or, respectively, it rewrites the symbol on worktape to $C' = \text{reject}$ and returns to the current parent in the state $\langle \text{res\_reject}, 0 \rangle$.

A' starts in the state $P_{INI} = \langle_{+1}q_0, 0\rangle$ with the head at the leftmost cell, labeled by $q_0 a_0 = q_0 \vdash$ and containing the read-write symbol $C = \alpha_{q_0, \vdash} =$ unknown. This starts the evaluation of the computation tree of A, rooted in $\langle q_0, 0\rangle$. When this evaluation is over, the outcome is written as the final value $C' \in \{accept, reject\}$ in the leftmost cell. Next, A' switches to the respective state $P_{ACC} = \langle res\_accept, \mu_{-1, q_0, q_0}\rangle = \langle res\_accept, \_{-2}q_0\rangle$ or $P_{REJ} = \langle res\_reject, \mu_{-1, q_0, q_0}\rangle = \langle res\_reject, \_{-2}q_0\rangle$, trying to back up with the result to a nonexistent parent $\langle q_0, 0-1\rangle$ to the left of the leftmost cell. At this moment, the computation is halted.

As the finishing touch, we can eliminate transitions in which A' does not move its head (see, e.g., [10, p. 319]). The idea is quite simple: each chain of transitions $\langle P_1, L, C_1\rangle \rightarrow \langle C_2, 0, P_2\rangle, \ldots, \langle P_{g-1}, L, C_{g-1}\rangle \rightarrow \langle C_g, 0, P_g\rangle, \langle P_g, L, C_g\rangle \rightarrow \langle C, d, P\rangle$, with $d \in \{-1, +1\}$, is replaced by $\langle P_1, L, C_1\rangle \rightarrow \langle C, d, P\rangle$.

It is trivial to see that the size of the read-only worktape alphabet can be bounded by $\|L\| = \|Q\| \cdot (\|\Sigma\| + 2) \leq O(n \cdot \|\Sigma\|)$, the size of the read-write worktape alphabet by $\|C\| = (3 \cdot \|Q\|) \cdot (3 \cdot \|Q\|) + 3 = 9n^2 + 3 \leq O(n^2)$, while for the number of finite control states we get $\|P\| = (3 \cdot \|Q\| + 2) \cdot (4 \cdot \|Q\| + 1) = 12n^2 + 11n + 2 \leq O(n^2)$. Another important complexity measure for A' is the number of visits at any worktape cell during the computation, bounded by

$$T = 5 \cdot n^2 - 4n + 1. \tag{3}$$

To sum it up, we have derived the following:

Lemma 1. *For each two-way alternating finite state automaton* A *with* n *states accepting a language* $\mathcal{L} \subseteq \Sigma^*$, *there exists a deterministic linear bounded automaton* A' *and a homomorphism* $\mathcal{H}$ *such that, for each* $w \in \Sigma^*$, *the machine* A' *accepts* $\mathcal{H}(\vdash w \dashv)$ *if and only if* A *accepts* w. *The machine* A' *starts in the initial state* P_{INI} *at the leftmost worktape cell and returns the outcome by trying to leave the worktape to the left, in the respective state* P_{ACC} *or* P_{REJ}.

The machine A' *uses a linear worktape with two tracks; the first track is read-only; the second track is used in the standard read-write way. The read-write alphabet is of size* $O(n^2)$ *and the number of finite control states is bounded by* $O(n^2)$. *Moreover,* A' *does not visit any cell along the worktape more than* $O(n^2)$ *times in the course of the entire computation.*

4   Two-Way Alternating Automata, Complemented

The deterministic two-way machine A' constructed in the previous section uses linear worktape space. However, A' does not visit any cell along the worktape more than $T \leq O(n^2)$ times, which is a value depending on n, the number of states in the original two-way alternating finite automaton A, but not increasing in the length of the input. Here we shall convert A' back into a two-way alternating finite automaton again, this time accepting the *complement* of $\mathcal{L}(A)$.

Recall that $w \in \Sigma^*$ is rejected by A if and only if the string $\mathcal{H}(\vdash w \dashv)$ is rejected by A', where $\mathcal{H}$ is the homomorphism introduced by (2). The machine

A′ starts in the initial state P_{INI} with the worktape head at the leftmost symbol and rejects by trying to leave the worktape to the left in the unique state P_{REJ}. Thus, to verify that A′ rejects, the alternating automaton A″ verifies whether there exists a backward path starting from P_{REJ} and ending in P_{INI}. A″ guesses the trajectory of this unique backward path existentially. Along this path, A″ keeps track of the current *local configuration* $S = \langle k, q_j a, t, P, C \rangle$, consisting of:

k, an auxiliary value, represented by the position of the head of A″ along the input tape containing $\vdash w \dashv = \vdash a_1 \cdots a_m \dashv$ (on the worktape of A′, this corresponds to a block of n cells, labeled by $q_0 a_k, q_1 a_k, \ldots, q_{n-1} a_k$),

$q_j a$, the label in the current cell under the head of A′ along its worktape, satisfying $a = a_k$ (the current worktape position of A′ is thus completely determined by $\langle q_j, k \rangle$),

t, the current number of visits at the current worktape position, made by A′ along the computation path by which it gets from the initial configuration to the current configuration,

P, the current finite control state of A′, and

C, the current read-write contents in the worktape cell under the head.

The local configuration S is *correct*, if all values in S agree with some real configuration along the computation path of A′ starting in the initial configuration on $\mathcal{H}(\vdash w \dashv)$. This real configuration is unique, since A′ is deterministic and loop-free, and hence each reachable configuration is completely determined by the worktape head position and by the number of visits at this position. Along the backward path, the current local configuration S is obtained by existential guessing. Now, the machine A″ has to verify whether S is correct. Before presenting the verification procedure, consider how the correctness of S can be derived from other local configurations in a "near neighborhood".

If S represents the initial configuration, then $S = \langle 0, q_0 \vdash, 1, P_{INI}, \text{unknown} \rangle$. Otherwise, there must exist $S' = \langle k', q_{j'} a', t', P', C' \rangle$, a correct local configuration describing the previous configuration, one step of A′ back in time. S' must be consistent with S, which means that: *(i)* S', S must agree with $d' \in \{-1, +1\}$, the direction of the latest move of A′ along the worktape—the structure of this worktape was presented by (2). Namely, if $d' = +1$, then $j' = (j-1) \bmod n$. Now, if $j > 0$, then $k' = k$, otherwise $k' = k-1$. The case of $d' = -1$ is symmetric: $j' = (j+1) \bmod n$ and, if $j < n-1$, then $k' = k$, otherwise $k' = k+1$. *(ii)* Moreover, all data must agree with $\mathbb{H}$, the set of transitions of A′. Namely, if $\langle P', q_{j'} a', C' \rangle \to \langle \widetilde{C}, \widetilde{d}, \widetilde{P} \rangle$ is a transition in $\mathbb{H}$, then $d' = \widetilde{d}$ and $P = \widetilde{P}$.

Second, if $t > 1$, the current worktape position had to be visited in the past. Thus, there must exist $S'_\square = \langle k'_\square, q_{j'_\square} a'_\square, t'_\square, P'_\square, C'_\square \rangle$, a correct local configuration describing the previous visit of A′ at the same worktape position, i.e., with $k'_\square = k$, $q_{j'_\square} a'_\square = q_j a$, and $t'_\square = t-1$. But then there must also exist $S_\square = \langle k_\square, q_{j_\square} a_\square, t_\square, P_\square, C_\square \rangle$, a correct local configuration describing the next configuration, one step of A′ forward in time from $S'_\square$. The local configuration $S'_\square$ must be consistent with $S_\square$ in the same way as S' with S, that is, they agree with $d'_\square$, the direction of the head movement from $S'_\square$ to $S_\square$, and with $\mathbb{H}$, the set of transitions. Moreover, $d'_\square = -d'$, since A' cannot leave the current worktape cell to the right and then return back from the left, or vice versa. For these reasons, $k_\square = k'$ and $q_{j_\square} a_\square = q_{j'} a'$. In addition, $t_\square \leq t'$, since A' cannot reach S' earlier than $S_\square$, not excluding the possibility that $S_\square = S'$, with $t_\square = t'$. Finally, if $\langle P'_\square, q_j a, C'_\square \rangle \to \langle \widetilde{C}, \widetilde{d}, \widetilde{P} \rangle$ is a transition in $\mathbb{H}$, then $\widetilde{C} = C$, since the contents in the current worktape cell did not change along the path connecting $S'_\square$ with S, not visiting the current worktape position in the meantime. This also gives $\widetilde{d} = d'_\square = -d'$.

Third, for each $\tau \in \{t_\square, \ldots, t'-1\}$, the computation path connecting $S_\square$ with S' must pass through a configuration visiting the same worktape position for the τ-th time. Thus, there must exist $S_\tau = \langle k_\tau, q_{j_\tau} a_\tau, \tau, P_\tau, C_\tau \rangle$, a correct local configuration with the head placed at the same position as in S', with $k_\tau = k'$ and $q_{j_\tau} a_\tau = q_{j'} a'$. (If $t_\square = t'$, the set $\{t_\square, \ldots, t'-1\}$ is empty.) In addition, A' must move the head from S_τ in the direction $-d'$, since the path connecting $S'_\square$ with S does not visit the same worktape position in the meantime. Thus, if $\langle P_\tau, q_{j'} a', C_\tau \rangle \to \langle \widetilde{C}, \widetilde{d}, \widetilde{P} \rangle$ is a transition in $\mathbb{H}$, then $\widetilde{d} = -d'$.

The situation is different for S with $t = 1$. That is, the current worktape position has not been visited before. Also here there must exist S' as specified above, but here the latest executed step, from S' to S, must move the head of A' in the direction $d' = +1$, since the first visit to any cell must arrive from the left. Moreover, C must be equal to $\alpha_{q_j, a_k} = \alpha_{q_j, a}$, the initial read-write contents for the current cell.

Next, for each $\tau \in \{1, \ldots, t'-1\}$, the computation path connecting the initial configuration with S' must pass through a configuration visiting the same work-tape position for the τ-th time. Thus, there must exist $S_\tau = \langle k_\tau, q_{j_\tau} a_\tau, \tau, P_\tau, C_\tau \rangle$, a correct local configuration with the head placed at the same position as in S', which gives $k_\tau = k'$ and $q_{j_\tau} a_\tau = q_{j'} a'$. (If $t' = 1$, the set $\{1, \ldots, t'-1\}$ is empty.) In addition, A' must move the head from S_τ in the direction $-d' = -1$, since A' does not visit the worktape segment on the right of S' before it reaches S'.

Summing up, a local configuration S with $t > 1$ is correct if and only if there exist correct local configurations $S', S'_\square, S_\square$ such that, for each $\tau \in \{t_\square, \ldots, t'-1\}$, there exists a correct local configuration S_τ such that $S', S'_\square, S_\square, S_\tau$ satisfy all requirements specified above. Thus, verifying the correctness of S can be reduced to the same kind of verification for $S', S'_\square, S_\square$ and $S_{t_\square}, \ldots, S_{t'-1}$, all of them backward in time along the path of A'. An analogous reduction works for S with $t = 1$, using S' and $S_1, \ldots, S_{t'-1}$ instead of $S', S'_\square, S_\square$ and $S_{t_\square}, \ldots, S_{t'-1}$.

By implementation of these ideas, we can construct a 2AFA A'' representing the local configuration $S = \langle k, q_j a, t, P, C \rangle$ by k, the head position along the input, and by $q_j a, t, P, C$, kept in the finite state control. Since $\|Q\| = n, T \leq O(n^2)$ by (3), and $\|\mathbb{P}\|, \|\mathbb{C}\|$ are bounded by $O(n^2)$, a local configuration can be represented by the use of $O(\|\Sigma\| \cdot n^7)$ finite control states. For the given S, the machine A'' verifies whether S is correct.

This is done as follows. First, branching existentially, A'' generates the local configurations $S', S'_\square, S_\square$. Next, branching universally, A'' generates a value

$\tau \in \{t_\square, \ldots, t'-1\}$. After that, in each branch of the computation tree running in parallel, A'' branches existentially again and generates the local configuration S_τ. Next, A'' checks whether S', $S'_\square$, $S_\square$, and S_τ are consistent, i.e., whether they satisfy all requirements, as specified above. Then, in parallel, the correctness of each of them is verified in the same way, i.e., it becomes a new current local configuration S, after "forgetting" all data that are no longer required and, if necessary, updating the position of the head along the input tape. Even though these parallel branches make existential guesses about the computation of A' independently of each other, the global consistency of all *correct* guesses is ensured by the fact that A' is deterministic, and hence the computation of A' is unique. Any wrong existential guess that contradicts this unique computation is overridden; such guess leads to an alternating subtree that is rejecting. Now, along each branch running in parallel and guessing correctly, A'' traces the unique computation of A' backward in time to the initial local configuration, where the correctness is decided deterministically, by comparing S with $\langle 0, q_0\vdash, 1, P_{\mathsf{INI}}, \text{unknown}\rangle$. (We skip narration for the case of $t = 1$, handled analogically.) Starting from a local configuration that represents a moment when A' is going to reject, A'' can verify whether the given input is rejected by the original machine A.

Recall that the local configuration S is represented by using $O(\|\Sigma\| \cdot n^7)$ finite control states, except for k, represented by the head position along the input. All values in S', $S'_\square$, $S_\square$ and in S_τ, where $\tau \in \{t_\square, \ldots, t'-1\}$, are also kept in the finite state control, except for k', $k'_\square$, $k_\square$, k_τ. However, since $k' \in \{k-1, k, k+1\}$, $k'_\square = k$, and $k_\square = k_\tau = k'$, also these values can be kept in the finite state control, relative to k. Moreover, utilizing $q_{j'_\square} a'_\square = q_j a$, $q_{j_\square} a_\square = q_{j_\tau} a_\tau = q_{j'} a'$, and $t'_\square = t-1$, we can bound the total number of states by $O(\|\Sigma\|^2 \cdot n^{30})$.

Theorem 2. *For each two-way alternating finite state automaton A with n states accepting a language $\mathcal{L} \subseteq \Sigma^*$, there exists a two-way alternating finite state automaton A'' with $O(\|\Sigma\|^2 \cdot n^{30})$ states accepting the complement of the original language.*

5   Concluding Remarks

The contribution of the paper is a technique of handling infinite loops in two-way alternating automata without the necessity of counting simulated steps. This gives a polynomial complementing for 2AFAs independently of whether they are halting, which solves a long-standing open problem. The cost of this complementing is $O(\|\Sigma\|^2 \cdot n^{30})$ states. This basic version can be improved, which reduces the number of states, down to $O(n^7)$. Because of the page limit, the improved version will appear in a full version of the paper only. The known lower bound is $\Omega(n \cdot \log n)$ states [9]. This raises another open problem, namely, the exact cost. At present, we have only a better cost for *halting* 2AFAs, by trivially inverting the roles of existential and universal decisions and the roles of accepting and rejecting states, using n states and preserving halting properties.

It would also be nice to obtain the complementary machine A'' by a direct construction, without taking a detour via $\mathsf{DSpace}(m)$.

Despite Theorem 2, it is still open whether we can transform each n-state 2AFA to a halting 2AFA with a polynomial number of states. The lower bound $\Omega(n \cdot \log n)$, derived in [9] for complementing, applies also to making the machine loop-free, because of the linear cost of complementing for loop-free machines.

Several other problems are still open, if the number of alternations is restricted. For a 2AFA making at most $k-1$ alternations between existential and universal states, starting in an existential or universal state ($2\Sigma_k\mathsf{FA}$ or $2\Pi_k\mathsf{FA}$, respectively), the cost of complementing is exponential [6]: for each $k \geq 2$, at least $2^{\Omega(n)-O(\log k)}$ states are required. But we still do not know whether, keeping the number of states polynomial, we can convert a $2\Sigma_k\mathsf{FA}$ into a $2\Pi_k\mathsf{FA}$ for the complement. The same problem is open the other way round, from $2\Pi_k\mathsf{FA}$ to $2\Sigma_k\mathsf{FA}$. The most important is the case of $k = 1$, i.e., the cost of complementing a two-way automaton making only existential choices by a two-way automaton making only universal choices, or vice versa.

References

1. Aho, A., Hopcroft, J., Ullman, J.: The Design and Analysis of Computer Algorithms. Addison-Wesley, Boston (1976)
2. Berman, L., Chang, J., Ibarra, O., Ravikumar, B.: Some observations concerning alternating Turing machines using small space. Inf. Process. Lett. **25**, 1–9 (1987). Corr. ibid. 27, p. 53 (1988)
3. Birget, J.: Partial orders on words, minimal elements of regular languages, and state complexity. Theor. Comput. Sci. **119**, 267–291 (1993)
4. Brassard, G., Bratley, P.: Fundamentals of Algorithmics. Prentice Hall, Upper Saddle River (1996)
5. Chandra, A., Kozen, D., Stockmeyer, L.: Alternation. J. Assoc. Comput. Mach. **28**, 114–133 (1981)
6. Geffert, V.: An alternating hierarchy for finite automata. Theor. Comput. Sci. **445**, 1–24 (2012)
7. Geffert, V.: Alternating space is closed under complement and other simulations for sublogarithmic space. Inf. Comput. **253**, 163–178 (2017)
8. Geffert, V., Mereghetti, C., Pighizzini, G.: Complementing two-way finite automata. Inf. Comput. **205**, 1173–1187 (2007)
9. Geffert, V., Okhotin, A.: Transforming two-way alternating finite automata to one-way nondeterministic automata. In: Csuhaj-Varjú, E., Dietzfelbinger, M., Ésik, Z. (eds.) MFCS 2014. LNCS, vol. 8634, pp. 291–302. Springer, Heidelberg (2014). https://doi.org/10.1007/978-3-662-44522-8_25
10. Hopcroft, J., Motwani, R., Ullman, J.: Introduction to Automata Theory, Languages, and Computation. Addison-Wesley, Boston (2001)
11. Kapoutsis, C.A.: Size complexity of two-way finite automata. In: Diekert, V., Nowotka, D. (eds.) DLT 2009. LNCS, vol. 5583, pp. 47–66. Springer, Heidelberg (2009). https://doi.org/10.1007/978-3-642-02737-6_4
12. Kapoutsis, C.: Minicomplexity. J. Autom. Lang. Comb. **17**, 205–224 (2012)
13. Kunc, M., Okhotin, A.: Reversibility of computations in graph-walking automata. In: Chatterjee, K., Sgall, J. (eds.) MFCS 2013. LNCS, vol. 8087, pp. 595–606. Springer, Heidelberg (2013). https://doi.org/10.1007/978-3-642-40313-2_53

14. Ladner, R., Lipton, R., Stockmeyer, L.: Alternating pushdown and stack automata. SIAM J. Comput. **13**, 135–155 (1984)
15. Liśkiewicz, M., Reischuk, R.: Computing with sublogarithmic space. In: Hemaspaandra, L., Selman, A. (eds.) Complexity Theory Retrospective II. Springer, New York (1997). ISBN 978-0-387-94973-4
16. Papadimitriou, C.H.: Computational Complexity. Addison-Wesley, Boston (1994)
17. Szepietowski, A.: Turing Machines with Sublogarithmic Space. LNCS, vol. 843. Springer, Heidelberg (1994). https://doi.org/10.1007/3-540-58355-6

Closure Under Reversal of Languages over Infinite Alphabets

Daniel Genkin[1], Michael Kaminski[2($\boxtimes$)], and Liat Peterfreund[2]

[1] Department of Computer and Information Science,
University of Pennsylvania, 3330 Walnut Street, Philadelphia, PA 19104, USA
[2] Department of Computer Science,
Technion – Israel Institute of Technology, 32000 Haifa, Israel
`kaminski@cs.technion.ac.il`

Abstract. It is shown that languages definable by weak pebble automata are not closed under reversal. For the proof, we establish a kind of periodicity of an automaton's computation over a specific set of words. The periodicity is partly due to the finiteness of the automaton description and partly due to the word's structure. Using such a periodicity we can find a word such that during the automaton's run on it there are two different, yet indistinguishable, configurations. This enables us to remove a part of that word without affecting acceptance. Choosing an appropriate language leads us to the desired result.

Keywords: Infinite alphabets · Weak pebble automata
Closure properties · Reversal

1 Introduction

While automata for words over finite alphabets are well-understood, a broad research activity began very recently on automata for words over infinite alphabets. Note, that for infinite alphabets, states alone are not sufficient, because an automaton should be able to check equality of input symbols. This can be done by (dynamically) marking a set of symbols of a fixed finite cardinality and allowing equality tests with these symbols.

Finite-Memory Automata (FMA) [2,3] keep marked symbols in a finite number of registers and Pebble Automata (PA) [5,6] keep marked symbols under a finite number of pebbles. Both are very restrictive models intended for recognizing an analog of regular languages over finite alphabets. The class of languages recognizable by FMA and PA enjoys many of the properties of regular languages. Languages recognizable by FMA are closed under standard language operations: intersection, union, concatenation, and iteration (Kleene star), whereas languages recognizable by PA are closed under all boolean operations (i.e., union, intersection and complementation), but are not closed under iteration. However, the emptiness problem for FMA is decidable, whereas it is decidable for weak 2-PA only.

© Springer International Publishing AG, part of Springer Nature 2018
F. V. Fomin and V. V. Podolskii (Eds.): CSR 2018, LNCS 10846, pp. 145–156, 2018.
https://doi.org/10.1007/978-3-319-90530-3_13

This paper deals with weak PA which, as mentioned above, are finite state automata equipped with a finite number k of pebbles, numbered from 1 to k. Each pebble can serve as the head of the automaton or point at a position in the input word. The pebbles are placed on the input word in the stack discipline: the first pebble placed is the last to be lifted. The first pebble placed is pebble 1. One pebble can only point at one position and the most recently placed pebble serves as the head of the automaton. The automaton moves from one state to another depending on the symbol under the head pebble, the equality tests among symbols under the pebbles, and the equality tests among the pebbles' positions.

There are two main variants of PA, weak and strong and we focus on weak PA (wPA). We show that, unlike languages accepted by strong PA, languages accepted by wPA are not closed under reversal. In this paper we deal with two languages - $L_\subseteq$ and $L_\supseteq$. Both languages consist of words of a special form $u\$v$, where $\$$ is a special symbol (the separator) not occurring in u or v and the symbols in each of these words are pairwise different. In $L_\subseteq$, each symbol occurring in u also occurs in v and, in $L_\supseteq$, each symbol occurring in v also occurs in u. Thus, $L_\subseteq$ and $L_\supseteq$ are the reversals of each other. We show that the language $L_\subseteq$ is accepted by wPA, whereas $L_\supseteq$ is not. For this, for each automaton accepting all words of $L_\supseteq$, we construct a special word $u\$v \notin L_\supseteq$, yet still accepted by the automaton. In that word a prefix of u is "properly" spread in v and its construction is based on a kind of periodicity of the sequence of the states in the run of PA. In particular, the head pebble behaves periodically when it sees symbols different from those under the other pebbles.

The paper is organized as follows. In Sect. 2 we present the definition of wPA and state the separation theorem whose proof, for the case of 2-wPA, is presented in Sect. 3.[1] In Sect. 4 we show how to modify the languages $L_\subseteq$ and $L_\supseteq$ to languages whose words do not contain a distinguished separator symbol. We conclude the paper with a short remark that concerns the technique applied for the proof of the separation theorem.

2   Weak Pebble Automata

As introduced in [5,6], Pebble Automata over infinite alphabets are finite state machines equipped with a finite set of numbered pebbles. The computation of an automaton on an input word starts when the lowest numbered pebble is located at the leftmost position of the input and acts as the head of the automaton. During the computation, an automaton can place (respectively, lift) a pebble on (respectively, from) the input. It can also move the pebble that acts as the head of the automaton. That pebble is the highest numbered pebble present on the input, whereas the other pebbles serve as pointers at the input symbols. The use of the pebbles is restricted by the stack discipline (pebble i can only be placed

---

[1] The proof of the general case can be found in [7], and, hopefully, will also appear elsewhere.

when pebble $i - 1$ is present on the input word and pebble i can only be lifted when pebble $i + 1$ is not present on the input word).

A transition depends on the current state, equality type of the symbols under the placed pebbles and equality among the pebbles' positions. The transition relation specifies change of state, the movement of the head and, possibly, whether the head pebble is lifted or a new pebble is placed.

Definition 1. *A deterministic one-way[2] k-wPA over an infinite alphabet Σ, is a tuple $\mathfrak{A} = \langle S, s_0, F, T \rangle$ whose components are as follows.*

- *S is a finite set of states,*
- *$s_0 \in S$ is the initial state,*
- *$F \subseteq S$ is a set of accepting states,*
- *T is a finite set of transitions of the form $\alpha \to \beta$, where*
  - *α is of the form (i, σ, P, V, s) or (i, P, V, s), $i \in \{1, \ldots, k\}$, $\sigma \in \Sigma$, $P, V \subseteq \{1, \ldots, i - 1\}$, and $s \in S$, and*
  - *β is of the form $(p, \texttt{action})$, where $p \in S$ and $\texttt{action} \in \{\texttt{move}, \texttt{place}, \texttt{lift}\}$,*

  such that $\alpha \to \beta$ and $\alpha \to \beta'$ imply $\beta = \beta'$.

For a word $\boldsymbol{w} \in \Sigma^*$, a configuration of $\mathfrak{A}$ on $\boldsymbol{w}$ is of the form $\gamma = [i, s, \theta]$, where $i \in \{1, \ldots, k\}$, $s \in S$, and $\theta : \{1, \ldots, i\} \to \{1, \ldots, |\boldsymbol{w}|\}$ indicates the pebble's positions on the input word. That is, $\theta(j)$ is the position of pebble j. In what follows, we identify θ with the i-tuple $(\theta(1), \ldots, \theta(i))$. Thus, i can be recovered from θ, but it is convenient to include it into a configuration explicitly.

The *initial configuration* is $\gamma_0 = [1, s_0, (1)]$. That is, the run starts in the initial state s_0 with pebble 1 placed at the beginning of the input word. An *accepting configuration* is of the form $[i, s, \theta]$, where $s \in F$.

Let $\boldsymbol{w} = w_1 \cdots w_n \in \Sigma^+$. A transition $(i, \sigma, P, V, s) \to \beta$ *applies* to a configuration $\gamma = [j, s', \theta]$ if

(1) $i = j$ and $s' = s$,
(2) $P = \{h < i : \theta(h) = \theta(i)\}$,
(3) $V = \{h < i : w_{\theta(h)} = w_{\theta(i)}\}$, and
(4) $w_{\theta(i)} = \sigma$.

In the above definition, P is the set of pebbles placed at the same position as the head pebble, V is the set of pebbles placed above the same symbol as the head pebble, and the current symbol under the head pebble is σ.

A transition $(i, P, V, s) \to \beta$ applies to a configuration $\gamma = [j, s', \theta]$, if the above conditions (1)–(3) are satisfied and no transition of the form $(i, \sigma, P, V, s) \to \beta$ applies to γ.

The transition relation $\vdash_{\boldsymbol{w}}$ on the set of all configurations is defined as follows:[3] $[i, s, \theta] \vdash [i', s', \theta']$ if and only if there is a transition $\alpha \to (p, \texttt{action})$ that applies to $[i, s, \theta]$ such that $s' = p$ and the following holds.

---

[2] It has been shown in [8] that alternating non-deterministic and deterministic one-way wPA have the same expressive power.

[3] We omit the subscript $\boldsymbol{w}$ of $\vdash$, if it is clear from the context.

- For all $j < i$, $\theta'(j) = \theta(j)$,
- if action is move, then $i' = i$ and $\theta'(i) = \theta(i) + 1$,
- if action is place, then $i' = i + 1$ and $\theta'(i + 1) = \theta'(i) = \theta(i)$,[4] and
- if action is lift, then $i' = i - 1$ and θ' is the restriction of θ on $\{1, \ldots, i-1\}$.

The *language* $L(\mathfrak{A})$ of $\mathfrak{A}$ consists of all words $\boldsymbol{w}$ such that $\gamma_0 \vdash^*_{\boldsymbol{w}} \gamma$ for an accepting configuration γ.

Remark 1. Note that the accepted languages are quite symmetric: they contain only finitely many "distinguished" symbols explicitly mentioned in the automaton transitions and are invariant under any permutation of all other symbols of the infinite alphabet Σ, cf. [3, Proposition 2].

Remark 2. It follows from the definition that wPA languages do not contain the empty word ϵ, but the languages we deal with in this paper do not contain ϵ either.

Next, we observe the following. To each configuration $\gamma = [i, s, \theta]$ of a deterministic one-way wPA corresponds the vector $\varphi^\gamma = (P_1, \ldots, P_i)$, where

$$P_j = \{h < j : \theta(h) = \theta(j)\}.$$

That is, P_j is the set of pebbles placed before pebble j which are at the same position as pebble j in configuration γ.[5]

If $\gamma \vdash \gamma'$, then $\varphi^{\gamma'}$ can be computed from φ^γ, according to the automaton transitions. Namely, if $\varphi^\gamma = (P_1, \ldots, P_i)$ and the transition applied to γ is $\alpha \to (p, \text{action})$, then

- if action is move, then $\varphi^{\gamma'} = (P_1, \ldots, P_{i-1}, \emptyset)$,
- if action is lift, then $\varphi^{\gamma'} = (P_1, \ldots, P_{i-1})$, and
- if action is place, then $\varphi^{\gamma'} = (P_1, \ldots, P_i, P_i \cup \{i\})$.

We can extend the set of states from S to

$$S \times \bigcup_{i=1}^{k} \{(P_1, \ldots, P_i) : P_j \subseteq \{1, \ldots, j-1\}, \ j = 1, \ldots, i\}$$

capturing in such a way the pebbles' positions by the state. This allows us to remove the P component from the left hand side of transitions. That is, we may assume that the left hand side of a transition is of the form (i, σ, V, s) or (i, V, s).

Finally, by adding some extra states and modifying the transitions appropriately, we can normalize the k-wPA behavior such that for each $i \in \{2, \ldots, k\}$ it acts as follows, cf. [6].

---

[4] That is, pebble $i + 1$ is placed at the position of pebble i, whereas in the strong PA model this pebble is placed at the beginning of the input word, i.e., at the leftmost position.

[5] By definition, $P_1 = \emptyset$ and, therefore, is redundant.

- A pebble is never lifted, but falls down when moving from the right end of the input. Thus, action `lift` is redundant.
- Only pebble 1 can enter a final state and only after it falls down from the right end of the input. In such a case, the accepting configuration consists of the corresponding accepting state only.
- Immediately after pebble i moves without falling down, pebble $i+1$ is placed.
- Immediately after pebble i falls down, pebble $i-1$ moves.

In what follows, we denote the set of letters occurring in a word u by $[u]$. That is, if $u = u_1 \cdots u_n$, then $[u] = \{u_1, \ldots, u_n\}$.

Example 1 (Cf. [4, Example 3.1]). This example deals with the language $L_{\mathtt{diff}}$ consisting of all words in which every symbol from Σ occurs at most one time:

$$L_{\mathtt{diff}} = \{\sigma_1 \cdots \sigma_n : n \geq 1, \sigma_i \neq \$, \text{ for each } i = 1, \ldots, n, \text{ and}$$

$$\sigma_i \neq \sigma_j, \text{ whenever } i \neq j\}.$$

This language is accepted by a 2-wPA that acts as follows. Pebble 1 advances through the input from left to right. At each step it verifies that the symbol under it is not $\$$, and then pebble 2 scans the suffix to the right of the position of pebble 1 to verify that the input symbol under pebble 1 differs from all symbols in that suffix.

Example 2. The language

$$L_{\mathtt{diff\$diff}} = \{u\$v : u, v \in L_{\mathtt{diff}}\}$$

is accepted by a 2-wPA that first, using the automaton from Example 1 scans u and then, using the same automaton, scans v.

Example 3. The language

$$L_{\subseteq} = \{u\$v : u, v \in L_{\mathtt{diff}} \text{ and } [u] \subseteq [v]\}$$

is accepted by a 2-wPA that acts as follows. Pebble 1 advances through the input to the separator $\$$. After each move of pebble 1 on u, pebble 2 moves to $\$$ and then scans the suffix v of the input to find the symbol under pebble 1. Verifying that both u and v are in $L_{\mathtt{diff}}$ can be done by the automaton from Example 2.

Theorem 1. *The language*

$$L_{\supseteq} = \{u\$v : u, v \in L_{\mathtt{diff}} \text{ and } [v] \subseteq [u]\}$$

is not accepted by wPA.

The proof of Theorem 1 for the case of 2-wPA (see footnote 1) is presented in the next section.

Since $L_{\supseteq}$ is the reversal of $L_{\subseteq}$, by Example 3 and Theorem 1, the languages accepted by wPA are not closed under reversal.

3   Proof of Theorem 1

As we have already mentioned above, the proof is restricted to the case of 2-wPA only. For the proof of the general case, see [7, Sects. 6 and 7].[6]

For the rest of this paper, $\mathfrak{A} = \langle S, s_0, F, T \rangle$ is a 2-wPA and the positions of pebble 1 in runs of $\mathfrak{A}$ will be denoted by p_1, possibly primed.

We construct a word $\boldsymbol{w} \in L_{\supseteq}$ such that the run of $\mathfrak{A}$ on its prefix is periodic. That is, the sequence of states in the run on the prefix is periodic. Using the periodicity we can shrink this prefix without affecting acceptance. It should be emphasized that periodicity alone is not sufficient for deleting a pattern from the input. This is because each move of pebble 1 depends not only on the prefix up to its position, but on the whole input word, see also the note in the end of this section.

We start with examining the run of pebble 2.

Proposition 1. *There exists a positive integer ℓ_2 such that for all $\boldsymbol{w} \in \Sigma^+$, $\boldsymbol{w} = w_1 \cdots w_n$, the following holds. If*

$$[2, s_{j_1}, (p_1, j_1)] \vdash [2, s_{j_1+1}, (p_1, j_1 + 1)] \vdash \cdots \vdash [2, s_{j_2}, (p_1, j_2)], \tag{1}$$

where $w_j \neq w_{p_1}$ for all $j_1 \leq j \leq j_2$, then the sequence of states $s_{j_1+\ell_2}, \ldots, s_{j_2}$, is periodic with period ℓ_2.[7]

Proof. Let j_1 and j_2 satisfy the prerequisites of the proposition. The transitions applied to the configurations in (1) are of the form $(2, \emptyset, s) \to (\mathtt{move}, s')$ for some $s, s' \in S$, because the moves of pebble 2 do not depend on w_{p_1}.

Since $\mathfrak{A}$ is deterministic, for some positive integers $m_{s_{j_1}}, \ell_{s_{j_1}} \leq |S|$, after $m_{s_{j_1}}$ steps from w_{j_1}, pebble 2 becomes periodic with a period $\ell_{s_{j_1}}$. Thus, the proposition holds for $\ell_2 = |S|!$, because $m_{s_{j_1}} \leq |S| \leq |S|!$ and $\ell_{s_{j_1}} \leq |S|$ implies that $\ell_{s_{j_1}}$ divides $|S|!$.

Corollary 1. *Let $\boldsymbol{z}' = \boldsymbol{xy}'$ and $\boldsymbol{z}'' = \boldsymbol{xy}''$, $\boldsymbol{x} = x_1 \cdots x_n$, where*

$$[\boldsymbol{x}] \cap ([\boldsymbol{y}'] \cup [\boldsymbol{y}'']) = \emptyset,$$

$$|\boldsymbol{y}'|, |\boldsymbol{y}''| \geq \ell_2,$$

and

$$|\boldsymbol{y}''| \equiv_{\ell_2} |\boldsymbol{y}'|.[8]$$

If

$$[2, s, (p, |\boldsymbol{x}|)] \vdash_{\boldsymbol{z}'} [2, t, (p, |\boldsymbol{xy}'|)], \tag{2}$$

then

$$[2, s, (p, |\boldsymbol{x}|)] \vdash_{\boldsymbol{z}''} [2, t, (p, |\boldsymbol{xy}''|)]. \tag{3}$$

---

[6] Note that in [7] the pebbles are placed in the reversed order, i.e., the computation start with pebble k and pebble i is placed *after* pebble $i + 1$, $i = 1, \ldots, k - 1$.

[7] Recall that we identify θ with the tuple of its values and, by the observation in the previous section, we omit the P-component of transitions.

[8] As usual, $\equiv_{\ell_2}$ is the congruence modulo l_2.

Proof. Let $\boldsymbol{y}' = \boldsymbol{y}'_1\boldsymbol{y}'_2$ and $\boldsymbol{y}'' = \boldsymbol{y}''_1\boldsymbol{y}''_2$, where

$$\boldsymbol{y}'_1 = \boldsymbol{y}''_1 = \ell_2. \tag{4}$$

It follows from (2) that for some state s_1

$$[2, s, (p, |\boldsymbol{x}|)] \vdash_{z'} [2, s_1, (p, |\boldsymbol{x}\boldsymbol{y}'_1|)] \tag{5}$$

and

$$[2, s_1, (p, |\boldsymbol{x}\boldsymbol{y}'_1|)] \vdash_{z'} [2, t, (p, |\boldsymbol{x}\boldsymbol{y}'_1\boldsymbol{y}'_2|)] = [2, t, (p, |\boldsymbol{x}\boldsymbol{y}'|)]. \tag{6}$$

It follows from (5) that

$$[2, s, (p, |\boldsymbol{x}|)] \vdash_{z''} [2, s_1, (p, |\boldsymbol{x}\boldsymbol{y}''_1|)], \tag{7}$$

because the moves of the automaton do not depend on x_p – the symbol under pebble 1, and it follows from (6) that

$$[2, s_1, (p, |\boldsymbol{x}\boldsymbol{y}''_1|)] \vdash_{z''} [2, t, (p, |\boldsymbol{x}\boldsymbol{y}''_1\boldsymbol{y}''_2|)] = [2, t, (p, |\boldsymbol{x}\boldsymbol{y}''|)], \tag{8}$$

because, by Proposition 1, (4) implies that the automaton is periodic with period ℓ_2 from state s_1 and

$$|\boldsymbol{y}'_2| = |\boldsymbol{y}'| - |\boldsymbol{y}'_1| \equiv_{\ell_2} |\boldsymbol{y}''| - |\boldsymbol{y}''_1| = |\boldsymbol{y}''_2|.$$

Combining (7) and (8), we obtain (3).

Corollary 2. *Let* $w, w' \in L_{\mathtt{diff\$diff}}$, $w = \boldsymbol{u}'v\$\boldsymbol{x}$ *and* $w' = \boldsymbol{u}'\boldsymbol{u}''v\$\boldsymbol{x}$ *be such that* $|\boldsymbol{u}''| \equiv_{\ell_2} 0$ *and* $|v| \geq \ell_2$. *If*

$$[1, s_0, (1)] \vdash^*_w [1, t, (|\boldsymbol{u}'|)],$$

then

$$[1, s_0, (1)] \vdash^*_{w'} [1, t, (|\boldsymbol{u}'|)].$$

Proof. It suffices to show that for any $i < |\boldsymbol{u}'|$

$$[1, s, (i)] \vdash^*_w [1, t, (i + 1)] \tag{9}$$

implies

$$[1, s, (i)] \vdash^*_{w'} [1, t, (i + 1)] \tag{10}$$

from which the corollary follows by a straightforward induction on the length of $\boldsymbol{u}'$.

We break the automaton run (9) into three parts:

$$[1, s, (i)] \vdash_w [2, s_1, (i, i)] \vdash^*_w [2, s_2, (i, |\boldsymbol{u}'|)], \tag{11}$$

$$[2, s_2, (i, |\boldsymbol{u}'|)] \vdash^*_w [2, s_3, (i, |\boldsymbol{u}'v\$|)], \tag{12}$$

and

$$[2, s_3, (i, |\boldsymbol{u}'\boldsymbol{v}\$|)] \vdash_{\boldsymbol{w}}^* [2, s_4, (i, |\boldsymbol{w}|)] \vdash_{\boldsymbol{w}} [1, s_5, (i)] \vdash_{\boldsymbol{w}} [1, t, (i+1)], \qquad (13)$$

where the automaton enters configuration $[1, s_5, (i)]$ after pebble 2 falls down from the right end of the input entering state s_5.

From (11), by the same moves of the automaton,

$$[1, s, (i)] \vdash_{\boldsymbol{w}'} [2, s_1, (i, i)] \vdash_{\boldsymbol{w}'}^* [2, s_2, (i, |\boldsymbol{u}'|)] \qquad (14)$$

and from (12), by Corollary 1 with p, $\boldsymbol{x}$, $\boldsymbol{y}'$, and $\boldsymbol{y}''$ being i, $\boldsymbol{u}'$, $\boldsymbol{v}$, and $\boldsymbol{u}''\boldsymbol{v}$, respectively,

$$[2, s_2, (i, |\boldsymbol{u}'|)] \vdash_{\boldsymbol{w}'}^* [2, s_3, (i, |\boldsymbol{u}'\boldsymbol{u}''\boldsymbol{v}\$|)]. \qquad (15)$$

Finally, from (13), by the same moves of the automaton,

$$[2, s_3, (i, |\boldsymbol{u}'\boldsymbol{u}''\boldsymbol{v}\$|)] \vdash_{\boldsymbol{w}'}^* [2, s_4, (i, |\boldsymbol{w}|)] \vdash_{\boldsymbol{w}'} [1, s_5, (i)] \vdash_{\boldsymbol{w}'} [1, t, (i+1)], \qquad (16)$$

because both $\boldsymbol{w}$ and $\boldsymbol{w}'$ have the same suffix $\boldsymbol{x}$ and in both runs pebble 1 is placed above the same symbol.

Combining (14)–(16), we obtain (10).

Definition 2. *Let ℓ be a positive integer and let $\boldsymbol{u}, \boldsymbol{v} \in L_{diff}$, $\boldsymbol{u} = u_1 \cdots u_m$ and $\boldsymbol{v} = v_1 \cdots v_n$, be such that $[\boldsymbol{u}] \subseteq [\boldsymbol{v}]$: $u_i = v_{j_i}$, $i = 1, \ldots, m$. We say that $\boldsymbol{u}$ is ℓ-spread in $\boldsymbol{v}$, if for all $i = 1, \ldots, m$, $j_i > j_{i-1}$ and $j_i \equiv_\ell j_{i-1}$, where $j_0 = 0$.*

Proposition 2. *Let $\boldsymbol{w} = \boldsymbol{u}\boldsymbol{v}\$\boldsymbol{x} \in L_{diff\$diff}$, where $\boldsymbol{u}$ is ℓ_2-spread in $\boldsymbol{x}$, and let $1 < p_1' < p_1'' \leq |\boldsymbol{u}|$. If*

$$[2, s, (p_1', |\boldsymbol{u}\boldsymbol{v}\$|)] \vdash^* [1, t, (p_1')], \qquad (17)$$

then

$$[2, s, (p_1'', |\boldsymbol{u}\boldsymbol{v}\$|)] \vdash^* [1, t, (p_1'')].^9 \qquad (18)$$

Proof. Let

- $\boldsymbol{u} = u_1 \cdots u_m$ and $\boldsymbol{x} = x_1 \cdots x_n$, and
- $u_{p_1'} = x_{j'}$ and $u_{p_1''} = x_{j''}$.

Then $j' < j''$ and it follows from (17) that for some states t' and t'',

$$[2, s, (p_1', |\boldsymbol{u}\boldsymbol{v}\$|)] \vdash^* [2, t', (p_1', |\boldsymbol{u}\boldsymbol{v}\$| + j')] \vdash [2, t'', (p_1', |\boldsymbol{u}\boldsymbol{v}\$| + j' + 1)] \qquad (19)$$

and

$$[2, t'', (p_1', |\boldsymbol{u}\boldsymbol{v}\$| + j' + 1)] \vdash^* [1, t, (p_1')]. \qquad (20)$$

---

[9] The automaton enters configurations $[1, t, (p_1')]$ and $[1, t, (p_1'')]$ after pebble 2 falls down from the right end of the input entering state t.

Since $\boldsymbol{u}$ is ℓ_2-spread in $\boldsymbol{x}$, both j' and j'' are divisible by ℓ_2. Thus, it follows from (19), by Proposition 1, that

$$[2, s, (p_1'', |\boldsymbol{uv}\$|)] \vdash^* [2, t', (p_1'', |\boldsymbol{uv}\$| + j'')] \vdash [2, t'', (p_1'', |\boldsymbol{uv}\$| + j'' + 1)], \quad (21)$$

because $x_{j'} = u_{p_1'}$ and $x_{j''} = u_{p_1''}$.

Finally, since

$$(|\boldsymbol{w}| - (|\boldsymbol{uv}\$| + j' + 1)) - (|\boldsymbol{w}| - (|\boldsymbol{uv}\$| + j'' + 1)) = j'' - j' \equiv_{\ell_2} 0,$$

it follows from (20), by Proposition 1, that

$$[2, t'', (p_1'', |\boldsymbol{uv}\$| + j'' + 1)] \vdash^* [1, t, (p_1'')]. \quad (22)$$

Combining (21) and (22), we obtain (18).

Corollary 3. *Let* $\boldsymbol{w} = \boldsymbol{uv}\$\boldsymbol{x} \in L_{\mathtt{diff}\$\mathtt{diff}}$ *be such that* $\boldsymbol{u}$ *is* ℓ_2-*spread in* $\boldsymbol{x}$ *and* $|\boldsymbol{v}| \geq \ell_2$, *and let* $p_1' < p_1'' \leq |\boldsymbol{u}|$ *be equivalent modulo* ℓ_2. *If*

$$[2, s, (p_1', p_1')] \vdash^* [2, t, (p_1' + 1, p_1' + 1)],$$

then

$$[2, s, (p_1'', p_1'')] \vdash^* [2, t, (p_1'' + 1, p_1'' + 1)].$$

Proof. Let s_1, s_2, and s_3 be the states such that

$$[2, s, (p_1', p_1')] \vdash^* [2, s_1, (p_1', |\boldsymbol{uv}\$|)] \vdash^* [1, s_2, (p_1')]$$
$$\vdash [1, s_3, (p_1' + 1)] \vdash [2, t, (p_1' + 1, p_1' + 1)].$$

Then, by Propositions 1 and 2,

$$[2, s, (p_1'', p_1'')] \vdash^* [2, s_1, (p_1'', |\boldsymbol{uv}\$|)] \vdash^* [1, s_2, (p_1'')]$$
$$\vdash [1, s_3, (p_1'' + 1)] \vdash [2, t, (p_1'' + 1, p_1'' + 1)].$$

Proposition 3 below shows that the behavior of pebble 1 on words $\boldsymbol{uv}\$\boldsymbol{x} \in L_{\mathtt{diff}\$\mathtt{diff}}$ such that $\boldsymbol{u}$ is ℓ_2-spread in $\boldsymbol{x}$ is also periodic.

Proposition 3. *For each* $\boldsymbol{w} = \boldsymbol{uv}\$\boldsymbol{x} \in L_{\mathtt{diff}\$\mathtt{diff}}$ *such that* $\boldsymbol{u}$ *is* ℓ_2-*spread in* $\boldsymbol{x}$ *and* $|\boldsymbol{v}| \geq \ell_2$, *there exist positive integers* m_w *and* ℓ_w *for which the following holds. If*

$$[2, s_{j_1}, (p_1, p_1)] \vdash^* [2, s_{j_2}, (p_1 + 1, p_1 + 1)] \vdash^* \cdots \vdash^* [2, s_{j_{|u|-p_1}}, (|\boldsymbol{u}|, |\boldsymbol{u}|)],$$

then the sequence of states $s_{p_1+m_w}, \ldots, s_{|u|-p_1}$ *is periodic with period* ℓ_w.

Proof. Let t_i, $i = 1, \ldots, |\boldsymbol{u}| - p_1$, be such that

$$[2, s_{j_i}, (p_1 + i, p_1 + i)] \vdash^* [2, t_i, (p_1 + i, |\boldsymbol{uv}\$|)].$$

That is, t_i is the state in which pebble 2 arrives at $\$$, when pebble 1 is placed above the $(p_1 + i)$th symbol of $\boldsymbol{u}$.

Let $m_w = \ell_2 |S| + 1$. Since there are ℓ_2 equivalence classes modulo ℓ_2 and the number of different states in the sequence is bounded by $|S|$, there are two indices j_1 and j_2,

$$p_1 \leq j_1 < j_2 \leq p_1 + m_w$$

such that $j_1 \equiv_{\ell_2} j_2$ and $t_{j_1} = t_{j_2}$.

We put $\ell_w = j_2 - j_1$. It follows from Corollary 3 by a straightforward induction on $i = 0, 1, \ldots$ (with $s = t_{j_1+i}$, $t = t_{j_1+i+1}$, $p_1' = j_1 + i$ and $p_1'' = j_2 + i$) that

$$[2, t_{j_1+i}, (j_1 + i, j_1 + i)] \vdash^* [2, t_{j_1+i+1}, (j_1 + i + 1, j_1 + i + 1)]$$

implies

$$[2, t_{j_2+i}, (j_2 + i, j_2 + i)] \vdash^* [2, t_{j_2+i+1}, (j_2 + i + 1, j_2 + i + 1)].$$

Thus, the proposition follows from the equality $t_{j_1} = t_{j_2}$.

Corollary 4. *There exist a positive integer ℓ_1 such that the following holds. Let $w = uv\$x \in L_{\mathtt{diff\$diff}}$, where u is ℓ_2-spread in x and $|v| \geq \ell_2$. If*

$$[2, s_{j_1}, (p_1, p_1)] \vdash^* [2, s_{j_2}, (p_1 + 1, p_1 + 1)] \vdash^* \cdots \vdash^* [2, s_{j_{|u|-p_1}}, (|u|, |u|)],$$

then the sequence of states $s_{p_1+\ell_1}, \ldots, s_{|u|-p_1}$ is periodic with period ℓ_1.

Proof. It follows from the proof of Proposition 3 that for each $w = uv\$x \in L_{\mathtt{diff\$diff}}$ such that u is ℓ_2-spread in x, $m_w, \ell_w \leq \ell_2 |S| + 1$. Thus, we can put $\ell_1 = (\ell_2 |S| + 1)!$, because the latter is divisible by both m_w and ℓ_w for all above w.

At last, we have arrived at the proof of Theorem 1.

Proof (of Theorem 1). Assume to the contrary that $L(\mathfrak{A}) = L_{\supseteq}$. Let

$$w' = u'u''v\$x \in L_{\supseteq} \cap L_{\mathsf{C}},^{10}$$

where u' is ℓ_2-spread in x, $|v| \geq \ell_2$, and $|u'| = |u''| = \ell_1$; and let $w = u'v\$x$.

Since $\ell_1 \equiv_{\ell_2} 0$, by Corollary 2,

$$[1, s_0, (1)] \vdash_w^* [1, t, (|u'|)]$$

implies

$$[1, s_0, (1)] \vdash_{w'}^* [1, t, (|u'|)]$$

and, since $|u'| = |u''| = \ell_1$, by Corollary 4 with $p_1 = 1$,

$$[1, s_0, (1)] \vdash_{w'}^* [1, t, (|u'u''|)].$$

In addition, the runs of $\mathfrak{A}$ from state t on the (same) suffix $v\$x$ of w and w' are the same. In particular, they terminate in the same state. However, w' belongs to $L_{\supseteq}$, whereas w does not.

Note that periodicity of pebble 1 alone (Corollary 4) without periodicity of pebble 2 (Corollary 2) is not sufficient for deleting the pattern u'' from w'. This is because pebble 1 has to arrive at position $|u'|$ in the same state in the runs of $\mathfrak{A}$ on w and w' and these runs depend on the whole inputs.

---

[10] Thus, both $u'u''v$ and x are in $L_{\mathtt{diff}}$ and $[u'u''v] = [x]$.

4 Removing the Distinguished Separator Symbol $

In this section we show how to modify the languages $L_\subseteq$ and $L_\supseteq$ to languages whose words do not contain a distinguished separator symbol.

Let

$$L'_\subseteq = \{\sigma u \sigma v : \sigma u, \sigma v \in L_{\mathtt{diff}} \text{ and } [u] \subseteq [v]\}$$

and

$$L'_\supseteq = \{\sigma u \sigma v : \sigma u, \sigma v \in L_{\mathtt{diff}} \text{ and } [v] \subseteq [u]\}.$$

Then $L'_\supseteq$ is the reversal of $L'_\subseteq$.

The language $L'_\subseteq$ is accepted by a 3-wPA that acts as follows. Pebble 1 at the leftmost position is used to distinguish the second σ of the input word. Pebble 2 advances through the input to the second σ. After each move of pebble 2 on u, pebble 3 moves to the second σ and then scans the suffix v of the input to find the symbol under pebble 2. At the end of the computation, pebble 1 moves to the end of the input to accept. Verifying that both u and v are in $L_{\mathtt{diff}}$ can be done by the automaton from Example 2, cf. Example 3.

It can be readily seen that each automaton accepting $L'_\supseteq$ modifies to an automaton accepting $L_\supseteq$. Thus, it follows from Theorem 1 that $L'_\supseteq$ is not accepted by wPA.

Alternatively, similarly to the proof of Theorem 1, one can show that $L'_\supseteq$ is not accepted by wPA that is normalized as follows.

- A pebble is never lifted, but falls down when moving from the right end of the input.
- Pebble 1 never leaves the leftmost position.
- Only pebble 2 can enter a final state and only after it falls down from the right end of the input.
- Immediately after pebble i moves without falling down, pebble $i + 1$ is placed.
- Immediately after pebble i falls down, pebble $i - 1$ moves.

In such a way, transitions of the form

$$(i, \$, V, s) \rightarrow \beta$$

in the proof of Theorem 1 are replaced with transitions of the form

$$(i + 1, \{1\} \cup \{j + 1 : j \in V\}, s) \rightarrow \beta.$$

5 Concluding Remark

It seems that the "shrinking" technique applied for the proof of Theorem 1 is quite appropriate for dealing with computations over infinite alphabets. For example, shrinking the input (by totally different tools) was used in [1] for proving decidability of languages accepted by certain variants of FMA.

References

1. Genkin, D., Kaminski, M., Peterfreund, L.: A note on the emptiness problem for alternating finite-memory automata. Theoret. Comput. Sci. **526**, 97–107 (2014)
2. Kaminski, M., Francez, N.: Finite-memory automata. In: Proceedings of the 31st Annual IEEE Symposium on Foundations of Computer Science, Los Alamitos, CA, pp. 683–688. IEEE Computer Society Press (1990)
3. Kaminski, M., Francez, N.: Finite-memory automata. Theoret. Comput. Sci. **134**, 329–363 (1994)
4. Kaminski, M., Tan, T.: A note on two-pebble automata over infinite alphabets. Fundam. Inform. **98**, 1–12 (2010)
5. Neven, F., Schwentick, T., Vianu, V.: Towards regular languages over infinite alphabets. In: Sgall, J., Pultr, A., Kolman, P. (eds.) MFCS 2001. LNCS, vol. 2136, pp. 560–572. Springer, Heidelberg (2001). https://doi.org/10.1007/3-540-44683-4_49
6. Neven, F., Schwentick, T., Vianu, V.: Finite state machines for strings over infinite alphabets. ACM Trans. Comput. Log. **5**, 403–435 (2004)
7. Peterfreund, L.: Closure under reversal of languages over infinite alphabets: a case study. Master's thesis, Department of Computer Science, Technion - Israel Institute of Technology (2015). http://www.cs.technion.ac.il/users/wwwb/cgi-bin/tr-get.cgi/2015/MSC/MSC-2015-20
8. Tan, T.: On pebble automata for data languages with decidable emptiness problem. J. Comput. Syst. Sci. **76**, 778–791 (2010)

Structural Parameterizations
of Dominating Set Variants

Dishant Goyal[1], Ashwin Jacob[2]([✉]), Kaushtubh Kumar[3],
Diptapriyo Majumdar[2], and Venkatesh Raman[2]

[1] Indian Institute of Technology Delhi, New Delhi, India
dishant.in@gmail.com
[2] The Institute of Mathematical Sciences, HBNI, Chennai, India
{ajacob,diptapriyom,vraman}@imsc.res.in
[3] Mentor Graphics, Noida, India
kaustubhkp10@gmail.com

Abstract. We consider structural parameterizations of the fundamental dominating set problem and its variants in the parameter ecology program. We give improved fixed-parameter tractable (FPT) algorithms and lower bounds under well-known conjectures for dominating set in graphs that are k vertices away from a cluster graph or a split graph. These are graphs in which there is a set of k vertices (called the modulator) whose deletion results in a cluster graph or a split graph. We also call k as the deletion distance (to the appropriate class of graphs). Specifically, we show the following results. When parameterized by the deletion distance k to cluster graphs,

- we can find a minimum dominating set in $\mathcal{O}^*(3^k)$ time ($\mathcal{O}^*$ notation ignores polynomial factors of input). Within the same time, we can also find a minimum independent dominating set (IDS) or a minimum efficient dominating set (EDS) or a minimum total dominating set. These algorithms are obtained through a dynamic programming approach for an interesting generalization of set cover which may be of independent interest.
- We complement our upper bound results by showing that at least for dominating set and total dominating set, $\mathcal{O}^*((2-\epsilon)^k)$ time algorithm is not possible for any $\epsilon > 0$ under, what is known as, Set Cover Conjecture. We also show that most of these variants of dominating set do not have polynomial sized kernel.

The standard dominating set and most of its variants are NP-hard or W[2]-hard in split graphs. For the two variants IDS and EDS that are polynomial time solvable in split graphs, we show that when parameterized by the deletion distance k to split graphs,

- IDS can be solved in $\mathcal{O}^*(2^k)$ time and we provide an $\Omega(2^k)$ lower bound under the strong exponential time hypothesis (SETH);
- the 2^k barrier can be broken for EDS by designing an $\mathcal{O}^*(3^{k/2})$ algorithm. This is one of the very few problems with a runtime better than $\mathcal{O}^*(2^k)$ in the realm of structural parameterization. We also show that no $2^{o(k)}$ algorithm is possible unless the exponential time hypothesis (ETH) is false.

© Springer International Publishing AG, part of Springer Nature 2018
F. V. Fomin and V. V. Podolskii (Eds.): CSR 2018, LNCS 10846, pp. 157–168, 2018.
https://doi.org/10.1007/978-3-319-90530-3_14

1 Introduction

1.1 Motivation

The DOMINATING SET problem is one of the classical NP-Complete graph theoretic problems. It asks for a minimum set of vertices in a graph such that every
vertex is either in that set or has a neighbor in that set. It, along with several
variations including *independent domination, total domination, efficient domination, connected domination, total perfect domination, threshold domination*
are well-studied in all algorithmic paradigms including parameterized complexity and approximation and in structural points of view. All of these versions
are hard for the parameterized complexity class W[1] in general graphs when
parameterized by solution size [12] and hence is unlikely to be fixed-parameter
tractable (See [8] for more details).

One of the goals in parameterized complexity is to identify parameters under
which (even hard) problems are fixed-parameter tractable. This is also of practical interest as often there are some small parameters (other than solution size)
that capture important practical inputs. This has resulted in the parameter ecology program where one studies problems under a plethora of parameters and
recently there has been a lot of active research [4,13,18] in this area. In particular, identifying a parameter as small as possible, under which a problem is
fixed-parameter tractable or has a polynomial sized kernel is an interesting direction of research. We continue this line of research and consider parameterizations
of DOMINATING SET variants that are more natural and functions of the input
graph. *Structural parameterization* of a problem is where the parameter is a
function of the input structure rather than the standard output size. To the best
of our knowledge, this is the first serious study of structural parameterization of
any version of the dominating set problem.

Our parameter of interest is the 'distance' of the graph from a natural class of
graphs. Here by distance we mean the number of vertices whose deletion results
in the class of graphs. Note that if dominating set is NP-hard in a graph class,
then it will continue to be NP-hard even on graphs that are k away from the
class, even for constant k (in particular for $k = 0$) and hence is unlikely to be
fixed-parameter tractable. Hence it is natural to consider graphs that are not
far from a class of graphs where DOMINATING SET is polynomial time solvable.
Our case study considers two such special graphs: cluster graphs where each
connected component is a clique and split graphs where the vertex set can be
partitioned into a clique and an independent set. In the former, all the variants
of dominating set we consider are polynomial time solvable, while in the latter
class of split graphs, we consider independent and efficient dominating set that
are polynomial time solvable. We call the set of vertices whose deletion results
in a cluster graph and split graph as *cluster vertex deletion set* (CVD) and *split
vertex deletion set* (SVD) respectively.

Finally, we remark that the size of minimum CVD and minimum SVD are
at most the size of a minimum vertex cover in a graph, which is a well-studied
parameterization in the parameter-ecology program [13].

1.2   Definitions, Our Results and Organization of the Paper

We start with describing the variants of dominating set we consider in the paper. A subset $S \subseteq V(G)$ is a *dominating set* if $N[S] = V(G)$. If S is an independent set, then S is an *independent dominating set*. It is called an *efficient dominating set* if for every vertex $v \in V$, $|N[v] \cap S| = 1$. Note that an efficient dominating set may not exist for a graph (for example, for a 4-cycle). If for every vertex v, $|N(v) \cap S| \geq r$, S is a *threshold dominating set* with threshold r. When $r = 1$, S is a *total dominating set*. Note that for dominating set, the vertices in S do not need other vertices to dominate them, but they do in a total dominating set. For more on these dominating set variants, see [15]. We will often denote *dominating set, efficient dominating set, independent dominating set, total dominating set* and *threshold dominating set* by DS, EDS, IDS, TDS and ThDS respectively in the rest of the article. When we say that a graph G is k-away from a graph in a graph class, what we mean is that there is a subset S of k vertices in the graph such that $G\backslash S$ belongs to the class.

Now we describe the main results in the paper (See Table 1 for a summary). When parameterized by the deletion distance k to cluster graphs,

- we can find a minimum dominating set in $\mathcal{O}^*(3^k)$ time. Within the same time, we can also find a minimum independent dominating set (IDS) or a minimum efficient dominating set (EDS) or a minimum total dominating set. We also give an $\mathcal{O}^*((r + 2)^k)$ algorithm for minimum threshold dominating set with threshold r. These algorithms are obtained through a dynamic programming approach for interesting generalizations of set cover which may be of independent interest. These results are discussed in Sect. 4.1.
- We complement our upper bound results by showing that for dominating set and total dominating set, $\mathcal{O}^*((2 - \epsilon)^k)$ algorithm is not possible for any $\epsilon > 0$ under what is known as Set Cover Conjecture. We also show that for IDS, $\mathcal{O}^*((2 - \epsilon)^k)$ algorithm is not possible for any $\epsilon > 0$ under the Strong Exponential Time Hypothesis (SETH) and for EDS no $2^{o(k)}$ algorithm is possible unless the Exponential Time Hypothesis (ETH) is false. It also follows from our reductions that dominating set, TDS and IDS do not have polynomial sized kernels unless NP $\subseteq$ coNP/poly. These results are discussed in Sect. 4.2.

The standard dominating set and most of its variants are NP-hard or W[2]-hard in split graphs [20]. For the two variants IDS and EDS that are polynomial time solvable in split graphs, we show that when parameterized by the deletion distance k to split graphs,

- IDS can be solved in $\mathcal{O}^*(2^k)$ time and provide an $\mathcal{O}^*((2-\epsilon)^k)$ lower bound for any $\epsilon > 0$ assuming SETH. We also show that IDS-SVD has no polynomial kernel unless NP $\subseteq$ coNP/poly.
- The 2^k barrier can be broken for EDS by designing an $\mathcal{O}^*(3^{k/2})$ algorithm. This is one of the very few problems with a runtime better than $\mathcal{O}^*(2^k)$ in the realm of structural parameterization. We also show that no $2^{o(k)}$ algorithm is possible unless the ETH is false. These results are discussed in Sect. 5.

Table 1. Summary of results. Results marked ⋆ indicate our results.

	Cluster deletion set		Split deletion set	
	Algorithms	Lower bounds	Algorithms	Lower bounds
DS, TDS	$\mathcal{O}^*(3^k)$ ⋆	$\mathcal{O}^*((2-\epsilon)^k)$ and No polynomial kernel ⋆		para-NP-hard
IDS	$\mathcal{O}^*(3^k)$ ⋆	$\mathcal{O}^*((2-\epsilon)^k)$ and No polynomial kernel	$\mathcal{O}(2^k)$ ⋆	$\mathcal{O}^*((2-\epsilon)^k)$ and No polynomial kernel
EDS	$\mathcal{O}^*(3^k)$ ⋆	$\mathcal{O}^*(2^{o(k)})$ ⋆	$\mathcal{O}^*(3^{k/2})$ ⋆	$\mathcal{O}^*(2^{o(k)})$ ⋆
ThDS	$\mathcal{O}^*((r+2)^k)$ ⋆	No polynomial kernel ⋆		para-NP-hard

2 Preliminaries and Notations

We use $[n]$ to denote the set $\{1,\ldots,n\}$. We use standard terminologies of graph theory book by Diestel [10]. For a graph $G = (V,E)$ we denote n as the number of vertices and m as the number of edges. For a vertex $v \in V(G)$, we denote $N_G(v) = \{u \in V(G)|(u,v) \in E(G)\}$ as the open neighborhood of v. When there is no confusion, we drop the subscript G. By $N[v]$ we denote the close neighborhood of v, i.e. $N[v] = N(v) \cup \{v\}$. For $S \subseteq V(G)$, we denote $N(S) = \{v \in V(G)|\exists u \in S$ such that $(u,v) \in E(G)\}\backslash S$. And we denote $N[S] = N(S)\cup S$. By $N^{=2}(v)$ we denote the set of vertices that are at minimum distance exactly two from v. For $S \subseteq V(G)$, we denote $G[S]$ to be the subgraph induced on S. We say that for vertices $u,v \in V$, u *dominates* v if $v \in N(u)$.

We give a general template of formal definition of problems as follows:

$\mathcal{P}$-$\mathcal{Q}$ **Parameter:** $|S|$
Input: An undirected graph $G = (V,E), S \subseteq V(G)$ which is a $\mathcal{Q}$ and an integer ℓ.
Question: Is there a $\mathcal{P}$ in G of size atmost ℓ?

---

where $\mathcal{P}$ represents an acronym of a dominating set variant among DS, EDS, IDS, TDS and ThDS and $\mathcal{Q}$ that of a modulator among CVD and SVD. For example, in the EDS-SVD problem we are interested in finding an EDS of size atmost ℓ given a k sized SVD where k is the parameter.

We use the following conjectures and theorems to prove some of our lower bounds.

Conjecture 1 (Strong Exponential Time Hypothesis (SETH)) ([17]). There is no $\epsilon > 0$ such that $\forall q \geq 3$, q-CNFSAT can be solved in $\mathcal{O}^*((2-\epsilon)^n)$ time where n is the number of variables in input formula.

Conjecture 2 (Exponential Time Hypothesis (ETH)) ([16,17]). 3-CNF-SAT cannot be solved in $\mathcal{O}^*(2^{o(n)})$ time where the input formula has n variables and m clauses.

Conjecture 3 (Set Cover Conjecture (SCC)) ([7]). There is no $\epsilon > 0$ such that SET COVER can be solved in $\mathcal{O}^*((2 - \epsilon)^n)$ time where n is the size of the universe.

Theorem 1 ([11]). SET-COVER *parameterized by the universe size does not admit any polynomial kernel unless* $\mathsf{NP} \subseteq \mathsf{coNP/poly}$.

Theorem 2 ([14]). CNF-SAT *parameterized by the number of variables admits no polynomial kernel unless* $\mathsf{NP} \subseteq \mathsf{coNP/poly}$.

3   Related Work

Clique-width [6] of a graph is a parameter that measures how close to a clique the graph is. Courcelle et al. [5] showed that for a graph with clique-width at most k, any problem expressible in MSO_1 (monadic second order logic of the first kind) has an FPT algorithm with k as the parameter if a k-expression for the graph (a certificate showing that the clique-width of the graph is at most k) is also given as input. The clique-width of a graph that is k away from a cluster graph can be shown to be $k + 1$ (with a k-expression) and all the dominating set variants discussed in the paper can be expressed in MSO_1 and hence can be solved in FPT time in such graphs. But the running time function $f(k)$ in Courcelle's theorem is huge (more than doubly exponential). Oum et al. [19] gave an $\mathcal{O}^*(k^{O(k)})$ algorithm to solve the minimum dominating set for clique-width k graphs without assuming that the k-expression is given. There is a $\mathcal{O}^*(4^k)$ algorithm by Bodlaender et al. [2] for finding minimum dominating set in graphs with clique-width k when the k-expression is given as input. It is easy to construct the k-expression for graphs k away from a cluster graph and hence we have a $\mathcal{O}^*(4^k)$ algorithm. The algorithms we give in Sect. 4, not only improve the running time but also are applicable for other variants of dominating set.

4   Dominating Set Variants Parameterized by CVD Size

4.1   Upper Bounds

In cluster graphs, a dominating set simply picks an arbitrary vertex from each clique. This dominating set is also efficient and independent. For threshold dominating set with threshold r, we arbitrarily pick $r + 1$ vertices from every clique if possible so that every vertex has r neighbors excluding itself.

We can assume that the CVD set S of size k is given with the input. If not, we can use the algorithm by Boral et al. [3] that runs in $\mathcal{O}^*(1.92^k)$ time and either outputs a CVD set of size at most k or says that no such set exists.

We first look at the problem DS-CVD which is NP-hard as any graph having an edge has a CVD set of at most $n - 2$.

Theorem 3. DS-CVD *can be solved in* $\mathcal{O}^*(3^k)$ *time.*

Proof. Our FPT algorithm starts with making a guess S' for the solution's intersection with S. We delete vertices in $N[S'] \cap S$ as they have been already dominated. We will keep the vertices of $N[S'] \cap (V \backslash S)$ as they can be used to cover the remaining vertices of S.

Let us denote the cliques in $G' = G \backslash S$ as $C_1, C_2, \ldots, C_q$ where $q \leq n - k$. We label the vertices of G' as $v_1, v_2, \ldots, v_{|V \backslash S|}$ such that the first l_1 of them belong to the clique C_1, the next l_2 of them belong to clique C_2 and so on for integers $l_1, l_2, \ldots, l_q$. Note that for some cliques, all the vertices of the clique gets dominated by S'. We are left with the problem of picking the minimum number of vertices from the cliques to dominate, the vertices of the cliques that are not yet dominated (by S'), and $S \backslash N[S']$. We abstract out the problem below.

DS-DISJOINTCLUSTER   **Parameter:** $|S|$
Input: An undirected graph $G = (V, E)$, $S \subseteq V$ such that every connected component of $G \backslash S$ is a clique, a $(0,1)$ vector $(f_1, f_2, \ldots, f_q)$ corresponding for the cliques $(C_1, \ldots, C_q)$ and an integer ℓ.
Question: Does there exist a subset $T \subseteq V \backslash S$ of size ℓ, that dominates all vertices of S and all vertices of all cliques C_i with flags $f_i = 1$?

For the problem we started off with, the set S in this new formulation is the remaining vertices of S after deleting $N[S'] \cap S$. Also f_i is set to 1 if the clique C_i has not been dominated by S' and is set to 0 otherwise.

Lemma 1. *DS-DISJOINTCLUSTER can be solved in $\mathcal{O}^*(2^{|S|})$ time.*

Proof. We formulate this problem instance as a variant of SET-COVER instance. Define the universe U as the set S. For each vertex $v \in V \backslash S$, we define a set $S_v = N(v) \cap S$. Define the family of sets $\mathcal{F} = \{S_v | v \in V \backslash S\}$. We say that a subfamily $\mathcal{F}' \subseteq \mathcal{F}$ covers a subset $W \subseteq U$ if for every element $w \in W$, there exist some set in $\mathcal{F}'$ containing w. Now a SET-COVER solution $\mathcal{F}' \subseteq \mathcal{F}$ for $(U, \mathcal{F})$ will cover all the elements of S. In the graph, the vertices corresponding to the sets in $\mathcal{F}'$ will dominate all the vertices in S. But DS-DISJOINTCLUSTER has the additional requirement of dominating the vertices of every clique C_i with $f_i = 1$ as well. This means from every such clique at least one vertex has to be picked. With this in mind, we define for each clique C_i a collection of sets $\mathcal{B}_i = \{S_v : v \in C_i\}$. We call these sets as *blocks*. Hence the number of blocks and the number of cliques in $G \backslash S$ are the same. We order the sets in the block in the order of the vertices $v_1, \ldots, v_{|V \backslash S|}$. We have the following problem which is a slight generalization of SET-COVER.

SET-COVER WITH PARTITION   **Parameter:** $|U| = k$
Input: A universe U, a family of sets $\mathcal{F} = \{S_1, \ldots, S_m\}$, a partition $\mathcal{B} = (\mathcal{B}_1, \mathcal{B}_2, \ldots, \mathcal{B}_q)$ of $\mathcal{F}$, a $(0,1)$ vector $(f_1, f_2, \ldots, f_q)$ corresponding to each block in the partition $(\mathcal{B}_1, \mathcal{B}_2, \ldots, \mathcal{B}_q)$ and an integer ℓ.
Question: Does there exist a subset $\mathcal{F}' \subseteq \mathcal{F}$ of size ℓ covering U and from each block $\mathcal{B}_i$ with flags $f_i = 1$ at least one set is picked?

Lemma 2 $(\star)$.[1] SET-COVER WITH PARTITION *can be solved in* $\mathcal{O}^*(2^{|U|})$ *time.*

We construct the SET-COVER WITH PARTITION instance from the DS-DISJOINTCLUSTER instance as discussed above. It can be easily seen that there exists a solution of size ℓ in DS-DISJOINTCLUSTER instance if and only if there exists a solution of size ℓ in SET-COVER WITH PARTITION instance. In Lemma 2, we solve SET-COVER WITH PARTITION via dynamic programming. And $|S| = |U|$. This completes the proof. □

Now for each guess $S' \subseteq S$ with $|S'| = i$, we construct the SET-COVER WITH PARTITION instance with $|U| \leq k - i$ and solve it with running time $\mathcal{O}^*(2^{k-i})$. Hence the total running time is $\sum_{i=1}^{k} \binom{k}{i} \mathcal{O}^*(2^{k-i})$ which is $\mathcal{O}(3^k n^{\mathcal{O}(1)})$. □

We show that with some careful modifications to the above dynamic programming algorithm, efficient FPT algorithms for minimum EDS, IDS, TDS and ThDS when parameterized by the size of cluster deletion set can be obtained.

Theorem 4 $(\star)$. EDS-CVD, TDS-CVD *and* IDS-CVD *can be solved in* $\mathcal{O}^*(3^k)$ *time.* THDS-CVD *can be solved in* $\mathcal{O}^*((r+2)^k)$ *time.*

4.2   Lower Bounds

Lemma 3 $(\star)$. *There is a polynomial time algorithm that takes an instance* $(U, \mathcal{F}, \ell)$ *of* SET-COVER *and outputs an instance* (G, ℓ) *of* DS-CVD *(or* TDS-CVD*) such that* G *has a cluster vertex deletion set with exactly* $|U|$ *vertices, such that* $(U, \mathcal{F}, \ell)$ *has a set cover of size* ℓ *if and only if* G *has a (total) dominating set of size* ℓ.

The following theorem follows from the above lemma and Conjecture 3.

Theorem 5 $(\star)$. DS-CVD *and* TDS-CVD *cannot be solved in* $O^*((2-\epsilon)^k)$ *running time for any* $\epsilon > 0$ *unless Set Cover Conjecture fails.*

The following theorem follows from Theorem 1 and Lemma 3.

Theorem 6 $(\star)$. DS-CVD, TDS-CVD *and* THDS-CVD *do not have polynomial sized kernels unless* $\mathsf{NP} \subseteq \mathsf{coNP/poly}$.

Note that the proof idea of Theorem 6 does not work for IDS-CVD. To show $\mathcal{O}^*((2-\epsilon)^k)$ lower bound for IDS-CVD under SETH, we use the following theorem and an observation. Here MMVC-VC problem refers to the problem of finding a maximum sized minimal vertex cover (MMVC) in a graph parameterized by the size of a given vertex cover (VC). Recall that a vertex cover in a graph is a subset of vertices that covers all edges.

---

[1] Due to lack of space, the proofs of Theorems, Lemmas, Observations, Safeness of Reduction Rules marked $\star$ and some omitted details will appear in the full version.

Theorem 7 ([21]).[2]   *Unless SETH fails, MMVC-VC cannot be solved in $\mathcal{O}^*((2-\epsilon)^k)$ time. Moreover, MMVC-VC does not admit polynomial sized kernel unless* NP $\subseteq$ coNP/poly.

Observation 1 (⋆). *If T is a minimal vertex cover of the graph G, then $V(G)\backslash T$ is an independent dominating set in G. Furthermore, if T is a maximum minimal vertex cover, then $V(G)\backslash T$ is a minimum independent dominating set.*

From Observation 1, we know that the complement of a maximum minimal vertex cover is a minimum independent dominating set. Also, any vertex cover is a cluster vertex deletion set. So, from Theorem 7, we have the following result.

Corollary 1 (⋆). *IDS-CVD cannot be solved in $O^*((2-\epsilon)^k)$ time for any $\epsilon > 0$ unless SETH fails. Moreover, IDS-CVD does not have any polynomial kernel unless* NP $\subseteq$ coNP/poly.

For EDS-CVD, we can only prove a weaker lower bound of $2^{o(k)}$ time assuming ETH, but we give the lower bound for EDS parameterized by even a larger parameter, i.e. the size of a vertex cover. We have the following results.

Theorem 8 (⋆). *EDS-VC cannot be solved in $2^{o(|S|)}$ time unless ETH fails.*

Corollary 2. *EDS-CVD cannot be solved in $2^{o(|S|)}$ time unless ETH fails.*

5   Dominating Set Variants Parameterized by SVD Size

In this section, we address the parameterized complexity of dominating set variants when parameterized by the size of a given SVD set S. Note that DS and TDS are NP-hard on split graphs [20]. Hence we focus only on EDS and IDS.

We assume that S is given with the input. Otherwise given (G, k), we use an $\mathcal{O}^*(1.27^{k+o(k)})$ algorithm due to Cygan and Pilipczuk [9] to find a set of vertices of size at most k whose removal makes G into a split graph.

5.1   EDS and IDS Parameterized by SVD Size

First, we provide a simple algorithm for IDS-SVD. The idea is to make a guess for the solution within the SVD and solve the resulting disjoint problem in polynomial time. It turns out that it works for EDS-SVD too.

Theorem 9 (⋆). *EDS-SVD and IDS-SVD can be solved in $O^*(2^k)$ time.*

---

[2] Note that the SETH based lower bound result and the result ruling out the existence of polynomial kernel in this paper use different constructions.

5.2   Lower Bounds for IDS and EDS

We know that any vertex cover is a split vertex deletion set. So, we have the following corollary as a consequence of Theorem 7.

Corollary 3 *(⋆).*[3]   IDS-SVD *cannot be solved in* $\mathcal{O}^*((2 - \epsilon)^k)$ *time unless* SETH *fails and it does not admit polynomial kernels unless* NP $\subseteq$ coNP/poly.

For EDS, as the size of the SVD set is always smaller than the size of the vertex cover, we have the following corollary of Theorem 8.

Corollary 4. EDS-SVD *cannot be solved in* $2^{o(|S|)}$ *time unless ETH fails.*

5.3   Improved Algorithm for EDS-SVD

In this section, we give an improved algorithm for EDS-SVD parameterized by the size of a given split vertex deletion set S breaking the barrier of $\mathcal{O}^*(2^k)$.

Let $F = G \backslash S$. As F is a split graph, $V(F) = C \uplus I$ where C induces a clique and I induces an independent set. The algorithm uses the standard branching technique. Consider any efficient dominating set D of a graph. Any two vertices $u, v \in D$ must have distance at least three. At any intermediate stage of the algorithm, we make a choice of not picking a vertex and we mark such vertices by coloring them red. Other vertices are colored blue. Hence all vertices of G are blue initially. We initialize $D = \emptyset$ which is the solution set we seek. Consider any pair of blue vertices $x, y \in S$. If the distance between x and y is at most two in G, then we use the following branching rule. And we measure the progress of the algorithm by $\mu(G)$ which is the number of blue vertices in S, which is k initially.

Branching Rule 1. *Consider a pair of blue vertices* $x, y \in S$ *such that the distance between* x *and* y *is at most two in* G. *In the first branch, we add* x *into* D, *delete* $N[x]$ *from* G, *color the vertices in* $N^{=2}(x)$ *by red. In the second branch, we add* y *into* D, *delete* $N[y]$ *from* G, *color the vertices in* $N^{=2}(y)$ *by red. In the third branch, we color* x, y *by red.*

Clearly the branches are exhaustive as both x and y cannot be in the EDS solution we seek. Furthermore, in the first branch, x is deleted from S and y is colored red. Symmetrically in the second branch, y is deleted from S and x is colored red. In the third branch, x and y are colored red. So in all the branches, $\mu(G)$ drops by at least two resulting in a $(2, 2, 2)$ branching rule. When this branching rule is not applicable, for every pair of blue vertices $x, y \in S$, $N[x] \cap N[y] = \emptyset$. Now, as C is a clique, we can have at most one vertex from C in the solution. When we decide to pick some vertex $v \in C$ into the solution, then we delete $N[v]$ and color $N^{=2}(v)$ as red. So all vertices of C get deleted. There are at most $|C|$ vertices in C. When we decide not to pick any vertex from C into the solution, then we color all vertices of C as red. So we have $(|C| + 1)$

---

[3] We provide an alternate proof in the full version.

choices from the vertices of C. Measure $\mu(G)$ does not increase in any of these choices. A multiplicative factor of $(|C| + 1)$ would come in the running time because of this one-time branching. Now, we are left with only the vertices of I. Now, we apply the following reduction rule to rule out some simple boundary conditions.

Reduction Rule 1. *If there exists a red vertex $x \in V(G)$ such that $N_G(x)$ has only one blue vertex y, then add y into D, delete $N[y]$ from G and color $N^{=2}(y)$ as red. Also if there exists a blue vertex $x \in V(G)$ such that $N_G(x)$ contains no blue vertex, then add x into D, delete $N[x]$ from G and color $N^{=2}(x)$ as red.*

It is easy to see that the above reduction rule is safe. Note that we have some blue vertices in I. Such vertices can only be dominated by themselves or a unique blue vertex in S, as otherwise Branching Rule 1 would have been applicable. Now, suppose that there exists a blue vertex $x \in S$ that has at least two blue neighbors $u, v \in I$. If we decide to pick u (or symmetrically v) into D, then we are not allowed to pick x or v (symmetrically u) in D but then u or v cannot be dominated. This forces x into D. We have the following reduction rule.

Reduction Rule 2 ($\star$). *If there exists a blue vertex $x \in S$ such that $N_G(x)$ contains at least two blue neighbors in I, then add x into D, delete $N[x]$ from G and color vertices in $N^{=2}(x)$ red.*

Lemma 4 ($\star$). *Reduction Rules 1 and 2 do not increase $\mu(G)$.*

Now if there are red vertices in I having no blue neighbor in S, then we move to the next branch as such a vertex cannot be dominated. Thus any blue vertex in I has only one blue neighbor in S and any blue vertex in S has only one blue neighbor in I. As Reduction Rule 1 is not applicable, any red vertex $x \in S \cup C$ has at least two blue neighbors in $u, v \in N_G(x)$. Clearly both $\{u, v\} \not\subseteq S$ as otherwise Branching Rule 1 would have been applicable. So, now we are left with the case that $u, v \in I$ or $u \in I, v \in S$ but (u, v) may or may not be an edge. Now we apply the following branching rule.

Branching Rule 2. *Let x be a red vertex in S with two blue neighbors u, v.*

1. *If $u, v \in I$, then we branch as follows. In one branch we add u into D, delete $N[u]$ from G , color $N^{=2}(u)$ as red. As $v \in N^{=2}(u)$ and v has only one blue neighbor $z \in S$, we add z also into D, delete $N[z]$ from G and color $N^{=2}(z)$ by red. In the second branch, we add v into D, delete $N[v]$ from G, color $N^{=2}(v)$ as red. As $u \in N^{=2}(v)$ and u has only one blue neighbor $y \in S$, we add y also into D, delete $N^{=2}(y)$ from G and color $N^{=2}(z)$ by red. In the third branch, color both u and v by red. Add the only blue neighbor y of u and z of v into D. Delete $N[y], N[z]$ from G and color the vertices in $N^{=2}(y) \cup N^{=2}(z)$ by red.*

2. *$u \in I, v \in S, (u, v) \notin E(G)$, then we branch as follows. In the first branch, we add u to D, color v as red. This forces us to pick the only blue neighbor z of v where $z \in I$. So, we add z to D. Delete $N[u], N[z]$ from G and color*

$N^{=2}(u), N^{=2}(z)$ as red. In the second branch, e color u as red. This forces us to pick the only neighbor y of u where $y \in S$. And we pick v into D as well as y into D. We delete $N[v], N[y]$ from G and color $N^{=2}(v), N^{=2}(y)$ by red. In the third branch, we color both u and v by red. This forces us to pick the only blue neighbor $z \in N_G(v) \cap I, y \in N_G(u) \cap S$ into D. So, we pick z into D, delete $N[z], N[y]$ from G and color $N^{=2}(y), N^{=2}(z)$ by red.

It is easy to see that $\mu(G)$ drops by at least two in all three branches as eventually two blue vertices of S get deleted in all the branches.

When none of the above rules are applicable, then we have $u \in S, v \in I$ and $(u, v) \in E(G)$. We know that either $u \in D$ or $v \in D$. Consider the red vertices in $N(u)$ and red vertices in $N(v)$. As Branching Rule 1, Reduction Rule 1 and Branching Rule 2 are not applicable, by the following lemma using which we can pick u or v arbitrarily.

Lemma 5 ($\star$). *If Branching Rule 1, Reduction Rule 1 and Branching Rule 2 are not applicable, then $N(u) \backslash \{v\} = N(v) \backslash \{u\}$.*

This completes the description of our algorithm that consists of a sequence of reduction rules and branching rules. The measure is k initially and the branching continues as long as k drops to 0. So, we have the following recurrence.

$$T(k) \leq 3T(k - 2) + \alpha \cdot (n + k)^c$$

Solving this recurrence, we get $\mathcal{O}(1.732^k \cdot n^{\mathcal{O}(1)})$ implying the following theorem.

Theorem 10. EDS-SVD *can be solved in* $\mathcal{O}^*(3^{k/2})$ *time.*

6 Concluding Remarks

We have initiated a study of structural parameterizations of some dominating set variants and complemented with lower bounds based on ETH and SETH. One immediate open problem is to narrow the gap between upper and lower bounds, especially for the dominating set variants parameterized by the size of CVD set.

We know that IDS is the complementary version of MAXIMUM MINIMAL VERTEX COVER problem. So a natural approach for an $\mathcal{O}^*(2^k)$ algorithm for IDS-CVD is to apply the ideas used in [21] to get $\mathcal{O}^*(2^k)$ algorithm for MMVC-VC. But this seems to require more work as there may not exist a minimal vertex cover that intersects the CVD set S in a particular subset.

Recently Bergougnoux and Kanté [1] have given an $\mathcal{O}^*(2^{O(k)})$ algorithm for connected dominating set (the dominating set induces a connected graph) for clique-width k graphs when the k-expression is given as input. An interesting open problem is whether connected dominating set has a simpler FPT algorithm as in the FPT algorithms in this paper, when parameterized by the CVD set size.

References

1. Bergougnoux, B., Kanté, M.M.: Fast exact algorithms for some connectivity problems parametrized by clique-width. arXiv preprint arXiv:1707.03584 (2017)
2. Bodlaender, H.L., van Leeuwen, E.J., van Rooij, J.M.M., Vatshelle, M.: Faster algorithms on branch and clique decompositions. In: Hliněný, P., Kučera, A. (eds.) MFCS 2010. LNCS, vol. 6281, pp. 174–185. Springer, Heidelberg (2010). https://doi.org/10.1007/978-3-642-15155-2_17
3. Boral, A., Cygan, M., Kociumaka, T., Pilipczuk, M.: A fast branching algorithm for cluster vertex deletion. Theory Comput. Syst. **58**(2), 357–376 (2016)
4. Cai, L.: Parameterized complexity of vertex colouring. Discrete Appl. Math. **127**(3), 415–429 (2003)
5. Courcelle, B., Makowsky, J.A., Rotics, U.: Linear time solvable optimization problems on graphs of bounded clique-width. Theory Comput. Syst. **33**, 125–150 (2000)
6. Courcelle, B., Olariu, S.: Upper bounds to the clique width of graphs. Discrete Appl. Math. **101**(1), 77–114 (2000)
7. Cygan, M., Dell, H., Lokshtanov, D., Marx, D., Nederlof, J., Okamoto, Y., Paturi, R., Saurabh, S., Wahlström, M.: On problems as hard as CNF-SAT. ACM Trans. Algorithms (TALG) **12**(3), 41 (2016)
8. Cygan, M., Fomin, F.V., Kowalik, L., Lokshtanov, D., Marx, D., Pilipczuk, M., Pilipczuk, M., Saurabh, S.: Parameterized Algorithms. Springer, Heidelberg (2015). https://doi.org/10.1007/978-3-319-21275-3
9. Cygan, M., Pilipczuk, M.: Split vertex deletion meets vertex cover: new fixed-parameter and exact exponential-time algorithms. Inf. Process. Lett. **113**(5–6), 179–182 (2013)
10. Diestel, R.: Graph Theory. Springer, Heidelberg (2006)
11. Dom, M., Lokshtanov, D., Saurabh, S.: Kernelization lower bounds through colors and IDs. ACM Trans. Algorithms (TALG) **11**(2), 13 (2014)
12. Downey, R.G., Fellows, M.R.: Fixed-parameter tractability and completeness II: on completeness for W[1]. Theor. Comput. Sci. **141**(1–2), 109–131 (1995)
13. Fellows, M.R., Jansen, B.M.P., Rosamond, F.A.: Towards fully multivariate algorithmics: parameter ecology and the deconstruction of computational complexity. Eur. J. Comb. **34**(3), 541–566 (2013)
14. Fortnow, L., Santhanam, R.: Infeasibility of instance compression and succinct PCPs for NP. J. Comput. Syst. Sci. **77**(1), 91–106 (2011)
15. Haynes, T.W., Hedetniemi, S., Slater, P.: Domination in Graphs: Advanced Topics. Marcel Dekker, New York (1997)
16. Impagliazzo, R., Paturi, R.: On the complexity of k-SAT. J. Comput. Syst. Sci. **62**, 367–375 (2001)
17. Impagliazzo, R., Paturi, R., Zane, F.: Which problems have strongly exponential complexity? J. Comput. Syst. Sci. **63**(4), 512–530 (2001)
18. Jansen, B.M.P., Raman, V., Vatshelle, M.: Parameter ecology for feedback vertex set. Tsinghua Sci. Technol. **19**(4), 387–409 (2014)
19. Oum, S.-I., Sæther, S.H, Vatshelle, M.: Faster algorithms parameterized by clique-width. arXiv preprint arXiv:1311.0224 (2013)
20. Raman, V., Saurabh, S.: Short cycles make W-hard problems hard: FPT algorithms for W-hard problems in graphs with no short cycles. Algorithmica **52**(2), 203–225 (2008)
21. Zehavi, M.: Maximum minimal vertex cover parameterized by vertex cover. SIAM J. Discrete Math. **31**(4), 2440–2456 (2017)

Complexity and Inapproximability Results for Parallel Task Scheduling and Strip Packing

Sören Henning, Klaus Jansen, Malin Rau$^{(\boxtimes)}$, and Lars Schmarje

Institut für Informatik, Christian-Albrechts-Universität zu Kiel, Kiel, Germany
{stu114708,kj,mra,stu115194}@informatik.uni-kiel.de

Abstract. We study Parallel Task Scheduling $Pm|size_j|C_{\max}$ with a constant number of machines. This problem is known to be strongly NP-complete for each $m \geq 5$, while it is solvable in pseudo-polynomial time for each $m \leq 3$. We give a positive answer to the long-standing open question whether this problem is strongly NP-complete for $m = 4$. As a second result, we improve the lower bound of $\frac{12}{11}$ for approximating pseudo-polynomial Strip Packing to $\frac{5}{4}$. Since the best known approximation algorithm for this problem has a ratio of $\frac{4}{3} + \varepsilon$, this result narrows the gap between approximation ratio and inapproximability result by a significant step. Both results are proven by a reduction from the strongly NP-complete problem 3-Partition.

1 Introduction

In the Parallel Task Scheduling problem denoted as $P|size_j|C_{\max}$ in the three-field-notation, a set of jobs J has to be scheduled on m machines minimizing the makespan T. Each job $j \in J$ has a processing time $p(j) \in \mathbb{N}$ and requires $q(j) \in \mathbb{N}$ machines. A schedule S is given by two functions $\sigma : J \to \mathbb{N}$ and $\rho : J \to 2^{\{1,\dots,m\}}$. The function σ maps each job to a start point in the schedule, while ρ maps each job to the set of machines it is processed on. We say a machine i contains a job $j \in J$ if $i \in \rho(j)$. A schedule is feasible if each machine processes at most one job at a time and each job is processed on the required number of machines (i.e. $|\rho(j)| = q(j)$). The objective is to find a feasible schedule S minimizing the makespan $T := \max_{j \in J}(\sigma(j) + p(j))$.

In 1989, Du and Leung [1] proved the Parallel Task Scheduling problem $P|size_j|C_{\max}$ to be strongly NP-complete for all $m \geq 5$, while $P|size_j|C_{\max}$ is solvable by a pseudo-polynomial algorithm for all $m \leq 3$. In this paper, we address the case of $m = 4$, which has been open since and prove:

Theorem 1. *Parallel Task Scheduling on 4 machines is strongly* NP-*complete.*

Building on this result, we can prove a lower bound for the absolute approximation ratio of pseudo polynomial algorithms for the Strip Packing problem. In the Strip Packing problem a set of rectangular items I has to be placed into

© Springer International Publishing AG, part of Springer Nature 2018
F. V. Fomin and V. V. Podolskii (Eds.): CSR 2018, LNCS 10846, pp. 169–180, 2018.
https://doi.org/10.1007/978-3-319-90530-3_15

a strip with width $W \in \mathbb{N}$ and infinite height. Each item $i \in I$ has a width $w_i \in \mathbb{N}_{\leq W}$ and a height $h_i \in \mathbb{N}$. A *packing* of the items I into the strip is a function $\rho : I \to \mathbb{Q}_0 \times \mathbb{Q}_0$, which assigns the left bottom corner of an item to a position in the strip, such that for each item $i \in I$ with $\rho(i) = (x_i, y_i)$ we have $x_i + w_i \leq W$. We say two items $i, j \in I$ *overlap* if they share an inner point. A packing is *feasible* if no two items overlap. The height of a packing is defined as $H := \max_{i \in I} y_i + h_i$. The objective is to find a feasible packing of the items I into the strip, that minimizes the packing height. If all item sizes are integral, we can transform feasible packings to packings where all positions are integral, without enlarging the packing height [2]. Therefore, we can assume that we have packings of the form $\rho : I \to \mathbb{N}_0 \times \mathbb{N}_0$.

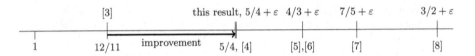

Fig. 1. The upper and lower bounds for the best possible approximation for pseudo-polynomial Strip Packing achieved so far

Lately, pseudo-polynomial algorithms for Strip Packing, where the width of the strip is allowed to appear polynomially in the input size gained high interest, see Fig. 1. In a series of papers [4–8], the best approximation ratio was improved to $\frac{5}{4} + \varepsilon$. On the other hand, it is not possible to find an algorithm with approximation ratio better than $\frac{12}{11}$, except P $=$ NP [3]. In this paper, we improve this lower bound to $\frac{5}{4}$, which almost closes the gap between lower bound and best algorithm.

Theorem 2. *For each $\varepsilon > 0$ it is NP-Hard to approximate Strip Packing with a ratio of $\frac{5}{4} - \varepsilon$ in pseudo-polynomial time.*

Related Work

Parallel Task Scheduling. In 1989, Du and Leung [1] proved Parallel Task Scheduling $Pm|size_j|C_{max}$ to be strongly NP-complete for all $m \geq 5$, while it is solvable by a pseudo-polynomial algorithm for all $m \leq 3$. Amoura et al. [9], as well as Jansen and Porkolab [10], presented a polynomial time approximation scheme (in short PTAS) for the case that m is a constant. A PTAS is a family of algorithms that finds a solution with an approximation ratio of $(1 + \varepsilon)$ for any given value $\varepsilon > 0$. If m is polynomially bounded by the number of jobs, a PTAS still exists [8]. Nevertheless, if m is arbitrarily large, the problem gets harder. By a simple reduction from the Partition problem, one can see that there is no polynomial algorithm with approximation ratio smaller than $\frac{3}{2}$. Parallel Task Scheduling with arbitrarily large m has been widely studied [11–14]. The algorithm with the best known absolute approximation ratio of $\frac{3}{2} + \varepsilon$ was presented by Jansen [15].

Strip Packing. The Strip Packing problem was first studied in 1980 by Baker et al. [16]. They presented an algorithm with an absolute approximation ratio of 3. This ratio was improved by a series of papers [17–21]. The algorithm with the best known absolute approximation ratio by Harren et al. [22] achieves a ratio of $\frac{5}{3} + \varepsilon$. By a simple reduction from the Partition problem, one can see that it is impossible to find an algorithm with better approximation ratio than $\frac{3}{2}$, unless P = NP.

The lower bound of $\frac{3}{2}$ does not hold for asymptotic approximation ratios and they have been studied in various papers [17,23,24]. Kenyon and Rémila [25] presented an asymptotic fully polynomial approximation scheme (in short AFPTAS) with additive term $\mathcal{O}(h_{\max}/\varepsilon^2)$, where h_{max} is the largest occurring item height. An approximation scheme is fully polynomial if its running time is polynomial in $1/\varepsilon$ as well. This algorithm was simultaneously improved by Sviridenko [26] and Bougeret et al. [27] to an algorithm with an additive term of $\mathcal{O}(h_{\max} \log(1/\varepsilon)/\varepsilon)$. Furthermore, at the expense of the running time, Jansen and Solis-Oba [28] presented an asymptotic PTAS with an additive term of $h_{\max}$.

Recently, the focus shifted to pseudo-polynomial algorithms. Jansen and Thöle [8] presented an pseudo-polynomial algorithm with approximation ratio of $\frac{3}{2} + \varepsilon$. Later Nadiradze and Wiese [7] presented an algorithm with ratio $\frac{7}{5} + \varepsilon$. Its approximation ratio was independently improved to $\frac{4}{3} + \varepsilon$ by Gálvez et al. [6] and by Jansen and Rau [5]. $5/4 + \varepsilon$ is the best approximation ratio so far, achieved by an algorithm by Jansen and Rau [4]. All these algorithms have a polynomial running time if the width of the strip W is bounded by a polynomial in the number of items.

In contrast to Parallel Task Scheduling, Strip Packing cannot be approximated arbitrarily close to 1, if we allow pseudo-polynomial running time. This was proved by Adamaszek et al. [3] by presenting a lower bound of $\frac{12}{11}$. As a consequence, Strip Packing admits no quasi-polynomial time approximation scheme, unless NP $\subseteq DTIME(2^{\text{polylog}(n)})$. For an overview on 2-dimensional packing problems and open questions regarding these problems, we refer to the survey by Christensen et al. [29].

2   Hardness of Scheduling Parallel Tasks

First, we introduce some notations. Let $j \in J$ and $J' \subseteq J$. We define the work of j as $w(j) := p(j) \cdot q(j)$ and the total work of J' as $w(J') := \sum_{j \in J'} w(j)$. We denote by $n_j(J')$ the number of jobs from the set J', which are finished before the start of the job j, i.e., $n_j(J') = |\{i \in J' : \sigma(i) + p(i) \le \sigma(j)\}|$. Furthermore, we will use a notation defined in [1] for swapping a part of the content of two machines; let $j \in J$ be a job, that is processed by at least two machines $\tilde{M}$ and $\tilde{M}'$ with start point $\sigma(j)$. We can swap the content of the machines $\tilde{M}$ and $\tilde{M}'$ after time $\sigma(j)$ without violating any scheduling constraint. We define this swapping operation as $SWAP(\sigma(j), \tilde{M}, \tilde{M}')$.

We will prove Theorem 1 by a reduction from the 3-Partition problem. In this problem, we are given a list $\mathcal{I} = (\iota_1, \ldots, \iota_{3z})$ of $3z$ positive integers, with

$\sum_{i=1}^{3z} \iota_i = zD$ and $D/4 < \iota_i < D/2$ for each $1 \le i \le 3z$. The problem is to decide whether there exists a partition of the set $I = \{1, \ldots, 3z\}$ into sets $I_1, \ldots I_z$, such that $\sum_{i \in I_j} \iota_i = D$ for each $1 \le j \le z$. 3-Partition is strongly NP-complete [30]. Hence, it cannot be solved in pseudo-polynomial time, unless P = NP.

The main idea of our reduction is to construct a set of *structure jobs*. These structure jobs have the property that each possible way to schedule them with the optimal makespan leaves z gaps each with processing time D, i.e., it happens exactly at z distinct times that a machine is idle, and the duration of each idle time is exactly D, see Fig. 2 at the hatched areas. As a consequence, *partition jobs* which have processing times equating the 3-Partition numbers can only be scheduled with the desired makespan if the 3-Partition instance is a Yes-instance.

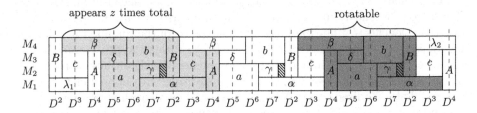

Fig. 2. Packing of structure jobs with gaps (hatched area) for 3-Partition items. The items in the green area (left) are repeated z times. With the current choice of processing times, the items in the red area (right) can be rotated by 180° such that α is scheduled on M_4 after the job in B and β is scheduled on M_1 before the job in A. (Color figure online)

Construction. Given a 3-Partition instance, we construct ten disjoint sets of jobs A, B, a, b, c, α, β, γ, δ, and λ, which will be forced to be scheduled as in Fig. 2 by choosing suitable processing times. First, we add a unique token to the processing time of each set of jobs processed simultaneous to ensure that these jobs have to be processed at the same time in every schedule. As this token, we choose D^x, where $x \in \{2, \ldots, 7\}$ and D is the required sum of the items in each partition set, see Fig. 2. For example jobs in B have processing time D^2, while jobs in α have processing time $D^7 + D^2 + D^3$.

Given a set of jobs, their total processing time has the form $\sum_{i=2}^{7} x_i D^i$, with $x_i \in \mathbb{N}$ for $i = 2, \ldots, 7$. We want the tokens D^i to be unique in the way that $x_i D^i < D^{i+1}$ for each possible occurring sum of processing times of structure jobs and each $i = 2, \ldots, 7$. Let $k_{\max}$ be the larges occurring coefficient in the sum of processing times of any given subset of the generated structure jobs, i.e., $k_{\max} = 3(z+1)$ with the current choice of processing times. We scale each number in the 3-Partition instance with $k_{\max}$ if $D \le k_{\max}$, resulting in $k_{\max} D^i < D^{i+1}$. If $k_{\max}$ depends polynomially on z, the input size of the scaled instance will still depend polynomially on the input size of the original instance.

Unfortunately, the tokens D^2 to D^7 are not enough to ensure that the schedule in Fig. 2 is the only possible one. Consider the jobs contained in the red area

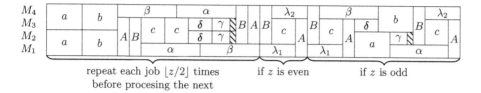

Fig. 3. A reordering we have to prohibit, since it fuses the areas for 3-partition items into two areas, one area on M_2 and one area on M_3 if z is even, and into three areas if z is odd.

(right) in Fig. 2. With the current choice of processing times, it is possible to rotate the red area by $180°$, such that α is scheduled on M_4 and β is scheduled on M_1. After rotating every second of these set of jobs, it is possible to reorder the jobs, and fusing the areas for the 3-Partition items into two or three areas, see Fig. 3. To prohibit this possibility to rotate, we introduce one further token D^8. This token is added to the processing time of some jobs such that the combined processing time of the jobs in the red area on M_1 differs from the one on M_4. To ensure this, we have to give up the property that in each of the sets $A, B, a, b, c, \alpha, \beta, \gamma, \delta$ all jobs have the same processing time. More precisely, each job in the sets c, δ, and γ receives a unique processing time.

In the following, we describe the jobs constructed for the reduction. We introduce two sets A and B of 3-processor jobs, three sets a, b and c of 2-processor jobs, and five sets α, β, γ, δ, and λ of 1-processor jobs. The description of the jobs inside these sets and their processing times can be found in Table 1. We call these jobs *structure jobs*. Additionally, we generate for each $i \in \{1, \ldots, 3z\}$ one 1-processor job, called *partition job*, with processing time ι_i and define P as the set containing all partition jobs. Last, we define $W := (z+1)(D^2 + D^3 + D^4) + z(D^5 + D^6 + D^7) + z(7z+1)D^8$. Note that the total work of the introduced jobs adds up to $4W$, i.e., a schedule without idle times has makespan W.

Table 1. Overview of the structure jobs

$$
\begin{array}{lll}
p(j) = D^4 & \text{if } j \in A & := \{A_0, \ldots, A_z\} \\
p(j) = D^2 & \text{if } j \in B & := \{B_0, \ldots, B_z\} \\
p(j) = D^5 + D^6 + 3zD^8 & \text{if } j \in a & := \{a_1, \ldots, a_z\} \\
p(j) = D^6 + D^7 + 3zD^8 & \text{if } j \in b & := \{b_1, \ldots, b_z\} \\
p(j) = D^3 + (z+i)D^8 & \text{if } j = c_i \in c & := \{c_0, \ldots, c_z\} \\
p(j) = D^2 + D^3 + D^7 + 4zD^8 & \text{if } j \in \alpha & := \{\alpha_1, \ldots, \alpha_z\} \\
p(j) = D^3 + D^4 + D^5 + (4z-1)D^8 & \text{if } j \in \beta & := \{\beta_1, \ldots, \beta_z\} \\
p(j) = D^7 + (3z-i)D^8 - D & \text{if } j = \gamma_i \in \gamma & := \{\gamma_1, \ldots, \gamma_z\} \\
p(j) = D^5 + (3z-i)D^8 & \text{if } j = \delta_i \in \delta & := \{\delta_1, \ldots, \delta_z\} \\
p(j) = D^2 + D^3 + zD^8 & \text{if } j = \lambda_1 & \\
p(j) = D^3 + D^4 + 2zD^8 & \text{if } j = \lambda_2 &
\end{array}
$$

If we add the processing times of all generated jobs, the largest coefficient is bounded by $4z(7z + 1)$. Let us assume that $D > 4z(7z + 1)$ in the given 3-Partition instance. Otherwise we scale each 3-Partition number with $4z(7z+1)$, to assure this property. Note that in a schedule with out idle times, a machine cannot contain a set of jobs, with processing times that add up to a value where one of the coefficients is larger than the corresponding one in W.

Partition to Schedule. Let $\mathcal{I}$ be a Yes-instance with partition $I_1, \ldots, I_z$. One can easily verify that the *structure jobs* can be scheduled as shown in Fig. 4. After each job γ_j, for each $1 \leq j \leq z$, we have a gap with processing time D. We schedule the *partition jobs* with indices out of I_j directly after γ_j. Their processing times add up to D, and therefore they fit into the gap. The resulting schedule has a makespan of W.

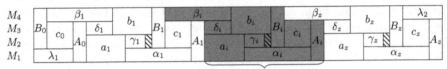

ith appearance of the repeated sequence

Fig. 4. An optimal schedule, for a Yes-instance.

Schedule to Partition. Let a schedule $S = (\sigma, \rho)$ with makespan W be given. We will now step by step describe why $\mathcal{I}$ has to be a Yes-instance. In the first step, we will show that we can transform the schedule, such that each machine contains a certain set of jobs.

Lemma 1. *We can transform the schedule S into a schedule, where M_1 contains the jobs $A \cup a \cup \alpha \cup \lambda_1$, M_2 contains the jobs $A \cup B \cup c \cup \breve{a} \cup \breve{b} \cup \breve{\gamma} \cup \breve{\delta}$, M_3 contains the jobs $A \cup B \cup c \cup \hat{a} \cup \hat{b} \cup \hat{\gamma} \cup \hat{\delta}$ and M_4 contains the jobs $B \cup b \cup \beta \cup \lambda_2$, with $\breve{a} \subseteq a$, $\hat{a} = a \backslash \breve{a}$, $\breve{b} \subseteq b$, $\hat{b} = b \backslash \breve{b}$, $\breve{\gamma} \subseteq \gamma$, $\hat{\gamma} = \gamma \backslash \breve{\gamma}$, and $\breve{\delta} \subseteq \delta$, $\hat{\delta} = \delta \backslash \breve{\delta}$. Furthermore, if the jobs are scheduled in this way, it holds that $|\breve{a}| = |\breve{\gamma}|$ and $|\breve{b}| = |\breve{\delta}|$.*

Proof. First, we will show that the content of the machines can be swapped without enlarging the makespan, such that M_2 and M_3 each contain all the jobs in $A \cup B$. Let $x \in A \cup B$ be the job with the smallest starting point in this set. We can swap the complete content of the machines such that M_2 and M_3 contain x. Let us suppose that, after some swapping operations, M_2 and M_3 contain the first i jobs in $A \cup B$ but not the $i + 1$th job. Let $\tilde{M} \in \{M_1, M_4\}$ be the third machine containing the i-th job $x_i \in A \cup B$ and $\tilde{M}' \in \{M_2, M_3\}$ be the machine not containing the $(i + 1)$-th job. We transform the schedule such that M_2 and M_3 contain the $(i + 1)$-th job, by performing one more swapping operation $SWAP(\sigma(x_i), \tilde{M}, \tilde{M}')$. Therefore, we can transform the given schedule such that M_2 and M_3 each contain all the jobs in $A \cup B$.

In the next step, we will determine the set of jobs contained by the machines M_1 and M_4 using the token D^8. Besides the jobs in $A \cup B$, M_2 and M_3 contain jobs with total processing time of $(z+1)D^3 + zD^5 + zD^6 + zD^7 + z(7z+1)D^8$. Hence, M_2 and M_3 cannot contain jobs in $\alpha \cup \beta \cup \lambda$, since their processing times contain D^2 or D^4. Therefore, each job in $A \cup B \cup \alpha \cup \beta \cup \lambda$ is either processed on M_1 or on M_4. In addition to these jobs, M_1 and M_4 together contain further jobs with a total processing time of $zD^5 + 2zD^6 + zD^7 + 6z^2D^8$. Exclusively jobs from the set $a \cup b$ have a processing time containing D^6. Therefore, each machine processes z of them. Hence corresponding to D^8, a total processing time of $3z^2D^8$ is used by jobs in the set $a \cup b$ on each machine. This leaves a processing time of $(4z^2 + z)D^8$ for the jobs in $\alpha \cup \beta \cup \lambda$ on M_1 and M_4. All the $2(z+1)$ jobs in $\alpha \cup \beta \cup \lambda$ contain D^3 in their processing time. Therefore, each machine M_1 and M_4 processes exactly $z+1$ of them. We will swap the content of M_1 and M_4 such that λ_1 is scheduled on M_1. As a consequence, M_1 processes z jobs from the set $\alpha \cup \beta \cup \{\lambda_2\}$, with processing times, which sum up to $4z^2D^8$ in the D^8 component. The jobs in α have with $4zD^8$ the largest amount of D^8 in their processing time. Therefore, M_1 has to process all of them since $z \cdot 4zD^8 = 4z^2D^8$, while M_4 contains the jobs in $\beta \cup \{\lambda_2\}$. Since we have $p(\alpha \cup \{\lambda_1\}) = (z+1)D^2 + (z+1)D^3 + zD^7 + z(4z+1)D^8$, jobs from the set $A \cup B \cup a \cup b$ with total processing time of $(z+1)D^4 + zD^5 + zD^6 + 3z^2D^8$ have to be scheduled on M_1. In this set, the jobs in A are the only jobs with processing times containing D^4, while the jobs in a are the only jobs with a processing time containing D^5. As a consequence, M_1 processes the jobs $A \cup a \cup \alpha \cup \{\lambda_1\}$. Analogously we can deduce that M_4 processes the jobs $B \cup b \cup \beta \cup \{\lambda_2\}$.

In the last step, we will determine which jobs are scheduled on M_2 and M_3. As shown before, each of them contains the jobs $A \cup B$. Furthermore, since no job in c is scheduled on M_1 or M_4, and they require two machines to be processed, machines M_2 and M_3 both contain the set c. Additionally, each job in $\gamma \cup \delta$ has to be scheduled on M_2 or M_3 since they are not scheduled on M_1 or M_4. Each job in $a \cup b$ occupies one of the machines M_1 and M_4. The second machine they occupy is either M_2 or M_3. Let $\breve{a} \subseteq a$ be the set of jobs, which is scheduled on M_2 and $\hat{a} \subseteq a$ be the set which is scheduled on M_3. Clearly $\breve{a} = a \backslash \hat{a}$. We define the sets $\hat{b}, \breve{b}, \hat{\delta}, \breve{\delta}, \hat{\gamma}$, and $\breve{\gamma}$ analogously. By this definition, M_2 contains the jobs $A \cup B \cup \breve{a} \cup \breve{b} \cup \breve{\delta} \cup \breve{\gamma} \cup c$ and M_3 contains the jobs $A \cup B \cup \hat{a} \cup \hat{b} \cup \hat{\delta} \cup \hat{\gamma} \cup c$.

We still have to prove that $|\breve{a}| = |\breve{\gamma}|$ and $|\breve{b}| = |\breve{\delta}|$. First, we notice that $|\breve{a}| + |\breve{b}| = z$ since these jobs are the only jobs with a processing time containing D^6. So besides the jobs in $A \cup B \cup c \cup \breve{a} \cup \breve{b}$, M_2 contains jobs with total processing time of $(z - |\breve{a}|)D^5 + (z - |\breve{b}|)D^7 + \sum_{i=1}^{z}(3z-i)D^8 = |\breve{b}|D^5 + |\breve{a}|D^7 + \sum_{i=1}^{z}(3z-i)D^8$. Since the jobs in δ are the only jobs in $\delta \cup \gamma$ having a processing time containing D^5, we have $|\breve{\delta}| = |\breve{b}|$ and analogously $|\breve{\gamma}| = |\breve{a}|$. $\square$

In the next steps, we will prove that it is possible to transform the order in which the jobs appear on the machines to the one in Fig. 4. Notice that, since there is no idle time in the schedule, each start point of a job i is given by the sum of processing times of the jobs on the same machine scheduled before i. So the start position $\sigma(i)$ of a job i has the form

$$\sigma(i) = x_0 + x_2 D^2 + x_3 D^3 + x_4 D^4 + x_5 D^5 + x_6 k D^6 + x_7 D^7 + x_8 D^8$$

for $-zD \leq x_0 \leq zD < D^2$ and $0 \leq x_j \leq 4z(7z+1) \leq D$ for each $2 \leq j \leq 8$. This allows us to make implications about the correlation between the number of jobs scheduled on different machines when a job from the set $A \cup B \cup a \cup b \cup c$ starts. For example, let us look at the coefficient x_4. This value is just influenced by jobs with processing times containing D^4. The only jobs with these processing times are the jobs in the set $A \cup \beta \cup \{\lambda_2\}$. The jobs in $\beta \cup \{\lambda_2\}$ are just processed on M_4, while the jobs in A each are processed on the three machines M_1, M_2, and M_3. Therefore, we know that at the starting point $\sigma(i)$ of a job i scheduled on machines M_1, M_2 or M_3 we have that $x_4 = n_i(A)$. Furthermore, if i is scheduled on M_4 we know that $x_4 = n_i(\beta) + n_i(\{\lambda_2\})$. In Table 2, we present which sets influences which coefficients in which way when job i is started on the corresponding machine.

Table 2. Overview of the values of the coefficients at the start point of a job i, if i is scheduled on machine M_j.

	M_1	M_2	M_3	M_4
x_2	$n_i(\alpha) + n_i(\{\lambda_1\})$	$n_i(B)$	$n_i(B)$	$n_i(B)$
x_3	$n_i(\alpha) + n_i(\{\lambda_1\})$	$n_i(c)$	$n_i(c)$	$n_i(\beta) + n_i(\{\lambda_2\})$
x_4	$n_i(A)$	$n_i(A)$	$n_i(A)$	$n_i(\beta) + n_i(\{\lambda_2\})$
x_5	$n_i(a)$	$n_i(\check{a}) + n_i(\check{\delta})$	$n_i(\hat{a}) + n_i(\hat{\delta})$	$n_i(\beta)$
x_6	$n_i(a)$	$n_i(\check{a}) + n_i(\check{b})$	$n_i(\hat{a}) + n_i(\hat{b})$	$n_i(b)$
x_7	$n_i(\alpha)$	$n_i(\check{b}) + n_i(\check{\gamma})$	$n_i(\hat{b}) + n_i(\hat{\gamma})$	$n_i(b)$

Let us consider the start point $\sigma(i)$ of a job i, which uses more than one machine. We know that $\sigma(i)$ is the same on all the used machines and therefore the coefficients are the same as well. In the following, we will study for each of the sets A, B, a, b, c what we can conclude for the starting times of these jobs. For each of the sets, we will present an equation, which holds at the start of each item in this set. These equations give us a strong set of tools for our further arguing.

First, we will consider the start points of the jobs in A. Each job $A' \in A$ is scheduled on machines M_1, M_2 and M_3. Therefore, we know that at $\sigma(A')$ we have $n_{A'}(B) =_{x_2} n_{A'}(\alpha) + n_{A'}(\{\lambda_1\}) =_{x_3} n_{A'}(c)$. Furthermore, we know that $n_{A'}(a) =_{x_6} n_{A'}(\check{a}) + n_{A'}(\check{b}) =_{x_6} n_{A'}(\hat{a}) + n_{A'}(\hat{b})$. Since $n_{A'}(a) = n_{A'}(\check{a}) + n_{A'}(\hat{a})$ and $n_{A'}(b) = n_{A'}(\check{b}) + n_{A'}(\hat{b})$, we can deduce that $n_{A'}(\hat{a}) = n_{A'}(\check{b})$ and $n_{A'}(\check{a}) = n_{A'}(\hat{b})$ and therefore $n_{A'}(a) = n_{A'}(b)$. Additionally, we know that $n_{A'}(\alpha) =_{x_7} n_{A'}(\check{b}) + n_{A'}(\check{\gamma}) =_{x_7} n_{A'}(\hat{b}) + n_{A'}(\hat{\gamma})$. Thanks to this equality, we can show that $n_{A'}(\alpha) = n_{A'}(b)$: First, we show $n_{A'}(\alpha) \geq n_{A'}(b)$. Let $b' \in b$ be the last job in b scheduled before A' if there is any. Let us w.l.o.g assume that $b' \in \hat{b}$. It holds that $n_{A'}(b) = n_{b'}(b) + 1 =_{x_7} n_{b'}(\hat{b}) + n_{b'}(\hat{\gamma}) + 1 \leq n_{A'}(\hat{b}) + n_{A'}(\hat{\gamma}) =_{x_7} n_{A'}(\alpha)$. If there is no such b' we have $n_{A'}(b) = 0 \leq n_{A'}(\alpha)$. Next, we show $n_{A'}(\alpha) \leq n_{A'}(b)$. Let $b'' \in b$ be the first job in b scheduled after A if there is any. Let us w.l.o.g assume that $b'' \in \check{b}$. It holds that $n_{A'}(b) = n_{b''}(b) =_{x_7} n_{b''}(\check{b}) + n_{b''}(\check{\gamma}) \geq n_{A'}(\check{b}) + n_{A'}(\check{\gamma}) =_{x_7} n_{A'}(\alpha)$. If there is no such b'', we have $n_{A'}(b) = z \geq n_{A'}(\alpha)$. As a consequence, we have $n_{A'}(\alpha) = n_{A'}(b)$. In summary, we can deduce that

$$n_{A'}(c) - n_{A'}(\{\lambda_1\}) = n_{A'}(B) - n_{A'}(\{\lambda_1\}) = n_{A'}(\alpha) = n_{A'}(b) = n_{A'}(a). \tag{1}$$

Analogously, we can deduce that at the start of each $B' \in B$ we have that

$$n_{B'}(c) - n_{B'}(\{\lambda_2\}) = n_{B'}(A) - n_{B'}(\{\lambda_2\}) = n_{B'}(\beta) = n_{B'}(a) = n_{B'}(b). \tag{2}$$

Each item $a' \in a$ is scheduled on machine M_1 and on one of the machines M_2 or M_3. For each possibility $a \in \hat{a}$ or $a \in \check{a}$, we can deduce the equation

$$n_{a'}(B) =_{x_2} n_{a'}(\alpha) + n_{a'}(\{\lambda_1\}) =_{x_3} n_{a'}(c). \tag{3}$$

Analogously, we deduce for each $b' \in b$ that

$$n_{b'}(A) =_{x_4} n_{b'}(\beta) + n_{b'}(\{\lambda_2\}) =_{x_3} n_{b'}(c). \tag{4}$$

Last, each item $c' \in c$ is scheduled on M_2 and M_3. Let $a' \in a$ be the job with the smallest $\sigma(a') \geq \sigma(c')$. Let us w.l.o.g assume that $a' \in \hat{a}$. It holds that $n_{c'}(\check{a}) + n_{c'}(\check{b}) =_{x_6} n_{c'}(\hat{a}) + n_{c'}(\hat{b}) \leq n_{a'}(\hat{a}) + n_{a'}(\hat{b}) =_{x_6} n_{a'}(a) = n_{a'}(\hat{a}) + n_{a'}(\check{a}) = n_{c'}(\hat{a}) + n_{c'}(\check{a})$. As a consequence, we have $n_{c'}(\check{b}) \leq n_{c'}(\hat{a})$ and $n_{c'}(\hat{b}) \leq n_{c'}(\check{a})$. Analogously, let $b' \in b$ be the job with the smallest $\sigma(b') \geq \sigma(c')$. Let us w.l.o.g assume that $b' \in \check{b}$. It holds that $n_{c'}(\hat{a}) + n_{c'}(\hat{b}) =_{x_6} n_{c'}(\check{a}) + n_{c'}(\check{b}) \leq n_{b'}(\check{a}) + n_{b'}(\check{b}) =_{x_6} n_{b'}(b) = n_{b'}(\hat{b}) + n_{b'}(\check{a}) = n_{c'}(\hat{b}) + n_{c'}(\check{b})$. Therefore, $n_{c'}(\check{a}) \leq n_{c'}(\hat{b})$ and $n_{c'}(\hat{a}) \leq n_{c'}(\check{b})$. As a consequence, we can deduce that

$$n_{c'}(b) = n_{c'}(a) \tag{5}$$

These equations give us the tools to analyze the given schedule with makespan W. First, we will show that in this schedule the first and last jobs have to be elements from the set $A \cup B$, (see Lemma 2). After that, we will prove that the jobs in A and jobs in B have to be scheduled alternating, (see Lemma 3). With the knowledge gathered in the proofs of Lemmas 2 and 3, we can prove that the given schedule can be transformed such that all jobs are scheduled contiguously, and that $\mathcal{I}$ has to be a Yes-instance (see Lemma 4).

Lemma 2. *The first and the last job on M_2 and M_3 are elements of $A \cup B$.*

Proof. Let $i := \arg\min_{i \in A \cup B} \sigma_i$ be the job with the smallest start point in $A \cup B$, (i.e. $n_i(A) = 0 = n_i(B)$). If $i \in A$ it holds that $0 = n_i(B) =_{(1)} n_i(\alpha) + n_i(\{\lambda_1\}) =_{(1)} n_i(a) + n_i(\{\lambda_1\})$ and therefore $n_i(a) = n_i(\alpha) = 0 = n_i(\{\lambda_1\})$. The jobs $a \cup \alpha \cup \{\lambda_1\} \cup A$ are the only jobs, which are contained on machine M_1. Since $n_i(A) = 0$ as well, it has to be that $\sigma_i = 0$, and therefore i is the first job on M_2 and M_3. If $i \in B$ we can prove $\sigma_i = 0$ analogously using equality (2).

Since the schedule stays valid, if we mirror the schedule such that the new start points are $s'(i) = W - \sigma(i) - p(i)$ for each job i, the last job has to be in the set $A \cup B$ as well. □

Next, we will show that the items in the sets A and B have to be scheduled alternating. Due to space limitations, the proofs of Lemmas 3 and 4 are not included in this extended abstract, but can be found in [31]. Let $(A_0, \ldots, A_z)$ be the set A and $(B_0, \ldots, B_z)$ be the set B each ordered by increasing size of the starting points. Simply swap the jobs if they do not have this order.

Lemma 3. *If the first job on M_2 is the job $B_0 \in B$ it holds for each item $i \in \{0, \ldots, z\}$ that*

$$n_{A_i}(B) - n_{A_i}(\{\lambda_1\}) = n_{A_i}(A) \qquad (6)$$

and $n_{A_i}(\{\lambda_1\}) = 1$.

A direct consequence of Lemma 3 is that the last job on M_2 is a job in A. Since the Eqs. (1) and (2), as well as (3) and (4), are symmetric, we can deduce an analogue statement if the first job on M_2 is in A. More precisely, we can show that $n_{B_i}(A) - n_{B_i}(\{\lambda_2\}) = n_{B_i}(B)$ and $n_{B_i}(\{\lambda_2\}) = 1$ for each $B_i \in B$ in this case. This would imply that the last job on M_2 is a job in B. Since we can mirror the schedule such that the last job is the first job, we can suppose that the first job on M_2 is a job in B. In this case a further direct consequence of Lemma 3 and Eq. (1) is the equation

$$i = n_{A_i}(A) = n_{A_i}(B) - 1 = n_{A_i}(c) - 1 = n_{A_i}(\alpha) = n_{A_i}(b) = n_{A_i}(a) \qquad (7)$$

Lemma 4. *$\mathcal{I}$ is a Yes-instance and we can transform the schedule such that all jobs are scheduled on continuous machines.*

3   Hardness of Strip Packing

In the transformed schedule, all jobs are scheduled on contiguous machines. As a consequence, we have proven that this problem is strongly NP-complete even if we restrict the set of feasible solutions to those where all jobs are scheduled on continuous machines. We will now describe how this insight delivers a lower bound of $\frac{5}{4}$ for the best possible approximation ratio for pseudo-polynomial Strip Packing and in this way prove Theorem 2.

To show our hardness result for Strip Packing, let us consider the following instance. We define $W := (z+1)(D^2+D^3+D^4)+z(D^5+D^6+D^7)+z(7z+1)D^8$ as the width of the considered strip, so it is the same as the considered makespan in the scheduling problem. For each job j defined in the reduction above, we define an item i with $w(i) = p(j)$ and height $h(i) = q(j)$. Now, we can show analogously that if the 3-Partition instance is a Yes-instance, there is a packing of height 4 (one example is the packing in Fig. 4); and on the other hand if there is a packing with height 4, the 3-Partition instance has to be a Yes-instance. If the 3-Partition instance is a No-instance, the optimal packing has a height of at least 5 since the optimal height for this instance is integral. Therefore, we cannot approximate Strip Packing in pseudo-polynomial time better than $\frac{5}{4}$.

Acknowledgements. The authors would like to thank the anonymous reviewers for their valuable comments and suggestions to improve the quality of the paper. This work was in part supported by German Research Foundation (DFG) project JA 612/14-2.

References

1. Du, J., Leung, J.Y.: Complexity of scheduling parallel task systems. SIAM J. Discrete Math. **2**(4), 473–487 (1989)
2. Bansal, N., Correa, J.R., Kenyon, C., Sviridenko, M.: Bin packing in multiple dimensions: inapproximability results and approximation schemes. Math. Oper. Res. **31**(1), 31–49 (2006)
3. Adamaszek, A., Kociumaka, T., Pilipczuk, M., Pilipczuk, M.: Hardness of approximation for strip packing. TOCT **9**(3), 14:1–14:7 (2017)
4. Jansen, K., Rau, M.: Closing the gap for pseudo-polynomial strip packing. CoRR abs/1705.04587 (2017)
5. Jansen, K., Rau, M.: Improved approximation for two dimensional strip packing with polynomial bounded width. In: Poon, S.-H., Rahman, M.S., Yen, H.-C. (eds.) WALCOM 2017. LNCS, vol. 10167, pp. 409–420. Springer, Cham (2017). https://doi.org/10.1007/978-3-319-53925-6_32
6. Gálvez, W., Grandoni, F., Ingala, S., Khan, A.: Improved pseudo-polynomial-time approximation for strip packing. In: 36th IARCS Annual Conference on Foundations of Software Technology and Theoretical Computer Science (FSTTCS), pp. 9:1–9:14 (2016)
7. Nadiradze, G., Wiese, A.: On approximating strip packing with a better ratio than 3/2. In: 27th Annual ACM-SIAM Symposium on Discrete Algorithms (SODA), pp. 1491–1510 (2016)
8. Jansen, K., Thöle, R.: Approximation algorithms for scheduling parallel jobs. SIAM J. Comput. **39**(8), 3571–3615 (2010)
9. Amoura, A.K., Bampis, E., Kenyon, C., Manoussakis, Y.: Scheduling independent multiprocessor tasks. Algorithmica **32**(2), 247–261 (2002)
10. Jansen, K., Porkolab, L.: Linear-time approximation schemes for scheduling malleable parallel tasks. Algorithmica **32**(3), 507–520 (2002)
11. Garey, M.R., Graham, R.L.: Bounds for multiprocessor scheduling with resource constraints. SIAM J. Comput. **4**(2), 187–200 (1975)
12. Turek, J., Wolf, J.L., Yu, P.S.: Approximate algorithms scheduling parallelizable tasks. In: 4th annual ACM symposium on Parallel algorithms and architectures (SPAA), pp. 323–332 (1992)
13. Ludwig, W., Tiwari, P.: Scheduling malleable and nonmalleable parallel tasks. In: 5th Annual ACM-SIAM Symposium on Discrete Algorithms (SODA), pp. 167–176 (1994)
14. Feldmann, A., Sgall, J., Teng, S.: Dynamic scheduling on parallel machines. Theor. Comput. Sci. **130**(1), 49–72 (1994)
15. Jansen, K.: A $(3/2+\varepsilon)$ approximation algorithm for scheduling moldable and non-moldable parallel tasks. In: 24th ACM Symposium on Parallelism in Algorithms and Architectures, (SPAA), pp. 224–235 (2012)
16. Baker Jr., B.S., Coffman, E.G., Rivest, R.L.: Orthogonal packings in two dimensions. SIAM J. Comput. **9**(4), 846–855 (1980)
17. Coffman Jr., E.G., Garey, M.R., Johnson, D.S., Tarjan, R.E.: Performance bounds for level-oriented two-dimensional packing algorithms. SIAM J. Comput. **9**(4), 808–826 (1980)
18. Sleator, D.D.: A 2.5 times optimal algorithm for packing in two dimensions. Inf. Proc. Lett. **10**(1), 37–40 (1980)
19. Schiermeyer, I.: Reverse-fit: A 2-optimal algorithm for packing rectangles. In: van Leeuwen, J. (ed.) ESA 1994. LNCS, vol. 855, pp. 290–299. Springer, Heidelberg (1994). https://doi.org/10.1007/BFb0049416

20. Steinberg, A.: A strip-packing algorithm with absolute performance bound 2. SIAM J. Comput. **26**(2), 401–409 (1997)

21. Harren, R., van Stee, R.: Improved absolute approximation ratios for two-dimensional packing problems. In: Dinur, I., Jansen, K., Naor, J., Rolim, J. (eds.) APPROX/RANDOM -2009. LNCS, vol. 5687, pp. 177–189. Springer, Heidelberg (2009). https://doi.org/10.1007/978-3-642-03685-9_14

22. Harren, R., Jansen, K., Prädel, L., van Stee, R.: A $(5/3 + \epsilon)$-approximation for strip packing. Comput. Geom. **47**(2), 248–267 (2014)

23. Golan, I.: Performance bounds for orthogonal oriented two-dimensional packing algorithms. SIAM J. Comput. **10**(3), 571–582 (1981)

24. Baker, B.S., Brown, D.J., Katseff, H.P.: A 5/4 algorithm for two-dimensional packing. J. Algorithms **2**(4), 348–368 (1981)

25. Kenyon, C., Rémila, E.: A near-optimal solution to a two-dimensional cutting stock problem. Math. Oper. Res. **25**(4), 645–656 (2000)

26. Sviridenko, M.: A note on the Kenyon-Remila strip-packing algorithm. Inf. Proc. Lett. **112**(1–2), 10–12 (2012)

27. Bougeret, M., Dutot, P., Jansen, K., Robenek, C., Trystram, D.: Approximation algorithms for multiple strip packing and scheduling parallel jobs in platforms. Discrete Math. Algorithms Appl. **3**(4), 553–586 (2011)

28. Jansen, K., Solis-Oba, R.: Rectangle packing with one-dimensional resource augmentation. Discrete Optim. **6**(3), 310–323 (2009)

29. Christensen, H.I., Khan, A., Pokutta, S., Tetali, P.: Approximation and online algorithms for multidimensional bin packing: a survey. Comput. Sci. Rev. **24**, 63–79 (2017)

30. Garey, M.R., Johnson, D.S.: Computers and Intractability: A Guide to the Theory of NP-Completeness. W. H. Freeman, New York (1979)

31. Henning, S., Jansen, K., Rau, M., Schmarje, L.: Complexity and inapproximability results for parallel task scheduling and strip packing. CoRR abs/1705.04587 (2017)

Operations on Boolean and Alternating Finite Automata

Michal Hospodár, Galina Jirásková, and Ivana Krajňáková[(⊠)]

Mathematical Institute, Slovak Academy of Sciences,
Grešákova 6, 040 01 Košice, Slovakia
hosmich@gmail.com, {jiraskov,krajnakova}@saske.sk

Abstract. We investigate the descriptional complexity of basic regular operations on languages represented by Boolean and alternating finite automata. In particular, we consider the operations of difference, symmetric difference, star, reversal, left quotient, and right quotient, and get tight upper bounds $m + n, m + n, 2^n, 2^n, m,$ and 2^m, respectively, for Boolean automata, and $m + n + 1, m + n, 2^n, 2^n, m + 1,$ and $2^m + 1$, respectively, for alternating finite automata. To describe witnesses for symmetric difference, we use a ternary alphabet. All the remaining witnesses are defined over binary or unary alphabets that are shown to be optimal.

1   Introduction

The Boolean finite automata (BFAs) are generalization of nondeterministic finite automata (NFAs). In an NFA, the transition function maps any pair of state and input symbol to a subset of states. This subset can be viewed as disjunction of its states. We obtain a BFA by considering other Boolean functions on states as a result of the transition function. Alternating finite automata (AFAs) start from the only one initial state, wheares Boolean automata may start their computation in any Boolean function designated as the initial function.

Boolean automata recognize the class of regular languages [2,4]. Every n-state Boolean automaton can be simulated by 2^{2^n}-state deterministic finite automaton (DFA), or by $(2^n + 1)$-state NFA, and both upper bounds are tight already in the binary case [2,10].

Some of the constructions and upper bounds for elementary operations on alternating automata were introduced in [5]. The upper bound $2^m + n + 1$ for concatenation from [5] has been shown to be tight in [8]. Detailed results for the square on alternating and Boolean automata can be found in [12]. Tight upper bounds for union and intersection were shown in [10]. For star and reversal, the upper and lower bound provided in [10] differed by one.

Research supported by grant VEGA 2/0084/15 and grant APVV-15-0091. This work was conducted as a part of PhD study of Michal Hospodár and Ivana Krajňáková at the Faculty of Mathematics, Physics and Informatics of the Comenius University.

F. V. Fomin and V. V. Podolskii (Eds.): CSR 2018, LNCS 10846, pp. 181–193, 2018.
https://doi.org/10.1007/978-3-319-90530-3_16

In this paper we continue the study of the operational complexity on Boolean and alternating finite automata. We improve the results on star and reversal from [10] and provide exact complexity of these two operations. We also examine other regular operations: complementation, difference, symmetric difference, left and right quotient on both Boolean and alternating automata. We get the exact complexity for each operation on both BFAs and AFAs. All our witness languages are defined over a small fixed alphabet which is optimal in most of the cases.

2   Preliminaries

Let Σ be a finite alphabet of symbols. Then Σ^* denotes the set of words over Σ including the empty word ε. A language is any subset of Σ^*. The cardinality of a finite set A is denoted by $|A|$, and its power-set by 2^A. The reader may refer to [7,17,18] for details.

A *nondeterministic finite automaton* (NFA) is a quintuple $A = (Q, \Sigma, \circ, I, F)$, where Q is a finite set of states, Σ is a finite non-empty alphabet, $\circ : Q \times \Sigma \to 2^Q$ is the transition function which is naturally extended to the domain $2^Q \times \Sigma^*$, $I \subseteq Q$ is the set of initial states, and $F \subseteq Q$ is the set of final states. The *language accepted by* A is the set $L(A) = \{w \in \Sigma^* \mid I \circ w \cap F \neq \emptyset\}$. For a symbol a, we say that (p, a, q) is a transition in NFA A if $q \in p \circ a$, and the state q has an in-transition on a. For a word w, we write $p \xrightarrow{w} q$ if $q \in p \circ w$.

An NFA A is *deterministic* (DFA) if $|I| = 1$ and $|q \circ a| = 1$ for each q in Q and each a in Σ; so all DFAs in this paper are assumed to be complete. We write $p \cdot a = q$ instead of $p \circ a = \{q\}$ in such a case. The *state complexity* of a regular language L, $\mathrm{sc}(L)$, is the smallest number of states in any DFA for L. A state q of a DFA is called *sink state* if $q \cdot a = q$ for each a in Σ.

For unary DFAs we use the Nicaud's notation [15]. For two integers ℓ and n such that $0 \leq \ell \leq n - 1$ and a subset F of $\{0, \ldots, n - 1\}$, $A = (n, \ell, F)$ is the unary automaton whose set of states is $Q = \{0, \ldots, n - 1\}$ and the transition function is given by $q \cdot a = q + 1$ if $0 \leq q \leq n - 2$ and $(n - 1) \cdot a = \ell$. The initial state of this automaton is 0 and its set of final states is F.

Every NFA $A = (Q, \Sigma, \circ, I, F)$ can be converted to an equivalent DFA $\mathcal{D}(A) = (2^Q, \Sigma, \cdot, I, F')$, where $S \cdot a = S \circ a$ for each S in 2^Q and a in Σ and $F' = \{R \in 2^Q \mid R \cap F \neq \emptyset\}$. We call the DFA $\mathcal{D}(A)$ the *subset automaton* of the NFA A. The subset automaton may not be minimal since some of its states may be unreachable or equivalent to other states.

To prove distinguishability of the states of the subset automaton, the following notions and observations are useful. A state q of an NFA A is called *uniquely distinguishable* if there is a word w which is accepted by A from and only from the state q, that is $p \circ w \cap F \neq \emptyset$ if and only if $p = q$. A transition (p, a, q) is called a *unique in-transition* if there is no state r such that $r \neq p$ and (r, a, q) is a transition in A. A state q is *uniquely reachable* from a state p if there exists a sequence of unique in-transitions (q_i, a, q_{i+1}) for $i = 0, 1, \ldots, k$ such that $q_0 = p$ and $q_{k+1} = q$.

Proposition 1 [1, Propositions 14 and 15]. *Let A be an NFA and $\mathcal{D}(A)$ be the corresponding subset automaton.*

(a) If two subsets of $\mathcal{D}(A)$ differ in a uniquely distinguishable state of A, then they are distinguishable.

(b) If a state q of A is uniquely distinguishable and uniquely reachable from a state p, then the state p is uniquely distinguishable as well.

(c) If there is a uniquely distinguishable state of A which is uniquely reachable from any other state of A, then every state of A is uniquely distinguishable.

(d) If every state of A is uniquely distinguishable, then the subset automaton $\mathcal{D}(A)$ does not have equivalent states.

$\square$

Let K and L be languages over an alphabet Σ. The *difference* and *symmetric difference* of K and L are the languages $K \backslash L = \{w \in K \mid w \notin L\}$ and $K \oplus L = \{w \in K \mid w \notin L\} \cup \{w \in L \mid w \notin K\}$, respectively. If languages K and L are accepted by DFAs $A = (Q_A, \Sigma, \cdot_A, s_A, F_A)$ and $B = (Q_B, \Sigma, \cdot_B, s_B, F_B)$, then the language $K \cap L$ is accepted by the *product automaton* $A \times B = (Q_A \times Q_B, \Sigma, \cdot, (s_A, s_B), F_A \times F_B)$ where $(p, q) \cdot a = (p \cdot_A a, q \cdot_B a)$. For the remaining Boolean operations we only need to change the set of final states in the product automaton. For union, difference, symmetric difference the set of final states is $(F_A \times Q_B) \cup (Q_A \times F_B)$, $F_A \times (Q_B \backslash F_B)$, $(F_A \times (Q_B \backslash F_B)) \cup ((Q_A \backslash F_A) \times F_B)$, respectively.

The reverse of a word is defined as $\varepsilon^R = \varepsilon$ and $(wa)^R = aw^R$ for each symbol a and word w. The reverse of a language L is the language $L^R = \{w^R \mid w \in L\}$. The reverse of an NFA A is an NFA A^R obtained from A by reversing all the transitions and by swapping the roles of initial and final states. The NFA A^R recognizes the reverse of $L(A)$.

The *concatenation* of K and L is the language $KL = \{uv \mid u \in K \text{ and } v \in L\}$. The *square* of a language L is the language $L^2 = LL$. The *right quotient* of K by L is the language $KL^{-1} = \{x \in \Sigma^* \mid xy \in K \text{ for some } y \in L\}$. The *left quotient* of K by L is the language $L^{-1}K = \{x \in \Sigma^* \mid yx \in K \text{ for some } y \in L\}$.

A *Boolean finite automaton* (BFA) is a quintuple $A = (Q, \Sigma, \delta, g_s, F)$, where Q is a finite non-empty set of states, $Q = \{q_1, \ldots, q_n\}$, Σ is an input alphabet, δ is the transition function that maps $Q \times \Sigma$ into the set $\mathcal{B}_n$ of Boolean functions with variables $\{q_1, \ldots, q_n\}$, $g_s \in \mathcal{B}_n$ is the initial Boolean function, and $F \subseteq Q$ is the set of final states. The transition function δ can be extended to the domain $\mathcal{B}_n \times \Sigma^*$ as follows: For all g in $\mathcal{B}_n$, a in Σ, and w in Σ^*, we have $\delta(g, \varepsilon) = g$; if $g = g(q_1, \ldots, q_n)$, then $\delta(g, a) = g(\delta(q_1, a), \ldots, \delta(q_n, a))$; $\delta(g, wa) = \delta(\delta(g, w), a)$. Next, let $f = (f_1, \ldots, f_n)$ be the Boolean vector with $f_i = 1$ iff $q_i \in F$. The language accepted by the BFA A is the set $L(A) = \{w \in \Sigma^* \mid \delta(g_s, w)(f) = 1\}$.

A Boolean finite automaton is called *alternating* (AFA) if the initial function is a projection $g(q_1, \ldots, q_n) = q_i$. For details, we refer to [2,5,10,13,17,18].

The *Boolean (alternating) state complexity* of L, bsc(L)(asc(L)), is the smallest number of states in any BFA (AFA) for L. It is known that a language L is accepted by an n-state BFA (AFA) if and only if the language L^R is accepted by an 2^n-state DFA (with 2^{n-1} final states). Since this is the crucial observation used later in the paper, we state it in the next two lemmas and provide proof ideas here.

Lemma 2 (cf. [5, Theorem 4.1, Corollary 4.2] and [10, Lemma 1]). *Let L be a language accepted by an n-state BFA (AFA). Then the reversal L^R is accepted by a DFA of 2^n states (of which 2^{n-1} are final).*

Proof (Proof Idea). Let $A = (\{q_1, q_2, \ldots, q_n\}, \Sigma, \delta, g_s, F)$ be an n-state BFA for L. Construct a 2^n-state NFA $A' = (\{0,1\}^n, \Sigma, \delta', S, \{f\})$, where

- for every $u = (u_1 \ldots, u_n) \in \{0,1\}^n$ and every $a \in \Sigma$,
  $\delta'(u, a) = \{u' \in \{0,1\}^n \mid \delta(q_{i,a})(u') = u_i \text{ for } i = 1, \ldots, n\}$;
- $S = \{(b_1, \ldots, b_n) \in \{0,1\}^n \mid g_s(b_1, \ldots, b_n) = 1\}$;
- $f = (f_1, \ldots, f_n) \in \{0,1\}^n$ with $f_i = 1$ iff $q_i \in F$.

Then $L(A) = L(A')$ and $(A')^R$ is deterministic. Moreover if A is an AFA then A' has 2^{n-1} initial states. It follows that L^R is accepted by a DFA with 2^n states, of which 2^{n-1} are final if A is an AFA. □

Lemma 3 (cf. [10, Lemma 2]). *Let L^R be accepted by a DFA A of 2^n states (of which 2^{n-1} are final). Then L is accepted by an n-state BFA (AFA).*

Proof (Proof Idea). Consider 2^n-state NFA A^R for L which has exactly one final state and the set of initial states S (and $|S| = 2^{n-1}$). Let the state set Q of A^R be $\{0, 1, \ldots, 2^n - 1\}$ with final state k and the initial set S ($S = \{2^{n-1}, \ldots, 2^n - 1\}$). Let δ be the transition function of A^R. Moreover, for every $a \in \Sigma$ and for every $i \in Q$, there is exactly one state j such that j goes to i on a in A^R. For a state $i \in Q$, let $\text{bin}(i) = (b_1, \ldots, b_n)$ be the binary n-tuple such that $b_1 b_2 \cdots b_n$ is the binary notation of i on n digits with leading zeros if necessary.

Let us define an n-state BFA $A' = (Q', \Sigma, \delta', g_s, F')$, where $Q' = \{q_1, \ldots, q_n\}$, $F' = \{q_\ell \mid \text{bin}(k)_\ell = 1\}$, and $g_s(\text{bin}(i)) = 1$ iff $i \in S$ ($g_s = q_1$). We define δ' to suffice the condition: for each i in Q and a in Σ, $(\delta'(q_1, a), \ldots, \delta'(q_n, a))(\text{bin}(i)) = \text{bin}(j)$ where $i \in \delta(j, a)$. Then $L(A') = L(A^R)$. □

As a corollary of the previous two lemmas, we get the following results.

Corollary 4. *If L is a regular language, then $\text{bsc}(L) \geq \lceil \log(\text{sc}(L^R)) \rceil$ and $\text{asc}(L) \geq \lceil \log(\text{sc}(L^R)) \rceil$.* □

Corollary 5. *Let L be a unary language. Then L is accepted by an n-state BFA (AFA) if and only if L is accepted by a 2^n-state DFA (with 2^{n-1} final states).* □

Now we prove several propositions which we use later in our paper.

Proposition 6. *If L is accepted by an n-state BFA, then L is accepted by an $(n+1)$-state AFA.*

Proof. Let a language L be accepted by an n-state BFA $(Q, \Sigma, \delta, g, F)$. Let $A = (Q \cup \{s\}, \Sigma, \delta', s, F')$ where $s \notin Q$, $\delta'(q, a) = \delta(q, a)$ if $q \in Q$ and $\delta'(q, a) = \delta(g, a)$ if $q = s$; $F' = F$ if $\varepsilon \notin L$ and $F' = F \cup \{s\}$ if $\varepsilon \in L$. Then A is an $(n+1)$-state AFA for L.   □

Proposition 7. *Let K and L be languages over Σ. Then*

(a) $(KL^{-1})^R = (L^R)^{-1} K^R$;
(b) $(L^{-1}K)^R = K^R (L^R)^{-1}$.

□

Proposition 8. *Let a non-empty language L be accepted by an n-state DFA. Then L^* is accepted by a 2^n-state DFA with half of the states final.*

Proof. Let L be accepted by an n-state DFA $A = (Q, \Sigma, \cdot, s, F)$. If the initial state is the only final state in A, then $L^* = L$, and we may add final and non-final unreachable sink states to get the desired automaton. Otherwise there is a final state q_F such that $q_F \neq s$. Construct an NFA N for L^* from A as follows:

(a) add the transition (q, a, s) whenever $q \cdot a \in F$;
(b) add a new initial and final state q_0;
(c) the initial states of N are s and q_0 and the set of final states is $F \cup \{q_0\}$.

In the corresponding subset automaton $\mathcal{D}(N)$ the initial subset is $\{q_0, s\}$ and any other reachable subset S is a non-empty subset of Q such that $S \cap F \neq \emptyset$ implies $s \in S$. By the construction above every set S such that $q_F \in S$ and $s \notin S$ is unreachable. That means that there are at most $1 + 2^n - 1 - 2^{n-2} = \frac{3}{4} 2^n$ reachable sets in $\mathcal{D}(N)$. Let us show that in the minimal DFA for L^* the number of non-final states as well as the number of final states is at most 2^{n-1}. The non-final subsets in $\mathcal{D}(N)$ must not contain the state q_F, so there are at most 2^{n-1} of them. Next the initial subset $\{q_0, s\}$ is final and any other final subset must contain the state s. This gives at most $1 + 2^{n-1}$ subsets. However, if $s \in F$ then $\{q_0, s\}$ and $\{s\}$ are equivalent, and if $s \notin F$ then $\{s\}$ is non-final. Therefore the minimal DFA for L^* has at most 2^{n-1} final states. To obtain 2^n-state DFA we may add some unreachable sink states. Since the number of final and non-final states are at most 2^{n-1} it is possible to achieve that exactly half of the states would be final and the other half non-final in the resulting 2^n-state DFA.   □

Proposition 9. *Let $m, n \geq 2$ and $\gcd(m, n) = 1$. Let K and L be unary regular languages accepted by deterministic finite automata $A = (m, 0, \{0\})$ and $B = (n, 0, \{1, 2, \ldots, n-1\})$, respectively. Then $\mathrm{sc}(K \oplus L) = mn$.*

Proof. Since symmetric difference is a commutative operation, we may assume that $m < n$. Denote $Q_A = \{0, 1, \ldots, m-1\}$, $Q_B = \{0, 1, \ldots, n-1\}$. Consider the product automaton $A \times B = (Q_A \times Q_B, \{a\}, \cdot, (0, 0), F)$ where the set of final states is $F = \{(0, 0)\} \cup \{1, 2, \ldots, m-1\} \times \{1, 2, \ldots, n-1\}$. Since $\gcd(m, n) = 1$, every state of the product automaton is reachable. To prove distinguishability, let p and q be two distinct states of the product automaton. Then there is an integer $k \geq 0$ such that $p \cdot a^k = (m-1, 0)$ and $q \cdot a^k = q'$ where $q' \neq (m-1, 0)$. We have three cases:

(a) $q' \in F$. Then a^k distinguishes p and q since $(m-1, 0) \notin F$.
(b) $q' = (0, n-1)$. Then $a^k a^m$ distinguishes p and q since

$$p \xrightarrow{a^k} (m-1, 0) \xrightarrow{a^m} (m-1, m) \in F,$$
$$q \xrightarrow{a^k} (0, n-1) \xrightarrow{a} (1, 0) \xrightarrow{a^{m-1}} (0, m-1) \notin F; \text{ recall that } m < n.$$

(c) q' is a non-final state different from $(0, n-1)$. Then $a^k a$ distinguishes p and q since $(m-1, 0) \cdot a \notin F$ and $q' \cdot a \in F$.

Hence all the states of the product automaton are reachable and pairwise distinguishable. This means that $\mathrm{sc}(K \oplus L) = mn$. □

3   Operations on Boolean and Alternating Automata

In this section we investigate the descriptional complexity of basic regular operations on languages represented by Boolean and alternating automata. We start with the complementation operation and we show that a language and its complement have the same complexity.

Theorem 10 (Complementation). *Let L be a regular language. Then we have* $\mathrm{asc}(L) = \mathrm{asc}(L^c)$ *and* $\mathrm{bsc}(L) = \mathrm{bsc}(L^c)$.

Proof. Let L be accepted by a minimal n-state BFA (AFA). Then the language L^R is accepted by a 2^n-state DFA (with half of the states final) by Lemma 2. This means that $(L^R)^c$ is accepted by a 2^n state DFA (with half of the states final) since we only interchange final and non-final states in the DFA for L^R. Next $(L^R)^c = (L^c)^R$. Therefore L^c is accepted by an n-state BFA (AFA) by Lemma 3. Hence $\mathrm{asc}(L^c) \le n$ and $\mathrm{bsc}(L^c) \le n$. Moreover we cannot have $\mathrm{asc}(L^c) < n$ because after another complementation we would get $\mathrm{asc}(L) < n$. The argument for $\mathrm{bsc}(L^c)$ is the same. □

We continue with the star operation. We improve the results from [10, Theorems 8, 9] where upper and lower bounds differed by one. We get tight upper bound 2^n for both BFAs and AFAs as a corollary of the next theorem.

Theorem 11 (Star). *Let $n \ge 2$.*

(a) If L is accepted by an n-state BFA, then L^* is accepted by a 2^n-state AFA.
(b) There exists a language L accepted by an n-state AFA such that every BFA for L^* has at least 2^n states.

Proof
(a) Let L be accepted by an n-state BFA. Then L^R is accepted by a 2^n-state DFA by Lemma 2. By Proposition 8, $(L^R)^*$ is accepted by a 2^{2^n}-state DFA with half of the states final. Next $(L^R)^* = (L^*)^R$. This means that L^* is accepted by a 2^n-state AFA by Lemma 3.

(b) Let L^R be the Palmovský's witness language for star [16] with 2^n states and 2^{n-1} final states shown in Fig. 1. By Lemma 3 the language L is accepted by an n-state AFA. By [16, Proof of Theorem 4.4] $\mathrm{sc}((L^R)^*) = 2^{2^n-1} + 2^{2^n-1-2^{n-1}} = 2^{2^n-1}(1 + 2^{-2^{n-1}})$. Since $(L^R)^* = (L^*)^R$ we get $\mathrm{bsc}(L^*) \ge \lceil \log(\mathrm{sc}((L^*)^R)) \rceil = 2^n$ by Corollary 4. □

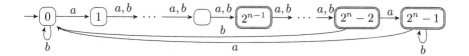

Fig. 1. The reverse of a binary witness for star on BFAs and AFAs.

In what follows we use Lemmas 2, 3 and Corollary 4 without citing them again and again. The next theorem provides tight upper bounds on the complexity of difference, symmetric difference, reversal, and right and left quotient on languages represented by Boolean finite automata.

Theorem 12 (Operations on BFAs). *Let K and L be (regular) languages over an alphabet Σ accepted by an m-state and n-state BFA, respectively. Then*

(a) $\mathrm{bsc}(K \setminus L) \leq m + n$, and the bound is tight if $|\Sigma| \geq 2$;
(b) $\mathrm{bsc}(K \oplus L) \leq m + n$, and the bound is tight if $|\Sigma| \geq 3$;
(c) $\mathrm{bsc}(L^R) \leq 2^n$, and the bound is tight if $|\Sigma| \geq 2$;
(d) $\mathrm{bsc}(KL^{-1}) \leq 2^m$, and the bound is tight if $|\Sigma| \geq 2$;
(e) $\mathrm{bsc}(L^{-1}K) \leq m$, and the bound is tight if $|\Sigma| \geq 1$.

Proof. Let $A = (Q_A, \Sigma, \delta_A, g_A, F_A)$ be an m-state BFA for the language K and $B = (Q_B, \Sigma, \delta_B, g_B, F_B)$ be an n-state BFA for L with $Q_A \cap Q_B = \emptyset$.

(a) The language $K \setminus L$ is accepted by BFA $(Q_A \cup Q_B, \Sigma, \delta, g_A \wedge \overline{g_B}, F_A \cup F_B)$, where $\delta = \delta_A$ on Q_A and $\delta = \delta_B$ on Q_B. Thus $\mathrm{bsc}(K \setminus L) \leq m+n$. For tightness, let K and L be binary witness languages for intersection on BFAs described in [10, Proof of Theorem 2]. Then K and L^c are witnesses for difference since $K \setminus L^c = K \cap L$.

(b) The symmetric difference $K \oplus L$ is accepted by BFA
$$(Q_A \cup Q_B, \Sigma, \delta, (g_A \wedge \overline{g_B}) \vee (\overline{g_A} \wedge g_B), F_A \cup F_B)$$
where $\delta = \delta_A$ on Q_A and $\delta = \delta_B$ on Q_B. Thus $\mathrm{bsc}(K \oplus L) \leq m+n$. For tightness, let K^R and L^R be the languages accepted by 2^m-state and 2^n-state DFAs with half of states final shown in Fig. 2. Then K and L are accepted by m-state and n-state BFAs. In the product automaton, each state (i, j) is reached by $a^i b^j$. Two (non-)final states are distinguished by c if they are in different quadrants and by a word in $a^* + b^*$ otherwise. So we get $\mathrm{sc}(K^R \oplus L^R) = 2^{m+n}$. Next $K^R \oplus L^R = (K \oplus L)^R$. Therefore $\mathrm{bsc}(K \oplus L) \geq m + n$.

(c) The language L^R is accepted by 2^n-state DFA, the special case of BFA. For tightness, let L^R be the Šebej's binary witness language for reversal [11] accepted by a DFA with 2^n states. Then L is accepted by an n-state BFA. By [11, Proof of Theorem 5] $\mathrm{sc}((L^R)^R) = 2^{2^n}$ and therefore $\mathrm{bsc}(L^R) \geq 2^n$.

(d) If K and L are accepted by an m-state and n-state BFA, respectively, then K^R and L^R are accepted by a 2^m-state and 2^n-state DFA, respectively. By Proposition 7 $(KL^{-1})^R = (L^R)^{-1}K^R$ and by [19, Theorem 4.1] $\mathrm{sc}((L^R)^{-1}K^R) \leq 2^{2^m} - 1$. It follows that $\mathrm{bsc}(KL^{-1}) \leq 2^m$. For tightness, let $L = \Sigma^*$ and K be the language accepted by the DFA shown in Fig. 3. Then $\mathrm{bsc}(K) \leq m$ and

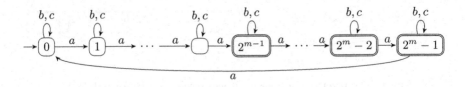

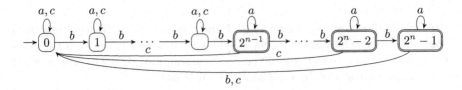

Fig. 2. The reverses of ternary witnesses for symmetric difference on BFAs.

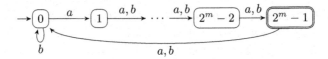

Fig. 3. The reverse of a binary witness for right quotient (by Σ^*) on BFAs.

$\mathrm{bsc}(L) \le n$. Next $(KL^{-1})^R = (\Sigma^*)^{-1}K^R$ and by [19, Proof of Theorem 4.1] $\mathrm{sc}((\Sigma^*)^{-1}K^R) = 2^{2^m} - 1$. Therefore $\mathrm{bsc}(KL^{-1}) \ge 2^m$.

(e) Since $(L^{-1}K)^R = K^R(L^R)^{-1}$ and $\mathrm{sc}(K^R(L^R)^{-1}) \le 2^m$ [19, p. 323], we get $\mathrm{bsc}(L^{-1}K) \le m$. For tightness, let $K = \{a^i \mid 2^{m-1}-1 \le i \le 2^m-2\}$ and $L = a^*$. Then $\mathrm{bsc}(K) \le m$ and $\mathrm{bsc}(L) \le n$. Next $K^R(a^*)^{-1} = \{a^i \mid 0 \le i \le 2^m - 2\}$, so $\mathrm{sc}(K^R(a^*)) = 2^m$. Therefore $\mathrm{bsc}(L^{-1}K) \ge m$.

In the next theorem we study the complexities of same operations on languages represented by alternating finite automata. Note that while the complexities of intersection, union, and difference on AFAs exceed those on BFAs by one, the complexity of symmetric difference on AFAs and BFAs is the same.

Theorem 13 (Operations on AFAs). *Let K and L be (regular) languages over an alphabet Σ accepted by an m-state and n-state AFA, respectively. Then*

(a) $\mathrm{asc}(K \setminus L) \le m + n + 1$, *and the bound is tight if $|\Sigma| \ge 2$;*
(b) $\mathrm{asc}(K \oplus L) \le m + n$, *and the bound is tight if $|\Sigma| \ge 3$;*
(c) $\mathrm{asc}(L^R) \le 2^n$, *and the bound is tight if $|\Sigma| \ge 2$;*
(d) $\mathrm{asc}(KL^{-1}) \le 2^m + 1$, *and the bound is tight if $|\Sigma| \ge 2$;*
(e) $\mathrm{asc}(L^{-1}K) \le m + 1$, *and the bound is tight if $|\Sigma| \ge 1$.*

Proof

(a) Since every AFA is BFA we get $\mathrm{bsc}(K \setminus L) \le m + n$ by Theorem 12(a). Therefore $\mathrm{asc}(K \setminus L) \le m+n+1$. For tightness, let K and L be the binary witness languages for intersection on AFAs described in [10, Proof of Theorem 3]. Then K and L^c are witnesses for difference since $\mathrm{asc}(K \setminus L^c) = \mathrm{asc}(K \cap L) = m + n + 1$.

(b) If K and L are accepted by m-state and n-state AFAs, then K^R and L^R are accepted by 2^m-state and 2^n-state DFAs with half of the states final. It follows that $K^R \oplus L^R$ is accepted by a product automaton of 2^{m+n} states and half of them are final. Therefore $K \oplus L$ is accepted by $(m+n)$-state AFA. For tightness, let K^R and L^R be the languages accepted by 2^m-state and 2^n-state DFAs with half of the states final shown in Fig. 2. Then K and L are accepted by m-state and n-state AFAs. As shown in Theorem 12(b) every BFA for $K \oplus L$ has at least $m+n$ states. Therefore $\mathrm{asc}(K \oplus L) \geq m+n$.

(c) If L is accepted by an n-state AFA, then L^R is accepted by 2^n-state DFA. Every DFA is a special case of AFA. Therefore AFA for language L^R has 2^n states. For tightness, let L^R be the language accepted by 2^n-state Šebej's automaton in which half of the states are final shown in Fig. 4. By [11, Proof of Theorem 5] we have $\mathrm{sc}((L^R)^R) = 2^{2^n}$; notice that any nontrivial number of final states does not matter since the subset automaton of NFA for $(L^R)^R$ does never have equivalent states [11, Proposition 3]. Hence $\mathrm{asc}(L^R) \geq 2^n$ by Corollary 4.

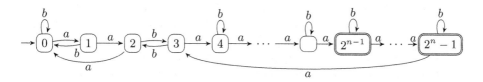

Fig. 4. The reverse of a binary witness for reversal on AFAs.

(d) By Propostion 6 and Theorem 12(d) we get $\mathrm{asc}(KL^{-1}) \leq \mathrm{bsc}(KL^{-1}) + 1 \leq 2^m + 1$. To prove tightness, let $L = \Sigma^*$ and K^R be the language accepted by the DFA A shown in Fig. 5 in which half of the states are final. Then $\mathrm{asc}(K) \leq m$ and $\mathrm{asc}(L) \leq n$. Next $(KL^{-1})^R = (\Sigma^*)^{-1} K^R$. Let us show that $\mathrm{sc}((\Sigma^*)^{-1} K^R) = 2^{2^m} - 1$. Construct an NFA N for $(\Sigma^*)^{-1} K^R$ from the DFA A by making all the states initial. Every non-empty subset in the corresponding subset automaton is reachable as it was shown in [19, Proof of Theorem 4.1]. To prove distinguishability, notice that the state 1 is uniquely distinguishable by the word b^{2^m-2}, and it is uniquely reachable in N from any other state through the unique in-transitions $2 \xrightarrow{a} 3 \xrightarrow{a} \cdots \xrightarrow{a} 2^m-1 \xrightarrow{a} 0 \xrightarrow{a} 1$. By Proposition 1, all states of the subset automaton are pairwise distinguishable. The number of final states in the subset automaton is $2^{2^m} - 2^{2^m-1}$, which is greater than 2^{2^m-1}. Therefore by Lemma 2 we get $\mathrm{asc}(KL^{-1}) \geq 2^m + 1$.

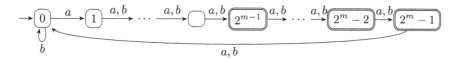

Fig. 5. The reverse of a binary witness for right quotient (by Σ^*) on AFAs.

(e) By Proposition 6 and Theorem 12(e) $\mathrm{asc}(L^{-1}K) \leq \mathrm{bsc}(L^{-1}K) + 1 \leq m + 1$. To get tightness, consider the same two languages as in Theorem 12(e). Notice that the minimal DFA for $K^R(a^*)^{-1}$ has more than 2^{m-1} final states. $\square$

In the next theorem we study the complexity of basic regular operations on unary languages represented by Boolean finite automata.

Theorem 14 (Unary BFAs). *Let $n \geq 2$ and K and L be unary languages accepted by an m-state and n-state BFA, respectively. Then*

(a) $\mathrm{bsc}(K \cap L) \leq m + n$, *and the bound is tight if* $\gcd(m, n) = 1$;
(b) $\mathrm{bsc}(K \cup L) \leq m + n$, *and the bound is tight if* $\gcd(m, n) = 1$;
(c) $\mathrm{bsc}(K \backslash L) \leq m + n$, *and the bound is tight if* $\gcd(m, n) = 1$;
(d) $\mathrm{bsc}(K \oplus L) \leq m + n$, *and the bound is tight if* $\gcd(m, n) = 1$;
(e) $\mathrm{bsc}(L^R) = \mathrm{bsc}(L)$;
(f) $\mathrm{bsc}(L^*) \leq 2n$ *and the bound is tight;*
(g) $\mathrm{bsc}(KL^{-1}) \leq m$, *and the bound is tight.*

Proof. Let unary languages K and L be accepted by m-state and n-state BFA, respectively. Then K and L are accepted by 2^m-state and 2^n-state DFA, respectively, by Corollary 5, and the languages $K \cap L$, $K \cup L$, $K \backslash L$, $K \oplus L$ are accepted by a $2^m 2^n$-state product automaton. This gives upper bounds $m + n$ in cases (a)–(d). To prove tightness for intersection, let $K = (a^{2^m})^*$ and $L = (a^{2^n-1})^*$. Then K and L are accepted by a 2^m-state and 2^n-state DFA, respectively, so by an m-state and n-state BFA, respectively. Since $\gcd(2^m, 2^n - 1) = 1$, we have $\mathrm{sc}(K \cap L) = 2^m(2^n - 1)$. This means that $\mathrm{bsc}(K \cap L) \geq \lceil \log(2^m(2^n-1)) \rceil = m+n$. For union, we may use the languages K^c and L^c, since $K^c \cup L^c = (K \cap L)^c$ and a language and its complement have the same Boolean state complexity. Similarly, for difference we use the languages K and L^c. For symmetric difference, let us consider unary languages K and L accepted by automata $A = (2^m, 0, \{0\})$ and $B = (2^n - 1, 0, \{1, 2, \ldots, 2^n - 2\})$. By Proposition 9 $\mathrm{sc}(K \oplus L) = 2^m(2^n - 1)$. It follows that $\mathrm{bsc}(K \oplus L) \geq \lceil \log(2^m(2^n - 1)) \rceil = m + n$.

(e) The equality follows from the fact that $L = L^R$ in the unary case.

(f) The state complexity of the star operation in the unary case is $(n-1)^2 + 1$ [3, 19]. If a unary language L is accepted by an n-state BFA then L is accepted by a 2^n-state DFA. This means that L^* is accepted by a DFA of at most $(2^n-1)^2 + 1$ states, so by a DFA of at most 2^{2n} states. Therefore $\mathrm{bsc}(L^*) \leq 2n$. For tightness, let L be the unary language accepted by the DFA $(2^n, 0, \{2^n - 1\})$ meeting the upper bound for star [19, Theorem 5.3]. Then L is accepted by an n-state BFA and $\mathrm{bsc}(L^*) \geq \lceil \log(\mathrm{sc}(L^*)) \rceil = \lceil \log((2^n - 1)^2 + 1) \rceil = 2n$.

(g) In the unary case, $KL^{-1} = L^{-1}K$. In Theorem 12(e) we proved that $\mathrm{bsc}(L^{-1}K) \leq m$ and we provided a unary witness. $\square$

Recall that by Proposition 6 $\mathrm{asc}(L) \leq \mathrm{bsc}(L) + 1$. Therefore as a corollary of the previous theorem we get the following upper bounds.

Corollary 15 (Unary AFAs). *Let $n \geq 2$ and K and L be unary languages accepted by an m-state and n-state AFA, respectively. Then*

(a) $\mathrm{asc}(K \cap L) \leq m + n + 1;$
(b) $\mathrm{asc}(K \cup L) \leq m + n + 1;$
(c) $\mathrm{asc}(K \setminus L) \leq m + n + 1;$
(d) $\mathrm{asc}(L^R) = \mathrm{asc}(L);$
(e) $\mathrm{asc}(L^*) \leq 2n + 1;$
(f) $\mathrm{asc}(KL^{-1}) \leq m + 1.$

We are not able to prove the tightness since the complexity of operations on unary DFAs with half of the states final is not known. The previous theorem and its corollary imply that a binary alphabet for some of our witness languages is optimal in the sense that it cannot be reduced to a unary alphabet.

4 Conclusions

We investigated the descriptional complexity of basic regular operations on languages represented by Boolean and alternating finite automata. We considered the operations of complementation, star, difference, symmetric difference, reversal, and left and right quotient. For each operation we obtained the tight upper bound on its complexity on both Boolean and alternating automata.

Our results are summarized in Table 1. The table also shows the size of alphabet used for describing witness languages, and compares our results to the known results for deterministic [11,14,19] and nondeterministic finite automata from [6,9]. The results for intersection and union on Boolean and alternating automata are from [10]. Notice that the complexity of intersection, union, and difference on alternating automata is $m + n + 1$ while the complexity of symmetric difference is $m + n$. Except for ternary witnesses for symmetric difference, all the other provided witnesses are defined over a binary or unary alphabets and, moreover, a binary alphabet for the witness languages for star, reversal, and right quotient on BFAs and AFAs is optimal in the sense that it cannot be reduced to a unary alphabet.

Table 1. The complexity of operations on languages represented by BFAs, AFAs, DFAs, NFAs. The results for DFAs are from [11,14,19], the results for NFAs are from [6,9], and the results for intersection and union on BFAs and AFAs are from [10].

	BFA	$|\Sigma|$	AFA	$|\Sigma|$	DFA	$|\Sigma|$	NFA	$|\Sigma|$
Complement	n	1	n	1	n	1	2^n	2
Intersection	$m + n$	2	$m + n + 1$	2	mn	2	mn	2
Union	$m + n$	2	$m + n + 1$	2	mn	2	$m + n + 1$	2
Difference	$m + n$	2	$m + n + 1$	2	mn	2	$\leq m2^n$	
Symmetric difference	$m + n$	3	$m + n$	3	mn	2	$\leq 2^{m+n}$	
Reversal	2^n	2	2^n	2	2^n	2	$n + 1$	2
Star	2^n	2	2^n	2	$\frac{3}{4}2^n$	2	$n + 1$	1
Left quotient	m	1	$m + 1$	1	$2^m - 1$	2	$m + 1$	2
Right quotient	2^m	2	$2^m + 1$	2	m	1	m	1

References

1. Brzozowski, J., Jirásková, G., Liu, B., Rajasekaran, A., Szykuła, M.: On the state complexity of the shuffle of regular languages. In: Câmpeanu, C., Manea, F., Shallit, J. (eds.) DCFS 2016. LNCS, vol. 9777, pp. 73–86. Springer, Cham (2016). https://doi.org/10.1007/978-3-319-41114-9_6

2. Brzozowski, J.A., Leiss, E.L.: On equations for regular languages, finite automata, and sequential networks. Theoret. Comput. Sci. **10**, 19–35 (1980). https://doi.org/10.1016/0304-3975(80)90069-9

3. Čevorová, K.: Kleene star on unary regular languages. In: Jurgensen, H., Reis, R. (eds.) DCFS 2013. LNCS, vol. 8031, pp. 277–288. Springer, Heidelberg (2013). https://doi.org/10.1007/978-3-642-39310-5_26

4. Chandra, A.K., Kozen, D., Stockmeyer, L.J.: Alternation. J. ACM **28**(1), 114–133 (1981). https://doi.org/10.1145/322234.322243

5. Fellah, A., Jürgensen, H., Yu, S.: Constructions for alternating finite automata. Int. J. Comput. Math. **35**(1–4), 117–132 (1990). https://doi.org/10.1080/00207169008803893

6. Holzer, M., Kutrib, M.: Nondeterministic descriptional complexity of regular languages. Int. J. Found. Comput. Sci. **14**(6), 1087–1102 (2003). https://doi.org/10.1142/S0129054103002199

7. Hopcroft, J.E., Ullman, J.D.: Introduction to Automata Theory, Languages and Computation. Addison-Wesley, Boston (1979)

8. Hospodár, M., Jirásková, G.: Concatenation on deterministic and alternating automata. In: Bordihn, H., Freund, R., Nagy, B., Vaszil, G. (eds.) NCMA 2016, vol. 321, pp. 179–194. Österreichische Computer Gesellschaft (2016). books@ocg.at

9. Jirásková, G.: State complexity of some operations on binary regular languages. Theoret. Comput. Sci. **330**(2), 287–298 (2005). https://doi.org/10.1016/j.tcs.2004.04.011

10. Jirásková, G.: Descriptional complexity of operations on alternating and Boolean automata. In: Hirsch, E.A., Karhumäki, J., Lepistö, A., Prilutskii, M. (eds.) CSR 2012. LNCS, vol. 7353, pp. 196–204. Springer, Heidelberg (2012). https://doi.org/10.1007/978-3-642-30642-6_19

11. Jirásková, G., Šebej, J.: Reversal of binary regular languages. Theoret. Comput. Sci. **449**, 85–92 (2012). https://doi.org/10.1016/j.tcs.2012.05.008

12. Krajňáková, I., Jirásková, G.: Square on deterministic, alternating, and Boolean finite automata. In: Pighizzini, G., Câmpeanu, C. (eds.) DCFS 2017. LNCS, vol. 10316, pp. 214–225. Springer, Cham (2017). https://doi.org/10.1007/978-3-319-60252-3_17

13. Leiss, E.L.: Succint representation of regular languages by Boolean automata. Theoret. Comput. Sci. **13**, 323–330 (1981). https://doi.org/10.1016/S0304-3975(81)80005-9

14. Maslov, A.N.: Estimates of the number of states of finite automata. Soviet Math. Doklady **11**(5), 1373–1375 (1970)

15. Nicaud, C.: Average state complexity of operations on unary automata. In: Kutyłowski, M., Pacholski, L., Wierzbicki, T. (eds.) MFCS 1999. LNCS, vol. 1672, pp. 231–240. Springer, Heidelberg (1999). https://doi.org/10.1007/3-540-48340-3_21

16. Palmovský, M.: Kleene closure and state complexity. RAIRO - Theor. Inf. Appl. **50**(3), 251–261 (2016). https://doi.org/10.1051/ita/2016024

17. Sipser, M.: Introduction to the theory of computation. Cengage Learn (2012)

18. Yu, S.: Regular languages. In: Rozenberg, G., Salomaa, A. (eds.) Handbook of Formal Languages. Volume 1: Word, Language, Grammar, pp. 41–110. Springer, Heidelberg (1997). https://doi.org/10.1007/978-3-642-59136-5_2
19. Yu, S., Zhuang, Q., Salomaa, K.: The state complexities of some basic operations on regular languages. Theoret. Comput. Sci. **125**(2), 315–328 (1994). https://doi.org/10.1016/0304-3975(92)00011-F

Conflict Free Version of Covering Problems on Graphs: Classical and Parameterized

Pallavi Jain[1]([✉]), Lawqueen Kanesh[1], and Pranabendu Misra[2]

[1] Institute of Mathematical Sciences, HBNI, Chennai, India
{pallavij,lawqueen}@imsc.res.in
[2] Department of Informatics, University of Bergen, Bergen, Norway
Pranabendu.Misra@uib.no

Abstract. Let Π be a family of graphs. In the classical Π-VERTEX DELETION problem, given a graph G and a positive integer k, the objective is to check whether there exists a subset S of at most k vertices such that $G - S$ is in Π. In this paper, we *introduce the conflict free version* of this classical problem, namely CONFLICT FREE Π-VERTEX DELETION (CF-Π-VD), and study these problems from the viewpoint of classical and parameterized complexity. In the CF-Π-VD problem, given two graphs G and H on the same vertex set and a positive integer k, the objective is to determine whether there exists a set $S \subseteq V(G)$, of size at most k, such that $G - S$ is in Π and $H[S]$ is edgeless. Initiating a systematic study of these problems is one of the main conceptual contribution of this work. We obtain several results on the conflict free version of several classical problems. Our first result shows that if Π is characterized by a finite family of forbidden induced subgraphs then CF-Π-VD is Fixed Parameter Tractable (FPT). Furthermore, we obtain improved algorithms for conflict free version of several well studied problems. Next, we show that if Π is characterized by a "well-behaved" infinite family of forbidden induced subgraphs, then CF-Π-VD is W[1]-hard. Motivated by this hardness result, we consider the parameterized complexity of CF-Π-VD when H is restricted to well studied families of graphs. In particular, we show that the conflict free versions of several well-known problems such as FEEDBACK VERTEX SET, ODD CYCLE TRANSVERSAL, CHORDAL VERTEX DELETION and INTERVAL VERTEX DELETION are FPT when H belongs to the families of d-degenerate graphs and nowhere dense graphs.

Keywords: Hereditary properties · Parameterized algorithms
Kernelization · Vertex cover · Conflict free

1 Introduction

Graph-modification by either deleting vertices, or deleting edges, or adding edges such that the resulting graph satisfies certain properties or becomes a member of some well-understood graph class is one of the basic problems in

F. V. Fomin and V. V. Podolskii (Eds.): CSR 2018, LNCS 10846, pp. 194–206, 2018.
https://doi.org/10.1007/978-3-319-90530-3_17

graph theory and computer science. However, most of these problems are NP-complete [18, 29], and therefore they have been extensively studied in various algorithmic paradigms that are meant to cope with NP-completeness [13, 14, 20], such as restricted classes of inputs, approximation algorithms and parameterized complexity. This paper introduces a new variant of these classical problems, called the *conflict free version*, and studies them from viewpoint of classical and parameterized complexity.

In the past, the conflict free versions of some classical problems have been studied, e.g. for SHORTEST PATH [16], MAXIMUM FLOW [24, 25], KNAPSACK [26], BIN PACKING [11], SCHEDULING [12], MAXIMUM MATCHING and MINIMUM WEIGHT SPANNING TREE [9, 10]. It is interesting to note that some of these problems are NP-hard even when their non-conflicting version is polynomial time solvable. The study of conflict free problems has also been recently initiated in computational geometry motivated by various applications (see [2–4]). Motivated by these works, we initiate the study of the conflict free versions of several well studied vertex deletion problems in parameterized complexity. This is the main conceptual contribution of this paper. A typical parameterized vertex deletion problem on graphs is of the following form. Let Π be a family of graphs (or property) – such as edgeless graphs, forests, cluster graphs, chordal graphs, interval graphs, bipartite graphs, split graphs or planar graphs. The vertex deletion problem corresponding to Π is formally stated as follows.

Π-VERTEX DELETION   **Parameter:** k
Input: An undirected graph G and a non-negative integer k.
Question: Does there exist $S \subseteq V(G)$, such that $|S| \leq k$ and $G - S$ is in Π?

That is, given a graph G, can we delete at most k vertices such that the resulting graph belongs to Π? The set S *is called* Π-*deletion set*. An algorithm for Π-VERTEX DELETION that runs in time $f(k) \cdot |V(G)|^{\mathcal{O}(1)}$ is called *fixed-parameter tractable* (FPT) algorithm and the problem itself is said to be FPT. We refer to [8] for more details on parameterized complexity. The study of parameterized graph deletion problems together with their various restrictions and generalizations has been an active area of research recently.

To formulate the conflict free version of these classical problems, let us begin with an example. Consider SET COVER, that has the following conflict free version. We are given a universe $\mathcal{U}$ and a family $\mathcal{S}$ of subsets of $\mathcal{U}$, a positive integer k and a graph H (with $V(H) = \mathcal{S}$). The objective is to check whether there exists a $\mathcal{S}' \subset \mathcal{S}$ of size at most k whose union is $\mathcal{U}$ and $H[\mathcal{S}']$ is edgeless. Now, we may similarly combine the classical vertex deletion problems on graphs, with the conflict free model described in [2–4] and arrive at the following generic conflict free problem. Let Π be a family of graphs. The conflict free vertex deletion problem corresponding to Π is formally stated as follows.

CONFLICT FREE Π-VERTEX DELETION (CF-Π-VD)   **Parameter:** k
Input: An undirected graph G, a conflict graph H on vertex set $V(G)$ and a non-negative integer k.
Question: Does there exist a set $S \subseteq V(G)$, such that $|S| \leq k$, $G - S$ is in Π and S is an independent set in H?

We define CF-Π-VD for hypergraphs and directed graphs, appropriately. In this paper, we focus on CF-Π-VD problems corresponding to several well studied problems in parameterized complexity, namely VERTEX COVER, d-HITTING SET, SPLIT VERTEX DELETION, FEEDBACK VERTEX SET IN TOURNAMENTS (FVST) and FEEDBACK VERTEX SET (FVS). Observe that when H is an edgeless graph, CF-Π-VD is same as Π-VERTEX DELETION and thus it generalizes the non-conflict free version of the problem. Furthermore, when H is same as G it corresponds to *independent* version of these problems which are also well studied, such as INDEPENDENT FEEDBACK VERTEX SET [19,23]. Thus, CF-Π-VD is a generalization of well studied problems in algorithms and complexity.

Our Results. Apart from introducing an interesting family of problems, we obtain the following results in the realm of parameterized and classical complexity. We note that several of these results are in sharp contrast to the non-conflict version of the same problem.

A *graph property* Π is a set of graphs, and a graph in Π is called a Π-*graph*. We say that Π is *hereditary* if for any graph G in Π, every induced subgraph of G is also in Π. A graph property Π has a forbidden set characterization if there is a set $\mathcal{F}$ of graphs such that a graph is a Π-graph if and only if it does not contain any graph in $\mathcal{F}$ as an induced subgraph, and further, it has a finite forbidden characterization if $\mathcal{F}$ is a finite set. We study the complexity of CF-Π-VD based on the forbidden set of the property Π.

Graph properties with finite forbidden characterization. The starting point of our results is a generic result by Cai [5] about graph properties which have a finite forbidden characterization. We show an analogous result for CF-Π-VD. In particular we show that CF-Π-VD is FPT whenever Π has a finite forbidden set characterization. Indeed, we show that this problem admits an algorithm with running time $\mathcal{O}(\alpha^k \cdot n \cdot T(m, n))$, where $T(m, n)$ is time to recognize a graph in Π and α is the size of largest graph in the finite forbidden set $\mathcal{F}$. Furthermore, it also admits a kernel with $\mathcal{O}(\alpha^2 \alpha! k^\alpha)$ vertices.

Next, we study the conflict free version of several well-studied cases of Π-VERTEX DELETION, where Π is characterized by the finite family of forbidden induced subgraphs. These results improve upon the generic result stated above.

1. CONFLICT FREE VERTEX COVER (CF-VC) admits a $2k$-vertex kernel, a factor 2-approximation algorithm, an $\mathcal{O}^\star(1.2738^k)$ FPT algorithm[1] and a $\mathcal{O}^\star(1.1996^n)$ exact algorithm. Further, CF-VC is NP-complete even when graph G is of degree at most 2. This holds even when G is disjoint union of P_3 (P_ℓ denotes path on ℓ vertices). Furthermore, CF-VC is polynomial

---

[1] $\mathcal{O}^\star$ suppresses the polynomial factor in the running time.

time solvable when G has degree at most one, or when both G and H have a perfect matching.

2. The CONFLICT FREE d-HITTING SET (d-CF-HS) problem can be solved in $\mathcal{O}^{\star}(((d-1) + .2738)^k) = \mathcal{O}^{\star}((d - 0.7262)^k)$ time.

3. CONFLICT FREE SPLIT VERTEX DELETION (CF-SVD) can be solved in $\mathcal{O}^{\star}(1.2738^k k^{\mathcal{O}(\log k)})$ time and polynomial space.

4. CONFLICT FREE FEEDBACK VERTEX SET IN TOURNAMENTS (CF-FVST) can be solved in $\mathcal{O}^{\star}(2^k)$ time.

Let us note that given an instance (G, H) of CF-VC, we can test whether there exists a conflict free vertex cover (of any size) in polynomial time. However, one can show that testing whether there exists a conflict free feedback vertex set is NP-complete.

Graph properties without finite forbidden characterization. Next, we consider those graph properties that are not characterized by a finite family of forbidden induced subgraphs. We show that if Π is characterized by a "well-behaved" infinite family of forbidden induced subgraphs, then CF-Π-VD is W[1]-hard. In particular, we show that CONFLICT FREE FEEDBACK VERTEX SET (CF-FVS) is W[1]-hard even when G is disjoint union of cycles. A similar result holds for CONFLICT FREE ODD CYCLE TRANSVERSAL (CF-OCT), CONFLICT FREE CHORDAL VERTEX DELETION (CF-CVD) and CONFLICT FREE INTERVAL VERTEX DELETION (CF-IVD).

This motivates us to restrict the families of conflict graphs. We show that conflict free versions of several well-known problems such as FEEDBACK VERTEX SET, ODD CYCLE TRANSVERSAL, CHORDAL VERTEX DELETION and INTERVAL VERTEX DELETION are FPT when H belongs to the family of d-degenerate graphs, or nowhere dense graphs. It is worth noting that the families of d-degenerate graphs and nowhere dense graphs include trees, graphs of bounded degree, planar graphs, graphs that exclude a fixed graph H as a minor (or a topological minor) and graphs of bounded expansion. These algorithms are based on the notion of "k-independence covering family" introduced in [19].

Due to space constraints, basic graph theoretic preliminaries and proofs of results marked ($\star$) have been omitted. These will appear in the full version of the paper.

2   Conflict Free Version of Properties with Forbidden Set Characterizations

2.1   Properties with Finite Forbidden Set Characterizations

In this subsection, we study the CF-FINITE Π-VD problem when Π is hereditary and admits a *finite forbidden set characterization*.

FPT Algorithm for CF-FINITE Π-VD. Let $\mathcal{F}$ be the finite forbidden set corresponding to the property Π. Cai [5] showed that the FINITE Π-D is FPT. That is, given a graph G, testing whether there exists a set $S \subseteq V(G)$ of size at most k such that $G - S$ is a Π-graph is FPT. The algorithm works as follows. It starts by finding a forbidden vertex set X in G; among which we know that at least one vertex must go in the solution set S. Therefore, we branch on this collection of vertices, and for each vertex $v \in X$, we recursively apply the algorithm to solve the instance $(G - v, k - 1)$. If one of these branches returns a Π-*deletion set* S, then clearly $S \cup \{v\}$ is of size at most k and it is a Π-*deletion set* in G. Else, we return that the given instance is a *no* instance. At every recursive call we decrease the parameter by 1, and thus the height of the search tree does not exceed k. At every step, we branch in at most α subproblems; where α is the size of largest graph in $\mathcal{F}$. Hence the number of nodes in the search tree does not exceed α^k. Observe that, the algorithm actually enumerates all the *minimal* Π-deletion sets of size at most k. Thus for CF-FINITE Π-VD, all we need to do in addition, is to check whether $H[S]$ is edgeless or not. We will also need the following result for the above algorithm.

Proposition 1 [5, Theorem 1]. *For any hereditary property Π, if Π is recognizable in time $T(m, n)$, then for any graph G that is not a Π-graph, a minimal forbidden induced subgraph of Π in G can be found in $\mathcal{O}(n \cdot T(m, n))$ time.*

With the above theorem in hand, we obtain the following theorem.

Theorem 1. *CF-FINITE Π-VD is FPT and admits an algorithm with running time $\mathcal{O}(\alpha^k \cdot n \cdot T(m, n))$, where $T(m, n)$ is the time to recognize a graph in Π and α is the size of largest graph in the finite forbidden set $\mathcal{F}$.*

We also obtain a polynomial kernel with at most $\mathcal{O}(\alpha^2 \alpha! k^\alpha)$ vertices for CF-FINITE Π-VD. The details will appear in the full version of the paper.

2.2 Properties that Do Not Admit Finite Forbidden Characterization

It is well know that a property Π is hereditary if and only if Π admits a forbidden set characterization [5]. Let $\mathcal{F}$ denote the forbidden set corresponding to Π. Following the previous section, a natural question that arises is what happens when $\mathcal{F}$ is *infinite*. We call the corresponding vertex deletion problem as CONFLICT FREE Π-VERTEX DELETION (CF-Π-VD). For example, suppose that Π is a family of forests, or chordal graphs, or interval graphs, or bipartite graphs. Then the corresponding classical problems of Π-VERTEX DELETION (Π-VD) problems are known as FEEDBACK VERTEX SET (FVS), CHORDAL VERTEX DELETION (CVD), INTERVAL VERTEX DELETION (IVD) and ODD CYCLE TRANSVERSAL (OCT) and these problems are known to be FPT [6,17,21,27]. However, we will show now that conflict free version of these problems is W[1]-hard. Indeed, CONFLICT FREE FEEDBACK VERTEX SET (CF-FVS) is W[1]-hard even when G is *disjoint union of cycles*.

Towards this, we present a parameter preserving reduction from the W[1]-hard MULTICOLORED INDEPENDENT SET (MCIS) problem to CF-FVS. See [8] for further details on the notion of W[1]-hardness and for the fact that MCIS is W[1]-hard. In MCIS, given a graph G, an integer k, and a partition of $V(G)$ into k sets, say $V_1, \ldots, V_k$, the objective is to check whether there exists a set $X \subseteq V(G)$ such that it contains exactly one element from every set V_i and is an independent set in G. We call such an independent set as *multicolored independent set*.

Theorem 2 ($\star$). CF-FVS *is* W[1]-*hard*.

Proof (Sketch). Let $(G, (V_1, \ldots, V_k), k)$ be an instance of MCIS. Given this, we construct an instance (G', H, k) of CF-FVS as follows. The vertices of G' and H are same as $V(G)$. For each set V_i, we construct a cycle $C_{|V_i|}$ in G' (C_ℓ denotes cycle on ℓ vertices) on vertex set V_i in G'. The graph H is identical to graph G. Now we can show that G has a multicolored independent set of size k, if and only if, (G', H) has a conflict free feedback vertex set of size k.   □

The proof of Theorem 2 requires nothing specific about CF-FVS, except that G is a disjoint union of forbidden sets where each forbidden set is identified with a color class V_i. If $\mathcal{F}$ is infinite and *well behaved* in the following sense: given an integer n we can output a forbidden set F of size polynomial in n (in fact size $f(k) \cdot n^{\mathcal{O}(1)}$ will also work for our purpose) in time $\tau(k) \cdot n^{\mathcal{O}(1)}$, then we can mimic the proof of Theorem 2 and show that the corresponding CF-Π-VD is W[1]-hard. Let us note that in certain cases, e.g. for bipartite graphs where the family of forbidden subgraphs are odd cycles, we may need to augment a color class V_i with additional vertices to obtain a forbidden set in $\mathcal{F}$ in the graph G. This is easily handled by making the additional vertices adjacent to all vertices in the conflict graph H, which ensures that they cannot be selected in any solution of cardinality greater than one. In particular, this holds for Π being the family of chordal graphs, or interval graphs, or bipartite graphs. Here, f and τ are computable functions.

2.3   Results on Properties Without Finite Forbidden Characterization

In Sect. 2.2, we have shown that if $\mathcal{F}$ is infinite, CF-Π-VD is W[1]-hard in general, even though the corresponding classical problem is FPT, e.g. CF-FVS, CF-OCT, CF-CVD etc. In light of this, a natural question that arises is what happens if H is restricted to certain graph classes. In this section, we show that CF-Π-VD is FPT when H is restricted to the class of *d-degenerate graphs* or *no-where dense graphs*.

   The *degeneracy* of an n-vertex graph $G^\star$ is defined as the minimum integer d such that there exists an ordering $\sigma : V(G^\star) \to \{1, \cdots, n\}$ where every vertex v has at most d neighbors u for which $\sigma(u) > \sigma(v)$. Such an ordering σ is called a *d-degeneracy sequence* of graph $G^\star$. We fix one such sequence, and then for any vertex $v \in V(G^*)$, we define its *forward* and *backward* neighbors in G^* with respect to this ordering. Our algorithm is based on the construction of a k-independence covering family of a graph, using the Independence Covering Lemma of [19]. For a graph $H^\star$ and an integer k, a *k-independence covering* *family*, denoted by $\mathscr{F}(H^\star, k)$, is a family of independent sets of graph $H^\star$ such that for any independent set X in $H^\star$ of size at most k there exists a set Y in $\mathscr{F}(H^\star, k)$ such that $X \subseteq Y$. We will use the following propositions to construct a k-independence covering family for H.

Proposition 2 [19, Lemma 1.1]. *There exists a linear time randomized algorithm, that given as input a d-degenerate graph $H^\star$ and $k \in \mathbb{N}$, outputs an independent set Y, such that for every independent set X in $H^\star$ of size at most k the probability that X is a subset of Y is at least $\left(\binom{k(d+1)}{k} k(d+1)\right)^{-1}$.*

Proposition 3 [19, Lemmas 3.2 and 3.3]. *There are two deterministic algorithms, that given a d-degenerate graph $H^\star$ and $k \in \mathbb{N}$, outputs independence covering families $\mathscr{F}_1(H^\star, k)$ of size at most $\binom{k(d+1)}{k} 2^{o(k(d+1))} \log n$ and $\mathscr{F}_2(H^\star, k)$ of size at most $\binom{k^2(d+1)^2}{k}(k(d+1))^{\mathcal{O}(1)} \log n$ respectively. These algorithms run in time $\mathcal{O}(|\mathscr{F}_1(H^\star, k)|(n+m))$ and $\mathcal{O}(|\mathscr{F}_1(H^\star, k)|(n+m))$, respectively.*

Now we present our algorithm for CF-Π-VD problems, when the conflict graph is d-degenerate. The algorithm is based on the observation that, given a independence covering family of conflict graph, the conflict free solution of the problem lies inside one of the sets in this family. By construction, each set in this family is an independent set in H, and therefore the problem of finding a solution to the given instance of CF-Π-VD boils down to finding a solution of Π-VD in the graph G that also lies in a chosen set in the family. In particular, it reduces to solving the following annotated version of CF-Π-VD.

ANNOTATED-Π-VD (A-Π-VD)   **Parameter:** k
Input: A graph G, $Y \subseteq V(G)$ and an integer k.
Question: Does there exist $S \subseteq Y$ of size at most k such that $G - S$ is a Π-graph?

Theorem 3 ($\star$). *Let Π be a property such that A-Π-VD admits an algorithm with running time $\tau(k)n^{\mathcal{O}(1)}$. Then CF-Π-VD admits a randomized algorithm with running time $\binom{k(d+1)}{k} k(d+1) \tau(k) n^{\mathcal{O}(1)}$, and a deterministic algorithm with running time $\min\left\{\binom{k(d+1)}{k} 2^{o(k(d+1))} \log n, \binom{k^2(d+1)^2}{k}(k(d+1))^{\mathcal{O}(1)} \log n\right\} \tau(k) n^{\mathcal{O}(1)}$, on the family of conflict graphs that are d-degenerate.*

Proof (Sketch). Given an instance (G, H, k) of CF-Π-VD we do as follows. Run the following two step procedure $\left(\binom{k(1+d)}{k} k(d+1)\right)$ times.

1. Run the algorithm in Proposition 2 on (H, k), and obtain the set Y.
2. Solve A-Π-VD on the instance (G, H, Y, k) using the algorithm running in time $\tau(k)n^{\mathcal{O}(1)}$.

The algorithm will output *yes* if, Step 2 returns *yes* at least once, else algorithm returns *no*. Now we prove the correctness of algorithm. Since in Step 1, the output set Y is an independent set in conflict graph H, if the algorithm returns *yes* then the input instance is a *yes* instance. Now suppose that the input instance is a *yes* instance and X be its solution. By Proposition 2, probability that $X \subseteq Y$ is at least $p = (\binom{k(d+1)}{k}k(d+1))^{-1}$. We repeat the procedure $1/p$ times, so the probability that in all executions $X \not\subseteq Y$ is at most $(1-p)^{1/p} \leq 1/e$. Therefore algorithm returns *yes* with probability at least $1 - 1/e$. Running time follows from Proposition 2 and the assumed running time of the algorithm for A-Π-VD.

Now we give the deterministic algorithm. Given an instance (G, H, k) of CF-Π-VD the algorithm works as follows. Algorithm first constructs k-independence covering family $\mathscr{F}(H, k)$ of conflict graph H, using Proposition 3. Now for all sets $Y \in \mathscr{F}(H, k)$, algorithm solve A-Π-VD on instance (G, H, Y, k) using the algorithm assumed in the statement of the theorem. The algorithm outputs *yes* if for some set $Y \in \mathscr{F}(H, k)$, the A-Π-VD returns *yes*, otherwise returns *no*. The correctness of algorithm follows from the definition of k-independence covering family. The running time follows from Proposition 3, and the assumed running time of the algorithm for A-Π-VD. This completes the proof. □

The above theorem naturally leads to the question that when can A-Π-VD be FPT. We give an affirmative answer for several cases when the integer weighted version (W-Π-VD) of the corresponding Π-VD is FPT.

Lemma 1. *Let Π be a property such that* W-Π-VD *admits an algorithm with running time* $\gamma(k)n^{\mathcal{O}(1)}$. *Then* A-Π-VD *also admits an algorithm with running time* $\gamma(k)n^{\mathcal{O}(1)}$.

Proof. We give a polynomial time reduction from A-Π-VD to W-Π-VD. Towards this given an instance (G, Y, k) of A-Π-VD we construct an instance (G', w, k) of W-Π-VD as follows. We take graph G' identical to graph G. We define weight function w as follows. We assign $w(v) = k + 1$ if $v \in V(G) \setminus Y$, otherwise $w(v) = 1$. We now show that (G, Y, k) is a yes instance of A-Π-VD if and only if (G', w, k) is a yes instance of W-Π-VD. Let $S \subseteq Y$ be a minimal vertex subset of size at most k such that $G - S$ is a Π-graph, then we claim that S is also solution G'. Since $S \subseteq Y$, we have that $w(v) = 1$ for all vertices in S. Therefore, the weight of S is at most k. Since G' is same as G, $G' - S$ is also a Π-graph.

Conversely, let S' be a set of weight at most k such that $G' - S'$ is a Π-graph. We claim that S' is a solution of G. Note that all the vertices in $V(G') \setminus Y$ have weight $k + 1$, therefore $S' \subseteq Y$. Since each vertex in S' has weight one, $|S'| \leq k$. Furthermore, since graph G is identical to graph G', we have that $G - S'$ is a Π-graph. This completes the proof. □

It is known that WEIGHTED FEEDBACK VERTEX SET (WFVS) can be solved in time $\mathcal{O}(3.618^k n^{\mathcal{O}(1)})$ [1] and thus by Lemma 1 we have that A-FVS can be solved in time $\mathcal{O}(3.618^k n^{\mathcal{O}(1)})$. Now by applying Theorem 3 we get the following.

Corollary 1. CF-FVS *either admits a randomized algorithm with running time* $\binom{k(d+1)}{k}k(d + 1)\tau(k)n^{\mathcal{O}(1)}$ *or a deterministic algorithm with running* *time* $\min\left\{\binom{k(d+1)}{k}2^{o(k(d+1))}\log n, \binom{k^2(d+1)^2}{k}(k(d+1))^{\mathcal{O}(1)}\log n\right\}\tau(k)n^{\mathcal{O}(1)}$, *on the family of conflict graphs that are d-degenerate. Here,* $\tau(k) = 3.618^k$.

We may similarly obtain the results for CONFLICT FREE ODD CYCLE TRANSVERSAL, CONFLICT FREE CHORDAL VERTEX DELETION and CONFLICT FREE INTERVAL VERTEX DELETION.

Corollary 2 (⋆). CF-OCT, CF-CVD *and* CF-IVD *admit a randomized algorithm with running time* $\binom{k(d+1)}{k}k(d+1)\tau(k)n^{\mathcal{O}(1)}$, *and a deterministic algorithm with   running*

time $\min\left\{\binom{k(d+1)}{k}2^{o(k(d+1))}\log n, \binom{k^2(d+1)^2}{k}(k(d+1))^{\mathcal{O}(1)}\log n\right\}\tau(k)n^{\mathcal{O}(1)}$, *on the family of conflict graphs that are d-degenerate. Here,* $\tau(k)$ *is* $4^k k^6$, $2^{\mathcal{O}(k\log k)}$ *and* 8^k *for each of these problems respectively.*

The above results can be also extended to the class of nowhere dense graphs. The details will appear in the full version of the paper.

3   Well Studied Special Cases of CF-FINITE Π-VD

We can obtain improved algorithms for the conflict free version of several well-studied cases of Π-VERTEX DELETION whenever Π is characterized by the finite family of forbidden induced subgraphs. In this section, we give improved algorithms for CONFLICT FREE VERTEX COVER, CONFLICT FREE d-HITTING SET, CONFLICT FREE SPLIT VERTEX DELETION and CONFLICT FREE FEEDBACK VERTEX SET IN TOURNAMENTS.

3.1   CONFLICT FREE VERTEX COVER

In this section, we study the conflict free version of the classical VERTEX COVER, namely CONFLICT FREE VERTEX COVER (CF-VC). In particular, we study the following problem.

CONFLICT FREE VERTEX COVER (CF-VC)   **Parameter:** k
Input: A graph $G = (V, E)$, a conflict graph H and an integer k.
Question: Does there exist $X \subseteq V(G)$ of size at most k such that X is a vertex cover of G and an independent set of H ?

We call the set X a *conflict free vertex cover*. Next, we show that CF-VC can be solved as fast as the classical VERTEX COVER problem. Towards this, we present a polynomial time reduction from CF-VC to MIN ONES 2-SAT which preserves both parameter k and number of variables n. In MIN ONES 2-SAT, we are given a formula Φ such that every clause consists of at most two literals and an integer k, and the aim is to check whether there exists a satisfying assignment τ of Φ where at most k variables are set to 1. Given a formula Φ, let $V(\Phi)$ and $C(\Phi)$ denote the set of variables and clauses of Φ, respectively.

Lemma 2. *There is a polynomial time parameter preserving reduction from* CF-VC *to* MIN ONES 2-SAT. *That is, given an instance* (G, H, k) *of* CF-VC, *in polynomial time, we can construct an instance* (Φ, k) *of* MIN ONES 2-SAT, *such that* (G, H, k) *is a yes instance of* CF-VC *if and only if* (Φ, k) *is a yes instance of* MIN ONES 2-SAT. *Furthermore,* $|V(\Phi)| = |V(G)| = |V(H)|$.

Proof. We begin with the construction of the formula. Let (G, H, k) be an instance of CF-VC. Given this instance, we construct an instance (Φ, k) of MIN ONES 2-SAT as follows. For every edge $uv \in E(G)$, introduce a clause $(u \vee v)$ and for every edge $uv \in E(H)$, introduce a clause $(\bar{u} \vee \bar{v})$ in Φ. More precisely, given the graphs G and H, the CF-VC is formulated as the following instance MIN ONES 2-SAT.

$$\Phi = \bigwedge_{uv \in E(G)} (u \vee v) \bigwedge_{uv \in E(H)} (\bar{u} \vee \bar{v}).$$

Now, let X be a conflict free vertex cover of G of size at most k. We construct a truth assignment τ of Φ as follows. If $x \in X$ then $\tau(x) = 1$, otherwise it is 0. Clearly this satisfies the formula Φ and it is of weight at most k. Conversely, let τ be a satisfying assignment of Φ of weight at most k. We construct a set X as follows. If $\tau(u) = 1$, add the vertex u to X. For the clause $(u \vee v)$, at least one of $\tau(u)$ or $\tau(v)$ is 1. This ensures that every edge of G is incident to some vertex $u \in X$. For the clause $(\bar{u} \vee \bar{v})$, at least one of $\tau(u)$ or $\tau(v)$ is 0. This ensures that $H[X]$ is edgeless. Clearly, the size of X is at most k. □

Lemma 2 implies the following result.

Lemma 3 ($\star$). *Let* G *be a graph and* H *be a conflict graph of* G. *Then in polynomial time we can test whether there exists a conflict free vertex cover of the instance* (G, H, k).

Misra et al. [22] have shown that MIN ONES 2-SAT can be solved as fast as VERTEX COVER. This implies that CF-VC can also be solved as fast as VERTEX COVER. Further, using results from [7,15,28], we obtain the following.

Theorem 4. CF-VC *admits a* $2k$-*vertex kernel, a factor 2-approximation algorithm, an* $\mathcal{O}^\star(1.2738^k)$ FPT *algorithm and a* $\mathcal{O}^\star(1.1996^n)$ *exact algorithm.*

Next, we consider some special cases of CF-VC. It is well known that VERTEX COVER is NP-complete in general and polynomial time solvable for graphs with maximum degree at most two. We can prove that CF-VC is NP-complete even when graph G is of degree at most 2. In fact, it is true even when G is disjoint union of P_3 (P_ℓ denotes path on ℓ vertices).

Theorem 5 ($\star$). CF-VC *is* NP-complete *when* G *is of degree at most 2.*

However, certain special cases of CF-VC are polynomial time solvable.

Theorem 6 ($\star$). CF-VC *is solvable in polynomial time in the following cases:*

(a) The graph G has degree at most one.
(b) Both the graphs G and H have a perfect matching.

Further Results. Due to space constraints, detailed description of the following results of CONFLICT FREE d-HITTING SET, CONFLICT FREE SPLIT VERTEX DELETION and CONFLICT FREE FEEDBACK VERTEX SET IN TOURNAMENTS have been deferred to the full version of the paper.

Theorem 7

(a) The CONFLICT FREE d-HITTING SET *problem can be solved in* $\mathcal{O}^\star(((d-1)+.2738)^k) = \mathcal{O}^\star((d-0.7262)^k)$ *time.*
(b) CONFLICT FREE SPLIT VERTEX DELETION *can be solved in* $\mathcal{O}^\star(1.2738^k k^{\mathcal{O}(\log k)})$ *time and polynomial space.*
(c) CONFLICT FREE FEEDBACK VERTEX SET IN TOURNAMENTS *can be solved in* $\mathcal{O}^\star(2^k)$ *time.*

4 Conclusion

In this paper, we introduced a new variant, called the conflict free version, of classical vertex deletion problems that are studied in graph algorithms. We studied these problems in the realm of parameterized complexity and obtain several results that classify the complexity of these problems in various graph classes. Our work opens up a whole new area of research in obtaining dichotomy results. For every property Π, where CONFLICT FREE Π-VERTEX DELETION is W[1]-hard, it is a natural question to ask for which family of graphs H does the problem becomes FPT. As a concrete question in this direction, for which family of graphs H does CONFLICT FREE FVS and CONFLICT FREE OCT admit FPT algorithms and polynomial kernels.

Acknowledgement. The first author acknowledges DST, India for SERB-NPDF fellowship [PDF/2016/003508].

References

1. Agrawal, A., Kolay, S., Lokshtanov, D., Saurabh, S.: A faster FPT algorithm and a smaller kernel for BLOCK GRAPH VERTEX DELETION. In: Kranakis, E., Navarro, G., Chávez, E. (eds.) LATIN 2016. LNCS, vol. 9644, pp. 1–13. Springer, Heidelberg (2016). https://doi.org/10.1007/978-3-662-49529-2_1
2. Arkin, E.M., Banik, A., Carmi, P., Citovsky, G., Katz, M.J., Mitchell, J.S.B., Simakov, M.: Choice is hard. In: Elbassioni, K., Makino, K. (eds.) ISAAC 2015. LNCS, vol. 9472, pp. 318–328. Springer, Heidelberg (2015). https://doi.org/10.1007/978-3-662-48971-0_28
3. Arkin, E.M., Banik, A., Carmi, P., Citovsky, G., Katz, M.J., Mitchell, J.S.B., Simakov, M.: Conflict-free covering. In: CCCG (2015)
4. Banik, A., Panolan, F., Raman, V., Sahlot, V.: Fréchet distance between a line and avatar point set. In: FSTTCS, pp. 32:1–32:14 (2016)

5. Cai, L.: Fixed-parameter tractability of graph modification problems for hereditary properties. Inf. Process. Lett. **58**(4), 171–176 (1996)
6. Cao, Y., Marx, D.: Interval deletion is fixed-parameter tractable. ACM Trans. Algorithms **11**(3), 21:1–21:35 (2015)
7. Chen, J., Kanj, I.A., Xia, G.: Improved upper bounds for vertex cover. Theor. Comput. Sci. **411**(40–42), 3736–3756 (2010)
8. Cygan, M., Fomin, F.V., Kowalik, Ł., Lokshtanov, D., Marx, D., Pilipczuk, M., Pilipczuk, M., Saurabh, S.: Parameterized Algorithms. Springer, Cham (2015). https://doi.org/10.1007/978-3-319-21275-3
9. Darmann, A., Pferschy, U., Schauer, J.: Determining a minimum spanning tree with disjunctive constraints. In: Rossi, F., Tsoukias, A. (eds.) ADT 2009. LNCS (LNAI), vol. 5783, pp. 414–423. Springer, Heidelberg (2009). https://doi.org/10.1007/978-3-642-04428-1_36
10. Darmann, A., Pferschy, U., Schauer, J., Woeginger, G.J.: Paths, trees and matchings under disjunctive constraints. Discret. Appl. Math. **159**(16), 1726–1735 (2011)
11. Epstein, L., Favrholdt, L.M., Levin, A.: Online variable-sized bin packing with conflicts. Discret. Optim. **8**(2), 333–343 (2011)
12. Even, G., Halldórsson, M.M., Kaplan, L., Ron, D.: Scheduling with conflicts: online and offline algorithms. J. Sched. **12**(2), 199–224 (2009)
13. Fomin, F.V., Lokshtanov, D., Misra, N., Saurabh, S.: Planar F-deletion: approximation, kernelization and optimal FPT algorithms. In: FOCS (2012)
14. Fujito, T.: A unified approximation algorithm for node-deletion problems. Discret. Appl. Math. **86**, 213–231 (1998)
15. Gusfield, D., Pitt, L.: A bounded approximation for the minimum cost 2-sat problem. Algorithmica **8**(2), 103–117 (1992)
16. Kann, V.: Polynomially bounded minimization problems which are hard to approximate. In: Lingas, A., Karlsson, R., Carlsson, S. (eds.) ICALP 1993. LNCS, vol. 700, pp. 52–63. Springer, Heidelberg (1993). https://doi.org/10.1007/3-540-56939-1_61
17. Kociumaka, T., Pilipczuk, M.: Faster deterministic feedback vertex set. Inf. Process. Lett. **114**(10), 556–560 (2014)
18. Lewis, J.M., Yannakakis, M.: The node-deletion problem for hereditary properties is NP-complete. J. Comput. Syst. Sci. **20**(2), 219–230 (1980)
19. Lokshtanov, D., Panolan, F., Saurabh, S., Sharma, R., Zehavi, M.: Covering small independent sets and separators with applications to parameterized algorithms. CoRR abs/1705.01414 (to appear in SODA 2018)
20. Lund, C., Yannakakis, M.: On the hardness of approximating minimization problems. J. ACM **41**, 960–981 (1994)
21. Marx, D.: Chordal deletion is fixed-parameter tractable. Algorithmica **57**(4), 747–768 (2010)
22. Misra, N., Narayanaswamy, N.S., Raman, V., Shankar, B.S.: Solving min ones 2-sat as fast as vertex cover. Theor. Comput. Sci. **506**, 115–121 (2013)
23. Misra, N., Philip, G., Raman, V., Saurabh, S.: On parameterized independent feedback vertex set. Theor. Comput. Sci. **461**, 65–75 (2012)
24. Pferschy, U., Schauer, J.: The maximum flow problem with conflict and forcing conditions. In: Pahl, J., Reiners, T., Voß, S. (eds.) INOC 2011. LNCS, vol. 6701, pp. 289–294. Springer, Heidelberg (2011). https://doi.org/10.1007/978-3-642-21527-8_34
25. Pferschy, U., Schauer, J.: The maximum flow problem with disjunctive constraints. J. Comb. Optim. **26**(1), 109–119 (2013)

26. Pferschy, U., Schauer, J.: Approximation of knapsack problems with conflict and forcing graphs. J. Comb. Optim. **33**(4), 1300–1323 (2017)
27. Reed, B.A., Smith, K., Vetta, A.: Finding odd cycle transversals. Oper. Res. Lett. **32**(4), 299–301 (2004)
28. Xiao, M., Nagamochi, H.: Exact algorithms for maximum independent set. Inf. Comput. **255**, 126–146 (2017)
29. Yannakakis, M.: Node-and edge-deletion NP-complete problems. In: STOC, pp. 253–264. ACM, New York (1978)

Quadratically Tight Relations for Randomized Query Complexity

Rahul Jain[1,2], Hartmut Klauck[1,3], Srijita Kundu[1], Troy Lee[1,3],
Miklos Santha[1,4], Swagato Sanyal[1,3], and Jevgēnijs Vihrovs[5(✉)]

[1] Centre for Quantum Technologies, National University of Singapore,
Block S15, 3 Science Drive 2, Singapore 117543, Singapore
[2] MajuLab, UMI 3654, Singapore, Singapore
[3] SPMS, Nanyang Technological University, 21 Nanyang Link,
Singapore 637371, Singapore
[4] IRIF, Université Paris Diderot, CNRS, 75205 Paris, France
[5] Faculty of Computing, Centre for Quantum Computer Science,
University of Latvia, Raiņa 19, Riga 1586, Latvia
jevgenijs.vihrovs@lu.lv

Abstract. In this work we investigate the problem of quadratically tightly approximating the randomized query complexity of Boolean functions $R(f)$. The certificate complexity $C(f)$ is such a complexity measure for the zero-error randomized query complexity $R_0(f)$: $C(f) \leq R_0(f) \leq C(f)^2$. In the first part of the paper we introduce a new complexity measure, expectational certificate complexity $EC(f)$, which is also a quadratically tight bound on $R_0(f)$: $EC(f) \leq R_0(f) = O(EC(f)^2)$. For $R(f)$, we prove that $EC^{2/3} \leq R(f)$. We then prove that $EC(f) \leq C(f) \leq EC(f)^2$ and show that there is a quadratic separation between the two, thus $EC(f)$ gives a tighter upper bound for $R_0(f)$. The measure is also related to the fractional certificate complexity $FC(f)$ as follows: $FC(f) \leq EC(f) = O(FC(f)^{3/2})$. This also connects to an open question by Aaronson whether $FC(f)$ is a quadratically tight bound for $R_0(f)$, as $EC(f)$ is in fact a relaxation of $FC(f)$.

In the second part of the work, we investigate whether the corruption bound $corr_\epsilon(f)$ quadratically approximates $R(f)$. By Yao's theorem, it is enough to prove that the square of the corruption bound upper bounds the distributed query complexity $D_\epsilon^\mu(f)$ for all input distributions μ. Here, we show that this statement holds for input distributions in which the various bits of the input are distributed independently. This is a natural and interesting subclass of distributions, and is also in the spirit of the input distributions studied in communication complexity in which the inputs to the two communicating parties are statistically independent. Our result also improves upon a result of Harsha et al. [2015], who proved a similar weaker statement. We also note that a similar statement in the communication complexity is open.

© Springer International Publishing AG, part of Springer Nature 2018
F. V. Fomin and V. V. Podolskii (Eds.): CSR 2018, LNCS 10846, pp. 207–219, 2018.
https://doi.org/10.1007/978-3-319-90530-3_18

1   Introduction

The query model is arguably the simplest model for computation of Boolean functions. Its simplicity is convenient for showing lower bounds for the amount of time required to accomplish a computational task. In this model, an algorithm computing a function $f : \{0,1\}^n \to \{0,1\}$ on n bits is given query access to the input $x \in \{0,1\}^n$. The algorithm can *query* different bits of x, possibly in an adaptive fashion, and finally produces an output. The complexity of the algorithm is the number of queries made; in particular, the algorithm does not incur additional cost for any computation other than the queries.

Unlike the more general models of computation (e.g. Boolean circuits, Turing machines), it is often possible to completely determine the query complexity of explicit functions using existing tools and techniques. The study of query algorithms can thus be a natural first step towards understanding the computational power and limitations of more general and complex models. Query complexity has seen a long line of research by computational complexity theorists. We refer the reader to the survey by Buhrman and de Wolf [6] for a comprehensive introduction to this line of work.

To understand query algorithms, researchers have defined many complexity measures of Boolean functions and investigated their relationship to query complexity, and to one another. For a summary of the current state of knowledge about these measures, see [2]. In this work, we focus on characterizing the bounded-error and zero-error randomized query complexity measures, denoted by $R(f)$ and $R_0(f)$, respectively. More specifically, we study measures that could quadratically approximate the randomized query complexity for all Boolean functions.

The following measures are known to lower bound $R_0(f)$: block sensitivity $bs(f)$, fractional certificate complexity $FC(f)$ (also known as fractional block sensitivity $fbs(f)$, [14]), and certificate complexity $C(f)$ (see Fig. 1). They are related as follows:
$$bs(f) \le fbs(f) = FC(f) \le C(f).$$

Let $D(f)$ denote the deterministic query complexity of f. It is known that $R_0(f) \le D(f) \le C(f)^2$, and the TRIBES function (an AND of $\sqrt{n}$ ORs on $\sqrt{n}$ bits) demonstrates that this relation is tight [11]. It is also known that $R_0(f) = O(bs(f)^3) = O(FC(f)^3)$ [4,13]. A quadratic separation between $R_0(f)$ and $FC(f)$ is also achieved by TRIBES. Aaronson posed a question whether $R_0(f) = O(FC^2(f))$ holds [1] (stated in terms of the randomized certificate complexity $RC(f)$, which later has been shown to be equivalent to $FC(f)$ [8]). A positive answer to this question would imply that $R_0(f) = O(\widetilde{deg}(f)^4) = O(Q(f)^4)$ [2], where $\widetilde{deg}(\cdot)$ and $Q(\cdot)$ stand for approximate polynomial degree and quantum query complexity respectively.

One approach to showing $R_0(f) \le FC(f)^2$ is to consider the natural generalization of the proof $D(f) \le C(f)^2$ to the randomized case; the analysis of this algorithm, however, has met some unresolved obstacles [12]. We define a new complexity measure *expectational certificate complexity* $EC(f)$ that is specifically designed to avert these problems and is of a similar form to $FC(f)$. We show that EC gives a quadratically tight bound for R_0:

Theorem 1. *For all total Boolean functions f,*

$$EC(f) \leq R_0(f) \leq O(EC(f)^2).$$

In fact, $FC(f)$ is a relaxation of $EC(f)$, and we show that $FC(f) \leq EC(f) \leq C(f)$. Moreover, we show that $EC(f)$ lies closer to $FC(f)$ than $C(f)$ does: $FC(f) \leq EC(f) \leq FC(f)^{3/2}$. While we don't know whether $EC(f)$ is a lower bound on $R(f)$, the last property gives $EC(f)^{2/3} \leq R(f)$.

As mentioned earlier, $C(f)^2$ bounds $R_0(f)$ from above. But for specific functions, $EC(f)^2$ can be an asymptotically tighter upper bound than $C(f)^2$. We demonstrate that by showing that the same example that provides a quadratic separation between $C(f)$ and $FC(f)$ [8] also gives $C(f) = \Omega(EC(f)^2)$. This is the widest separation possible between $EC(f)$ and $C(f)$, because $C(f) \leq R_0(f) = O(EC(f)^2)$.

In the second part of the paper, we investigate whether the query corruption bound $corr_\epsilon(f)$ quadratically approximates $R(f)$. By Yao's Minimax Principle (see Fact 2), it is sufficient to show that the distributional query complexity $D_\epsilon^\mu(f)$ is upper bounded by the product of $corr_\epsilon(f)$ for all distributions μ. We show that this holds for the bitwise product distributions, where the distributional query complexity can be upper bounded by the product of the minimum product query corruption bound $corr_{\min,\epsilon}^\times(f)$ and the block sensitivity $bs(f)$ (see Definition 10 and Sect. 2).

Theorem 2. *Let $\epsilon \in [0, 1/2)$ and μ a product distribution over the inputs. Then*

$$D_{4\epsilon}^\mu(f) = O(corr_{\min,\epsilon}^\times(f) \cdot bs(f)).$$

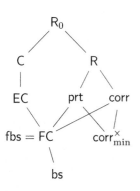

Fig. 1. Lower bounds on $R_0(f)$ and $R(f)$.

We then show that $bs(f) \leq corr_\epsilon(f)$, thus $D_{4\epsilon}^\mu(f) = O(corr_\epsilon(f)^2)$, as we have $corr_{\min,\epsilon}^\times(f) \leq corr_\epsilon(f)$.

We contrast Theorem 2 with the past work by Harsha et al. [9], who showed that for product distributions, the distributional query complexity is bounded above by the square of the smooth corruption bound corresponding to inverse polynomial error. Theorem 2 improves upon their result, firstly by upper bounding the distributional complexity by minimum query corruption bound, which is an asymptotically smaller measure than the smooth corruption bound, and secondly by losing a constant factor in the error as opposed to a polynomial worsening in their work.

Theorem 7, a consequence of Theorem 2, shows that for product distribution over the inputs, the distributional query complexity is asymptotically bounded above by the square of the query corruption bound.

Thus Theorem 7 resolves a question that was open after the work of Harsha et al. The analogous question in communication complexity is still open.

Theorem 2 also bounds distributional query complexity in terms of the *partition bound* $\mathsf{prt}(\cdot)$ of Jain and Klauck [11]. The following theorem follows from Theorems 2 and 4 .

Theorem 3. *Let $\epsilon \in \left[0, \frac{1}{8}\right]$ and μ a product distribution over the inputs. Then*

$$\mathsf{D}^{\mu}_{8\epsilon}(f) = O(\mathsf{prt}_{\epsilon}(f)^2).$$

Jain and Klauck showed that $\mathsf{prt}(f)$ is a powerful lower bound on $\mathsf{R}(f)$. In the same work, $\mathsf{prt}(f)$ was used to give a tight $\Omega(n)$ lower bound on $\mathsf{R}(f)$ for the TRIBES function on n bits. The authors proved that $\mathsf{prt}(f)$ is asymptotically larger than $\mathsf{FC}(f)$. This implies that $\mathsf{R}(f) = O(\mathsf{prt}(f)^3)$, since $\mathsf{R}(f) = O(\mathsf{bs}(f)^3)$. While a quadratic separation between $\mathsf{R}(f)$ and $\mathsf{prt}(f)$ is known [3], it is open whether $\mathsf{R}(f) = O(\mathsf{prt}(f)^2)$. Theorem 3 proves a distributional version of this quadratic relation, for the special case in which the input is sampled from a product distribution, i.e., a distribution where the input bits are independently distributed. We remark here that Harsha et al. [9] proved in their work that $\mathsf{D}^{\mu}_{1/3}(f) = O(\mathsf{prt}_{1/3}(f)^2 \cdot (\log \mathsf{prt}_{1/3}(f))^2)$; Theorem 3 achieves polylogarithmic improvement over this bound. Once again, an analogous statement for an arbitrary distribution together with the Minimax Principle will imply that $\mathsf{R}(f) = O(\mathsf{prt}(f)^2)$.

The paper is organized as follows. In Sect. 2, we give the definitions for some of the complexity measures. In Sect. 3, we define the expectational certificate complexity and prove the results concerning this measure, starting with Theorem 1. In Sect. 4, we define the minimum query corruption bound and prove Theorems 2 and 3. In Sect. 5, we list some open problems concerning our measures.

2   Preliminaries

In this section we recall the definitions of some known complexity measures. For detailed introduction on the query model, see the survey [6]. For the rest of this paper, f is any total Boolean function on n bits, $f : \{0, 1\}^n \to \{0, 1\}$.

Definition 1 (Randomized Query Complexity). *Let $\mathcal{A}$ be a randomized algorithm that as an input takes $x \in \{0, 1\}^n$ and returns a Boolean value $\mathcal{A}(x, r)$, where r is any random string used by $\mathcal{A}$. With one query $\mathcal{A}$ can ask the value of any input variable x_i, for $i \in [n]$. The complexity $C(\mathcal{A}, x, r)$ of $\mathcal{A}$ on x is the number of queries the algorithm performs under randomness r, given x. The worst-case complexity of $\mathcal{A}$ is $C(\mathcal{A}) = \max_{r, x \in \{0,1\}^n} C(\mathcal{A}, x, r)$.*

The zero-error randomized query complexity $\mathsf{R}_0(f)$ is defined as

$$\min_{\mathcal{A}} \max_{x} \mathbb{E}_r[C(\mathcal{A}, x, r)],$$

where $\mathcal{A}$ is any randomized algorithm such that for all $x \in \{0,1\}^n$, we have $\Pr_r[\mathcal{A}(x,r) = f(x)] = 1$.

The one-sided error randomized query complexity $R_\epsilon^0(f)$ is defined as $\min_{\mathcal{A}} C(\mathcal{A})$, where $\mathcal{A}$ is any randomized algorithm such that for every x such that $f(x) = 0$, we have $\Pr_r[\mathcal{A}(x,r) = 1] \leq \epsilon$, and for all x such that $f(x) = 1$, we have $\Pr_r[\mathcal{A}(x,r) = 1] = 1$. Similarly we define $R_\epsilon^1(f)$.

The two-sided error randomized query complexity $R_\epsilon(f)$ is defined as $\min_{\mathcal{A}} C(\mathcal{A})$, where $\mathcal{A}$ is any randomized algorithm such that for every $x \in \{0,1\}^n$, we have $\Pr_r[\mathcal{A}(x,r) \neq f(x)] \leq \epsilon$. We denote $R_{1/3}(f)$ simply by $R(f)$.

Definition 2 (Distributional Query Complexity). *Let μ be a probability distribution over $\{0,1\}^n$, and $\epsilon \in [0,1/2)$. The distributional query complexity $D_\epsilon^\mu(f)$ is the minimum number of queries made in the worst case (over inputs) by a deterministic query algorithm $\mathcal{A}$ for which $\Pr_{x \sim \mu}[\mathcal{A}(x) = f(x)] \geq 1 - \epsilon$.*

The *Minimax Principle* relates the randomized query complexity and distributional query complexity measures of Boolean functions.

Fact 4 (Minimax Principle). *For any Boolean function f, $R_\epsilon(f) = \max_\mu D_\epsilon^\mu(f)$.*

Definition 3 (Product Distribution). *A probability distribution μ over $\{0,1\}^n$ is a* product distribution *if there exist n functions $\mu_1, \ldots, \mu_n : \{0,1\} \to [0,1]$ such that $\mu_i(0) + \mu_i(1) = 1$ for all i and for all $x \in \{0,1\}^n$,*

$$\mu(x) = \prod_{i \in [n]} \mu_i(x_i).$$

Definition 4 (Assignment). *An* assignment *is a map $A : \{1, \ldots, n\} \to \{0,1,*\}$. All inputs consistent with A form a subcube $\{x \in \{0,1\}^n \mid \forall i \in [n] : x_i = A(i) \text{ or } A(i) = *\}$. The length or size of an assignment A, denoted by $|A|$, is defined to be the co-dimension of the subcube it corresponds to. Let $Q_A := \{j : A(j) \neq *\}$ be the set of variables fixed by A.*

Definition 5 (Certificate Complexity). *For $b \in \{0,1\}$, a b-certificate for f is an assignment A such that $x \in A \Rightarrow f(x) = b$. The* certificate complexity $C(f,x)$ *of f on x is the size of the shortest $f(x)$-certificate that is consistent with x. The certificate complexity of f is defined as $C(f) = \max_{x \in \{0,1\}^n} C(f,x)$. The b-certificate complexity of f is defined as $C^b(f) = \max_{x:f^{-1}(b)} C(f,x)$.*

Definition 6 (Sensitivity and Block Sensitivity). *For $x \in \{0,1\}^n$ and $S \subseteq [n]$, let x^S be x flipped on locations in S. The* sensitivity $s(f,x)$ *of f on x is the number of different $i \in [n]$ such that $f(x) \neq f(x^{\{i\}})$. The sensitivity of f is defined as $s(f) = \max_{x \in \{0,1\}^n} s(f,x)$.*

The block sensitivity $bs(f,x)$ *of f on x is the maximum number k of disjoint subsets $B_1, \ldots, B_k \subseteq [n]$ such that $f(x) \neq f(x^{B_i})$ for each $i \in [k]$. The block sensitivity of f is defined as $bs(f) = \max_{x \in \{0,1\}^n} bs(f,x)$.*

Definition 7 (Fractional Certificate Complexity). *The* fractional certificate complexity $\mathsf{FC}(f, x)$ *of* f *on* $x \in \{0, 1\}^n$ *is defined as the optimal value of the following linear program:*

$$minimize \sum_{i \in [n]} v_x(i) \quad subject\ to \quad \forall y \quad s.t.\ f(x) \neq f(y) : \sum_{i:x_i \neq y_i} v_x(i) \geq 1.$$

Here $v_x \in \mathbb{R}^n$ *and* $v_x(i) \geq 0$ *for each* $x \in \{0, 1\}^n$ *and* $i \in [n]$. *The fractional certificate complexity of* f *is defined as* $\mathsf{FC}(f) = \max_{x \in \{0,1\}^n} \mathsf{FC}(f, x)$.

Definition 8 (Fractional Block Sensitivity). *Let* $\mathcal{B} = \{B \mid f(x) \neq f(x^B)\}$ *be the set of sensitive blocks of* x. *The* fractional block sensitivity $\mathsf{fbs}(f, x)$ *of* f *on* x *is defined as the optimal value of the following linear program:*

$$maximize \sum_{B \in \mathcal{B}} u_x(B) \quad subject\ to \quad \forall i \in [n] : \sum_{\substack{B \in \mathcal{B} \\ i \in B}} u_x(B) \leq 1.$$

Here $u_x \in \mathbb{R}^{|\mathcal{B}|}$ *and* $u_x(B) \leq 1$ *for each* $x \in \{0, 1\}^n$ *and* $B \in \mathcal{B}$. *The fractional block sensitivity of* f *is defined as* $\mathsf{fbs}(f) = \max_{x \in \{0,1\}^n} \mathsf{fbs}(f, x)$.

Fractional certificate complexity and fractional block sensitivity were discovered independently by Tal [14] and Gilmer et al. [8]. The linear programs $\mathsf{FC}(f, x)$ and $\mathsf{fbs}(f, x)$ are duals of each other, hence their optimal solutions are equal and $\mathsf{FC}(f) = \mathsf{fbs}(f)$.

3   Expectational Certificate Complexity

In this section, we give the results for the expectational certificate complexity. The measure is motivated by the well-known $\mathsf{D}(f) \leq \mathsf{C}^0(f)\mathsf{C}^1(f)$ deterministic query algorithm which was independently discovered several times [5,10,15]. In each iteration, the algorithm queries the set of variables fixed by some consistent 1-certificate. Either the query answers agree with the fixed values of the 1-certificate, in which case the input must evaluate to 1, or the algorithm makes progress as the 0-certificate complexity of all 0-inputs still consistent with the query answers is decreased by at least 1. The latter property is due to the crucial fact that the set of fixed values of any 0-certificate and 1-certificate must intersect.

In hopes of proving $\mathsf{R}(f) \leq \mathsf{FC}^0(f)\mathsf{FC}^1(f)$, a straightforward generalization to a randomized algorithm would be to pick a consistent 1-input x and query each variable independently with probability $v_x(i)$, where v_x is a fractional certificate for x. To show that such an algorithm makes progress, one needs a property analogous to the fact that 0-certificates and 1-certificates overlap. Kulkarni and Tal give a similar intersection property for the fractional certificates:

Lemma 1 ([12], Lemma 6.2). *Let* $f : \{0, 1\}^n \to \{0, 1\}$ *be a total Boolean function and* $\{v_x\}_{x \in \{0,1\}^n}$ *be a feasible solution for the* $\mathsf{FC}(f)$ *linear program. Then for any two inputs* $x, y \in \{0, 1\}^n$ *such that* $f(x) \neq f(y)$, *we have*

$$\sum_{i:x_i \neq y_i} \min\{v_x(i), v_y(i)\} \geq 1.$$

However, it is not clear whether the algorithm makes progress in terms of reducing the fractional certificates of the 0-inputs. We get around this problem by replacing $\min\{v_x(i), v_y(i)\}$ with the product $v_x(i)v_y(i)$ and putting that the sum of these terms over i where $x_i \neq y_i$ is at least 1 as a constraint:

Definition 9 (Expectational Certificate Complexity). *The* expectational certificate complexity $\mathsf{EC}(f)$ *of f is defined as the optimal value of the following program:*

$$minimize \max_x \sum_{i=1}^{n} w_x(i) \quad s.t. \quad \sum_{i:x_i \neq y_i} w_x(i)w_y(i) \geq 1 \ \forall x, y \ s.t. f(x) \neq f(y),$$

$$0 \leq w_x(i) \leq 1 \ for \ all \ x \in \{0,1\}^n, i \in [n].$$

We use the term "expectational" because the described algorithm on expectation queries at least weight 1 in total from input y, when querying the variables with probabilities being the weights of x. While the informally described algorithm shows a quadratic upper bound on the worst-case expected complexity, in the next section we show a slight modification that directly makes a quadratic number of queries in the worst case.

3.1 Quadratic Upper Bound on Randomized Query Complexity

In this section we prove Theorem 1 (restated below).

Theorem 1. $\mathsf{EC}(f) \leq \mathsf{R}_0(f) \leq O(\mathsf{EC}(f)^2)$.

Proof. The first inequality follows from Lemma 4 and $\mathsf{C}(f) \leq \mathsf{R}_0(f)$.

To prove the second inequality, we give randomized query algorithms for f with 1-sided error ϵ.

Proposition 1. *For any $b \in \{0,1\}$, we have $\mathsf{R}_\epsilon^b(f) \leq \lceil \mathsf{EC}(f)^2/\epsilon \rceil$.*

The second inequality of Theorem 1 follows from Proposition 1 by standard arguments of $\mathsf{ZPP} = \mathsf{RP} \cap \mathsf{coRP}$.

Proof of Proposition 1. We prove the proposition for $b = 0$. The case $b = 1$ is similar.

Let $\{w_x\}_{x \in \{0,1\}^n}$ be an optimal solution to the $\mathsf{EC}(f)$ program. We say that an input y is consistent with the queries made by $\mathcal{A}$ on x if $y_i = x_i$ for all queries $i \in [n]$ that have been made. Also define a probability distribution $\mu_y(i) = w_y(i)/\sum_{i \in [n]} w_y(i)$ for each input $y \in \{0,1\}^n$.

Algorithm 1. The randomized query algorithm $\mathcal{A}$.

Input: $x \in \{0,1\}^n$

1. Repeat $\lceil \mathsf{EC}(f)^2/\epsilon \rceil$ many times:
   (a) Pick the lexicographically first consistent 1-input y.
       If there is no such y, return 0.
   (b) Sample a position i from μ_y and query x_i.
   (c) If the queried values form a c-certificate, return c.
2. Return 1.

The complexity bound is clear as $\mathcal{A}$ always performs at most $\lceil EC(f)^2/\epsilon \rceil$ queries. We prove the correctness of the algorithm in Sect. 3.1 of the full version of the paper [7].   □

    □

3.2 Relation with the Fractional Certificate Complexity

Lemma 2. $FC(f) \leq EC(f)$.

Proof. We show that a feasible solution $\{w_x\}_x$ for $EC(f)$ is also feasible for $FC(f)$. Since $0 \leq w_x(i) \leq 1$ for any x, i, $\sum_{i:x_i \neq y_i} w_x(i) \geq \sum_{i:x_i \neq y_i} w_x(i)w_y(i) \geq 1$, and we are done.   □

Lemma 3. $EC(f) = O(FC(f)\sqrt{s(f)})$.

Proof. Let $\{v_x\}_x$ be an optimal solution to the fractional certificate linear program for f. We first modify each v_x to a new feasible solution v'_x by eliminating the entries $v_x(i)$ that are very small, and boosting the large entries by a constant factor. Namely, let

$$v'_x(i) = \begin{cases} \min\left\{\frac{3}{2}v_x(i), 1\right\}, & \text{if } v_x(i) \geq \frac{1}{3s(f)}, \\ 0, & \text{otherwise.} \end{cases}$$

We first claim that $\{v'_x\}_x$ is still a feasible solution. Fix any $x \in \{0,1\}^n$, and let B be a minimal sensitive block for x. As v_x is part of a feasible solution, we have

$$1 \leq \sum_{i \in B} v_x(i) = \sum_{\substack{i \in B, \\ v_x(i) < 1/3s(f)}} v_x(i) + \sum_{\substack{i \in B, \\ v_x(i) \geq 1/3s(f)}} v_x(i) \leq \frac{1}{3} + \sum_{\substack{i \in B, \\ v_x(i) \geq 1/3s(f)}} v_x(i).$$

The second line follows because $|B| \leq s(f)$, as B is a minimal sensitive block and therefore every index in B is sensitive. Rearranging the last inequality, we have $\sum_{\substack{i \in B \\ v_x(i) \geq 1/3s(f)}} v_x(i) \geq \frac{2}{3}$, and therefore, $\sum_{i \in B} v'_x(i) \geq 1$.

Next, $w_x(i) := \sqrt{v'_x(i)}$ is a feasible solution to the expectational certificate program, as

$$\sum_{i:x_i \neq y_i} w_x(i)w_y(i) = \sum_{i:x_i \neq y_i} \sqrt{v'_x(i)v'_y(i)} \geq \sum_{i:x_i \neq y_i} \min\{v'_x(i), v'_y(i)\} \geq 1.$$

The second inequality holds by Lemma 1.

Now that we have shown that $\{w_x\}_x$ forms a feasible solution to the expectation certificate program, it remains to bound its objective value:

$$\sum_{i \in [n]} w_x(i) = \sum_{i \in [n]} \sqrt{v'_x(i)} = \sum_{i:v'_x(i) \neq 0} \frac{v'_x(i)}{\sqrt{v'_x(i)}} \leq \sqrt{3s(f)} \sum_{i \in [n]} v'_x(i) \leq \sqrt{3s(f)}\frac{3}{2}FC(f),$$

where the first inequality follows from $v'_x(i) \geq v_x(i) \geq 1/3s(f)$ for $v'_x(i) \neq 0$.

    □

Since $s(f) \leq FC(f)$ and $FC(f) \leq R(f)$, we immediately get

Corollary 1. $EC(f) = O(FC(f)^{3/2}) = O(R(f)^{3/2})$.

3.3 Relation with the Certificate Complexity

Lemma 4. $\mathsf{EC}(f) \leq \mathsf{C}(f)$.

Proof. We construct a feasible solution $\{w_x\}_x$ for $\mathsf{EC}(f)$ from $\mathsf{C}(f)$. Let A_x be the shortest certificate for x. Assign $w_x(i) = 1$ iff $i \in A_x$, otherwise let $w_x(i) = 0$. Let x, y be any two inputs such that $f(x) \neq f(y)$. There is a position i where $A_x(i) \neq A_y(i)$, otherwise there would be an input consistent with both A_x and A_y, which would give a contradiction. Therefore, $w_x(i) w_y(i) \geq 1$. The value of this solution is $\max_x \sum_{i \in [n]} w_x(i) = \max_x \mathsf{C}(f, x) = \mathsf{C}(f)$. □

As $\mathsf{FC}(f) \leq \mathsf{EC}(f) \leq \mathsf{C}(f) \leq \mathsf{FC}(f)^2$, there can be at most quadratic separation between $\mathsf{EC}(f)$ and $\mathsf{C}(f)$. We show that this is achieved by the example of Gilmer et al. that separates $\mathsf{FC}(f)$ and $\mathsf{C}(f)$ quadratically:

Theorem 4 ([8], Theorem 32). *For every $n \in \mathbb{N}$ sufficiently large, there is a function $f : \{0,1\}^{n^2} \to \{0,1\}$ such that $\mathsf{FC}(f) = O(n)$ and $\mathsf{C}(f) = \Omega(n^2)$.*

Their construction for f is as follows. First a function $g : \{0,1\}^n \to \{0,1\}$ is exhibited such that $\mathsf{FC}^0(g) = \Theta(1)$, $\mathsf{C}^0(g) = \Theta(n)$ and $\mathsf{FC}^1(g) = \mathsf{C}^1(f) = n$. The function $f : \{0,1\}^{n^2} \to \{0,1\}$ is defined as a composition $\mathrm{OR}(g(x^{(1)}), \ldots, g(x^{(n)}))$. This gives $\mathsf{FC}(f) = \max\{n\mathsf{FC}^0(g), \mathsf{FC}^1(g)\} = \Theta(n)$ and $\mathsf{C}(f) \geq n\mathsf{C}^0(g) = \Theta(n^2)$ (both properties follow by Proposition 31 in their paper).

Let us construct a feasible solution w for $\mathsf{EC}(f)$. For any $x = x^{(1)} \ldots x^{(n)}$ such that $f(x) = 1$, let j be the first index such that $g(x^{(j)}) = 1$. Let $S \subseteq [n^2]$ be the set of positions that correspond to $x^{(j)}$. Let $w_x(i) = 1$ for each position i in S, and $w_x(i) = 0$ for all other positions. Then $\sum_{i=1}^{n^2} w_x(i) = n$.

On the other hand, let $\{v_x\}_{x \in \{0,1\}^n}$ be an optimal solution to $\mathsf{FC}(f)$. For any $x \in \{0,1\}^{n^2}$ such that $f(x) = 0$, let $w_x(i) = v_x(i)$ for all $i \in [n^2]$. Then $\sum_{i=1}^{n^2} w_x(i) = \mathsf{FC}(f, x) = O(n)$.

Now, for any two inputs x, y such that $f(x) = 1$ and $f(y) = 0$, let j be the smallest index such that $g(x^{(j)}) = 1$, then we have $g(y^{(j)}) = 0$. By construction,

$$\sum_{i:x_i \neq y_i} w_x(i) w_y(i) = \sum_{i:x_i \neq y_i} w_y(i) \geq 1.$$

Hence $\{w_x\}_x$ is a feasible solution to the expectational certificate and $\mathsf{EC}(f) = n$.

4 Minimum Query Corruption Bound and Partition Bound

In this section, we prove upper bounds on the distributional query complexity D_ϵ^μ, where μ is bitwise product distribution on the inputs. We first consider the *query corruption bound* and *minimum query corruption bound*.

Definition 10 (Query Corruption Bound and Minimum Query Corruption Bound for product distributions). *Let $\epsilon \in [0, 1/2)$ and $\mu :$ $\{0,1\}^n \to [0,1]$ be a probability distribution over the inputs. For a $b \in \{0,1\}$, let an assignment A be an ϵ-error b-certificate under μ, if*

$$\Pr_{x \sim \mu} [f(x) \neq b \mid x \in A] \leq \epsilon.$$

Define the query corruption bound *for b, distribution μ and error ϵ as*

$$\mathrm{corr}_{\epsilon}^{b,\mu}(f) = \min\{|A| \mid A \text{ is an } \epsilon - \text{error } b - \text{certificate under } \mu\}.$$

The query corruption bound *of f is defined as $\mathrm{corr}_{\epsilon}(f) = \max_{\mu} \max_b \mathrm{corr}_{\epsilon}^{b,\mu}(f)$, where μ ranges over all distributions on $\{0,1\}^n$. Define the* minimum query corruption bound *of f for product distributions by $\mathrm{corr}_{\min,\epsilon}^{\times}(f) = \max_{\mu} \min_b \mathrm{corr}_{\epsilon}^{b,\mu}(f)$, where μ ranges over all product distributions on $\{0,1\}^n$.*

4.1 Upper Bound in Terms of the Corruption Bound and Block Sensitivity

In this subsection, we give a deterministic algorithm that achieves the bound of Theorem 2 (restated below).

Theorem 2. *Let $\epsilon \in [0, 1/2)$ and μ a product distribution over the inputs. Then*

$$\mathsf{D}_{4\epsilon}^{\mu}(f) = O(\mathrm{corr}_{\min,\epsilon}^{\times}(f) \cdot \mathsf{bs}(f)).$$

In the algorithm, we will work with restrictions of probability distributions. Let η be a probability distribution over $\{0,1\}^n$, $x \in \{0,1\}^n$ be an n-bit string, and $Q \subseteq \{1, \ldots, n\}$ be a set of indices. The restriction of x to the indices of Q, $(x_j : j \in Q)$, will be denoted by x_Q. Then the distribution $\eta \mid_{x_Q}$ is the distribution obtained by conditioning η on the event that the bits in the locations in Q agree with x. Formally, for each $y \in \{0,1\}^n$

$$\eta \mid_{x_Q} (y) = \begin{cases} \dfrac{\eta(y)}{\sum_{z : \forall i \in Q, z_i = x_i} \eta(z)} & \text{if } \forall i \in Q, y_i = x_i, \\ 0 & \text{otherwise.} \end{cases}$$

Algorithm 2. The deterministic query algorithm $\mathcal{B}$.

Input: $x \in \{0,1\}^n$
1. Set $t_0, t_1 \leftarrow 0, i \leftarrow 1, \eta^{(1)} \leftarrow \mu$.
2. Repeat:
   - (a) Pick a shortest ϵ-error certificate A under $\eta^{(i)}$.
   - (b) Query all the variables in Q_A that are still unknown.
   - (c) Let A be an ϵ-error b-certificate for some $b \in \{0,1\}$. Set $t_b \leftarrow t_b + 1$.
   - (d) If the results of the queries are consistent with A, return b.
   - (e) If $t_b = 2\mathsf{bs}(f)$, return b.
   - (f) $\eta^{(i+1)} \leftarrow \eta^{(i)} \mid_{x_{Q_A}}$.
   - (g) $i \leftarrow i + 1$.

We analyze the correctness and performance of the algorithm in Sect. 4 of the full version of the paper [7].

4.2 Quadratic Upper Bound in Terms of the Partition Bound

In this subsection, we show that the partition bound is a quadratic upper bound on the distributional query complexity. We prove Theorem 3 (restated below).

Theorem 3. Let $\epsilon \in \left[0, \frac{1}{8}\right]$ and μ a product distribution over the inputs. Then

$$\mathsf{D}_{8\epsilon}^{\mu}(f) = O(\mathsf{prt}_\epsilon(f)^2).$$

We reproduce the definition of the partition bound by Jain and Klauck [11]. Here, ϵ is an error parameter between 0 and 1, A stands for subcubes, or equivalently, partial assignments, z stands for a bit, i.e., a 0 or a 1, and x stands for an input to f from $\{0,1\}^n$.

Definition 11 (Partition Bound). *The ϵ-partition bound* bound *of f, denoted* $\mathsf{prt}_\epsilon(f)$, *is given by the logarithm to base 2 of the optimal value of the following linear program*[1]:

$$minimize \sum_{z,A} w_{z,A} \cdot 2^{|A|} \qquad subject\ to\quad \forall x : \sum_{A \ni x} w_{f(x),A} \geq 1 - \epsilon,$$

$$\forall x : \sum_{z, A \ni x} w_{z,A} = 1,$$

$$\forall z, A : w_{z,A} \geq 0.$$

Jain and Klauck showed that the partition bound bounds randomized query-complexity from below. They also showed that randomized query complexity is bounded above by the third power of the partition bound.

Theorem 5 ([11] Theorem 3).

1. $\mathsf{R}_\epsilon(f) \geq \frac{1}{2}\mathsf{prt}_\epsilon(f)$.
2. $\mathsf{R}_{1/3}(f) \leq \mathsf{D}(f) = O(\mathsf{prt}_{1/3}(f)^3)$.

The best known separation between $\mathsf{D}(f)$ and $\mathsf{prt}(f)$ is quadratic [3]. Theorem 3 proves that this is tight for product distributions. As stated in Sect. 1, Theorem 3 improves upon the result of Jain et al. by a polylogarithmic factor.

Jain and Klauck showed that the partition bound is bounded below by the block sensitivity.

Theorem 6 ([11],Theorem 3). *For any error parameter $\epsilon \in [0, 1/2)$,*

$$\mathsf{prt}_{\epsilon/4}(f) \geq \epsilon \cdot \mathsf{bs}(f) + \log \epsilon - 2.$$

---

[1] Jain and Klauck in their paper defined $\mathsf{prt}_\epsilon(f)$ to be the value of the linear program, instead of the logarithm of the value of the program.

We show that the minimum query corruption bound lower-bounds the partition bound (for the proof, see Appendix A of the full version of the paper [7]). Our proof closely follows the proof that the corruption bound is asymptotically bounded above by square of the partition bound shown in [11].

Lemma 5. *For any error parameter $\epsilon \in [0, 1/2)$,*

$$\mathsf{corr}^{\times}_{\min, 2\epsilon}(f) \leq \mathsf{prt}_\epsilon(f) + \log(1/\epsilon).$$

Theorem 3 now follows, combining Theorems 2, 6 and Lemma 5 together.

4.3 Quadratic Upper Bound in Terms of the Corruption Bound

We conclude by showing that the query corruption bound is a quadratic upper bound on the distributional query complexity.

Theorem 7. *Let $\epsilon \in [0, 1/2)$ and μ a product distribution over the inputs. Then*

$$\mathsf{D}^{\mu}_{4\epsilon}(f) = O\left(\mathsf{corr}_\epsilon(f)^2\right).$$

The result follows by combining Theorem 2 with the following lemma (for the proof, see Appendix B of the full version of the paper [7]).

Lemma 6. *For any $\epsilon \in [0, 1)$, $\mathsf{fbs}(f) \leq \mathsf{corr}_\epsilon(f)$.*

5 Open Problems

Expectational vs. Fractional Certificate. What is the largest separation between the two measures? Is the upper bound $\mathsf{EC}(f) \leq \mathsf{FC}(f)^{3/2}$ tight? Any smaller upper bound would improve the $\mathsf{R}(f) \leq \mathsf{FC}(f)^3$ upper bound. Our attempts in finding a function where $\mathsf{EC}(f)$ is asymptotically larger than $\mathsf{FC}(f)$ so far have been unsuccessful. As evident by the proof of the quadratic separation between $\mathsf{EC}(f)$ and $\mathsf{C}(f)$, such an example would need to have $\mathsf{FC}^z(f) = o(\mathsf{C}^z(f))$ for both $z \in \{0, 1\}$. Examples of separations between $\mathsf{FC}(f)$ and $\mathsf{C}(f)$ given in [1] and [8] do not satisfy these properties.

Corruption and Partition Bounds. Can the proof of Theorem 2 be extended to non-product distributions? The definition of the corruption bound is in some sense a relaxation of the certificate complexity. Can the argument of $\mathsf{D}(f) \leq \mathsf{C}(f)^2$ be extended to the randomized setting in terms of the corruption bound?

Acknowledgements. This work is supported in part by the Singapore National Research Foundation under NRF RF Award No. NRF-NRFF2013-13, the Ministry of Education, Singapore under the Research Centres of Excellence programme by the Tier-3 grant. Grant "Random numbers from quantum processes" No. MOE2012-T3-1-009.

M.S. is partially funded by the ANR Blanc program under contract ANR-12-BS02-005 (RDAM project).

J.V. is supported by the ERC Advanced Grant MQC. Part of this work was done while J.V. was an intern at the Centre for Quantum Technologies at the National University of Singapore.

We thank Anurag Anshu for helpful discussions.

References

1. Aaronson, S.: Quantum certificate complexity. J. Comput. Syst. Sci. **74**(3), 313–322 (2008)
2. Aaronson, S., Ben-David, S., Kothari, R.: Separations in query complexity using cheat sheets. In: Proceedings of the Forty-eighth Annual ACM Symposium on Theory of Computing, STOC 2016, pp. 863–876. ACM, New York (2016)
3. Ambainis, A., Kokainis, M., Kothari, R.: Nearly optimal separations between communication (or query) complexity and partitions. In: Proceedings of the 31st Conference on Computational Complexity, CCC 2016, pp. 4:1–4:14. Schloss Dagstuhl-Leibniz-Zentrum fuer Informatik, Germany (2016)
4. Beals, R., Buhrman, H., Cleve, R., Mosca, M., de Wolf, R.: Quantum lower bounds by polynomials. J. ACM **48**(4), 778–797 (2001)
5. Blum, M., Impagliazzo, R.: Generic oracles and oracle classes. In: Proceedings of the 28th Annual Symposium on Foundations of Computer Science, SFCS 1987, pp. 118–126. IEEE Computer Society, Washington (1987)
6. Buhrman, H., de Wolf, R.: Complexity measures and decision tree complexity: a survey. Theor. Comput. Sci. **288**(1), 21–43 (2002)
7. Gavinsky, D., Jain, R., Klauck, H., Kundu, S., Lee, T., Santha, M., Sanyal, S., Vihrovs, J.: Quadratically tight relations for randomized query complexity. CoRR, abs/1708.00822 (2017)
8. Gilmer, J., Saks, M., Srinivasan, S.: Composition limits and separating examples for some Boolean function complexity measures. Combinatorica **36**(3), 265–311 (2016)
9. Harsha, P., Jain, R., Radhakrishnan, J.: Relaxed partition bound is quadratically tight for product distributions. CoRR, abs/1512.01968 (2015)
10. Hartmanis, J., Hemachandra, L.A.: One-way functions, robustness, and non-isomorphism of NP-complete sets. In: Proceedings of 2nd Structure in Complexity Theory, pp. 160–173 (1987)
11. Jain, R., Klauck, H.: The partition bound for classical communication complexity and query complexity. In: Proceedings of the 2010 IEEE 25th Annual Conference on Computational Complexity, CCC 2010, pp. 247–258. IEEE Computer Society, Washington (2010)
12. Kulkarni, R., Tal, A.: On fractional block sensitivity. Chicago J. Theor. Comput. Sci. **8**, 1–16 (2016)
13. Nisan, N.: CREW PRAMS and decision trees. In: Proceedings of the Twenty-First Annual ACM Symposium on Theory of Computing, STOC 1989, pp. 327–335. ACM, New York (1989)
14. Tal, A.: Properties and applications of boolean function composition. In: Proceedings of the 4th Conference on Innovations in Theoretical Computer Science, ITCS 2013, pp. 441–454. ACM, New York (2013)
15. Tardos, G.: Query complexity or why is it difficult to separate $\mathbf{NP}^A \cap \mathbf{coNP}^A$ from $\mathbf{P}^A$ by a random oracle. Combinatorica **9**, 385–392 (1990)

On Vertex Coloring Without Monochromatic Triangles

Michał Karpiński[✉] and Krzysztof Piecuch

Institute of Computer Science, University of Wrocław,
Joliot-Curie 15, 50-383 Wrocław, Poland
{karp,kpiecuch}@cs.uni.wroc.pl

Abstract. We study a certain relaxation of the classic vertex coloring problem, namely, a coloring of vertices of undirected, simple graphs, such that there are no monochromatic triangles. We give the first classification of the problem in terms of classic and parametrized algorithms. Several computational complexity results are also presented, which improve on the previous results found in the literature. We propose the new structural parameter for undirected, simple graphs – the triangle-free chromatic number χ_3. We bound χ_3 by other known structural parameters. We also present two classes of graphs with interesting coloring properties, that play pivotal role in proving useful observations about our problem.

1 Introduction

Graph coloring is probably the most popular subject in graph theory. It is an interesting topic from both algorithmic and combinatoric points of view. The coloring problems have many practical applications in areas such as operations research, scheduling and computational biology. For a recent survey one can turn to [10]. In this paper we study a variation of the classic coloring – we call it the *triangle-free coloring* problem, in which we ask for an assignment of colors to the vertices of a given graph, such that the number of colors used is minimum and that each cycle of length 3 has at least two vertices colored differently. We show that our problem has interesting graph-theoretical properties and we also present some evidence that this problem might be easier than the classic vertex coloring. This suggests that studying our variation, new results can be achieved in the field of classic vertex coloring, which is known to be one of the hardest known optimization problems. Apart from theoretical motivation, there is also a practical one – vertex coloring without monochromatic cycles can be used in the study of consumption behavior [6].

1.1 Related Work

Some researchers have already considered coloring problems that are similar to our variation. The class of planar graphs has been of particular interest, for example, Angelini and Frati [2] study planar graphs that admit an acyclic

© Springer International Publishing AG, part of Springer Nature 2018
F. V. Fomin and V. V. Podolskii (Eds.): CSR 2018, LNCS 10846, pp. 220–231, 2018.
https://doi.org/10.1007/978-3-319-90530-3_19

3-coloring – a proper coloring in which every 2-chromatic subgraph is acyclic. Algorithms for acyclic coloring can be used to solve/approximate a triangle-free coloring, although we do not explore this possibility in this paper. Another result is of Kaiser and Škrekovski [12], where they prove that every planar graph has a 2-coloring such that no cycle of length 3 or 4 is monochromatic. Thomassen [20], on the other hand, considers list-coloring of planar graphs without monochromatic triangles. Few hardness results for our problem are known – Karpiński [13] showed that verifying whether a graph admits a 2-coloring without monochromatic cycles of fixed length is $\mathcal{NP}$-complete. His proof was then simplified by Shitov [18], who also proposed and proved the hardness of an extension of our problem, where additional restriction is imposed on the coloring in the form of the set of *polar* edges – edges that must not be monochromatic in the resulting coloring. Another result is by Jain [11] who also consider a triangle-free 2-coloring problem. He shows that the problem is $\mathcal{NP}$-complete, even on classically 5-colorable graphs with maximum degree 8. It is also worth noting that our problem is a special case of 3-uniform hypergraph coloring problem, which was studied in the past (see, for example, [7]).

1.2   Our Contribution

Several novel results are presented in this paper. First, we explore the graph-theoretical side of our problem. We propose the new structural parameter $\chi_3(G)$ which is the minimum number of colors needed to label the vertices of an undirected, simple graph G, such that there are no monochromatic triangles. We then bound this new parameter by $\omega(G)$, $\chi(G)$ and $\Delta(G)$, which are clique number, chromatic number and the largest vertex degree of G, respectively. We also show that $\chi_3(G)$ cannot be upper-bounded by any function of $\omega(G)$. Additionally we present two gadgets that have interesting coloring properties and are also used in proving hardness results later in the paper.

For the positive side, several known graph classes are presented, for which our problem can be solved efficiently, for example, we can find χ_3 on planar graphs in polynomial time, whereas finding χ is $\mathcal{NP}$-complete, even on planar graphs with maximum degree 4 [5]. We use the fact that when χ is small (less than 5), then we can reduce our problem to the problem of deciding if a given graph is triangle-free. In general, the time needed for listing all triangles of a graph is $O(m\sqrt{m})$ [1], but in the presented graphs, we can find if a graph is triangle-free in time $O(n)$. We also prove that our problem is fixed-parameter tractable, when the parameter is the vertex cover number.

We present several hardness results, which improve on the work of Karpiński [13] and Shitov [18]. We show that given any fixed number $q \geq 2$, determining if graph is triangle-free q-colorable is $\mathcal{NP}$-hard. This improves on the result given in [13], where the author shows hardness only for $q = 2$. Another improvement to Karpiński's result [13] is $\mathcal{NP}$-hardness proof of triangle-free 2-coloring problem for the graphs that do not contain clique of size 4 as a subgraph. In [18], author formulates and proves the $\mathcal{NP}$-hardness of triangle-free 2-coloring problem where additional set of polar edges – edges that must not be monochromatic – are given on input. We show that this variation of our problem remains $\mathcal{NP}$-hard, even on graphs with maximum degree 3 – the sub-cubic graphs.

1.3   Structure of the Paper

In Sect. 2 we give formulation of all considered problems. We give several bounds on χ_3 in Sect. 3 and we construct gadgets used later in the paper. In Sect. 4 we present efficient algorithms for the triangle-free coloring problem for certain classes of graphs. A single FPT result is also shown there. Hardness results are presented in Sect. 5. We finish the paper with some concluding remarks in Sect. 6, where we also present the reader several open problems. Standard notation and properties from graph theory are used throughout this paper, therefore we moved them to the extended version of the paper due to space restriction (see [14]).

2   Problems

In this paper we study coloring of vertices, and to avoid confusion, we distinguish two types of coloring. A *classic k-coloring* of a graph is a function $c : V \rightarrow \{1, \ldots, k\}$, such that there are no two adjacent vertices u and v, for which $c(u) = c(v)$. Given G, the smallest k for which there exists a classic k-coloring for G is called the *chromatic number* and is denoted as $\chi(G)$. A *triangle-free k-coloring* of a graph is a function $c : V \rightarrow \{1, \ldots, k\}$, such that there are no three mutually adjacent vertices u, v and w, for which $c(u) = c(v) = c(w)$. If such vertices exist, then the induced subgraph (K_3) is called a *monochromatic* triangle. Given G, the smallest k for which there exists a triangle-free k-coloring for G we call the *triangle-free chromatic number* and we denote it as $\chi_3(G)$.

We now formulate the decision problems investigated in this paper, for any fixed integer $q > 0$:

TRIANGLEFREE-q-COLORING
Input: A finite, undirected, simple graph G.
Question: Is there a triangle-free q-coloring of G?

TRIANGLEFREEPOLAR-q-COLORING
Input: A finite, undirected, simple graph $G = (V, E)$ and a subset $S \subseteq E$.
Question: Is there a triangle-free q-coloring of G, such that no edge in S is monochromatic?

We denote $\lceil x \rceil$ to be the smallest integer not less than x. Let $r \in \mathbb{N}$, we define a binary operator $+_r$ on the set $\mathbb{Z}_r = \{0, 1, 2, \ldots, r-1\}$, which is addition modulo r, i.e., for every $n, m \in \mathbb{Z}_r$, $n +_r m = n + m \pmod{r}$. The operator $-_r$ can be defined in a similar way.

3   Bounds on the Triangle-Free Chromatic Number

We first give simple bounds on $\chi_3(G)$ in terms of $\omega(G)$ and $\chi(G)$:

Theorem 1. *For any graph G:* $\left\lceil \frac{\omega(G)}{2} \right\rceil \leq \chi_3(G) \leq \left\lceil \frac{\chi(G)}{2} \right\rceil$.

Proof. To see that the lower bound holds, take any clique in G of maximum cardinality. We can use one color for at most two vertices of that clique, otherwise we would create a monochromatic triangle. Therefore we need at least $\lceil \omega(G)/2 \rceil$ colors in order to make the coloring triangle-free in this clique, and therefore at least $\lceil \omega(G)/2 \rceil$ colors are needed to triangle-free color the entire graph.

The upper bound can be justified by the following argument. Let $k = \chi(G)$ and take any classic k-coloring of G. Let V_i be the set of vertices colored i, where $1 \le i \le k$. For each $0 \le j \le \lceil k/2 \rceil - 1$ recolor sets $V_{2j+1} \cup V_{2j+2}$ with j (we may need to add empty set V_{k+1}, if k is odd). Since every V_i is an independent set, then after recoloring, any monochromatic cycle is of even length. Therefore the resulting coloring is triangle-free.   □

We call the recoloring procedure from the above proof the *standard recoloring strategy*. It will be used in the construction of algorithms in the next section. The fundamental question we ask is how tight are the bounds in Theorem 1 and what kind of algorithmic consequences are implied by these observations. The following theorem shows that the upper bound can be arbitrarily large, which means that we should not expect to find an algorithm for general graphs that would acceptably approximate $\chi_3(G)$ based on a chromatic number alone.

Theorem 2. *For any $k \ge 1$, there exists a graph G for which $\chi_3(G) = 1$ and $\chi(G) = k$.*

Proof. The class of graphs called Mycielski graphs [16] meet this property, as they are triangle-free and can have arbitrarily large chromatic number.   □

Moreover, we cannot bound $\chi_3(G)$ from above by any function of $\omega(G)$.

Theorem 3. *For every k, there exist a graph G without K_4, such that $\chi_3(G) > k$.*

Proof. It is well known [8] that for every k, t and g there exists t-uniform hypergraph with girth at least g that cannot be colored using only k colors. Let's take such a hypergraph $H = (V, E)$ with $t = 3$ and $g = 4$. Let's create graph $G = (V, E')$ in such a way that $\{u, v\} \in E'$ if and only if $\{u, v\} \subset e$ and $e \in E$. We have $\chi_3(G) > k$ because for every k-coloring of hypergraph H there exists an edge $e = \{u, v, w\} \in E$ that is monochromatic. Since $\{u, v\}, \{u, w\}, \{v, w\} \in E'$ there exists monochromatic triangle in graph G. Now we need proof that G is K_4-free. Proof by contradiction. Let's say that v_1, v_2, v_3 and v_4 form a K_4. It means there need to be at least 3 edges from E that were involved in creating this clique. Moreover there need to be a triangle in this clique such that each edge was from different edges in hypergraph H. This means there was a cycle of size 3 in hypergraph H, but hypergraph H should have girth at least 4.   □

From the positive side, we can give upper bound on $\chi_3(G)$ using $\Delta(G)$. We recall the well-known Brook's Theorem [3] which states that for any graph G, we have $\chi(G) \le \Delta(G)$, unless G is a complete graph or an odd cycle. Using this theorem we can prove the following.

Theorem 4. *Let $G = (V, E)$ be any graph where $|V| > 3$, where G is not a complete graph of odd number of vertices. Then $\chi_3(G) \leq \left\lceil \frac{\Delta(G)}{2} \right\rceil$.*

Proof. If G is an odd cycle of length at least 5, then $\chi_3(G) = 1$ and $\left\lceil \frac{\Delta}{2} \right\rceil = 1$, so the inequality holds. If G is a complete graph (a clique) of n vertices, then $\chi_3(G) = \left\lceil \frac{n}{2} \right\rceil$ and $\Delta = n - 1$. The inequality $\left\lceil \frac{n}{2} \right\rceil \leq \left\lceil \frac{n-1}{2} \right\rceil$ holds iff n is even. If the above cases do not occur, then the application of the Brook's Theorem combined with the upper bound of Theorem 1 completes the proof.   □

3.1   Gadgets

We now present two auxiliary graphs with useful coloring properties. First one we call a *cycle-clique*.

Definition 1 (cycle-clique). *Let $k \in \mathbb{N}$. The k-cycle-clique is a graph with exactly $5 \cdot k$ vertices $v_{i,j}$ for $i \in \{0, \dots, 4\}$ and $j \in \{0, \dots, k-1\}$ and there is an edge between $v_{i,j}$ and $v_{i',j'}$ if and only if $|i -_5 i'| \leq 1$. Each set $J_i = \{v_{i,j} : j \in \{0, \dots, k-1\}\}$ is called a* joint.

Observation 1. *For any k-cycle-clique it holds that $J_i \sim K_k$ and $J_i \cup J_{i+_51} \sim K_{2k}$, for $i \in \{0, \dots, 4\}$.*

Example 1. In Fig. 1a and b we present k-cycle-cliques for $k \in \{1, 2\}$. In both graphs joints are marked with dashed lines. The 1-cycle-clique is simply a cycle of length 5. Notice how $J_i \cup J_{i+_51}$ forms a clique of size $2k$, for $i \in \{0, \dots, 4\}$, whereas each J_i is a clique of size k.

Observation 2. *Let $k \geq 1$. In every triangle-free k-coloring of graph K_{2k} every color is used exactly twice.*

Lemma 1. *Let $k \geq 1$. In every triangle-free k-coloring of k-cycle-clique all color are met on the vertices of J_i, for $i \in \{0, \dots, 4\}$.*

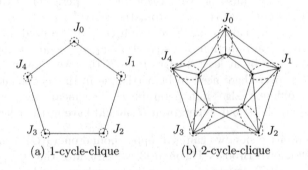

(a) 1-cycle-clique   (b) 2-cycle-clique

Fig. 1. Examples of cycle-cliques.

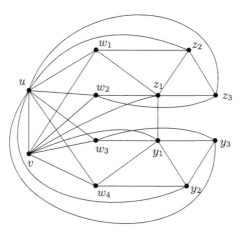

Fig. 2. The K_4-free polar gadget. It consists of 12 vertices, 30 edges and 20 triangles.

Proof. Proof by contradiction. Let's say that there is a triangle-free k-coloring of k-cycle-clique and $i \in \{0, \ldots, 4\}$ such that there exists two vertices in J_i that have the same color. Using Observation 2 we know that this color cannot appear in $J_{i+_5 1}$ or $J_{i-_5 1}$. Using Observation 2 again, this color need to be used twice in each of $J_{i+_5 2}$ and $J_{i-_5 2}$. Since $J_{i+_5 2} \cup J_{i-_5 2} \sim K_{2k}$ from the definition and that four vertices of $J_{i+_5 2} \cup J_{i-_5 2}$ are colored with the same color, we reach a contradiction with the assumption that the initial coloring was triangle-free. □

Second gadget is presented below.

Definition 2. *The K_4-free polar gadget is the graph $G = (V, E)$ with $V = \{u, v, w_1, w_2, w_3, w_4, z_1, z_2, z_3, y_1, y_2, y_3\}$, and edge set defined as in Fig. 2.*

We say that egde xy is *polar*, if in every triangle-free 2-coloring c, $c(x) \neq c(y)$.

Lemma 2. *Let G be a K_4-free polar gadget. Then the following are true: (i) G is triangle-free 2-colorable, (ii) uv is a polar edge, (iii) G is K_4-free.*

Proof. Ad. (i): Let $S = \{u, z_1, y_1\}$. Set $c(S) = 1$ and $c(V \setminus S) = 2$.

Ad. (ii): Assume by contradiction, that $c(u) = c(v) = 1$ and G is triangle-free 2-colorable. Then $c(\{w_1, w_2, w_3, w_4\}) = 2$, which is forced. Since $c(v) = 1$, then either $c(z_1) = 2$, or $c(y_1) = 2$. Assume the former, without loss of generality. Then $c(z_2) = 1$, otherwise $\{w_1, z_1, z_2\}$ would be monochromatic. But this means that we cannot color z_3, because if $c(z_3) = 1$, then $\{u, z_2, z_3\}$ is monochromatic, and if $c(z_3) = 2$, then $\{w_2, z_1, z_3\}$ is monochromatic – a contradiction.

Ad. (iii): Let $V_1 = \{u, v\}$, $V_2 = \{w_1, w_2, w_3, w_4\}$, $V_3 = \{z_1, z_2, z_3, y_1, y_2, y_3\}$. If there exists $H \subset G$ that is isomorphic to K_4, then we can observe that:

– H contains at most one vertex from V_2, since V_2 is an independent set, and
– H contains at most one vertex from V_1, because otherwise H would contain at least one vertex from V_3, and there is no vertex (or pair of vertices) in V_3 that is connected to both vertices in V_1.

Therefore we are left with the following four cases:

- H uses only vertices from V_3. But in $G(V_3)$ only z_1 and y_1 have degree at most 3.
- H uses 3 vertices from V_3 and one vertex from V_2. But each vertex from V_2 is adjacent to exactly two vertices from V_3.
- H uses 3 vertices from V_3 and one vertex from V_1. Then $v \notin H$ because it is adjacent to only z_1 and y_1. Thus $u \in H$, but it easy to verify that $N(u)$ does not contain a triangle.
- H uses 2 vertices from V_3, one vertex from V_2 and one vertex from V_1. But for each $\{s,t\} \subset V_3$, there exists exactly one triangle $\{s,t,r\}$, such that $r \in V_1 \cup V_2$.

We have reached contradictions in all possible cases, thus we conclude that G is K_4-free. □

We believe that the gadgets that we presented in this section are interesting from the graph-theoretic point of view. It would be instructive to find answers to the following questions.

Question 1. For any $k \geq 2$, what is the smallest graph (in terms of vertices and/or edges) for which the following are true: (i) $\chi_3(G) = k$, and (ii) for any triangle-free k-coloring of G, there exists a *polar clique* of size k in G, i.e., a clique of size k where each vertex has a unique color?

Question 2. Is K_4-free polar gadget the smallest (in terms of vertices, edges or triangles) graph that satisfies the properties of Lemma 2?

4   Tractable Classes of Graphs

Here we present several classes of graphs for which efficient algorithms for triangle-free coloring problem exist. We use the bounds derived in the previous section to show that for small values of structural parameters polynomial time complexity can be achieved in many cases.

Given graph $G = (V, E)$, we distinguish between two optimization variants of the triangle-free coloring problem: (i) finding the number $\chi_3(G)$, and (ii) finding the triangle-free coloring that uses exactly $\chi_3(G)$ colors, i.e., the mapping from V to $\{1, \ldots, \chi_3(G)\}$ is the required output. As we will soon see, complexities for these two variants can be different. To this end we propose the following notion of time complexity for our problem. For convenience, we say that n is the number of vertices of graph G and m is the number of edges of G, unless stated otherwise.

Definition 3. *Let $\mathcal{G}$ be a class of graphs. We say that the triangle-free coloring problem is solvable in time $O(f(n,m), g(n,m))$ on $\mathcal{G}$, iff there exist algorithms $\mathcal{A}$ and $\mathcal{B}$, such that for every input $G \in \mathcal{G}$, (i) algorithm $\mathcal{A}$ outputs $\chi_3(G)$ in $O(f(n,m))$ time, and (ii) algorithm $\mathcal{B}$ outputs the triangle-free coloring that uses exactly $\chi_3(G)$ colors, in $O(g(n,m))$ time.*

4.1 Graphs with Bounded Chromatic Number

Here we present a single theorem that presents the complexity of triangle-free coloring problem on several popular classes of graphs. We exploit the fact that if we know in advance that $\chi(G) \leq 4$ – and therefore $\chi_3(G) \leq 2$, by Theorem 1 – then $\chi_3(G)$ can be found easily and is one of the following: (a) $\chi_3(G) = 0$ iff G is empty, (b) $\chi_3(G) = 1$ iff G is triangle-free (K_3-free), (c) $\chi_3(G) = 2$ iff the above two cases do not hold. Checking the first case is trivial. Therefore the challenge of finding $\chi_3(G)$ lies in checking whether G contains a triangle.

For finding the actual coloring, we use the fact that in the given graphs we can find "good-enough" classic coloring, and then recolor vertices so that there are no monochromatic triangles using the *standard recoloring strategy*.

Theorem 5. *The triangle-free coloring problem is solvable:*

- *in time $O(n, n^2)$ on planar graphs,*
- *in time $O(n, n)$ on: outerplanar graphs, chordal graphs, graphs with bounded maximum degree Δ, with $\Delta \leq 4$.*

Proof. See extended paper [14].   □

We do not know if there exists a sub-quadratic algorithm for finding a triangle-free coloring in planar graphs, so we leave it as an open problem. It is worth noting that the situation would not improve even if we knew that the the input graph is classically 3-colorable, as the best known algorithm for classically 4-coloring planar graphs with this property still runs in $O(n^2)$ time [15]. Theorem 5 serves as an evidence, that the triangle-free coloring problem is easier than the classic coloring problem. We have efficient algorithms for finding χ_3 on planar graphs and graphs with $\Delta = 4$, but it is $\mathcal{NP}$-hard to find the χ even on 4-regular planar graphs [5]. This suggests that the above open problem might have a positive answer.

4.2 FPT Results

In this subsection we look at our problem from the parametrized complexity point of view. We prove that TRIANGLEFREE-q-COLORING problem is fixed-parameter tractable when parametrized by vertex cover number. We can also prove the stronger result with treewidth as a parameter (since treewidth is at most vertex cover number) by using the celebrated Courcelle's Theorem [4], which states that all graph properties definable in Monadic Second Order logic can be decided in linear time on graphs of bounded treewidth. Since the proof is elementary and the corresponding time complexity of the algorithm is big, we decided to move the proof to the extended version of the paper [14] and give an algorithm with vertex cover number as a parameter.

In the article by Fiala, Golovach and Kratochvíl [9], authors explore differences in the complexity of several coloring problems when parametrized by either vertex cover number or treewidth. We use techniques similar to theirs in the proof of the following theorem.

Theorem 6. *The* TRIANGLEFREE-*q*-COLORING *problem is* $\mathcal{FPT}$ *when parametrized by the vertex cover number.*

Proof. Let W be a minimum vertex cover of G, and let $|W| = k$. Then $I = V \setminus W$ is an independent set. The goal is to find a triangle-free coloring $c : V \to \{1, \ldots, q\}$.

The algorithm works in two steps: first, we find the triangle-free q-coloring of W by exhaustive search. Then, since I is an independent set, we can use greedy strategy to color its vertices, once we know the coloring of W. By greedy strategy we mean taking each $v \in I$, checking its colored neighborhood $N(v)$ (notice that $N(v) \subseteq W$), and coloring v in a way which does not create a monochromatic triangle in G. Clearly this procedure will find a triangle-free q-coloring iff such coloring exists. The analysis of running time now follows.

Since $\chi_3(W) \leq \lceil \chi(W)/2 \rceil \leq \lceil k/2 \rceil$, it is natural to consider the following two cases:

(i) If $q \leq \lceil k/2 \rceil$, then we consider all q-colorings of W and their extensions to I by the greedy algorithm. Since W has at most $k^{\lceil k/2 \rceil}$ colorings, the running time is $O(k^{\lceil k/2 \rceil + 1}n)$.

(ii) If $q > \lceil k/2 \rceil$, then assuming that $W = \{w_0, \ldots, w_{k-1}\}$, we can set $c(w_{2i}) = c(w_{2i+1}) = i + 1$, for each $0 \leq i < \lceil k/2 \rceil$, which clearly uses $\lceil k/2 \rceil$ colors, then set $c(I) = \lceil k/2 \rceil + 1$. Thus, in this case, the triangle-free q-coloring always exists, and we can find it in $O(n)$ time.

We conclude that the total running time of the algorithm is $O(k^{\lceil k/2 \rceil + 1}n)$, which proves the claim. □

5   Hardness Results

Here we improve the results of Karpiński [13] and Shitov [18]. In their papers they show that the generalized version of TRIANGLEFREE-2-COLORING problem is $\mathcal{NP}$-complete – they not only consider monochromatic triangles, but monochromatic cycles of arbitrary (fixed) length. Karpiński first presented the proof, which was then simplified by Shitov, by introducing an auxiliary problem, which we call the TRIANGLEFREEPOLAR-2-COLORING problem. In Shitov's proof, this new problem serves as a connection between NOTALLEQUAL 3-SAT and TRIANGLEFREE-2-COLORING.

Our contribution is as follows: we first prove that the triangle-free coloring remains $\mathcal{NP}$-hard even if we can use any fixed number of colors. Then we show that the TRIANGLEFREE-2-COLORING problem is $\mathcal{NP}$-hard in the restricted class of graphs, which do not contain K_4 as a subgraph. These improve the results of Karpiński [13]. Lastly, we show that the TRIANGLEFREEPOLAR-2-COLORING remains $\mathcal{NP}$-hard even in graphs of maximum degree 3, which improves the result of Shitov [18].

Theorem 7. *For any fixed* $q \geq 2$, *the* TRIANGLEFREE-*q*-COLORING *problem is* $\mathcal{NP}$-*hard.*

Proof. We prove the theorem by induction on q. For $q = 2$ the hardness has already been proved by Karpiński [13]. What remains is to show the polynomial reduction from the TRIANGLEFREE-q-COLORING problem to TRIANGLEFREE-$(q+1)$-COLORING problem, for any fixed $q \geq 2$. Take any graph $G = (V, E)$, where $|V| = n$ and $|E| = m$, and assume that the vertices are arbitrarily ordered, i.e., $V = \{v_1, \ldots, v_n\}$. We construct the graph $G' = (V', E')$ in the following way:

1. Let $V' = V \cup (\bigcup_{i=1}^{n} V_i)$ and $E' = E \cup (\bigcup_{i=1}^{n} E_i)$, where $V_i = V_i^0 \cup V_i^1 \cup V_i^2 \cup V_i^3 \cup V_i^4$ and (V_i, E_i) is a $(q+1)$-cycle-clique with the set $\{V_i^0, V_i^1, V_i^2, V_i^3, V_i^4\}$ being its joints.
2. For every $1 \leq i \leq n$, pick arbitrary vertex from V_i^0 and call it u_i. Identify the pairs of vertices in order: $\langle u_1, u_2 \rangle, \ldots, \langle u_1, u_n \rangle$. Rename u_1 to u.
3. For every $1 \leq i \leq n$, take any $s \in V_i^0$ such that $s \neq u$ and identify vertices $\langle v_i, s \rangle$.

Observe that $v_i u$ is an edge in V_i^0 (for $1 \leq i \leq n$) and therefore in any triangle-free $(q + 1)$-coloring of G', v_i has a different color than u, by Lemma 1. It is now easy to verify that G is triangle-free q-colorable iff G' is triangle-free $(q+1)$-colorable. Also, since q is part of the problem (and therefore a constant), the reduction is polynomial w.r.t. n and m.   □

Theorem 8. *The* TRIANGLEFREE-*2*-COLORING *problem is* $\mathcal{NP}$-*hard on* K_4-*free graphs.*

Proof. We create a polynomial reduction from the following problem, which is known to be $\mathcal{NP}$-complete [17]:

NOTALLEQUAL 3-SAT
Input: Boolean formula ϕ in conjunctive normal form, where each clause consists of exactly three literals.
Question: Does ϕ have a nae-satisfying assignment, i.e., in each clause at least one literal is true and at least one literal is false?

Let ϕ be a Boolean formula as described above, with n variables X and m clauses $C_1, \ldots, C_m$. We assume, without loss of generality, that each clause consists of at least two unique literals, otherwise the instance is trivially false. We show a construction of K_4-free graph $G = (V, E)$, such that ϕ is nea-satisfiable iff there exist a triangle-free 2-coloring of G.

1. For every clause $C_i \equiv (l_1^i \vee l_2^i \vee l_3^i)$, $1 \leq i \leq m$, add vertices $l_1^i, l_2^i,$ and l_3^i to V, and add edges $\{l_1^i, l_2^i\}$, $\{l_2^i, l_3^i\}$ and $\{l_3^i, l_1^i\}$ to E.
2. For every variable $x \in X$, add vertices t_x and f_x to V, and edge $\{t_x, f_x\}$ to E.
3. For every variable x and every occurrence of literal $l \equiv \neg x$ in ϕ, add edge $\{t_x, l\}$ to E. For every variable x and every occurrence of literal $l' \equiv x$ in ϕ, add edge $\{f_x, l'\}$ to E.
4. For every edge $e \in E$, except for edges added in step 1, create K_4-free polar gadget (Definition 2), with e as the polar edge.

Observe that introducing polar gadgets in step 4 forces every vertex corresponding to the literal l to have the same color, and also every vertex corresponding to literal $\neg l$ to have the same color, but different than the color of l's. Therefore the nae-satisfying assignment of each clause can be obtained from the coloring of each triangle introduced in step 1. Note that K_4-free polar gadget is 2-colorable, by Lemma 2. From the above observations, we can conclude that ϕ has a nae-satisfying assignment iff G is triangle-free 2-colorable. Also, the reduction is polynomial w.r.t. n and m, and G is K_4-free, since every polar gadget used is K_4-free, by Lemma 2. □

Theorem 9. *The* TRIANGLEFREEPOLAR-2-COLORING *problem is $\mathcal{NP}$-hard on graphs with maximum degree at most 3, but it is linear-time solvable on graphs with maximum degree at most 2.*

Proof. See extended paper [14]. □

6   Conclusions and Future Work

In this paper we have shown several results regarding vertex coloring without monochromatic triangles. Many new interesting problems can be derived from this new coloring variant. We have already asked a few questions throughout this paper. Apart from that, we propose a handful of ways to extend our research:

- For positive results one can look for an algorithm that finds $\chi_3(G)$ in graphs with $\Delta(G) \geq 5$. Using an algorithm from [19] when $\Delta(G) = 5$ can only get us classic 5-coloring of G, and therefore standard recoloring strategy may not produce an optimal solution. This issue requires more complicated algorithmic approach.
- For negative side, one can look for the smallest $\Delta(G)$, for which the TRIANGLEFREE-q-COLORING problem is $\mathcal{NP}$-hard.
- In the context of parametrized complexity we have shown two results. Constructing algorithms for other choices of parameters is a fine research direction. It would also be good to know where the problems proved to be $\mathcal{NP}$-hard in Sect. 5 reside in W-hierarchy.

Finally, we note that it is possible to further generalize the notion of χ_3 to get the parameter χ_r, for any $r \geq 3$ (notice that $\chi_2 = \chi$). This new parameter can restrict the coloring in two ways: either not allow monochromatic K_r, or not allow monochromatic $C_{r'}$ (cycle of length r'), for any $3 \leq r' \leq r$. Both extensions are interesting.

References

1. Alon, N., Yuster, R., Zwick, U.: Finding and counting given length cycles. Algorithmica **17**(3), 209–223 (1997)
2. Angelini, P., Frati, F.: Acyclically 3-colorable planar graphs. J. Comb. Optim. **24**(2), 116–130 (2012)

3. Brooks, R.: On colouring the nodes of a network. Math. Proc. Camb. Philos. Soc. **37**(2), 194–197 (1941)
4. Courcelle, B.: The monadic second-order logic for graphs I: recognizable of fiite graphs. Inf. Comput. **85**, 12–75 (1990)
5. Dailey, D.P.: Uniqueness of colorability and colorability of planar 4-regular graphs are NP-complete. Discret. Math. **30**(3), 289–293 (1980)
6. Deb R.: An efficient nonparametric test of the collective household model (2008). SSRN: http://ssrn.com/abstract=1107246
7. Dinur, I., Regev, O., Smyth, C.: The hardness of 3-uniform hypergraph coloring. In: The 43rd Annual IEEE Symposium on Foundations of Computer Science, pp. 33–40 (2002)
8. Erdos, P.: Graph thoery and probability. Canad. J. Math. **11**, 34–38 (1959)
9. Fiala, J., Golovach, P., Kratochvíl, J.: Parametrized complexity of coloring problems: treewidth versus vertex cover. Theor. Comput. Sci. **412**, 2513–2523 (2011)
10. Formanowicz, P., Tanaś, K.: A survey of graph coloring - its types, methods and applications. Found. Comput. Decis. Sci. **37**(3), 223–238 (2012)
11. Jain, P.: On a variant of Monotone NAE-3SAT and the Triangle-Free Cut problem. Pre-print: arXiv:1003.3704 [cs.CC] (2010)
12. Kaiser, T., Škrekovski, R.: Planar graph colorings without short monochromatic cycles. J. Graph Theory **46**(1), 25–38 (2004)
13. Karpiński, M.: Vertex 2-coloring without monochromatic cycles of fixed size is NP-complete. Theor. Comput. Sci. **659**, 88–94 (2017)
14. Karpiński, M., Piecuch, K.: On vertex coloring without monochromatic triangles. Pre-print, arXiv:1710.07132 [cs.DS] (2017)
15. Kawarabayashi, K., Ozeki, K.: A simple algorithm for 4-coloring 3-colorable planar graphs. Theor. Comput. Sci. **411**(26–28), 2619–2622 (2010)
16. Mycielski, J.: Sur le coloriage des graphs. Colloquium Mathematicae **3**(2), 161–162 (1955)
17. Schaefer, T.J.: The complexity of satisfiability problems. In: STOC 1978, pp. 216–226 (1978)
18. Shitov, Y.: A tractable NP-completeness proof for the two-coloring without monochromatic cycles of fixed length. Theor. Comput. Sci. **674**, 116–118 (2017)
19. Skulrattanakulchai, S.: Delta-list vertex coloring in linear time. Inf. Process. Lett. **98**(3), 101–106 (2006)
20. Thomassen, C.: 2-List-coloring planar graphs without monochromatic triangles. J. Comb. Theory Ser. B **98**, 1337–1348 (2008)

Recognizing Read-Once Functions
from Depth-Three Formulas

Alexander Kozachinskiy$^{(\boxtimes)}$

National Research University Higher School of Economics, Moscow, Russia
akozachinskiy@hse.ru

Abstract. Consider the following decision problem: for a given monotone Boolean function f decide, whether f is read-once. For this problem, it is essential how the input function f is represented. On a negative side we have the following results. Elbassioni et al. [1] proved that this problem is coNP-complete when f is given by a depth-4 read-2 monotone Boolean formula. Gurvich [2] proved that this problem is coNP-complete even when the input is the following expression: $C \vee D_n$, where $D_n = x_1 y_1 \vee \ldots \vee x_n y_n$ and C is a monotone CNF over the variables $x_1, y_1, \ldots, x_n, y_n$ (note that this expression is a monotone Boolean formula of depth 3; in [2] nothing is said about the readability of C, but the proof is valid even if C is read-2 and thus the entire formula is read-3).

On a positive side, from [3] we know that there is a polynomial time algorithm to recognize read-once functions when the input is a monotone depth-2 formula (that is, a DNF or a CNF). There are even very fast algorithms for this problem [4].

Our contribution consists of the following two results. We show that we can test in polynomial-time whether a given expression $C \vee D$ computes a read-once function, provided that C is a read-once monotone CNF and D is a read-once monotone DNF and all the variables of C occur also in D (recall that due to Gurvich, the problem is coNP-complete when C is read-2). The second result states that this is a coNP-complete problem to decide whether the expression $A \wedge D_n$ computes a read-once function, where D_n is as above and A is the $\bigwedge - \bigvee - \bigwedge$ depth-3 read-once monotone Boolean formula (so that the entire expression $A \wedge D_n$ is depth-3 read-2). This result improves the result of [1] in the depth and the result of [2] in the readability of the input formula.

Keywords: Read-once functions · Monotone Boolean functions
coNP-completeness

1 Introduction

In this paper we study the following decision problem: decide, for a given monotone Boolean function f, whether f is a read-once function (the latter means

A. Kozachinskiy—Supported in part by the RFBR grant 16-01-00362 and by the Russian Academic Excellence Project '5-100'.

F. V. Fomin and V. V. Podolskii (Eds.): CSR 2018, LNCS 10846, pp. 232–243, 2018.
https://doi.org/10.1007/978-3-319-90530-3_20

that the function can be computed by a monotone formula in which every variable occurs only once). Of course, to specify the problem, we need to specify the representation of f. For some representations this problem turns out to be tractable. For example, it is known (see [3]) that if f is given by a monotone DNF (or, equivalently, CNF), then the corresponding problem can be solved in polynomial time. Golumbic et al. [4] gave quite fast algorithm for this problem which works in time $O(nm)$, where m is the length of the given DNF and n is the number of variables. In [4] they also asked how hard is this problem when f is represented in some other way, for example, when f is given by an arbitrary Boolean formula.

First of all, let us note that if the input is a monotone Boolean formula, then the problem belongs to coNP. This follows from the following theorem by Gurvich:

Theorem 1 ([5]). *A monotone Boolean function f is read-once iff every minterm S of f and every maxterm T of f intersect in exactly one point.*

Thus to show that f is *not* read-once it is enough to demonstrate a minterm S and a maxterm T with $|S \cap T| > 1$ (it is not hard to show that given a formula S, T, we can decide in polynomial time whether S is a minterm and T is a maxterm).

Soon after Golumbic, Mintz and Rotics raised their question, Elbassioni et al. [1] proved that the read-once recognition problem is coNP-complete when the input function f is given by a depth-4 read-2 monotone Boolean formula. The same authors also proved that it is NP-hard to approximate the readability of the monotone Boolean function $f : \{0,1\}^n \to \{0,1\}$, given by a depth-4 monotone Boolean formula, within a factor of $O(n)$.

Later, Gurvich [2] proved, that the problem of recognizing read-once functions is coNP-complete even when the input is the following expression: $C \vee D_n$, where $D_n = x_1y_1 \vee \ldots \vee x_ny_n$ and C is a monotone CNF over the variables $x_1, y_1, \ldots, x_n, y_n$. Note that the entire formula $C \vee D_n$ is the depth-3 formula. The paper [2] says nothing about the readability of C, but the proof is valid even if C is read-2[1] (so that the whole expression is read-3).

Our first result shows that this problem becomes tractable, if C is read-once even when D_n is any monotone read-once DNF:

Theorem 2. *There is a polynomial-time algorithm which decides, whether a given expression $C \vee D$ computes a read-once function, provided that C is a monotone read-once CNF, D is a monotone read-once DNF, and every variable of C occurs also in D.*

We don't know whether the last restriction can be removed. However, we find this theorem interesting due to its connection to the result of Gurvich.

Our second result shows that the problem of recognizing read-once functions is coNP-complete when the input formula is depth-3 read-2 (note that [1] requires

---

[1] This is due to the fact that the SAT-problem is NP-complete even for read-3 (non-monotone) CNFs.

depth at least 4 and [2] requires readability at least 3 for the input formula). Moreover, we may consider only formulas which are conjunctions of two read-once formulas:

Theorem 3. *The problem to decide whether a given expression $A \wedge D_n$ computes a read-once function is coNP-complete. Here $D_n = x_1 y_1 \vee \ldots \vee x_n y_n$ and A is a $\wedge - \vee - \wedge$ depth-3 read-once monotone Boolean formula over $\{x_1, y_1, \ldots, x_n, y_n\}$.*

The reduction we establish for the Theorem 3 is from the clique-problem (while the reductions from [1,2] are from the SAT-problem).

Remark. Theorem 3 can be used to show that inapproximability result of Elbassioni, Makino and Raur is also true for depth-3 formulas. This is, however, can be done with the use of the result of Gurvich as well.

2   Preliminaries

A monotone Boolean formula (i.e., a $\wedge, \vee$-formula) Φ is called a *read-k formula* if every variable occurs at most k times in Φ. A monotone Boolean function f is called a *read-k function* if there is a monotone read-k formula, computing f. *Readability* of a Boolean function f (formula Φ) is the minimal k such that function f (formula Φ) is read-k.

Assume that f is a monotone Boolean function over the variables $x_1, \ldots, x_n$ and S is a subset of $\{x_1, \ldots, x_n\}$. To simplify notation below let $f(S \to i)$ (here $i \in \{0, 1\}$) denote the value of f when all the variables from S are set to i and all the variables from $\{x_1, \ldots, x_n\} \backslash S$ are set to $1 - i$.

A subset $S \subset \{x_1, \ldots, x_n\}$ is called a *minterm* of f if $f(S \to 1) = 1$ but for every proper subset S' of S it holds that $f(S' \to 1) = 0$. Similarly, a subset $T \subset \{x_1, \ldots, x_n\}$ is called a *maxterm* of f if $f(T \to 0) = 0$ but for every proper subset T' of T it holds that $f(T' \to 0) = 1$.

Obviously, every minterm of f intersects every f's maxterm.

3   Proof of Theorem 2

Our algorithm uses the following lemma:

Lemma 1. *There exists a polynomial-time algorithm which for any given read-once monotone CNF C and for any given read-once monotone DNF D decides, whether $C \to D$ is a tautology.*

Proof. It is known (see, e.g., [6]) that there is a polynomial-time algorithm to decide, whether a given read-2 CNF is satisfiable. Apply this algorithm to $\neg(C \to D) = C \wedge \neg D$ (the latter can be re-written as a read-2 CNF in polynomial time). □

Let $\{x_1, \ldots, x_n\}$ be variables occurring in D. Let $C_1, \ldots, C_m$ denote the clauses of C. Since C is read-once, we may identify $C_1, \ldots, C_m$ with m disjoint subsets of $\{x_1, \ldots, x_n\}$. The same thing can be done for D. Let

$$D = D_1 \vee D_2 \vee \ldots \vee D_l,$$

where $D_1, \ldots, D_l \subset \{x_1, \ldots, x_n\}$ are disjoint conjunctions. Note also that $D_1 \cup \ldots \cup D_l = \{x_1, \ldots, x_n\}$.

We provide first a description of minterms of $C \vee D$. Let S be a subset of $\{x_1, \ldots, x_n\}$. We say that S is a *right set* if for some $j \in \{1, \ldots, l\}$ we have $S = D_j$. We say that S is a *left set* if $S \subset C_1 \cup \ldots \cup C_m$ and for every $i \in \{1, \ldots, m\}$ it holds that $|C_i \cap S| = 1$. The following lemma is straightforward.

Lemma 2. *A set S is a minterm of $C \vee D$ if and only if S is a left set that does not properly include any right set or S is a right set that does not properly include any left set.*

A minterm S of $C \vee D$ is called *a left minterm* if S is a left set. Similarly, we call S *a right minterm* if S is a right set.

Now we are ready to present the algorithm for Theorem 2. In the description of the algorithm we will state several auxiliary lemmas whose proofs are deferred to Appendix (with exception of Lemma 6; its proof is omitted due to space constraints).

Algorithm. The algorithm works in four steps.

Step 1. Check, using Lemma 1, whether $C \to D$ is a tautology. If it is, then $C \vee D$ is equivalent to D and hence $C \vee D$ computes a read-once function and the algorithm halts. Otherwise proceed to Step 2.

Step 2. Obviously every maxterm T includes at least one clause of C. For every pair of distinct clauses C_u, C_v check whether there is a maxterm T of $C \vee D$ such that $C_u, C_v \subset T$. This can be done in polynomial time by the following

Lemma 3. *Let C_u, C_v be two distinct clauses of C. Then there exists a maxterm T of $C \vee D$ such that $C_u, C_v \subset T$ if and only if for every $j \in \{1, \ldots, l\}$ it holds that $|(C_u \cup C_v) \cap D_j| \leq 1$.*

If there is such T, then $C \vee D$ is not read-once. Indeed, consider any minimal $S_0 \subset \{x_1, \ldots, x_n\}$ such that $C(S_0 \to 1) = 1$ and $D(S_0 \to 1) = 0$. Such a set exists, since $C \to D$ is not a tautology. As $D(S_0 \to 1) = 0$, the set S_0 does not include any right set, and hence is a left minterm of $C \vee D$. As, $C(S_0 \to 1) = 1$, the set S_0 intersects both C_u and C_v. Recall that $C_u, C_v \subset T$ and hence $|T \cap S_0| \geq 2$. By Theorem 1, $C \vee D$ does not compute a read-once function.

Otherwise (if there is no maxterm T of $C \vee D$ that includes distinct clauses of C) we proceed to Step 3.

Step 3. For every clause C_u and for every pair of distinct variables p and q from C_u we check:

- whether there is a right minterm S such that $\{p, q\} \subset S$ (this can be done in polynomial time since there are only polynomially many right minterms);
- whether there is a maxterm T containing C_u.

The second check can be done in polynomial time using the following

Lemma 4. *Assume that $C \to D$ is not a tautology and no maxterm of $C \vee D$ contains two distinct clauses of C. Then for any clause C_i we can decide in polynomial-time whether there exists a maxterm T of $C \vee D$ such that $C_i \subset T$.*

If for some C_u, p, q both questions answer in positive, then $C \vee D$ is not a read-once function. Indeed, in this case both the minterm S and the maxterm T include distinct variables p and q, and by Theorem 1, $C \vee D$ does not compute a read-once function. Otherwise we proceed to Step 4.

Step 4. For all $u \in \{1, \ldots, m\}$ and $v \in \{1, \ldots, l\}$ with $C_u \cap D_v = \varnothing$, and for all $p \in D_v$ and $q \in C_u$ we check:

- whether there is a left minterm S such that $\{p, q\} \subset S$,
- whether there is a maxterm T such that $C_u \cup \{p\} \subset T$.

Both checks can be performed in polynomial time by the following lemmas.

Lemma 5. *There exists a polynomial-time algorithm which for any given pair of distinct variables $a, b \in \{x_1, \ldots, x_n\}$ decides, whether there is a left minterm S of $C \vee D$ such that $\{a, b\} \subset S$.*

Lemma 6. *Assume that there is no maxterm of $C \vee D$ which contains two distinct clauses of C. Then for any given C_u, D_v with $C_u \cap D_v = \varnothing$ and for any given $p \in D_v$ we can decide in polynomial time whether there exists a maxterm T of C such that $C_u \cup \{p\} \subset T$.*[2]

If for some C_u, D_v, p, q both questions answer in positive, then $C \vee D$ does not compute a read-once function. Indeed, in this case both the maxterm T and the minterm S include distinct variables p, q and by Theorem 1, $C \vee D$ does not compute a read-once function.

Otherwise the algorithm outputs the positive answer and halts. We have to show that in this case $C \vee D$ indeed computes a read-once function. For the sake of contradiction, assume that $C \vee D$ is not read-once. By Theorem 1 there is a maxterm T of $C \vee D$ and a minterm S of $C \vee D$ that have distinct common variables p, q. By Lemma 2 S is either a left or a right minterm. We will consider these two cases separately.

Case 1: S is a right set, say $S = D_j$. Then T contains some clause C_u of C. We claim that $S \cap T \subset C_u$. Indeed, otherwise $S \cap T$ would include a variable $y \notin C_u$. Then $C(T \setminus \{y\} \to 0) = 0$, as $C_u \subset T \setminus \{y\}$. And $D(T \setminus \{y\} \to 0) = 0$,

---

[2] The proof of this lemma is almost identical to the proof of Lemma 4. Actually, it is possible to formulate a single lemma which implies both of them, but then the formulation of the lemma becomes immense.

since D is read-once and $T \cap D_j$ has at least 2 variables. We obtain contradiction, as T is a maxterm of $C \vee D$.

Thus $T \cap S \subset C_u$. Hence there are two distinct variables p and q from C_u such that $\{p, q\} \subset S$. Hence the algorithm must have halted on Step 3, a contradiction.

Case 2: S is a left set. Again there is a clause $C_u \subset T$. Since S is a left set, S and C_u have exactly one common variable q. By our assumption $T \cap S$ includes another variable $p \neq q$. Note that $p \notin C_u$ because otherwise $S \cap C_u$ has more than one variable.

Let D_v be the (unique) right set that includes p (here we use the assumption that every variable is in some right set). We claim that D_v and C_u are disjoint. For the sake of contradiction assume that there is a variable $y \in C_u \cap D_v$. Then $(C \vee D)(T\backslash\{p\} \to 0) = 0$. Indeed, $C(T\backslash\{p\} \to 0) = 0$, as $p \notin C_u$ and hence $C_u \subset T\backslash\{p\}$. And $D(T\backslash\{p\} \to 0) = 0$, as $y \neq p$ (since y is in C_u and p is not) and hence $T\backslash\{p\}$ still intersects D_v.

Thus T is not a maxterm, and the contradiction shows that D_v and C_u are disjoint. Therefore, we have found C_u, p, q, D_v such that $q \in C_u$, $p \in D_v$, C_u and D_v are disjoint and there is a left minterm S and a maxterm T with $\{p, q\} \subset S, C_u \cap \{p\} \subset T$. Hence the algorithm just have halted on Step 4, a contradiction. **(End of Algorithm.)**

4   Proof of Theorem 3

Let $G = (V, E)$ be undirected graph and $k \leq |V|$ a positive integer. Let us define two auxiliary sets $\mathcal{A}(G, k)$ and $\mathcal{B}(G, k)$. Namely, let $\mathcal{A}(G, k)$ be the set of all quadruples (i, u, j, v) such that $i, j \in \{1, \ldots, k\}$, $u, v \in V$ and $i \neq j$, $\{u, v\} \notin E$. Further, let $\mathcal{B}(G, k)$ be the set of all unordered pairs $\{(i, u), (j, v)\}$ such that $(i, u, j, v) \in \mathcal{A}(G, k)$. Note that $\mathcal{A}(G, k)$ is two times larger than $\mathcal{B}(G, k)$. Note also that sizes of $\mathcal{A}(G, k)$ and $\mathcal{B}(G, k)$ are polynomial in size of G.

For each $(i, u, j, v) \in \mathcal{A}(G, k)$ introduce a variable $x_{j,v}^{i,u}$. Consider the following two formulas over variables $x_{j,v}^{i,u}$:

$$A(G, k) = \bigwedge_{i=1}^{k} \bigvee_{u \in V} \bigwedge_{\substack{(j,v): \\ (i,u,j,v) \in \mathcal{A}(G,k)}} x_{j,v}^{i,u},$$

$$B(G, k) = \bigvee_{\{(i,u),(j,v)\} \in \mathcal{B}(G,k)} (x_{j,v}^{i,u} \wedge x_{i,u}^{j,v}).$$

Observe that $A(G, k)$ is $\bigwedge - \bigvee - \bigwedge$ depth-3 read-once monotone Boolean formula and $B(G, k)$ is equal to D_n for $n = |\mathcal{B}(G, k)|$. Note also that these formulas can be obtained from G, k in polynomial time.

The following Lemma motivates this construction.

Lemma 7. *There is no clique of size k in G iff $A(G, k) \to B(G, k)$ is a tautology.*

Proof. Denote for brevity $A = A(G, k)$, $B = B(G, k)$.

Assume that $w_1, \ldots, w_k \in V$ form a clique in G. Let us define an assignment of variables for which A is true and B is false. Namely, set $x^{i,u}_{j,v}$ to 1 iff $u = w_i$. Clearly, A is true, since for every i from 1 to k there is a vertex w_i such that all variables with superscript i, w_i are set to 1. On the other hand, B is false, since $(x^{i,u}_{j,v} \wedge x^{j,v}_{i,u}) = 1$ implies that $u = w_i$, $v = w_j$, and this contradicts either $i \neq j$ or $\{u, v\} \notin E$.

Now assume that $A \to B$ is not a tautology. Hence there is an assignment of variables on which A is true and B is false. Since A is true, this means that for every i from 1 to k there exists $w_i \in V$ such that all the variables with superscript i, w_i are set to 1. Let us show that $w_1, \ldots, w_k$ form a clique in G. Assume for contradiction that there are $i \neq j$ such that w_i and w_j are not connected by an edge in G, that is, $\{(i, w_i), (j, w_j)\}$ belongs to $\mathcal{B}(G, k)$. This contradicts the assumption that B is false, because both x^{i,w_i}_{j,w_j} and x^{j,w_j}_{i,w_i} are set to 1. $\square$

To show Theorem 3 we reduce from coNP-complete language CO-CLIQUE, consisting of all pairs (G, k) such that G is an undirected graph, k is an integer and there is no clique of size k in G. It is enough to show the following

Lemma 8. *Assume that $3 \leq k < |V|$. Then there is no clique of size k in G iff $A(G, k) \wedge B(G, k)$ computes a read-once function.*

A technical restriction $3 \leq k < |V|$ is not essential; if we a have a pair (G, k) and k is either less than 3 or bigger than $|V| - 1$, then we can solve a clique problem for (G, k) without any reduction in polynomial time.

Proof (of Lemma 8).

Once again, denote for brevity $A = A(G, k)$, $B = B(G, k)$.

Indeed, if there is no clique of size k in G, then by Lemma 7 implication $A \to B$ is a tautology. The latter implies that $A \wedge B = A$. Since A is a read-once formula, we are done.

Now assume that $w_1, \ldots, w_k$ form a clique in G. It is enough to show that in this case $A \wedge B$ is not read-once. To do this we will use Theorem 1.

Define the following maxterm T of $A \wedge B$. Add a variable $x^{i,u}_{j,v}$ to T iff it satisfies at least one of the following two conditions:

- $v = w_j$;
- $u \neq w_i$, $v \neq w_j$, and $i < j$.

Let us verify that T is a maxterm. Note that $B(T \to 0) = 0$. Indeed, assume for contradiction that $B(T \to 0) = 1$. Then there is $\{(i, u), (j, v)\} \in \mathcal{B}(G, k)$ such that $x^{i,u}_{j,v}$ and $x^{j,v}_{i,u}$ are both set to 1, that is, $x^{i,u}_{j,v}$ and $x^{j,v}_{i,u}$ are both not from T. On the other hand, if $u = w_i$, then $x^{j,v}_{i,u}$ falls into T. Similarly, if $v = w_j$, then $x^{i,u}_{j,v}$ falls into T. This means that $u \neq w_i, v \neq w_j$. By definition of $\mathcal{B}(G, k)$ it holds that $i \neq j$. This leads us to a contradiction: either $x^{i,u}_{j,v} \in T$ (if $i < j$) or $x^{j,v}_{i,u} \in T$ (if $j < i$).

We have shown that $(A \wedge B)(T \to 0) = B(T \to 0) = 0$. It remains to show that $(A \wedge B)(T' \to 0) = 1$ for any proper subset T' of T. Consider any variable $x^{i',u'}_{j',v'}$ from $T \backslash T'$. Note that for every i all the variables with superscript i, w_i are not from T (if $x^{i,w_i}_{j,v} \in T$, then by definition of T it holds that $v = w_j$, but this either contradicts $i \neq j$ or $\{w_i, v\} \notin E$). This shows that $A(T \to 0) = 1$ and hence $A(T' \to 0) = 1$. On the other hand, $x^{i',u'}_{j',v'} \wedge x^{j',v'}_{i',u'}$ is true on the assignment $T' \to 0$ and so is B. To see this note that $x^{j',v'}_{i',u'}$ is not from T. Indeed, if $x^{j',v'}_{i',u'}$ is from T, then there are two cases:

- *the first case:* $u' = w_{i'}$. Since (i', u', j', v') is a quadriple from $\mathcal{A}(G, k)$, then $i' \neq j'$ and $u' = w_{i'}$ and v' are not connected by an edge. In particular, this means that $v' \neq w_{j'}$ and hence $x^{i',u'}_{j',v'}$ is not an element of T.
- *the second case:* $v' \neq w_{j'}, u' \neq w_{i'}$ and $j' < i'$. The first and the third inequalities again make it impossible for $x^{i',u'}_{j',v'}$ to be an element of T.

Thus we have showed that T is a maxterm. Let us define a minterm S which intersects with T in more than one point. Take any vertex y from $V \backslash \{w_1, \ldots, w_k\}$ (it is possible since by assumption $k < |V|$). Include in S all variables $x^{i,u}_{j,v}$ with $u = y$. First of all, let us show that $|S \cap T| > 1$. Consider variables $x^{1,y}_{2,y}$ and $x^{2,y}_{3,y}$. There are such variables since $\{y, y\} \notin E$ and by assumption $k \geq 3$. These two variables are clearly from S. Moreover, they are from T (since $y \notin \{w_1, w_2, w_3\}$, they satisfy the second condition from the definition of T).

It only remains to verify that S is a minterm. Set all the variables from S to 1 and set all the other variables to 0. Note that B is true on such assignment (because $\{\{1, y\}, \{2, y\}\} \in \mathcal{B}(G, k)$ and $x^{1,y}_{2,y} \wedge x^{2,y}_{1,y}$ is true). Clearly, A is also true, since for every i every variable with superscript i, y is set to 1.

On the other hand, consider any variable $x^{i,y}_{j',v'}$ from S. Set all the variables from S, except $x^{i,y}_{j',v'}$, to 1 and set all the other variables (including $x^{i,y}_{j',v'}$) to 0. Observe that A is false on such assignment. Indeed, for every $u \in V$ the conjunction

$$\bigwedge_{\substack{(j,v): \\ (i,u,j,v) \in \mathcal{A}(G,k)}} x^{i,u}_{j,v}$$

is false. For $u = y$ this conjunction is false since $x^{i,y}_{j',v'}$ is set to 0; if $u \neq y$, then every variable in this conjunction is set to 0). $\square$

Acknowledgments. I would like to thank Alexander Shen and Nikolay Vereshchagin for help in writing this paper.

Appendix

A   Proof of Lemma 3

Let T be a maxterm of $C \vee D$ and $C_u, C_v \subset T$. For the sake of contradiction assume that there is $j \in \{1, \ldots, l\}$ such that $|(C_u \cup C_v) \cap D_j| \geq 2$. Pick any two distinct $p, q \in (C_u \cup C_v) \cap D_j$. Let us show that $(C \vee D)(T \backslash \{q\} \rightarrow 0) = 0$. To show that $C(T \backslash \{q\} \rightarrow 0) = 0$ observe that C_u or C_v does not contain q and hence $C_u \subset T \backslash \{q\}$ or $C_v \subset T \backslash \{q\}$. To show that $D(T \backslash \{q\} \rightarrow 0) = 0$ observe that $D(T \rightarrow 0) = 0$ and hence T intersects all sets $D_1, \ldots, D_l$. Since $q \in D_j$ and hence $q \notin D_i$ for all $i \neq j$, the set $T \backslash \{q\}$ still intersects D_i for all $i \neq j$. And it intersects D_j since $p \in C_u \cup C_v \subset T$ and p was not removed from T. Since T is a maxterm, this is a contradiction.

On the other hand, assume that for every $j \in \{1, \ldots, l\}$ it holds that $|(C_u \cup C_v) \cap D_j| \leq 1$. We have to find a maxterm T that includes both C_u, C_v. Start with $T = C_u \cup C_v$. Then for all j such that D_j does not intersect $C_u \cup C_v$ pick a variable from D_j and include it in T. In this way we make T intersect every D_j in exactly one point. In particular, $D(T \rightarrow 0) = 0$ and $C(T \rightarrow 0) = 0$. On the other hand, every proper subset T' of T is disjoint with at least one D_j and hence $D(T' \rightarrow 0) = 1$. This shows that T is a maxterm.

B   Proof of Lemma 4

Since $C \rightarrow D$ is not a tautology, C_i is non-empty and intersects with some D_j. Further, without loss of generality we may assume that:

- $i = 1$;
- C_1 intersects with $D_1, \ldots D_r$ and C_1 is disjoint with $D_{r+1}, \ldots D_l$ for some $1 \leq r \leq l$;
- $C_2, \ldots, C_s$ all intersect with $D_1 \cup D_2 \cup \ldots \cup D_r$ and $C_{s+1}, \ldots, C_m$ are all disjoint with $D_1 \cup D_2 \cup \ldots \cup D_r$ for some $1 \leq s \leq m$.

From the fact that $D_1 \cup D_2 \cup \ldots \cup D_l = \{x_1, \ldots, x_n\}$ we may derive that:

$$C_{s+1}, \ldots, C_m \subset D_{r+1} \cup \ldots \cup D_l. \tag{1}$$

Define an auxiliary CNF $\widehat{C} = C_{s+1} \wedge \ldots \wedge C_m$ and an auxiliary DNF $\widehat{D} = D_{r+1} \vee \ldots \vee D_l$. Note that $\widehat{C}$ and $\widehat{D}$ are over variables from $D_{r+1} \cup \ldots \cup D_l$ (this follows from (1)).

We claim that there exists T such that T is a maxterm of $C \vee D$ and $C_1 \subset T$ if and only if $\widehat{C} \rightarrow \widehat{D}$ is not a tautology (the latter by Lemma 1 can be verified in polynomial time).

($\Leftarrow$) Assume that $\widehat{C} \rightarrow \widehat{D}$ is not a tautology. Then there exists $\widehat{T} \subset D_{r+1} \cup \ldots \cup D_l$ such that

$$C_{s+1} \not\subset \widehat{T}, \ldots, C_m \not\subset \widehat{T}; \tag{2}$$

$$|D_{r+1} \cap \widehat{T}| = 1, \ldots, |D_l \cap \widehat{T}| = 1; \tag{3}$$

(take minimal $\widehat{T} \subset D_{r+1} \cup \ldots \cup D_l$ such that $\widehat{C}(\widehat{T} \rightarrow 0) = 1$ and $\widehat{D}(\widehat{T} \rightarrow 0) = 0$).

Let us show that $T = \widehat{T} \cup C_1$ is the maxterm of $C \vee D$. First of all, let us verify that $(C \vee D)(T \rightarrow 0) = 0$. Indeed,

- $C(T \to 0) = 0$ because $C_1 \subset T$;
- $D(T \to 0) = 0$ because every $D_1, \ldots, D_l$ intersects with T; namely, $D_1, \ldots D_r$ intersect C_1 and $D_{r+1}, \ldots, D_l$ intersect $\widehat{T}$.

Now, assume that $T' \subset T$ and $(C \vee D)(T' \to 0) = 0$. Let us show that this is possible only when $T' = T$.

Since $D(T' \to 0) = 0$, we have that T' intersects with every $D_1, \ldots, D_l$. From the fact that C_1 is disjoint with $D_{r+1}, \ldots, D_l$ and from (3) it follows that $\widehat{T} \subset T'$.

It remains to show that $C_1 \subset T'$. This follows from the assumption that $C(T' \to 0) = 0$. Indeed, then at least one clause of C should be the subset of T'. Assume that this clause is C_u. If $C_u \neq C_1$, then $C_u \subset \widehat{T}$. There are two cases:

- *The first case.* Assume that $C_u \in \{C_2, \ldots, C_s\}$. Then $C_u \subset \widehat{T} \subset D_{r+1} \cup \ldots \cup D_l$ intersects with $D_1 \cup D_2 \cup \ldots \cup D_r$, but the latter is impossible.
- *The second case.* Assume that $C_u \in \{C_{s+1}, \ldots, C_m\}$. This case contradicts (2).

$(\Rightarrow)$ Assume that T is the maxterm of $C \vee D$ such that $C_1 \subset T$. Define $\widehat{T} = T \backslash C_1$. Later we will show that $\widehat{C}(\widehat{T} \to 0) = 1$, $\widehat{D}(\widehat{T} \to 0) = 0$ and hence $\widehat{C} \to \widehat{D}$ is not a tautology. But at first we should verify that $\widehat{T} \subset D_{r+1} \cup \ldots \cup D_l$ (recall that $\widehat{C}, \widehat{D}$ are over variables from $D_{r+1} \cup \ldots \cup D_l$).

To show that $\widehat{T} \subset D_{r+1} \cup \ldots \cup D_l$ assume for contradiction that $\widehat{T}$ intersects $D_1 \cup D_2 \cup \ldots \cup D_r$ and let q be the variable which lies in their intersection. Note that $q \notin C_1$ (this is because $q \in \widehat{T} = T \backslash C_1$). Let us demonstrate that for such q we have that $(C \vee D)(T \backslash \{q\} \to 0) = 0$ (this is already a contradiction since T is a maxterm). Indeed, $C(T \backslash \{q\} \to 0) = 0$ since $C_1 \subset T \backslash \{q\}$. Further, we should show that $T \backslash \{q\}$ intersects every $D_1, \ldots D_l$ and hence $D(T \backslash \{q\} \to 0) = 0$. Indeed:

- $T \backslash \{q\}$ intersects $D_1, \ldots, D_r$ because of C_1;
- $T \backslash \{q\}$ intersects $D_{r+1}, \ldots, D_l$ because of the following two reasons: (a) $D(T \to 0) = 0$ and hence T intersects every $D_{r+1}, \ldots, D_l$; (b) q is not in $D_{r+1} \cup \ldots \cup D_l$.

Thus it remains to show that $\widehat{C}(\widehat{T} \to 0) = 1$ and $\widehat{D}(\widehat{T} \to 0) = 0$. To show that the first equality is true assume for contradiction that there is $C_u \in \{C_{s+1}, \ldots, C_m\}$ such that $C_u \subset \widehat{T} = T \backslash C_1$. But then $C_u \subset T$. This contradicts the assumption that there is no maxterm of $C \vee D$ which contains two distinct clauses of C.

To show that the second equality ($\widehat{D}(\widehat{T} \to 0) = 0$) is true, observe that $\widehat{T}$ intersects every $D_{r+1}, \ldots, D_l$. This is true because $D(\widehat{T} \to 0) = 0$ and hence T intersects $D_{r+1}, \ldots, D_l$; but C_1 by assumption is disjoint with $D_{r+1}, \ldots, D_l$ and hence $T \backslash C_1$ still intersects them.

C   Proof of Lemma 5

If there is C_i such that $a, b \in C_i$ or $\{a, b\} \not\subset C_1 \cup \ldots \cup C_m$, then no left minterm S can contain both a and b. From now we assume that this is not the case, i.e. there is no C_i which contains both a and b and $a, b \in C_1 \cup \ldots \cup C_m$. Let $\widehat{C} \vee \widehat{D}$ be obtained from $C \vee D$ by setting a, b to 1. In other words, $\widehat{C}$ is obtained from C by erasing all clauses containing a or b and $\widehat{D}$ is obtained from D by erasing a and b. Assume without loss of generality that C_1, C_2 are erased clauses, $a \in C_1, b \in C_2$ and $D_1, \ldots, D_r$ are conjunctions containing a or b (note that r is either 1 or 2). Then $\widehat{C}$ and $\widehat{D}$ can be written as

$$\widehat{C} = C_3 \wedge \ldots \wedge C_m,$$

$$\widehat{D} = (D_1 \backslash \{a, b\}) \vee \ldots \vee (D_r \backslash \{a, b\}) \vee D_{r+1} \vee \ldots \vee D_l.$$

We assert that there is a left minterm S containing a and b iff $\widehat{C} \rightarrow \widehat{D}$ is not a tautology. The latter by Lemma 1 can be verified in polynomial times.

($\Leftarrow$) Assume that $\widehat{C} \rightarrow \widehat{D}$ is not a tautology. Take minimal $\widehat{S} \subset \{x_1, \ldots, x_n\} \backslash \{a, b\}$ such that $\widehat{C}(\widehat{S} \rightarrow 1) = 1, \widehat{D}(\widehat{S} \rightarrow 1) = 0$. Obviously, such $\widehat{S}$ satisfies the following two conditions:

$$\widehat{S} \subset C_3 \cup \ldots \cup C_m, \qquad |\widehat{S} \cap C_3| = 1, \ldots, |\widehat{S} \cap C_m| = 1, \tag{4}$$

$$D_1 \backslash \{a, b\} \not\subset \widehat{S}, \ldots, D_r \backslash \{a, b\} \not\subset \widehat{S}, \; D_{r+1} \not\subset \widehat{S}, \ldots, D_l \not\subset \widehat{S}. \tag{5}$$

Now, define $S = \widehat{S} \cup \{a, b\}$. Let us show that S is a left minterm of $C \vee D$. From (5) it follows that there is no $j \in \{1, \ldots, l\}$ such that $D_j \subset S$. Hence S contains no right set as a proper subset. Thus it remains to show by Lemma 2 that S is a left set. Since a, b are from $C_1 \cup \ldots \cup C_m$, we have that

$$S \subset (\{a, b\} \cup C_3 \cup \ldots \cup C_m) \subset C_1 \cup \ldots \cup C_m.$$

Moreover, S intersects every clause of C in exactly one point. For $C_3, \ldots, C_m$ this follows from (4) and from the fact that $C_3, \ldots, C_m$ contain neither a nor b. For C_1, C_2 this is true because: (a) $\widehat{S}$ is disjoint with C_1, C_2; (b) $a \in C_1, b \in C_2$.

($\Rightarrow$) Assume that there is a left minterm S of $C \vee D$ containing a and b. Define $\widehat{S} = S \backslash \{a, b\}$. Let us show that $\widehat{S}$ intersects every $C_3, \ldots C_m$. Indeed, this is true for S and a, b are not from $C_3 \cup \ldots \cup C_m$. Hence $\widehat{C}(\widehat{S} \rightarrow 1) = 1$. On the other hand, $\widehat{D}(\widehat{S} \rightarrow 1) = 0$, since:

- $D_1 \backslash \{a, b\} \not\subset \widehat{S}, \ldots D_r \backslash \{a, b\} \not\subset \widehat{S}$ because otherwise at least one $D_1, \ldots, D_r$ is the subset of $S = \widehat{S} \cup \{a, b\}$;
- $D_{r+1} \not\subset \widehat{S}, \ldots, D_l \not\subset \widehat{S}$ because it is true even for S.

Thus $\widehat{C} \rightarrow \widehat{D}$ is not a tautology.

References

1. Elbassioni, K., Makino, K., Rauf, I.: On the readability of monotone Boolean formulae. J. Comb. Optim. **22**(3), 293–304 (2011)
2. Gurvich, V.: It is a coNP-complete problem to decide whether a positive ∨-∧ formula of depth 3 defines a read-once or respectively quadratic Boolean function. RUTCOR Research Report (2010)
3. Golumbic, M.C., Gurvich, V.: Read-once functions. In: Boolean Functions: Theory Algorithms and Applications (2009)
4. Golumbic, M.C., Mintz, A., Rotics, U.: An improvement on the complexity of factoring read-once Boolean functions. Discrete Appl. Math. **156**(10), 1633–1636 (2008)
5. Gurvich, V.A.: Repetition-free Boolean functions. Usp. Mat. Nauk **32**(1), 183–184 (1977)
6. Büning, H., Lettmann, T.: Propositional Logic: Deduction and Algorithms, vol. 48. Cambridge University Press, Cambridge (1999)

MAX-CUT ABOVE SPANNING TREE
is Fixed-Parameter Tractable

Jayakrishnan Madathil[1]([✉]), Saket Saurabh[1,2], and Meirav Zehavi[3]

[1] The Institute of Mathematical Sciences, HBNI, Chennai, India
{jayakrishnanm,saket}@imsc.res.in
[2] Department of Informatics, University of Bergen, Bergen, Norway
saket.saurabh@uib.no
[3] Department of Computer Science, Ben-Gurion University, Beersheba, Israel
meiravze@bgu.ac.il

Abstract. Every connected graph on n vertices has a cut of size at least $n-1$. We call this bound the 'spanning tree bound'. In the MAX-CUT ABOVE SPANNING TREE (MAX-CUT-AST) problem, we are given a connected n-vertex graph G and a non-negative integer k, and the task is to decide whether G has a cut of size at least $n-1+k$. We show that MAX-CUT-AST admits an algorithm that runs in time $\mathcal{O}(8^k n^{\mathcal{O}(1)})$, and hence it is fixed parameter tractable with respect to k. Furthermore, we show that MAX-CUT-AST has a polynomial kernel of size $\mathcal{O}(k^5)$.

1 Introduction

The classic MAX-CUT is one of Karp's original 21 NP-complete problems [8]. Since the 1960s, this problem has been studied extensively in the areas of graph theory, combinatorics and graph algorithms (see, e.g., the well-known survey [11]). A cut in a graph G is defined by a bipartition of the vertex-set of G. Specifically, the cut corresponding to a bipartition (A, B) is the set of edges with one endpoint in A and the other endpoint in B. The input for MAX-CUT consists of a graph G and a non-negative integer k, and the task is to determine whether G has a cut of size at least k. It is now textbook knowledge that every graph with n vertices and m edges has a cut of size at least $m/2$, and that such a cut can be found in polynomial time. This result was first shown by Erdős in 1965 [6]. Edwards [4,5] showed that this lower bound of $m/2$ can be improved to $m/2 + (n-1)/4$ if the graph is connected. Nowadays, this famous lower bound is called the Erdős-Edwards bound. Furthermore, it is known that this bound is tight for infinitely many graphs.

In the area of parameterized complexity, the most obvious parameterization for MAX-CUT is by the cut size k. More precisely, we would like to determine in time $f(k)n^{\mathcal{O}(1)}$ whether the input graph G has a cut of size at least k. In light of the lower bound $m/2$, if $k \leq m/2$, then we can immediately conclude that G has a cut of size at least k, and otherwise $m < 2k$, i.e., the total number of edges in the graph is bounded by a function of k, and we can by brute force find the maximum sized cut in G. The entire problem, thus, becomes trivial.

F. V. Fomin and V. P. Podolskii (Eds.): CSR 2018, LNCS 10846, pp. 244–256, 2018.
https://doi.org/10.1007/978-3-319-90530-3_21

In order to address this triviality, Mahajan and Raman [9] introduced the idea of parameterizing "above guaranteed" lower bounds. Instead of asking if G has a cut of size at least k, they considered the question whether G has a cut of size at least $m/2 + k$. In particular, they showed that the latter question can be decided in time $2^{\mathcal{O}(k)}n^{\mathcal{O}(1)}$. In light of this result as well as the higher lower bound above, Crowston et al. [3] further asked the natural question of MAX-CUT ABOVE ERDŐS-EDWARDS. This paper, originally published in 2012 [2], has already become one of the most well known results in Parameterized Complexity concerning above guarantee parameterization. Specifically, Crowston et al. [3] showed that we can determine whether a given connected graph G has a cut of size at least $m/2 + (n-1)/4 + k$ in time $2^{\mathcal{O}(k)}n^{\mathcal{O}(1)}$, and furthermore, this problem admits a kernel of size $\mathcal{O}(k^5)$. These bounds were improved by Etscheid and Mnich [7], as well as extended by Mnich et al. [10].

While the important results above received significant attention, another central natural bound has so far been overlooked from the perspective of parameterized complexity: every connected graph on n vertices has a cut of size at least $n - 1$. To see this, first note that every connected graph G has a spanning tree T (containing $n - 1$ edges). Second, note that a tree is 2-colorable. Hence, it is simple to partition the vertex set of G into two sets in such a way that all edges of T fall into the cut. We refer to this folklore lower bound as the *spanning tree bound*. We stress that this bound is incomparable to the Erdős-Edwards bound. Specifically, whenever the average degree of the input graph is smaller than 3, it holds that $n - 1 > m/2 + (n-1)/4$, i.e., spanning tree bound gives a better guarantee on the cut size.

In this paper, we show that MAX-CUT ABOVE SPANNING TREE (MAX-CUT-AST) is also fixed parameter tractable and admits a polynomial sized kernel. Formally, our problem is stated as follows.

MAX-CUT ABOVE SPANNING TREE (MAX-CUT-AST)   **Parameter:** k
Input: A connected graph G and a non-negative integer k.
Question: Does G have a cut of size at least $n - 1 + k$?

Specifically, our contribution is twofold. First, we prove that MAX-CUT-AST admits a parameterized algorithm that runs in time $8^k n^{\mathcal{O}(1)}$. Second, we prove that MAX-CUT-AST admits a kernel of size $\mathcal{O}(k^5)$. We remark that our algorithm is essentially optimal, since unless the Exponential Time Hypothesis (ETH) fails, MAX-CUT-AST cannot be solved in time $2^{o(k)}n^{\mathcal{O}(1)}$. Indeed, as observed in [3] (from a reduction by [1]), unless the ETH fails, MAX-CUT ON CONNECTED GRAPHS, where the input consists of a connected graph G on n vertices a non-negative integer t and the question is to decide whether G has a cut of size at least t, cannot be solved in time $2^{o(t)}n^{\mathcal{O}(1)}$. Note that an instance (G, t) MAX-CUT ON CONNECTED GRAPHS is equivalent to the instance (G, k) of MAX-CUT-AST, where $k = t - (n - 1)$. Therefore, if MAX-CUT-AST were to have an algorithm with running time $2^{o(k)}n^{\mathcal{O}(1)}$, then MAX-CUT ON CONNECTED GRAPHS would admit an algorithm with running time $2^{o(t)}n^{\mathcal{O}(1)}$.

Due to lack of space, proofs of statements marked by stars ($\star$) are deferred to the full version of this paper.

Overview of Our Technique. On a high-level, our technique adapts the strategy of [3]. However, our arguments significantly differ in several basic aspects stemming from the fact that the spanning tree bound is high (in comparison to the maximum size of a cut) when the graph is *sparse* but low when the graph is *dense*, while the behavior of Erdős-Edwards bound is the complete opposite. We thus believe that our results and those of [3] complement each other in the sense that together they present a more complete picture of the complexity of MAX-CUT parameterized above a guarantee. We start our analysis with new structural results, which are particularly relevant to our kernel. In particular, we introduce the notion of an *even sunflower*. This notion is a curious modification of a classic sunflower (with a core of size at most 2) where every petal is of even length and an additional restriction concerning the interaction between the core and the petal is satisfied. The rationale behind the introduction of such a notion is that even cycles are more advantageous than odd cycles. As a simple illustrative example, say we only have one even cycle. Then, it is easy to see that we can choose a spanning tree that excludes exactly one edge of that even cycle, and then "win 1 over" the spanning tree bound as the excluded edge will necessarily lie across the cut. Delicate differentiation of the analysis of even and odd cycles is present throughout our paper. We remark that in [3], cycles do not play any such role.

For our algorithm, we begin by applying a set of one-way reduction rules, so that in polynomial time we can either conclude that G has a cut of size at least $n - 1 + k$ or find a set $S \subseteq V(G)$ such that $|S| \leq 3k$ and every 2-connected component of $G - S$ is a clique or a cycle. In comparison to [3], we introduce one new rule as well as modify those that we do adapt, and present a new analysis. In particular, we can only guarantee that every 2-connected component of $G - S$ is either a clique or a cycle, while Crowston et al. [3] could guarantee that every such component is simply a clique. We proceed by "guessing" to which side each vertex in S should belong in some bipartition defining a solution (if one exists). This step is captured by the definition of a weighted variant of MAX-CUT as in [3]. In our case, we need to solve the weighted problem in polynomial time on a broader class of graphs, as some blocks can be cycles.

For our kernelization algorithm, almost all of the reduction rules to bound $G - S$ are completely new. In comparison to [3], we face the following difficulties. First, while [3] can easily get rid of vertices of degree 2, for us, the analysis of such vertices is non-trivial. To see an example of the kind of "trouble" degree-2 vertices cause, we remark that we are able to devise a reduction rule (Rule 10) that gets rid of some degree-2 vertices if three such vertices appear consecutively on a path on five vertices in $G - S$, but if we only demanded that two such vertices appear consecutively on a path on four vertices in $G - S$, the argument would already fail. This particular situation is related to parity arguments. The second difficulty in comparison to [3] that we would like to highlight is rooted in the following fact: the measure of [3] consists of both m and n. Thus, Crowston et al. [3] have freedom in their analysis to, say, add edges to the graph, an operation that ensures that their measure would not increase when applying a reduction rule. We do not have this freedom—say, if we delete one vertex, then we must directly ensure that k would be decreased by at least 1 without being able to compensate for a further decrease in k by modifying m appropriately.

2 Preliminaries and Structural Results

Given $A' \subseteq A$ and a function $f : A \to B$, $f|_{A'}$ denotes the restriction of f to A'.

Graph Notation. All graphs in this paper are simple and undirected. For a graph G, we denote by $V(G)$ and $E(G)$ the vertex set and the edge set of G, respectively. A *cut* of G is a function $f : V(G) \to \{0, 1\}$. An edge $uv \in E(G)$ is *satisfied* by a cut f if $f(u) \neq f(v)$. The *size* of a cut f, denoted by $||f||$, is the number of edges satisfied by f, i.e., $||f|| = |\{uv \in E(G)|f(u) \neq f(v)\}|$. A *block* of G is a 2-connected component of G. Note that if $v \in V(G)$ is contained in two different blocks, then v is a cut vertex of G. For a graph G and $x, y \in V(G)$, $d_G(x)$ denotes the degree of x in G, and $\text{dist}_G(x, y)$ denotes the distance (i.e., number of edges on a shortest path) between x and y in G.

We define a class of graphs called clique-cycle-forest as follows.

Definition 1. *The class of clique-cycle-forests is defined as follows. A clique is a clique-cycle-forest, and so is a cycle. The disjoint union of two clique-cycle-forests is a clique-cycle-forest. In addition, a graph formed from a clique-cycle-forest by identifying two vertices, each from a different (connected) component, is also a clique-cycle-forest.*

Note that clique-cycle-forests are exactly the graphs in which every block is a clique or a cycle.

Simple Bound. Recall that a connected graph G on n vertices has a cut of size at least $n - 1$. We now claim that if G contains an even cycle, then G has a cut of size $(n - 1) + 1 = n$. To see this, let C be an even cycle in G and uv be an edge of C. Consider a spanning tree T of G that contains all edges of C except uv. Let f be a cut of G that satisfies all edges of T. Since $C - uv$ is an odd length path, and since f satisfies all edges of $C - uv$, it must be the case that $f(u) \neq f(v)$. This means, in addition to the $n - 1$ edges of T, f satisfies uv as well. Therefore, $||f|| \geq n - 1 + 1 = n$. The key point here is that "one even cycle means one additional edge in the cut." This modest combinatorial observation immediately leads to the following lemma.

Lemma 1. *Let G be a connected graph and k be a non-negative integer. If G contains k vertex disjoint even cycles, then (G, k) is a yes-instance of* Max-Cut-AST.

Proof. Take a spanning tree T of G that contains all but one edge of each of the k even cycles. Such a spanning tree exists because the cycles are vertex disjoint. Then any cut of G that satisfies all edges of T will satisfy *all* edges of each of the k even cycles.   □

The proof of Lemma 1 relies only on the existence of k even cycles and a spanning tree T with the property that a cut of G that satisfies all edges of T should automatically satisfy all edges of each of the k cycles. This observation leads us to the definition below.

Sunflower-Based Bound. We define a family of even cycles, which we call an *even sunflower*, that can guarantee a spanning tree with the property above.

Definition 2. *Let C be a family of cycles in a graph G and let $X \subseteq V(G)$ be of size at most 2. The family C is a* sunflower with core X *if for every distinct $C, C' \in C$, $V(C) \cap V(C') = X$. The* size *of the sunflower C is the number of cycles in C.*

We further need a special kind of a sunflower, defined as follows.

Definition 3. *A sunflower C with core X is said to be an* even sunflower *if*

(i) *every cycle in C is of even length, and*
(ii) *if $x, y \in X$, then in every cycle in C, the distance between x and y is even (i.e., $dist_C(x, y)$ is even for every $C \in C$).*

Let C be an even sunflower with core X. First, note that if $x, y \in X$, then condition (ii) in the above definition implies that x and y are not adjacent in any of the cycles in C (i.e., $xy \notin E(C)$ for every $C \in C$). Moreover, if $X = \emptyset$, then C is a collection of vertex disjoint even cycles, and if $|X| = 1$, then $x = y$, and C is a collection of even cycles that intersect only at x. Second, note that every $C \in C$ can be partitioned into two internally vertex disjoint $x - y$ paths, and $dist_C(x, y)$ is the length of the shorter of those two paths. Since C is an even sunflower, $dist_C(x, y)$ is even (i.e., the length of the shorter $x - y$ path is even), and since C is of even length, the length of the longer $x - y$ path in C must also be even.

Lemma 2. *Let G be a connected graph and let C be an even sunflower in G with core X. Then, G has a spanning tree T with the following property: any cut of G that satisfies all edges of T, also satisfies all edges of all the cycles in C.*

Proof. If $X = \emptyset$, then the cycles in C are vertex disjoint, and the lemma follows from Lemma 1. Next, assume that $X \neq \emptyset$. Let $x \in X$. Grow a spanning tree T as follows. Start with $E(T) = \emptyset$. Fix one cycle, say $\widetilde{C} \in C$. Add to T all edges of $\widetilde{C}$, except exactly one of the two edges incident on x. Now, for each $C \in C$, $C \neq \widetilde{C}$, do the following: if $|X| = 1$, then add to T all edges of C, except exactly one of the two edges incident on x; if $|X| = 2$, say $X = \{x, y\}$, then add to T all edges of C except the two edges incident on x. Finally, extend T to a spanning tree of G.

To see that T has the required property, consider a cut f of G that satisfies all edges of T. Assume $f(x) = 1$. For $C \in \mathcal{C}$, let u_C and v_C be the neighbors of x in C. Suppose $|X| = 1$. Then, for every $C \in \mathcal{C}$, exactly one of the two edges xu_C and xv_C is missing from T. Suppose xu_C is the missing edge. Then, T contains $C - xu_C$, and hence f satisfies all the edges of the odd length path $C - xu_C$. Therefore, it must be the case that $f(x) \neq f(u_C)$, i.e., f satisfies the edge xu_C. Using similar arguments, because of the discussion before the lemma, we can also show that f satisfies all edges of the cycles in $\mathcal{C}$ when $|X| = 2$.   □

Consequently, we have the following result.

Lemma 3. *Let G be a connected graph and k be a non-negative integer. If G contains an even-sunflower of size at least k, then (G, k) is a yes-instance of* MAX-CUT-AST.

We can further generalize the above lemma. It is not necessary that all the k cycles should come from the same even sunflower. It is not difficult to see that the k cycles may belong to different even sunflowers, and (G, k) will still be a yes-instance, provided no two cycles belonging to different even sunflowers share a vertex. We will use this observation in Sect. 4 to bound the kernel size. For future reference, we summarize this discussion in the following proposition.

Proposition 1. *Let G be a connected graph and k be a non-negative integer.*

1. *If G has even sunflowers $\mathcal{C}_1, \mathcal{C}_2, \ldots, \mathcal{C}_r$ such that for all distinct $i, j \in [r]$, $\sum_{i=1}^{r} |\mathcal{C}_i| \geq k$ and $\left(\bigcup_{C \in \mathcal{C}_i} V(C) \right) \cap \left(\bigcup_{C' \in \mathcal{C}_j} V(C') \right) = \emptyset$, then (G, k) is a yes-instance of* MAX-CUT-AST.
2. *If G contains k even cycles (not necessarily disjoint) and a spanning tree that contains all but one edge of each of the k even cycles, then (G, k) is a yes-instance of* MAX-CUT-AST.

Structural Results for Kernelization. We conclude this section with the following two lemmas that will be used in Sect. 4 to bound the kernel size.

Lemma 4 ($\star$). *Let $\widetilde{G}$ be a connected graph and let $s \in V(\widetilde{G})$ be such that s is not a cut vertex of $\widetilde{G}$. Let $x, y, z \in V(\widetilde{G})$ be three distinct neighbors of s. Then there exists an even cycle containing s in $\widetilde{G}$.*

Lemma 5 ($\star$). *Let $\widetilde{G}$ be a connected graph on n vertices and let k be a non-negative integer. Let $s \in V(\widetilde{G})$ be a vertex that is not a cut vertex of $\widetilde{G}$. If $d_{\widetilde{G}}(s) \geq 2k + 1$, then $\widetilde{G}$ has a cut of size at least $n - 1 + k$.*

3 FPT Algorithm for MAX-CUT ABOVE SPANNING TREE

In this section, we show that MAX-CUT ABOVE SPANNING TREE admits an algorithm that runs in time $2^{3k} n^{\mathcal{O}(1)}$. The following lemma is the first building block of our algorithm.

Lemma 6. *Given an instance (G, k) of* MAX-CUT-AST, *in polynomial time we can either conclude that (G, k) is a yes-instance or find a set $S \subseteq V(G)$ such that $|S| \leq 3k$ and $G - S$ is a clique-cycle-forest.*

In order to prove the lemma, we apply a set of one-way reduction rules (Rules 1–5) to (G, k). By a one-way reduction rule, we mean a rule that, when applied to (G, k), produces an instance (G', k') of MAX-CUT-AST such that if (G', k') is a yes-instance, then (G, k) is a yes-instance. However, the converse need not hold, i.e., if (G', k') is a no-instance, then (G, k) may or may not be a no-instance. While applying these rules, we ensure that G' is connected and $k' \leq k$. We then show that after an exhaustive application of these rules, we will have $G' = K_1$, (where K_1 is the clique on one vertex). Moreover, if we also have $k' \leq 0$, then (G', k') is yes-instance, and because the rules we applied were one-way safe, we can conclude that (G, k) is a yes-instance as well. Otherwise, the rules will have marked a set of vertices S such that $|S| \leq 3k$ and $G - S$ is a clique-cycle-forest, i.e., every block of $G - S$ is a clique or a cycle.

Let (G, k) be an instance of MAX-CUT-AST. Then G is a connected graph. If $a, b, c \in V(G)$ are such that (i) $ab, bc \in E(G)$ and $ac \notin E(G)$, and (ii) $G - \{a, b, c\}$ is connected, then the path abc is called a *connected P_3*. Let abc be a connected P_3 in G. If degree of at least one of a, b or c is at least 3, then we call abc a *good P_3*; otherwise we call abc a *bad P_3*. In the rules below, as X is a connected component, it is implicitly assumed to be non-empty.

Exhaustively apply the following reduction rules to (G, k).

Rule 1:	Apply to a connected graph G with $v \in V(G)$, and $X \subseteq V(G)$ such that X is a connected component of $G - \{v\}$ and $X \cup \{v\}$ is a clique.
Remove:	All vertices in X.
Mark:	Nothing.
Parameter:	Reduce k by $|X|^2/4 - |X|/2 + 1/4$ if $|X|$ is odd, and by $|X|^2/4 - |X|/2$ if $|X|$ is even.
Rule 2:	Apply to a connected graph G with $v \in V(G)$, and $X \subseteq V(G)$ such that X is a connected component of $G - \{v\}$, and $G[X \cup \{v\}]$ is an induced cycle.
Remove:	All vertices in X.
Mark:	Nothing.
Parameter:	Reduce k by 1 if $|X|$ is odd, and leave k the same if $|X|$ is even.
Rule 3:	Apply to a connected graph G reduced by Rules 1 and 2 with $v \in V(G)$, and $X \subseteq V(G)$ such that X is a connected component of $G - \{v\}$, and X is a clique.
Remove:	All vertices in X.
Mark:	The vertex v.
Parameter:	Reduce k by $\lfloor |X|^2/4 \rfloor + \min\{d_{G[X \cup \{v\}]}(v), \lceil |X|/2 \rceil\} - |X|$.

Rule 4:	Apply to a connected graph G reduced by Rules 1 and 2 with $a, b, c \in V(G)$ such that abc is a good P_3. Let $x \in \{a, b, c\}$ be such that $d(x) = \max\{d(a), d(b), d(c)\}$.
Remove:	The vertices a, b, c.
Mark:	The vertices a, b, c.
Parameter:	Reduce k by $\lceil (d(x) - 2)/2 \rceil$.

Rule 5:	Apply to a connected graph G reduced by Rules 1 and 2 with $a, c \in V(G)$ such that $ac \notin E(G)$, $G - \{a, c\}$ has exactly two connected components, X and Y, where $|Y| \geq 2$, and $Y \cup \{a\}$ and $Y \cup \{c\}$ are cliques.
Remove:	All vertices in $Y \cup \{a, c\}$.
Mark:	The vertices a, c.
Parameter:	Reduce k by $\lceil |Y|/2 \rceil \cdot \lfloor |Y|/2 \rfloor + \lceil d_{G[X \cup \{a\}]}(a)/2 \rceil + \lceil d_{G[X \cup \{c\}]}(c)/2 \rceil - 2$.

Lemma 7 ($\star$). *Rules 1–5 are one-way safe.*

Proof (Proof Sketch). Here we consider only Rule 1. The proofs of the other rules can be found in the full version. Let (G', k') be the instance obtained from (G, k) by a single application of Rule 1. Observe first that G' is connected. In G, v is a cut vertex, removal of which disconnects $G - X$ from X. Hence for every pair of vertices $s, t \in V(G) - X$, no $s - t$ path passes through X. So $G - X$ remains connected even after the removal of X. Now we show that if (G', k') is a yes-instance, then so is (G, k).

Suppose $|X|$ is odd. (The case when $|X|$ is even is similar.) Then G' is a graph on $n - |X|$ vertices and $k' = k - |X|^2/4 + |X|/2 - 1/4$. Note that since $|X|$ is odd, $|X|^2/4 - |X|/2 + 1/4$ is an integer, and hence k' is also an integer. Suppose (G', k') is a yes-instance, and let $f' : V(G') \rightarrow \{0, 1\}$ be a cut of size at least $n - |X| - 1 + k'$. Assume without loss of generality that $f'(v) = 1$, where v is the vertex referred to in Rule 1. Since $G[X]$ is a clique, its vertices can be partitioned into two parts, say X_0 and X_1 of sizes $(|X| + 1)/2$ and $(|X| - 1)/2$, respectively, so that the resulting cut will have size $((|X| + 1)(|X| - 1))/4$. Extend f' to a cut f of G by defining $f(u) = f'(u)$, if $u \in V(G) \backslash X$; $f(u) = 0$ if $u \in X_0$; and $f(u) = 1$, if $u \in X_1$. Then,

$$\|f\| \geq \underbrace{n - |X| - 1 + k'}_{\text{from the cut } f'} + \underbrace{(|X| + 1)/2}_{\text{edges between } v \text{ and } X_0} + \underbrace{((|X| + 1)(|X| - 1))/4}_{\text{edges between } X_0 \text{ and } X_1}$$

$$= n - |X| - 1 + k - |X|^2/4 + |X|/2 - 1/4 + |X|/2 + 1/2 + |X|^2/4 - 1/4$$

$$= n - 1 + k.$$

$\square$

After an exhaustive application of these rules, we are left with a reduced instance, say (G', k'), where G' is a connected graph and $k' \leq k$. Let S be the set of the marked vertices. Each application of Rules 3, 4 and 5 marks at most three vertices, and reduces the parameter k by at least 1. Therefore, if $|S| > 3k$

then $k' \leq 0$, and since G' is connected, we can conclude that (G', k') is a yes-instance, which in turn implies (G, k) is a yes-instance. So assume $|S| \leq 3k$.

Lemma 8 ($\star$). *Let G be a connected graph. If $G \neq K_1$, then at least one of Rules 1–5 is applicable to G.*

Lemma 9. *Let (G', k') be the instance obtained from (G, k) by exhaustively applying Rules 1–5, and let S be the set of vertices marked during the construction of (G', k'). Then every block of $G - S$ is either a clique or a cycle.*

Proof. Let ℓ be the length of the reduction from G to G'. Proof is by induction on ℓ. If $\ell = 0$, then $G = G' = K_1$ and $S = \emptyset$, and hence the statement of the lemma holds. Let G'' be a graph obtained from G by a single application of Rule 1, 2, 3, 4 or 5. It is enough to show that if every block of $G'' - S$ is a clique or a cycle, then so is every block of $G - S$.

If G'' is obtained by an application of Rule 1 or 2, then $G - S$ can be formed from $G'' - S$ by adding either (i) a disjoint clique or cycle, and identifying one of its vertices with the vertex v in G'' if $v \notin S$, where v is vertex referred to in Rules 1 and 2; or (ii) a disjoint clique or a path if $v \in S$. For Rule 4, observe that $G - S = G'' - S$. For Rule 3 and 5, observe that $G - (G'' \cup S)$ is a clique, and that S disconnects $G - (G'' \cup S)$ from G'. Therefore $G - S$ can be formed from $G'' - S$ by adding a disjoint clique. $\qquad\square$

Our goal is to show that any given cut of $G[S]$ can be optimally extended to a cut of G. Towards that end, consider the following problem, introduced by Crowston et al. in [2].

MAX-CUT-WITH-WEIGHTED-VERTICES
Input: A graph G with weight functions $w_0 : V(G) \to \mathbb{N} \cup \{0\}$ and $w_1 : V(G) \to \mathbb{N} \cup \{0\}$, and an integer $t \in \mathbb{N}$.
Question: Does there exist an assignment $f : V(G) \to \{0, 1\}$ such that $\sum_{xy \in E(G)} |f(x) - f(y)| + \sum_{f(x)=0} w_0(x) + \sum_{f(x)=1} w_1(x) \geq t$?

Crowston et al. [2] showed that MAX-CUT-WITH-WEIGHTED-VERTICES is polynomial time solvable when the input graph is a clique-forest (a graph in which every 2-connected component is a clique). For the sake of completeness and better understanding, we copy their result in its entirety.

Lemma 10 ([2]). MAX-CUT-WITH-WEIGHTED-VERTICES *is solvable in polynomial time if the input graph is a clique-forest.*

Proof. The algorithm consists of a polynomial time transformation that replaces an instance (G, w_0, w_1, t) with an equivalent instance (G', w_0', w_1', t') such that G' has fewer vertices than G. By applying this transformation at most $|V(G)|$ times, we reach a trivial instance, and thus have a polynomial time algorithm to decide MAX-CUT-WITH-WEIGHTED-VERTICES on (G, w_0, w_1, t). We may assume that G is connected, as otherwise we can handle each component of G separately. Let $X \cup \{r\}$ be the vertices of a leaf-block in G, with r a cut-vertex of G (unless G consists of a single block, in which case let r be an arbitrary vertex and $X = V(G) \setminus \{r\}$.) Recall that by definition of a clique-forest, $X \cup \{r\}$ is a clique. For each possible assignment to r, we will in polynomial time calculate the optimal extension to the vertices in X. (This optimal extension depends only on the assignment to r, since no other vertices are adjacent to vertices in X.) We can then remove all the vertices in X, and change the values of $w_0(r)$ and $w_1(r)$ to reflect the optimal extension for each assignment. Suppose we assign r the value 1. Let $\varepsilon(x) = w_1(x) - w_0(x)$ for each $x \in X$. Now arrange the vertices of X in order $x_1, x_2, \ldots, x_{n'}$, where $n' = |X|$, such that if $i < j$ then $\varepsilon(x_i) \geq \varepsilon(x_j)$. Observe that there is an optimal assignment for which x_i is assigned 1 for every $i \leq t$, and x_i is assigned 0 for every $i > t$, for some $0 \leq t \leq n'$. (Consider an assignment for which $f(x_i) = 1$ and $f(x_j) = 0$, for $i < j$, and observe that switching the assignments of x_i and x_j will increase $\sum_{f(x)=0} w_0(x) + \sum_{f(x)=1} w_1(x)$ by $\varepsilon(x_i) - \varepsilon(x_j)$.) Therefore we only need to try $n' + 1$ different assignments to the vertices in X in order to find the optimal cut when $f(r) = 1$. Let A be the value of this optimal assignment (over $X \cup \{r\}$.) By a similar method we can find the optimal assignment when r is assigned 0. Let the number of satisfied edges in this cut be B. Now remove the vertices in X and incident edges, and let $w_1(r) = A$, and let $w_0(r) = B$.   □

We now extend this result to clique-cycle-forests.

Lemma 11. MAX-CUT-WITH-WEIGHTED-VERTICES *is solvable in polynomial time if the input graph is a clique-cycle-forest.*

Proof. Let (G, w_0, w_1, t) be the given instance of MAX-CUT-WITH-WEIGHTED-VERTICES. Assume G is connected. We shall show that by repeated applications of Lemma 10, the given instance can be reduced to an equivalent instance (G', w_0', w_1', t') such that $|V(G')| \leq |V(G)|$. The algorithm is as follows. Let $X \cup \{r\}$ be the vertices of a leaf-block of G with cut vertex r. Let $H = G[X \cup \{r\}]$. If H is a clique, then follow the steps in Lemma 10.

Suppose H is an odd cycle. Then at least one of the edges of H cannot be part of any (optimal) cut. For each edge e of H, note that $H - e$ is a path, and hence is a clique-forest. For each of the two values of $f(r)$, proceed as in Lemma 10 on $H - e$. Let A_e and B_e be the sizes of the optimal cuts of H, obtained by setting $f(r) = 1$ and $f(r) = 0$, respectively. Take $A = \max\{A_e | e \in E(H)\}$ and $B = \max\{B_e | e \in E(H)\}$. Now delete X and update $w_1(r) = A$, and let $w_0(r) = B$.

Suppose H is an even cycle. There are two possibilities. (i) All edges of H are in an optimal cut. There are only two such cuts, and they are determined by setting $f(r) = 1$ or $f(r) = 0$. Let A_1 and B_1 be their corresponding sizes. (ii) At least one edge of H is missing from an optimal cut. For each edge e of H, proceed as we did in the case when H was an odd cycle. Take $A = \max\{A_1, A_e | e \in E(H)\}$ and $B = \max\{B_1, B_e | e \in E(H)\}$. Delete X and update $w_1(r) = A$, and let $w_0(r) = B$.   □

Theorem 1. MAX-CUT ABOVE SPANNING TREE *is fixed parameter tractable.*

Proof. Let (G, k) be an instance of MAX-CUT-AST, where $n = |V(G)|$. Apply Lemma 6. Then, in polynomial time we have either concluded that (G, k) is a yes-instance or found a set $S \subseteq V(G)$ of size at most $3k$ such that $G - S$ is a clique-cycle-forest. Assume we have found such a set S. We shall show that every possible cut of $G[S]$ can be optimally extended to cut of G, in polynomial time. Corresponding to every one of the $2^{|S|} \leq 2^{3k}$ cuts $f : S \to \{0, 1\}$ of $G[S]$, we create an instance of MAX-CUT-WITH-WEIGHTED-VERTICES on $G - S$ such that the original instance (G, k) of MAX-CUT-AST is a yes-instance if and only if at least one of the $2^{|S|}$ instances of MAX-CUT-WITH-WEIGHTED-VERTICES is a yes-instance. By Lemma 11, each such instance of MAX-CUT-WITH-WEIGHTED-VERTICES can be solved in polynomial time, and thus we have an algorithm for MAX-CUT-AST that runs in time $\mathcal{O}(2^{3k} n^{\mathcal{O}(1)})$.

For each $f : S \to \{0, 1\}$, construct an instance of MAX-CUT-WITH-WEIGHTED-VERTICES as follows. Let ℓ be the number of edges of $G[S]$ that are satisfied by f. For every $x \in V(G) - S$, let $w_0(x) = |\{s \in S | sx \in E(G), \}|$ and $f(s) = 1$, and let $w_1(x) = |\{s \in S | sx \in E(G), \text{ and } f(s) = 0\}|$. Let $t = n - 1 + k - \ell$. $G - S$ is a clique-cycle-forest; apply Lemma 11 to solve the instance $(G-S, w_0, w_1, t)$ of MAX-CUT-WITH-WEIGHTED-VERTICES in polynomial time. And let $f' : V(G - S) \to \{0, 1\}$ be the optimal solution thus obtained. Define a cut $g : V(G) \to \{0, 1\}$ of G by setting $g(x) = f(x)$ if $x \in S$ and $g(x) = f'(x)$ if $x \in V(G - S)$. Then,

$$||g|| = \ell + \sum_{xy \in E(G-S)} |g(x) - g(y)| + \sum_{\substack{x \in V(G - S) \\ g(x) = 0}} w_0(x) + \sum_{\substack{x \in V(G - S) \\ g(x) = 1}} w_1(x).$$

Note that $||g|| = \ell + ||f'||$. Therefore, $||g|| \geq n - 1 + k$ if and only if $||f'|| \geq n - 1 + k - \ell = t$. $\square$

4   Polynomial Kernel for MAX-CUT-AST

In this section, we show that MAX-CUT-AST admits a polynomial kernel. Specifically, we prove the following result.

Theorem 2 ($\star$). *Given an instance (G, k) of MAX-CUT-AST, in polynomial time, either we can conclude that (G, k) is a yes-instance or construct an equivalent instance (G', k') such that $k' \leq k$ and $|V(G')| \leq 12528k^5 - 6696k^4 - 63k^3 + 198k^2 - 3k - 13 = \mathcal{O}(k^5)$.*

Due to space constraints, we defer all technical details of the proof of Theorem 2 to the full version of the paper. Here we briefly outline our proof strategy. Let (G, k) be an instance of MAX-CUT-AST. By Lemma 6, in polynomial time we can either conclude that (G, k) is a yes-instance or find a set $S \subseteq V(G)$ such that $|S| \leq 3k$ and $G - S$ is a clique-cycle-forest, i.e., every block of $G - S$ is either a clique or a cycle. For the rest of this section, assume that we have found such a set S. The graph $G - S$ need not be connected.

First, we apply a set of two-way safe reduction rules (Rules 6–13) on (G, k) to obtain an equivalent instance (G', k'). While applying these rules we ensure that the set S remains untouched, the reduced graph G' is connected and $G' - S$ is a clique-cycle-forest. Thus each block C of $G' - S$ will be exactly one of the following: (i) a clique of size at least 4 (and thus contains at least one even cycle,) or (ii) an even cycle, or (iii) an odd cycle, or (iv) a K_2, or (v) a K_1.

Then we show that either (G', k') is a yes-instance or $|V(G')| = O(k^5)$. To achieve this, we bound the number of blocks of $G' - S$ as well as the size of each block. In order to bound the number of blocks in $G - S$, we exploit the structural property of $G - S$'s being a clique-cycle-forest. Thus, $G - S$ is a "forest of blocks," in which every block is a clique or a cycle. First, we classify the components of $G - S$ into different types, and bound the number of each type of components separately. Then we bound the number of leaf-blocks (blocks that contain at most one cut vertex) in each type of components. This in turn bounds the number of blocks of $G - S$ that contain at least three cut vertices. (The number of vertices of degree at least 3 in a forest is upper bounded by the number of leaves; and $G - S$ is a forest of blocks.) Finally, we bound the number of blocks that contain exactly two cut vertices. We also bound the number of cut vertices in each type of components, which, then will be used to bound the size of each block.

References

1. Cai, L., Juedes, D.W.: On the existence of subexponential parameterized algorithms. J. Comput. Syst. Sci. **67**(4), 789–807 (2003)
2. Crowston, R., Jones, M., Mnich, M.: Max-Cut parameterized above the Edwards-Erdős bound. In: Czumaj, A., Mehlhorn, K., Pitts, A., Wattenhofer, R. (eds.) ICALP 2012. LNCS, vol. 7391, pp. 242–253. Springer, Heidelberg (2012). https://doi.org/10.1007/978-3-642-31594-7_21
3. Crowston, R., Jones, M., Mnich, M.: Max-Cut parameterized above the Edwards-Erdős bound. Algorithmica **72**(3), 734–757 (2015)
4. Edwards, C.S.: An improved lower bound for the number of edges in a largest bipartite subgraph. In: Proceedings of the 2nd Czechoslovak Symposium on Graph Theory, Prague, pp. 167–181 (1975)
5. Edwards, C.: Some extremal properties of bipartite subgraphs. Can. J. Math. **25**(3), 475–483 (1973)
6. Erdös, P.: On some extremal problems in graph theory. Isr. J. Math. **3**(2), 113–116 (1965)
7. Etscheid, M., Mnich, M.: Linear kernels and linear-time algorithms for finding large cuts. In: 27th International Symposium on Algorithms and Computation, ISAAC 2016, Sydney, Australia, 12–14 December 2016, pp. 31:1–31:13 (2016)
8. Karp, R.M.: Reducibility among combinatorial problems. In: Miller, R.E., Thatcher, J.W., Bohlinger, J.D. (eds.) Complexity of Computer Computations. IRSS, pp. 85–103. Springer, Boston (1972). https://doi.org/10.1007/978-1-4684-2001-2_9
9. Mahajan, M., Raman, V.: Parameterizing above guaranteed values: MaxSat and MaxCut. J. Algorithms **31**(2), 335–354 (1999)

10. Mnich, M., Philip, G., Saurabh, S., Suchý, O.: Beyond Max-Cut: λ-extendible properties parameterized above the Poljak-Turzík bound. J. Comput. Syst. Sci. **80**(7), 1384–1403 (2014)
11. Poljak, S., Tuza, Z.: Maximum cuts and largest bipartite subgraphs. In: Proceedings of a DIMACS Workshop on Combinatorial Optimization, New Brunswick, New Jersey, USA, 1992/1993, pp. 181–244 (1993)

Slopes of 3-Dimensional Subshifts
of Finite Type

Etienne Moutot[1] and Pascal Vanier[2(✉)]

[1] LIP, ENS de Lyon – CNRS – INRIA – UCBL – Université de Lyon,
6 all ée d'Italie, 69364 Lyon Cedex, France
etienne.moutot@ens-lyon.org
[2] Laboratoire d'Algorithmique, Complexité et Logique Université de Paris-Est,
LACL, UPEC, Paris, France
pascal.vanier@lacl.fr

Abstract. In this paper we study the directions of periodicity of three-dimensional subshifts of finite type (SFTs) and in particular their slopes. A configuration of a subshift has a slope of periodicity if it is periodic in exactly one direction, the slope being the angles of the periodicity vector. In this paper, we prove that any Σ_2^0 set may be realized as a a a set of slopes of an SFT.

A d-dimensional subshift of finite type (SFT for short) is a set of colorings of $\mathbb{Z}^d$ by a finite number of colors containing no pattern from a finite family of forbidden patterns. Subshifts may be seen as discretizations of continuous dynamical systems: if X is a compact space and there are d commuting continuous actions $\phi_1, \ldots, \phi_d$ on X, one can partition X in a finite number of parts indexed by an alphabet Σ. The orbit of a point $x \in X$ maps to a coloring y of $\mathbb{Z}^d$ where $y(v)$ corresponds to the partition where $\phi^v(x)$ lies.

In dimension 1, most problems on SFTs are easy in a computational sense, since SFTs correspond to bi-infinite walks on finite automata. For instance, in dimension 1, detecting whether an SFT is non-empty is decidable since it suffices to detect if there exists a cycle in the corresponding automaton [13], which corresponds to the existence of a periodic configuration.

In higher dimensions however, the situation becomes more involved, and knowing whether an SFT is non-empty becomes undecidable [2,3]. The proof uses two key results on SFTs: the existence of an aperiodic SFT and an encoding of Turing machine's space time diagrams. The fact that there exists aperiodic SFTs is not straightforward, and the converse was first conjectured by Wang [20]. Had this conjecture been true, it would have meant the decidability of the emptiness problem for SFTs. Berger [2,3] proved however that there does exist SFTs containing only aperiodic configurations. Subsequently, many other aperiodic SFTs were constructed [4,8,12,16–18]. Note that the existence in itself of aperiodic SFTs does not suffice to prove that the emptiness problem is undecidable, one needs in addition to encode some computation in them, usually in the form of Turing machines.

© Springer International Publishing AG, part of Springer Nature 2018
F. V. Fomin and V. V. Podolskii (Eds.): CSR 2018, LNCS 10846, pp. 257–268, 2018.
https://doi.org/10.1007/978-3-319-90530-3_22

Periodicity has thus been central in the study of SFTs from the beginning, and it has been proved very early that knowing whether an SFT is aperiodic is undecidable [6]. In fact, sets of periods constitute a classical conjugacy/isomorphism invariant for subshifts in any dimension. As such, they have been studied extensively and even characterized: algebraically in dimension 1, see [13] for more details, and computationally in dimension 2. In fact it seems that computability theory is the right tool to study dynamical aspects of higher dimensional symbolic dynamical systems [1,5,7,14].

In dimensions $d \geq 2$, one may investigate periodicity from different angles. Denote $\Gamma_x = \{v \in \mathbb{Z}^d \mid x(z + v) = x(z), \forall z \in \mathbb{Z}^d\}$ the lattice of vectors of periodicity of configuration x: Γ_x may be of any dimension below d and some cases are particularly interesting:

- When it is of dimension 0, then x does not have any vector of periodicity and is hence aperiodic.
- When it is of dimension d, then x is somehow finite, this case has been studied and partly characterized in terms of complexity classes by Jeandel and Vanier [10].
- When $d = 1$, then there exists some vector v such that $\Gamma_x = v\mathbb{Z}$. In this case, one may talk about the direction or slope of the configuration.

In this paper, we are interested in this last case. In [9], this case was studied and characterized for 2-dimensional SFTs through the arithmetical hierarchy:

Theorem (Jeandel and Vanier). *The sets of slopes of 2-dimensional SFTs are exactly the Σ_1^0 subsets of $\mathbb{Q} \cup \{\infty\}$.*

In the end of [9] it was conjectured that slopes of higher dimensional SFTs are the Σ_2^0 subsets of $(\mathbb{Q} \cup \{\infty\})^{d-1}$. This gap between dimension 2 and dimension 3 for decidability of periodicity questions is similar to the gap between dimension 1 and 2 for decidability of emptiness questions: the subset of periodic configurations of a d-dimensional subshift along some periodicity vector may be seen as a $(d-1)$-dimensional subshift (see *e.g.* [9]), hence the jump in complexity. This is the idea that led to the conjecture. However, in dimension higher than 2, the construction of [9] cannot be reused.

In this article, we prove one direction of the aforementioned conjecture: we show how to realize any Σ_2^0 subset of $(\mathbb{Q} \cup \{\infty\})^2$ as a set of slopes of a 3D subshift:

Theorem 1. *Any Σ_2^0 subset of $(\mathbb{Q} \cup \{\infty\})^2$ may be realized as the set of slopes of some 3D SFT.*

In order to do this, we introduce a new way to synchronize computations between different dimensions, inspired partly by what is done by Durand et al. [5]. Note that our construction can be easily generalized to realize any Σ_2^0 subset of $(\mathbb{Q} \cup \{\infty\})^{d-1}$ as a set of slopes of a d-dimensional subshift for $d \geq 3$.

However, we did not manage to prove the other part of the conjecture, that is the fact that the sets of slopes of d-dimensional SFTs are in Σ_2^0 (for $d \geq 3$).

The paper is organized as follows: in Sect. 1 we recall the useful definitions about subshifts and the arithmetical hierarchy, and in Sect. 2 we prove Theorem 1.

1 Definitions and Properties

1.1 Subshifts and Tilesets

We give here some standard definitions and facts about subshifts, one may consult [13] for more details.

Let Σ be a finite alphabet, a *configuration* (or tiling) is a function $c : \mathbb{Z}^d \longrightarrow \Sigma$. A *pattern* is a function $p : N \longrightarrow \Sigma$, where $N \subseteq \mathbb{Z}^d$ is a finite set, called the *support* of p. A pattern p *appears* in another pattern p' if there exists $v \in \mathbb{Z}^d$ such that $\forall x \in N, p(x) = p'(x + v)$. We write then $p \sqsubseteq p'$. Informally, a configuration (or tiling) is a coloring of $\mathbb{Z}^d$ with elements of Σ. A *subshift* is a closed, shift-invariant subset of $\Sigma^{\mathbb{Z}^d}$, the d-dimensional *full shift*. For a subshift X we will sometimes note Σ_X its alphabet. The full shift is a compact metric space when equipped with the distance $d(x,y) = 2^{-\min\{\|v\|_\infty \mid v \in \mathbb{Z}^d, x(v) \neq y(v)\}}$ with $\|v\|_\infty = \max_i |v_i|$.

It is well known that subshifts may also be defined via collections of forbidden patterns. Let F be a collection of forbidden patterns, the subset X_F of $\Sigma^{\mathbb{Z}^d}$ defined by

$$X_F = \left\{ x \in \Sigma^{\mathbb{Z}^d} \mid \forall p \in F, p \not\sqsubseteq x \right\}$$

is a subshift. Any subshift may be defined via an adequate collection of forbidden patterns. A *subshift of finite type (SFT)* is a subshift which may be defined via a finite collection of forbidden patterns. A configuration of a subshift is also called a *point* of this subshift and is said to be valid with respect to the family of forbidden patterns F. Remark that F being finite, we can define a subshift of finite type either by a set of forbidden or authorized patterns.

Wang tiles are unit squares with colored edges which may not be flipped or rotated, a *tileset* is a finite set of Wang tiles. Tiles of a tileset maybe placed side by side on the $\mathbb{Z}^2$ plane only when the matching borders have the same color, thus forming a tiling of the plane. The set of all tilings by some tileset is an SFT, and conversely, any SFT may be converted into an isomorphic tileset. From a computability point of view, both models are equivalent and we will use both indiscriminately. In 3D, Wang tiles can be straightforwardly generalized to Wang cubes.

A subshift is **North-West-deterministic** if, for any position, and for any two colors placed above it and to its left (Fig. 1), there exists at most one valid color at this position. Likewise, we call a subshift **West-deterministic** if it is the case with the colors to its left and top-left (Fig. 2).

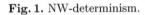

Fig. 1. NW-determinism.   **Fig. 2.** W-determinism.

1.2   Periodicity and Aperiodicity

The notion of periodicity being central in this paper, we will define it in this section.

Definition 1 (Periodicity). *A configuration c is* periodic *of period v if there exists $v \in \mathbb{Z}^d \setminus \{(0,0)\}$ such that $\forall x \in \mathbb{Z}^d, c(x) = c(x+v)$. If c has no period, then it is said to be* aperiodic. *A subshift is* aperiodic *if all its points are aperiodic.*

From now on, we will focus on dimension 3 in this paper. As seen in the introduction, the lattice of vectors of periodicity may be of any dimension between 0 and d and we are interested here in the case where it is 1-dimensional. In this case we can define the slope periodicity:

Definition 2 (Slope of periodicity). *Let c be a configuration periodic along $v = (p,q,r)$. We call* slope of v *the pair $\theta = (\theta_1, \theta_2)$ with $\theta_1 = \frac{p}{r}$ and $\theta_2 = \frac{p}{q}$. If all vectors of periodicity of c have slope θ, we say that θ is the* slope of periodicity *or* slope *of c. We write $Sl(X) = \{\theta \mid \exists x \in X, \theta \text{ is the slope of } x\}$ the set of slopes of X.*

1.3   Arithmetical Hierarchy

We give now some basic definitions used in computability theory and in particular about the arithmetical hierarchy. More details may be found in [19].

Usually the arithmetical hierarchy is seen as a classification of sets according to their logical characterization. For our purpose we use an equivalent definition in terms of computability classes and Turing machines with oracles:

- $\Delta_0^0 = \Sigma_0^0 = \Pi_0^0$ is the class of recursive (or computable) problems.
- Σ_n^0 is the class of recursively enumerable (RE) problems with an oracle Π_{n-1}^0.
- Π_n^0 the complementary of Σ_n^0, or the class of co-recursively enumerable (coRE) problems with an oracle Σ_{n-1}^0.
- $\Delta_n^0 = \Sigma_n^0 \cap \Pi_n^0$ is the class of recursive (R) problems with an oracle Π_{n-1}^0.

In particular, Σ_1^0 is the class of recursively enumerable problems and Π_1^0 is the class of co-recursively enumerable problems.

2   Proof of Theorem 1

Theorem 1. *Let $R \in \Sigma_2^0 \cap \mathcal{P}((\mathbb{Q} \cup \{\infty\})^2)$, there exists a 3D SFT X such that $Sl(X) = R$.*

Proof. Let M be a Turing machine accepting R with an oracle $O \in \Pi_1^0$. One can suppose that this machine takes as input 3 integers $(p, q, r) \in \mathbb{N}^3$ and that its output depends only on $\theta_1 = \frac{p}{r}$ and $\theta_2 = \frac{p}{q}$.

We only explain the case $0 < r < q < p$, the others are symmetric or quite similar and it suffices to take the disjoint union of the obtained SFTs to get the full characterization.

Let us construct a 3D SFT X_M that has a periodic configuration along θ if and only if $\theta \in R$. To do so, X_M will be such that "good" configurations (i.e. valid and 1-periodic) are formed of large cubes, shifted with an offset to allow periodicity along some slope. Then we encode M inside all the cubes, and give to it the slope as input. The machine halts (i.e the slope is in R) implies that the cubes are of finite size. Which means that the configuration is 1-periodic only when the slope actually corresponds to some element of R.

For that, we separate the construction in different layers, in order to make it clearer. We define $X_M = B \times B' \times B'' \times C \times W \times P \times S \times T_O \times T_M \times A$, with the following layers:

- B creates (yz) black planes, separated by an aperiodic tiling.
- B' and B'' create planes orthogonal to the ones of B, forming rectangular parallelepipeds.
- C forces the parallelepipeds to become cubes.
- W forces the aperiodicity vector to appear between cubes, and writes the input of the Turing machine in the cubes.
- P reduces the size of the input.
- S synchronizes the aperiodic backgrounds of the cubes.
- T_M encodes the "Σ_2^0" Turing machine M in the cubes.
- T_O encodes the Π_1^0 oracle O that is used by M.
- A ensures the existence of configurations with a unique direction of periodicity.

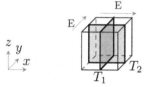

Fig. 3. T_1, T_2 are tiles of the 2D aperiodic tileset and form a Wang cube of the 3D aperiodic tileset. The "E" shows the east direction of the two planes.

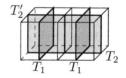

Fig. 4. Duplication of parallel backgrounds.

Aperiodic background. We first need an aperiodic background in order to ensure that there is no other directions of periodicity that the one we create later on. We even make a 3D West-deterministic aperiodic background since some layers will need that deterministic property to work. For that we cross two 2D West-deterministic aperiodic backgrounds: the set of aperiodic cubes are the sets of cubes of the form shown in Fig. 3. We also impose that all parallel planes are identical (Fig. 4). The 2D aperiodic tiling is from Kari [11], which is a NW-aperiodic tiling, and can be easily transformed into a West-deterministic SFT. With such a superposition, one can easily show that the resulting 3D tiling is aperiodic.

Layer B. The first layer is made with two types of cubes: a white cube ($\square$), which is a meta-cube that corresponds to any cube of the aperiodic background and a black cube ($\blacksquare$) which will serve to break the aperiodicity brought by the white cubes. The rules of this layer are:

- In coordinates $z + 1$ and $z - 1$ of $\blacksquare$, only a $\blacksquare$ can appear.
- In coordinates $y + 1$ and $y - 1$ of $\blacksquare$, only $\blacksquare$ can appear.
- In coordinates $x + 1$ and $x - 1$ of $\blacksquare$, only $\square$ can appear.

With only this layer, the valid periodic configurations are thick aperiodic (yz) planes separated by infinite black (yz) planes. At this stage, there may be several aperiodic planes "inside" a period.

Layer B'. For every Wang cube of this layer, we impose that the cube at $y + 1$ is the same. So we can describe the layer B' by a set of 2D Wang tiles in the (xz) plane, duplicated on the y axis. The tiles are:

The first four can only be superimposed with black tiles of layer B and the last two only with white ones.

With layers B and B' the periodic configurations are formed of infinite planes along (yz) linked by infinite (xy) strips infinite along y, see Fig. 5.

Layer B''. This layer is identical to B' but tiles are duplicated along the z axis. It creates portions of infinite planes along the z axis, also delimited by the black planes the layer B.

With these three layers, periodic configurations are formed of parallelepipeds delimited by black cubes and $\blacksquare$ with portions of aperiodic background inside them.

Layer C. This layer forces the parallelepipeds to be cubes, by forcing rectangles of (xz) to be squares, and same for rectangles of (xy).

Like the B layer this layer is created by duplicating 2D Wang tiles along y axis for rectangles of (xz) and z axis for rectangles of (xy):

These tiles are superimposed once on the B' tiles with rules on the (xz) plane and on the B'' tiles with rules on the (xy) plane. The superimpositions allowed are the following:

- $\blacksquare$ can only be on $\boxtimes$ and $\boxtimes$.
- $\blacksquare$ can only be on $\blacksquare$.

- ▨ can only be on ◪ and ▨ only on ◩.
- ◩, ◼ and ☐ can only be on white tiles ⊠.

Figure 6 shows how this layer forces squares to appear.

Layer W. This layer uses signals to synchronize the offsets of different cubes, and to force cubes to have the same size. In order to visualize the different offsets you can refer to Fig. 7. The construction is done in several parts.

The first one forces the offsets along y (denoted by r) to be the same in each cube. Here again, everything is duplicated along z. It creates signals (see Fig. 8), that have to correspond with the extension of the neighboring cubes. This also writes the number r in unary in the border of each cube. This number will be used by the Turing machine encoded later in the tiling. The second part is identical to the first one, but on the (xz) plane and rotated $90°$. It forces the offset along z (denoted by q) to be the same everywhere.

Finally, the cubes are forced to be of same size. For that we add the two signals shown on Fig. 9, which have to link a corner to the extension of a square. It has the effect to force each square (and hence each cube) to be of same size as its neighbors.

Layer P. This layer reduces the size of the input, in order to allow us to construct valid configurations as large as we want for the same input (p', q', r'). Starting from an unary input (p, q, r), this layer writes into cubes what the input of the Turing machine will be: (p', q', r'), with $(p, q, r) = 2^k(p', q', r')$, and $\gcd(p', q', r')$ not divisible by 2.

For that, we use a transducer to convert numbers in binary. Such a transducer can be easily encoded into tilings (see for example [12] or [8]). Then it only remains to remove the final 0's they have in common, which can easily be done through local rules.

Layer S. Aperiodic backgrounds of different "slices" may be different ("slices" are the thick planes in the (yz) plane). They must be synchronized in order to ensure the existence of a periodic configuration along (p, q, r). To do this synchronization in 2D, we use the following arrow tiles:

➡ ◿ ➡ ◿

with the following rules:

- To the left of ◪ (layer B') there is ◿ and bottom left neighbor of ◿ is ◿ or a ➡.
- Bottom left tile of a square is ➡. On the right of ➡ there is only ➡ or ◿.
- On the right, left and bottom of ◿ there is only ◿ or ◿.
- The *breaking lines* can only have ➡ on them.

We obtain the tiling shown on Fig. 10. If we impose that the background is the same at the beginning and at the end of the arrow with gray background, its West-periodicity ensures that it is repeated along the global periodicity vector.

We now use this 2D construction to build the 3D transmission of the background in X_M. We create two layers of 2D arrows. One in the (xy) plane, that are repeated along z (*front* arrows) and the other in (xz) and repeated along y (*top* arrow). We then create our real layer using these two 2D layers, with 3D arrows:

The superimposition of two 2D arrows gives directly which 3D arrow is on each tile (see Fig. 11). Like in 2D we impose that the background is the same at the beginning and at the end of the gray arrow. Thanks to the double West-periodicity of the background, this ensures that the background has a periodicity vector of (p, q, r) in valid configurations.

Layer T_M. This layer encodes the Turing machine M in the tiling. In our definition of the arithmetical hierarchy, the machine M being in Σ_2^0, it has access to a Π_1^0 oracle. This oracle will be represented by a tape R_O filled with zeros and ones, such that position i of R_O is a 1 if and only if the oracle O accepts i (i.e. the Turing machine of index i runs indefinitely). For the moment, we will encode M with an additional read-only arbitrary tape, and the next layer will ensure that the content of the tape is valid for O. This additional tape is supposed to be infinite, but since M has to halt in the periodic configurations, we can restrict the construction to a finite but arbitrarily large portion of it. The tape is a line along axis y duplicated along axes x and z (see Fig. 12). We add the two rules:

1. Inside a cube, a number at position x is the same as the number at position $x - 1$ and a number at position z must be equal to the number at position $z - 1$.
2. The first line of the tape is transmitted through black cubes like the aperiodic background.

The first rule duplicates the first line everywhere inside a cube, and the second one ensures that the same R_O tape is duplicated along the direction of periodicity.

Then, we encode M in the (xz) plane using the usual encoding of Turing machines in tilings. Let us say that the time is along the z axis and the working tape along x. In order to access the entire R_O tape, we add the spacial dimension y to the TM encoding: while doing a transition, the machine can move its head along the y axis and read the value of the R_O tape in it; rule 1 above prevents M to modify this extra tape.

Note that because O is a Π_1^0 oracle, it can only ensure that the 1s of R_O are correct. M has to check that the 0s are correct. But checking the 0s, i.e. checking if a TM halts in a computation does not add any complexity to the problem, because we are only interested in the periodic configurations, where M actually halts (and so do all its checks).

Layer T_O. This layer is the core of this proof, and it is where 3D actually comes to play: in the thick aperiodic (yz) planes we will compute the Π_1^0 oracle by encoding an infinite computation that checks simultaneously all possible inputs of M_O, the Turing machine checking the Π_1^0 oracle O (M_O halts if and only if there is a wrong 1 in the portion of R_O written in all the cubes). The key idea of this layer is the use of the previously constructed cubes as *macro-tiles* in order to encode computations of M_O. Each cube will thus represent one tile and the thick aperiodic planes will contain, more sparsely, another 2D tiling. See Fig. 14 to see how the cubes store this macro-tileset.

For this macro-tileset, we may use a construction of Myers [15] which modifies Robinson's aperiodic tileset in order to synchronize the input tapes on all of the partial computations. So each of our cubes contains/represents one tile of Myer's tileset, and the thick aperiodic planes thus also contain a Myers tiling checking some input that for the moment is not synchronized with the oracle written inside these cubes.

We now have a valid macro-tiling for the cubes if and only if the machine M_O never halts on R_O.

The one remaining thing to do is to explain the R_O tape that M_O accesses is synchronized with the R_O which is stored inside the large cubes. We add to the set of numbered tiles the same tiles, but in red, representing the head of the M_O Turing machine on the tape R_O. We impose that there is only one red number in every large cube (see Fig. 13).

The red tile of a cube must be synchronized with the cell of the oracle R_O currently contained in the Myers tile. Every time a new partial computation is started in the macro-tiles, the red tile must be placed at the beginning of R_O, whenever the macro-tile moves the head to the right/left, the red tile must also be moved, if the red tile reaches the border of the cube, in which case it reaches a special state of non-synchronization, since the beginning has already been synchronized.

To do that, we must allow two new transitions. These new transitions do not change the state of the working tape, thus we only move to the next time along z. But in the new position, the red-marked cell in the large cube must have changed. To do this, we again use signals between the bottom-cube (previous state), the cell doing the transition and the upper-cube (next state), see Fig. 15.

Layer A. This last layer forces the apparition of 1-periodic configurations. Using two cubes (▨ and ◧), superimposed only with ▢ and borders of big cubes. We impose that blue/red neighbors have the same color. It is easy to see that the color is uniform inside a cube and spread to two opposite corners of cubes. Thus all the cubes along (p, q, r) have the same color and there exists at least one 1-periodic configuration.

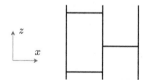

Fig. 5. Projection on the (xz) plane of a valid configuration with layers B and B'.

Fig. 6. Valid tiling with rules of layer C.

Now we prove that this construction does what we claim, finishing the proof of Theorem 1.

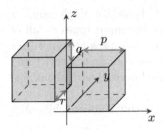

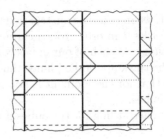

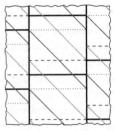

Fig. 7. Names of the offsets.

Fig. 8. Signals making the offsets identical.

Fig. 9. Signals making the cubes of same size.

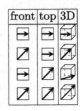

Fig. 10. Transmission of a 2D background.

Fig. 11. Rules for layer S (transmission in 3D).

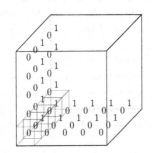

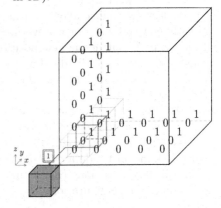

Fig. 12. Tape R_O of the oracle in a cube.

Fig. 13. The red tiles on R_O and the transmission of its value. (Color figure online)

2.1 Every Slope θ of X_M Is Accepted by M

Let $\theta = (\frac{p}{r}, \frac{q}{r})$ be a slope, by construction every periodic configuration along this slope is formed with cubes of the same size p, shifted with the same offset (q, r). Every cube has the same content, which corresponds to an execution of M. Cubes being of finite size, every execution is a halting execution of M. Let's take $(p, q, r) = 2^k(p', q', r')$, with p', q', r' odds. Thanks to the layer P, the input of M is (p', q', r'), then M accepts $(\frac{p'}{r'}, \frac{q'}{r'}) = (\frac{p}{r}, \frac{q}{r}) = \theta$.

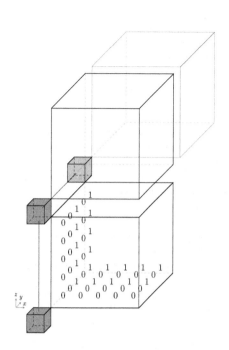 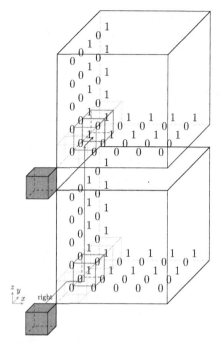

Fig. 14. Meta-tiles of large cubes, with the adjacency rules represented by the arrows. Myers' tiles are placed in the darker cubes, in the (yz) plane.

Fig. 15. Moving the red cube when the head of R_O moves. (Color figure online)

2.2 Accepting Inputs of M Are Slopes of X_M

If M accepts the input (p, q, r), there exists a time t and a space a on the working tape, b on the oracle tape, in which the machine M halts. Then, the cube of size $m = 2^{\lceil \log t \rceil} p \geq a, b, t$ can contain the computation of M. The configuration formed by cubes of size m and of offset $(n, o) = 2^{\lceil \log t \rceil}(q, r)$ is of slope $(\frac{m}{o}, \frac{n}{o}) = (\frac{p}{r}, \frac{q}{r})$. □

3   Open Problems

The problem of deciding if all configurations of a 2D SFT are aperiodic is well-known to be Π_1^0. Proving the other direction of the conjecture would require the study of a very similar problem: deciding if there **exists** a periodic configuration in a given SFT. Four our purpose, one needs to prove that the problem of the existence of an aperiodic configuration is Π_1^0 or Σ_2^0. However, we aren't aware of any study of this problem, not even a simpler bound like Π_2^0. Our quick look at it suggests that this could be a very challenging problem to tackle. Yet, it seems interesting by itself, as it would likely lead to a better understanding of periodicity and aperiodicity in SFTs.

Acknowledgements. The authors would like to thank anonymous reviewers who pointed out a mistake in a previous version of the paper.

This work was supported by grant TARMAC ANR 12 BS02 007 01.

References

1. Aubrun, N., Sablik, M.: Simulation of effective subshifts by two-dimensional subshifts of finite type. Acta Applicandae Math. **126**(1), 35–63 (2013)
2. Berger, R.: The Undecidability of the Domino Problem. Ph.D. thesis, Harvard University (1964)
3. Berger, R.: The Undecidability of the Domino Problem. No. 66 in Memoirs of the American Mathematical Society, The American Mathematical Society (1966)
4. Culik II, K., Kari, J.: An aperiodic set of Wang cubes. J. Univers. Comput. Sci. **1**(10), 675–686 (1995)
5. Durand, B., Romashchenko, A., Shen, A.: Fixed-point tile sets and their applications. J. Comput. Syst. Sci. **78**(3), 731–764 (2012)
6. Gurevich, Y., Koryakov, I.: Remarks on Berger's paper on the domino problem. Siberian Math. J. **13**(2), 319–320 (1972)
7. Hochman, M., Meyerovitch, T.: A characterization of the entropies of multidimensional shifts of finite type. Ann. Math. **171**(3), 2011–2038 (2010)
8. Jeandel, E., Rao, M.: An aperiodic set of 11 Wang tiles. CoRR abs/1506.06492 (2015). http://arxiv.org/abs/1506.06492
9. Jeandel, E., Vanier, P.: Slopes of tilings. In: Kari, J. (ed.) JAC, pp. 145–155. Turku Center for Computer Science (2010)
10. Jeandel, E., Vanier, P.: Characterizations of periods of multi-dimensional shifts. Ergod. Theor. Dyn. Syst. **35**(2), 431–460 (2015). http://journals.cambridge.org/article_S0143385713000606
11. Kari, J.: The nilpotency problem of one-dimensional cellular automata. SIAM J.Comput. **21**(3), 571–586 (1992)
12. Kari, J.: A small aperiodic set of Wang tiles. Discrete Math. **160**(1–3), 259–264 (1996)
13. Lind, D.A., Marcus, B.: An Introduction to Symbolic Dynamics and Coding. Cambridge University Press, New York (1995)
14. Meyerovitch, T.: Growth-type invariants for $\mathbb{Z}^d$ subshifts of finite type and arithmetical classes of real numbers. Inventiones Math. **184**(3), 567–589 (2010)
15. Myers, D.: Non recursive tilings of the plane II. J. Symbolic Log. **39**(2), 286–294 (1974)
16. Ollinger, N.: Two-by-two substitution systems and the undecidability of the domino problem. In: Beckmann, A., Dimitracopoulos, C., Löwe, B. (eds.) CiE 2008. LNCS, vol. 5028, pp. 476–485. Springer, Heidelberg (2008). https://doi.org/10.1007/978-3-540-69407-6_51
17. Poupet, V.: Yet another aperiodic tile set. In: Journées Automates Cellulaires (JAC), pp. 191–202. TUCS (2010)
18. Robinson, R.M.: Undecidability and nonperiodicity for tilings of the plane. Inventiones Math. **12**(3), 177–209 (1971)
19. Rogers Jr., H.: Theory of Recursive Functions and Effective Computability. MIT Press, Cambridge (1987)
20. Wang, H.: Proving theorems by pattern recognition I. Commun. ACM **3**(4), 220–234 (1960)

Facility Location on Planar Graphs
with Unreliable Links

N. S. Narayanaswamy, Meghana Nasre, and R. Vijayaragunathan[✉]

Department of Computer Science and Engineering,
Indian Institute of Technology Madras, Chennai, India
{swamy,meghana,vijayr}@cse.iitm.ac.in

Abstract. Hassin et al. [9] consider the MAX-EXP-COVER-R problem
to study the facility location problem on a graph in the presence of *unreliable links* when the link failure is according to the Linear Reliability
Order (LRO) model. They showed that for unbounded R the problem
is polynomial time solvable and for $R = 1$ and planar graphs the problem is NP-Complete. In this paper, we study the MAX-EXP-COVER-1
problem under the LRO edge failure model. We obtain a fixed parameter tractable algorithm for MAX-EXP-COVER-1 problem for bounded
treewidth graphs, parameterized by the treewidth. We extend the Baker's
technique (Baker, J. ACM 1994) to obtain PTAS for MAX-EXP-COVER-1
problem under the LRO model on planar graphs. We observe that the
coverage function of the MAX-EXP-COVER-R problem is submodular and
the problem admits a $(1 - 1/e)$-approximation for any failure model in
which the expected coverage of a set by another set can be computed in
polynomial time.

1 Introduction

We consider the facility location problem on a graph in the presence of *unreliable* links (edges). The input to our problem is an undirected graph $G = (V, E)$, a
survival probability $p(e)$ is associated with each edge $e \in E$, a positive valued
demand function w associated with each vertex, a budget $\mathcal{B}$ on the number
of facilities that can be opened to service demand, and a positive integer R.
The survival probability function models the probability with which the edge
survives if a disaster occurs. Each vertex is also a potential location for opening
a facility which can serve the demands and we can open at most $\mathcal{B}$ facilities.
The value R is a radius of coverage of a facility located at a vertex, that is,
a facility located at a vertex v can cover any vertex which is reachable by a
path of length at most R. The goal is to identify a $\mathcal{B}$ sized vertex set $F \subseteq V$
where facilities can be located such that the *expected demand* that is satisfied
by F is maximized. The expectation is over all possible graphs (realizations)
that can arise due to the failure of an edge set in the event of disaster. The
way in which edges fail are represented in a *failure model*. A natural failure
model is the *random failure model* (RFM) in which the failure of any edge is
independent of the failure of other edges in the graph. It is easy to note that due

© Springer International Publishing AG, part of Springer Nature 2018
F. V. Fomin and V. V. Podolskii (Eds.): CSR 2018, LNCS 10846, pp. 269–281, 2018.
https://doi.org/10.1007/978-3-319-90530-3_23

to the independence assumption, there are 2^m many realizations of the network that are possible under the RFM model. In some cases, failures on edges may not be independent. To study such scenarios, the *linear reliability order* (LRO) failure model was introduced by Daskin [5] and later studied by Hassin et al. [9]. The LRO model is defined as follows: if an edge e with a particular survival probability fails then all edges with smaller survival probability than e surely fail. That is, for any two edges e_i and e_j with survival probabilities $p(e_i) < p(e_j)$, the $Pr(e_i$ fails $\mid e_j$ fails$) = 1$. Clearly, the number of different realizations that are possible under the LRO model is exactly $m + 1$.

We now formally define the MAX-EXP-COVER-R problem. Let $G = (V, E)$ be the input graph with a demand function on vertices $w : V \to \mathbb{R}_{\geq 0}$ and a survival probability function on the edges $p : E \to [0, 1]$. Let $\mathcal{D}$ be the failure model that associates a probability to each realization, R be the radius of coverage, and $\mathcal{B}$ be the budget on the number of facilities. Let $\mathcal{Q}$ be the set of all realizations that can have a non-zero probability of occurrence in the failure model $\mathcal{D}$. Let $P : \mathcal{Q} \to [0, 1]$ be a function denotes the probability of a realization in the failure model $\mathcal{D}$. For any realization $G_i \in \mathcal{Q}$, the coverage of a set $F \subseteq V$ is the sum of demands of the nodes in the Rth closed neighborhood of F in G_i. For any $v \in V$, let $I(G_i, F, v, R)$ be an indicator variable which is set to 1 if v is within a radius of R from some node in F in the realization G_i. Otherwise $I(G_i, F, v, R) = 0$. Given a graph G, and an integer R, for any two sets $T, F \subseteq V$, define $\mathcal{C}_R(G, T, F) = \sum_{G_i \in \mathcal{Q}} P(G_i) \cdot \sum_{v \in T} w_v \cdot I(G_i, F, v, R)$ is the R-hop coverage of the *target* set T by the *source* F in the graph G. We always write $\mathcal{C}(T, F)$ instead of $\mathcal{C}_R(G, T, F)$ as R and G will be unambiguous from the context. Further, if T or F is a singleton set, we just write the element of the set instead of the set notation. The MAX-EXP-COVER-R problem is the following maximization problem:

$$\max_{F \subseteq V, |F| \leq \mathcal{B}} \sum_{G_i \in \mathcal{Q}} P(G_i) \sum_{v \in V} I(G_i, F, v, R) \cdot w(v) = \max_{F \subseteq V, |F| \leq \mathcal{B}} \mathcal{C}(V, F)$$

For any two sets $T, F \subseteq V$, the coverage function $\mathcal{C}(T, F) = \sum_{u \in T} \mathcal{C}(u, F)$.

Our Results: We present the following new results in the paper:

- We present a fixed parameter tractable (FPT) algorithm for the MAX-EXP-COVER-1 problem with the LRO failure model on bounded treewidth graphs. Our algorithm runs in time $O^*(8^{tw})$. The notation O^* hides the $poly(n)$ in the running time.
- Using the FPT algorithm on bounded treewidth graphs, we obtain a PTAS for the same on planar graphs by applying Baker's technique [1].
- We conclude by observing that the objective function of the MAX-EXP-COVER-R problem is monotone submodular. As a consequence, for all distributions where for any two sets $T, F \subseteq V$, the expected coverage of set T by F is computable in polynomial time, the MAX-EXP-COVER-R problem admits a $(1 - \frac{1}{e})$-approximation.

At the high level, our FPT algorithm resembles the standard dynamic programming (DP) algorithm on the *tree decomposition* of the graph. However, we crucially make use of the fact that the failure model is an LRO model. We show that under the LRO model, the expected coverage of a vertex by any set F depends only on a *single* vertex in F. This allows us to use the tree-decomposition meaningfully and obtain an efficient algorithm parameterized by the treewidth.

Related Work. Daskin [5] formulated the maximum expected coverage problem where nodes have a survival probability (unlike our case where edges have survival probability). For a fixed network G with demands on vertices, a surviving probability $p > 0$ common for all the vertices and the radius of coverage R, Daskin introduced the problem and LP-based heuristic algorithms. He also considered the dependencies in the vertex failure in the generalized version, which has been subsequently studied as the LRO failure model [9,10].

The MAX-EXP-COVER-R problem is studied as *most reliable source* problem when restricted to the budget $\mathcal{B} = 1$ and $R = \infty$ and the RFM failure model. Melachrinoudis and Helander [13] considered the most reliable source problem on trees, and gave a polynomial time algorithm for the same. Later, Ding and Xue [8] gave a linear time algorithm for most reliable source problem on trees. Colbourn and Xue [4] gave an $O(n^2)$-time algorithm for most reliable source problem on series-parallel graphs. Ding [6] gave an $O(n^2)$-time algorithm for most reliable source problem on ring graphs. Later, he generalized the result for most reliable source on general graphs and gave an $O(mn + n^2 \log n)$-time algorithm [7].

Hassin et al. [9] studied the MAX-EXP-COVER-R problem with assumption that the edge failure follows the LRO model. They proved that the MAX-EXP-COVER-R problem is NP-hard even under the LRO failure model. The MAX-EXP-COVER problem is a distance relaxed version of MAX-EXP-COVER-R problem in which $R = \infty$. Hassin et al. [9] showed that MAX-EXP-COVER problem is linear time solvable.

2   An FPT Algorithm for $R = 1$ Under LRO Model

In this section we present our FPT algorithm for the MAX-EXP-COVER-1 problem under the LRO failure model for *bounded tree-width graphs*. We start by recalling the LRO model and the preliminaries and then present our dynamic programming algorithm.

LRO model [9,10]: We assume that the input given to us is a graph $G = (V, E)$, a budget $\mathcal{B}$, the weight function on the vertices and the survival probability on the edges. For an edge $e = (u, v) \in E$, the survival probability is denoted by $p(e)$ or $p(uv)$. Throughout the paper we assume that the failure model is the LRO model. In the LRO model the failure of an edge e_i fails implies that all the edges of lower survival probability also fail. We assume that the survival probability of the edges are distinct.

We visualize the model as a set of all edge subgraphs that could survive in the event of a failure. Each graph in this set is referred to as a realization. For the LRO model there are exactly $m + 1$ realizations of the graph G. Each realization also has an associated probability of survival that can be calculated for the LRO model as follows: Let us order the edges $E = \{e_0, e_1 \ldots e_m, e_{m+1}\}$ in descending order of the surviving probability from the most reliable edge to the least reliable edge. The edges e_0 and e_{m+1} are dummy edges added in the ordering with survival probabilities $p(e_0) = 0$ and $p(e_{m+1}) = 1$, respectively. We consider the linear ordering $\preceq$ on the $m + 1$ realizations of G due to LRO failure model: for $0 \leq i < j \leq m$, $G_i \preceq G_j$. For $0 \leq i \leq m$ let G_i be a realization in which e_i is the weakest edge that survives. The probability of the realization is $P(G_i) = p(e_i) - p(e_{i+1})$.

Preliminaries: For every non empty vertex subset $U \subseteq V$, the subgraph induced by U is denoted by $G[U]$. For a vertex v let $N_G(v)$ denote the closed neighborhood of v. For any subset $U \subseteq V$, closed neighborhood of the set U in G is denoted by $N_G[U]$, that is, $N_G[U] = \cup_{u \in U} N_G(u)$. Suppose the graph parameter is not given to the neighborhood function, then the input graph G is the default parameter.

A tree decomposition [3,11] of a graph G is a pair (X, H) such that H is a tree and $X = \{X_i \subseteq V : i \in H\}$. For each node $i \in H$, X_i is referred as *bag* of i. In addition, the following three conditions hold:

(a) For each vertex $v \in V$, there is a node $i \in H$ such that $v \in X_i$.
(b) For each edge $(u, v) \in E$, there is a node $i \in H$ such that $u, v \in X_i$.
(c) For each vertex $v \in V$, the induced subtree of the nodes in H that contains v is connected.

The *width* of a tree decomposition is the $\max_{i \in H}(|X_i| - 1)$. The *tree-width* of a graph, denoted by tw, is the minimum width over all possible tree decompositions of G. For our algorithm, we require a special kind of decomposition, called the nice tree decomposition which we define below.

Definition 1 (Nice tree decomposition [11]). *A nice tree decomposition is a tree decomposition, rooted by a special node r with $X_r = \{\}$ and each node in the tree decomposition is one of the following four type of nodes.*

1. **Leaf node.** *A node $i \in H$ with no child and $X_i = \{\}$.*
2. **Introduce node.** *A node $i \in H$ with one child j such that $X_i = X_j \cup \{v\}$ for some $v \notin X_j$.*
3. **Forget node.** *A node $i \in H$ with one child j such that $X_i = X_j \setminus \{v\}$ for some $v \in X_j$.*
4. **Join node.** *A node $i \in H$ with two children j and k such that $X_i = X_j = X_k$.*

Given a tree decomposition (X, H) of a graph G with width k, a nice tree decomposition (X', H') with same width and $O(nk)$ nodes can be computed in time $O(nk^2)$ [11]. Hereafter we will assume the tree decomposition is nice. For each node $i \in H$, H_i denotes the subtree rooted at i. Let V_i be the set of all vertices in the bag of nodes in the subtree H_i.

$$V_i = \begin{cases} X_i & \text{if } i \text{ is a leaf node} \\ X_i \cup \bigcup_{j \in ch(i)} V_j & \text{if } i \text{ is a non-leaf node} \end{cases}$$

where $ch(i)$ is the set of all children of i in H. For a join node i, we denote by j and k its left and right children respectively. In case i has only one child, as in the case of introduce and forget nodes, only the left child j is well-defined and k does not exist. In this case, V_k is assumed to be the empty set.

2.1 Dynamic Programming Formulation for Bounded Treewidth

We now present an FPT algorithm for the MAX-EXP-COVER-1 problem under LRO model on bounded treewidth graphs. Our main observation is that the expected coverage of a vertex u by a set $S \subseteq V$ is dependent on a single vertex $v \in S$ in case of LRO model. We refer to v as the *best neighbour* of u in S. This LRO specific intuition is the key idea in our dynamic programming algorithm when G is represented by a nice tree decomposition of a fixed treewidth.

Definition 2 [Best Neighbour]. *For a vertex $u \in V$ and a set $S \subseteq V$, the best neighbor of u in S, $bn(u,S)$ is defined as follows:*

$$bn(u,S) = \begin{cases} u & \text{if } u \in S \\ \arg\max_{v \in S \cap N(u)} p(uv) & \text{if } u \notin S \text{ and } S \cap N(u) \neq \emptyset \\ undefined & \text{if } S \cap N[u] = \emptyset \end{cases}$$

We know that for any two sets $S, T \subseteq V$, $\mathcal{C}(T,S) = \sum_{u \in T} \mathcal{C}(u,S)$.

Lemma 1. *For $u \in V$ and $S \subseteq V$, if the coverage $\mathcal{C}(u,S) > 0$, then there is a vertex $v \in S$ such that $\mathcal{C}(u,S) = \mathcal{C}(u,v)$.*

Proof. The value of $\mathcal{C}(u,S)$ is same as $\mathcal{C}(u, S \cap N[u])$. Let $S' = S \cap N[u]$. If $u \in S'$ then the coverage of u depends on itself and we are done. Hence assume $u \notin S'$. Consider $S' = \{v_1, v_2, \ldots, v_l\}$ ordered according to the decreasing order of $p(uv_i)$, that is, $p(uv_i) > p(uv_{i+1})$ for $1 \leq i < l$. In the LRO model, if the edge (u, v_1) fails then all the edges $(u, v_i), i \geq 2$ would also fail. Now consider the ordering of the realizations $G_0, G_1, \ldots, G_m$ according to the LRO model. We note that $\mathcal{C}(u, S')$ can be written as:

$$\mathcal{C}(u, S') = w(u) \cdot \sum_{G_i \in \mathcal{Q}} P(G_i) \cdot I(G_i, S', u, 1)$$

In the linear ordering of the graph let G_k be the first graph such that the edge (u, v_1) survives. Note that for all realizations G'_k where $k' < k$, we have

$I(G_{k'}, S', u, 1) = 0$, implying that the coverage of u by S' in $G_{k'}$ is zero. On the other hand, for all realizations $G_{k'}$ where $k' \geq k$, we have $I(G'_k, S', u, 1) = 1$, implying that the coverage of u by S' in $G_{k'}$ is $w(u)$. Thus, we have

$$C(u, S') = w(u) \cdot \sum_{G_{k'}:k' \geq k} P(G_{k'})$$

Now we note that, $\sum_{G_{k'}:k' \geq k} P(G_{k'}) = p(uv_1)$. Therefore, the expected coverage of u by S is dependent only on v_1, that is $C(u, S) = C(u, S') = C(u, v_1) = w(u) \cdot p(uv_1)$.   □

As a consequence of Lemma 1, we write the following equation for $C(u, S)$ for any $u \in V$ and $S \subseteq V$.

Observation 1. *The expected coverage of a vertex $u \in V$ by a set $S \subseteq V$ is given by:*

$$C(u, S) = \begin{cases} w(u) & \text{if } u \in S \\ w(u) \cdot p(uv) & \text{if } v \notin S \text{ and } v = bn(u, S) \\ 0 & \text{if } bn(u, S) \text{ is undefined} \end{cases}$$

Structure of a solution S for V_i: We now present a recursive formulation of $C(V_i, S)$ for a node i in the tree decomposition H. For any $S \subseteq V_i$ of size at most B, we define the following partition of X_i:

- $A = S \cap X_i$ and $Z = S \backslash A$.
- Let Y_A be the set of vertices in $X_i \backslash A$ whose best neighbour in S is from A.
- Let Y_Z be the set of vertices in $X_i \backslash A$ whose best neighbour in S is from Z.
- Let U be the vertices in $X_i \backslash A$ whose best neighbour in S is undefined.

We refer to $\mathsf{P} = (A, Y_A, Y_Z, U)$ as the S-compatible partition of X_i which is unique for S. Note that for any set S and the S-compatible partition P we can rewrite the coverage of V_i by S as follows:

$$C(V_i, S) = C(V_i \backslash X_i, S) + \sum_{v \in X_i} C(v, S)$$

$$= C(V_i \backslash X_i, S) + \sum_{v \in A} C(v, v) + \sum_{v \in Y_A} C(v, A) + \sum_{v \in Y_Z} C(v, Z) + \sum_{v \in U} C(v, S)$$

$$= C(V_i \backslash X_i, S) + C(A, A) + C(Y_A, A) + C(Y_Z, Z) + C(U, S)$$

$$= C(V_i \backslash X_i, S) + C(A \cup Y_A, A) + C(Y_Z, Z)$$

Dynamic programming formulation based on H: Let i be a node in H. For each $S \subseteq V_i$ and $0 \leq |S| \leq B$, there is a unique S-compatible partition $\mathsf{P} = (A, C_A, C_Z, U)$ of X_i. While we do not know the set $S \subseteq V_i$ that maximizes $C(V_i, S)$, we know that each S gives a unique 4-way partition of X_i. Therefore, our DP formulation seeks to finds the optimal set S for each budget value b. In order to find the set S, our algorithm maintains at each node $i \in H$ a table T_i. Each row in T_i corresponds to a 2-tuple consisting of an integer $0 \le b \le B$ and a four way partition $P = (A, C_A, C_Z, U)$ of the set X_i. We compute two entries for each row (b, P) which are denoted by $T_i[b, P].\mathbf{solution}$ and $T_i[b, P].\mathbf{value}$. Let $T_i[b, P].\mathbf{solution}$ be a set $A \cup Z$ satisfying the following conditions:

1. $Z \subseteq V_i \backslash X_i$ and $|A \cup Z| = b$
2. $A \cup Z$ is such that $\mathcal{C}(V_i \backslash X_i, A \cup Z) + \mathcal{C}(A \cup C_A, A) + \mathcal{C}(C_Z, Z)$ is maximized.

For Z defined above, let

$$T_i[b, P].\mathbf{value} = \mathcal{C}(V_i \backslash X_i, A \cup Z) + \mathcal{C}(A \cup C_A, A) + \mathcal{C}(C_Z, Z)$$

Since we consider all possible values of b, it may be the case that, $b > |V_i|$. In such cases, the table entry $T_i[b, P].\mathbf{solution} = $ invalid and $T_i[b, P].\mathbf{value} = -\infty$ for all possible partitions P of X_i. Our DP algorithm will compute the set Z efficiently assuming that the table entries for the children of the node i have been correctly computed.

Table at the root of H: The root of the nice tree decomposition H a node with the empty bag. Let r be the root node with $X_r = \{\}$. From the structure of the nice tree decomposition it is clear that r is a forget node. The only valid four-way partition of X_r is $P = (\{\}, \{\}, \{\}, \{\})$. We will prove that the set $T_r[\mathcal{B}, P].\mathbf{solution}$ attains the optimum value $T_r[\mathcal{B}, P].\mathbf{value}$ for the MAX-EXP-COVER-1 problem on G.

2.2   Recursive Formulation at the Nodes of H

We now formally state the table entries for all the four types of nodes in the nice tree decomposition H, show how they are computed in a bottom-up approach (from leaf to the root), and prove the structure of an optimum solution for each type of node.

Leaf node. Let i be a leaf node with bag $X_i = \{\}$. Since the bag of the leaf node is an empty set, the T_i has only one valid entry in which the partition of X_i is $P = (\{\}, \{\}, \{\}, \{\})$ and $b = 0$.

$$T_i[b, P] = \begin{cases} \mathbf{solution} = \{\}, \mathbf{value} = 0 & \text{if } b = 0 \\ \mathbf{solution} = \text{invalid}, \mathbf{value} = -\infty & 1 \le b \le B \end{cases}$$

It is clear that the empty set is the set that achieves the maximum for the MAX-EXP-COVER-1 problem on the empty graph with budget $b = 0$. Therefore, at the leaf nodes in H, the table T_i maintains the optimum solution for each row.

Introduce node. Let i be an introduce node with child j such that $X_i = X_j \cup \{v\}$ for some $v \notin X_j$. Let $P = (A, C_A, C_Z, U)$ be a four way partitioning of X_i. There are two cases in the recurrence at an introduce node, depending on whether v belongs to A.

1. Let $v \notin A$. Then v belongs to exactly one of C_A, C_Z or U. Let $\mathsf{P}_j = (A, C_A \backslash v, C_Z \backslash v, U \backslash v)$ be the partition of X_j obtained by removing v from the corresponding set in the partition P. In this case

$$T_i[b, \mathsf{P}].\textbf{solution} = T_j[b, \mathsf{P}_j].\textbf{solution}$$

$$T_i[b, \mathsf{P}].\textbf{value} = \begin{cases} T_j[b, \mathsf{P}_j].\textbf{value} & \text{if } v \notin C_A \\ T_j[b, \mathsf{P}_j].\textbf{value} + C(v, A) & \text{if } v \in C_A \end{cases}$$

2. Let $v \in A$. Then, there may exist some vertices $u \in C_A$ such that the "best neighbour" of u in the set A is the vertex v, $bn(u, A) = v$. Let C_{Av} be the set of such vertices, that is, $C_{Av} = \{u \in C_A \mid bn(u, A) = v\}$. Let $\mathsf{P}_j = (A \backslash v, (C_A \backslash C_{Av}), C_Z, U \cup C_{Av})$ be a partitioning of $X_j = X_i \backslash v$. In this case

$$T_i[b, \mathsf{P}].\textbf{solution} = T_j[b - 1, \mathsf{P}_j].\textbf{solution} \cup \{v\}$$
$$T_i[b, \mathsf{P}].\textbf{value} = T_j[b - 1, \mathsf{P}_j].\textbf{value} + C(\{v\} \cup C_{Av}, v)$$

In both the cases, we spent $O(n)$ time to find compatible partition and $O(m+n)$ time to find the additional coverage. Then time to update an entry is $O(m+n)$.

Forget node. Let i be a forget node with child j such that $X_i = X_j \backslash \{v\}$ for some vertex $v \in X_j$. Let $\mathsf{P} = (A, C_A, C_Z, U)$ be a four way partitioning of X_i. Given P we consider the following four partitions of X_j.

- $\mathsf{P}_1 = (A \cup \{v\}, C_A, C_Z, U)$
- $\mathsf{P}_2 = (A, C_A \cup \{v\}, C_Z, U)$
- $\mathsf{P}_3 = (A, C_A, C_Z \cup \{v\}, U)$
- $\mathsf{P}_4 = (A, C_A, C_Z, U \cup \{v\})$

Define $\mathsf{P}_j = \underset{\mathsf{P}' \in \{\mathsf{P}_1, \mathsf{P}_2, \mathsf{P}_3, \mathsf{P}_4\}}{\arg\max} T_j[b, \mathsf{P}'].\textbf{value}$. Then $T_i[b, \mathsf{P}]$ is updated as follows:

$$T_i[b, \mathsf{P}].\textbf{solution} = T_j[b, \mathsf{P}_j].\textbf{solution}$$
$$T_i[b, \mathsf{P}].\textbf{value} = T_j[b, \mathsf{P}_j].\textbf{value}$$

In case of forget node, we spend $O(1)$ time to find the four compatible partitions and update a table entry.

Join node. Let i be a join node with children j and k. We know that $X_i = X_j = X_k$. Let $\mathsf{P} = (A, C_A, C_Z, U)$ be a partition of X_i. The recurrence for $T_i[b, \mathsf{P}]$ is based on the observation that if a vertex u in C_Z contributes a non-zero value to the expected coverage by $T_i[b, \mathsf{P}].\textbf{solution}$, then it is adjacent to a vertex v which is exclusively in one of $V_j \backslash X_j$ or $V_k \backslash X_k$. If such a $v \in T_i[b, \mathsf{P}].\textbf{solution}$ was to be in both $V_j \backslash X_j$ and $V_k \backslash X_k$, then it would also be in X_i by the definition of a tree decomposition. This would mean $v \in A$ and would then imply that $u \in C_A$. However, this contradicts the premise that $u \in C_Z$. Therefore, the value of $T_i[b, \mathsf{P}]$ is computed by considering all partitions of C_Z into $C_{Z1} \cup C_{Z2}$

as follows: For any partition $C_{Z1} \cup C_{Z2}$ of C_Z, we define the partitions on X_j and X_k as $\mathsf{P}'_1 = (A, C_A, C_{Z1}, U \cup C_{Z2})$ and $\mathsf{P}'_2 = (A, C_A, C_{Z2}, U \cup C_{Z1})$. We consider the following triple:

$$(b', C_{Zj}, C_{Zk}) = \underset{\substack{0 \le b_1 \le b-|A|, \\ C_{Z1} \cup C_{Z2} = C_Z}}{\arg \max} \quad T_j[b_1 + |A|, \mathsf{P}'_j].\mathbf{value} + T_k[b - b_1, \mathsf{P}'_k].\mathbf{value}$$

Based on the triple (b', C_{Zj}, C_{Zk}) we consider the partitions $\mathsf{P}_j = (A, C_A, C_{Zj}, U \cup C_{Zk})$ and $\mathsf{P}_k = (A, C_A, C_{Zk}, U \cup C_{Zj})$ of X_j and X_k, respectively. Then the $T_i[b, \mathsf{P}]$ is defined recursively as follows:

$$T_i[b, \mathsf{P}].\mathbf{solution} = T_j[b' + |A|, \mathsf{P}_j].\mathbf{solution} \cup T_k[b - b', \mathsf{P}_k].\mathbf{solution}$$
$$T_i[b, \mathsf{P}].\mathbf{value} = T_j[b' + |A|, \mathsf{P}_j].\mathbf{value} + T_k[b - b', \mathsf{P}_k].\mathbf{value} - \mathcal{C}(A \cup C_A, A)$$

Since we consider all possible partitions of the set C_Z and all possible budget values, each table entry for a join node is computed in $O(\mathcal{B} \cdot 2^{|X_i|})$ time.

Correctness of the update rules. We now show that the update rules correctly compute T_i at each node i in H.

Theorem 1. *Let i be a node in H. For a non-negative integer b and a partition $\mathsf{P} = (A, C_A, C_Z, U)$ of X_i, the update step for $T_i[b, \mathsf{P}]$ described in the equations given above is correct. In other words, $T_i[b, \mathsf{P}].\mathbf{solution}$ is the set $A \cup Z$ such that $(A \cup Z) \cap X_i = A$, $T_i[b, \mathsf{P}].\mathbf{value} = \mathcal{C}(V_i \backslash X_i, Z) + \mathcal{C}(A \cup Y_A, A) + \mathcal{C}(Y_Z, Z)$, and this value is the maximum over all $A \cup Z'$ such that $(A \cup Z') \cap X_i = A$.*

Proof. The proof is by induction on the height of a node in H. The height of a node i in a rooted tree H is the distance to the farthest leaf in the subtree rooted at i. The base case is when i is a leaf node in H, and clearly its height is 0. For a leaf node i with $X_i = \{\}$, the row with $b = 0$, and $\mathsf{P} = (\{\}, \{\}, \{\}, \{\})$ is the only valid row entry and its value is 0. This completes the proof of the base case. Let us assume that the claim is true for all nodes of height at most $h - 1 \ge 0$. We prove that if the claim is true at all nodes of height at most $h - 1$, then it is true for a node of height h. Let b be a budget and $\mathsf{P} = (A, C_A, C_Z, U)$ be a partition of X_i. Let $T_i[b, \mathsf{P}] = S = A \cup Z$. We prove that the table entry updated at $T_i[b, \mathsf{P}]$ is correct by contradiction. Let $S' = A \cup Z'$ be the solution optimal than S for $T_i[b, \mathsf{P}]$. That is, $T_i[b, \mathsf{P}].\mathbf{value} < \mathcal{C}(V_i \setminus X_i, S') + \mathcal{C}(A \cup C_A, A) + \mathcal{C}(C_Z, Z')$. The proof is by considering the 3 cases for the type of node i. In the interest of space we give the proof for the introduce node only.

Introduce node. Let i be an introduce node with child j such that $X_i = X_j \cup \{v\}$ for some $v \notin X_j$. We now consider two cases depending on whether v belongs to S.

1. Let $v \notin S$. The table entry $T_i[b, \mathsf{P}]$ is computed from the table entry $T_j[b, \mathsf{P}_j]$ of j where $\mathsf{P}_j = (A, C_A{}', C_Z{}', U')$ is a partition of X_j corresponding to P. In other words, the partition P_j is obtained by removing v from some set in the partition P. Since v is introduced at i, the neighbours of v in $G[V_i]$ are in X_i only. Then the coverage of v by S depends on A. We consider the coverage of V_j by S corresponding to the partition P_j.

$$
\begin{aligned}
T_j[b, \mathsf{P}_j].\textbf{value} &= \mathcal{C}(V_j \backslash X_j, S) + \mathcal{C}(A \cup C_A{}', A) + \mathcal{C}(C_Z{}', Z) \\
&= T_i[b, \mathsf{P}].\textbf{value} - \mathcal{C}(v, A) \\
&= \mathcal{C}(V_i \backslash X_i, S) + \mathcal{C}(A \cup C_A, A) + \mathcal{C}(C_Z, Z) - \mathcal{C}(v, A) \\
&< \mathcal{C}(V_i \backslash X_i, S') + \mathcal{C}(A \cup C_A, A) + \mathcal{C}(C_Z, Z') - \mathcal{C}(v, A)
\end{aligned}
$$

This shows that the set S' is optimal than S for the entry $T_j[b, \mathsf{P}_j]$. This contradicts the optimality of table constructed in j which is at height $h' < h$.

2. Let $v \in S$. The entry $T_j[b, \mathsf{P}]$ is computed using the entry $T_j[b-1, \mathsf{P}_j]$ in j where $\mathsf{P}_j = (A \backslash \{v\}, C_A \backslash C_{Av}, C_Z, U \cup C_{Av})$ is the partition of X_j corresponding to P. Let C_{Av} be the set of vertices in C_A such that their best neighbour in A is v. The coverage of C_{Av} by S is depends on v. We consider the coverage of V_j by $S \backslash \{v\}$ corresponding to the partition P_j.

$$
\begin{aligned}
T_j[b, \mathsf{P}_j].\textbf{value} &= \mathcal{C}(V_j \backslash X_j, S \backslash \{v\}) + \mathcal{C}((A \cup C_A) \backslash (\{v\} \cup C_{Av}), A \backslash \{v\}) \\
&\quad + \mathcal{C}(C_Z, Z) \\
&= T_i[b, \mathsf{P}].\textbf{value} - \mathcal{C}(\{v\} \cup C_{Av}, v) \\
&< \mathcal{C}(V_i \backslash X_i, S') + \mathcal{C}(A \cup C_A, A) + \mathcal{C}(C_Z, Z') - \mathcal{C}(\{v\} \cup C_{Av}, v)
\end{aligned}
$$

This shows that the set $S' \backslash \{v\}$ is optimal than $S \backslash \{v\}$ for the table entry $T_j[b-1, \mathsf{P}_j]$. This contradicts the optimality of table constructed in node j which is at height $h' < h$.

The proofs for the forget and the join node use similar arguments to show that if the table entry of T_i is not optimal then there exists a table entry in one of the child node T_j which is sub-optimal. □

Running time. At each node $i \in H$, the table T_i consists of $(\mathcal{B} + 1) \cdot 4^{X_i}$ rows. The time to compute the table entries depends on the type of node. Among the four type of nodes, the join node requires $O(\mathcal{B} \cdot 2^{|X_i|})$ time to update each entry. Then, the update operation on each node i requires $O(\mathcal{B}^2 \cdot 8^{|X_i|})$. For a graph G with width tw, the nice tree-decomposition has $O(n \cdot tw)$ many nodes. For any node $i \in H$, $|X_i| \leq tw$. Therefore, the time to complete the update operation on H requires $O(\mathcal{B}^2 \cdot tw \cdot n \cdot 8^{tw})$.

We conclude the following theorem.

Theorem 2. *Given an instance G of the* MAX-EXP-COVER-1 *problem where the graph is input as a nice tree decomposition of width at most tw, the Dynamic Programming based algorithm described above solves* MAX-EXP-COVER-1 *in time $O(\mathcal{B}^2 \cdot tw \cdot n \cdot 8^{tw})$. By hiding the polynomial factors, the running time is $O^*(8^{tw})$.*

3   A PTAS for MAX-EXP-COVER-1 on planar graphs

In this section, we design a PTAS for MAX-EXP-COVER-1 for planar graphs using the $O^*(8^{tw})$-time algorithm in Sect. 2.1 to solve the MAX-EXP-COVER-1 problem on bounded treewidth graphs.

A *planar graph* is a graph that can be drawn in the plane without edge crossings, such a drawing is called an *planar embedding*. An embedding of a planar graph divides the plane into regions, called faces. A face is called exterior or outer, if the region is outside the graph embedding on the plane. An outerplanar or 1-outerplanar graph is a planar graph, such that all vertices lie in the exterior face. Given a planar embedding of G, the vertices on the exterior face are level 1 vertices. The level i vertices of G is the set of vertices lie that on the exterior face of the G after removing the vertices of levels $1, \ldots, i - 1$. A graph G is k-outerplanar if there exists a planar embedding of G, such that all the vertices of G are level i vertices for some $i \leq k$. Given a planar graph G, a k-outerplanar embedding of G for which k is minimal can be found in polynomial time using the algorithm in [2].

Our approach is similar to the Baker's technique – each planar graph is a k-outerplanar graph for some k, and each k-outerplanar graphs has a treewidth of at most $3k + 1$. The Baker's technique uses dynamic programming on a tree decomposition of treewidth $3k + 1$ to design PTASs for different optimization problems [1]. Additionally, we use the property that a k-outerplanar graph has a nice tree decomposition with a treewidth of at most $3k + 1$.

For an input $0 < \epsilon < 1$ to the PTAS, Let $k = \lceil \frac{1}{\epsilon} \rceil$. As mentioned earlier, in an l-outerplanar embedding of a planar graph for each vertex $v \in V$ there is a unique i, $1 \leq i \leq l$ such that v belongs to level i. Therefore, for each vertex there is a corresponding level number given an l-outerplanar embedding. For $1 \leq i \leq k$, let G_i be an edge subgraph of G obtained by deleting all the edges (u, v) for which u has a level number congruent to $((i - 1) \mod k)$ and v has a level number congruent to $(i \mod k)$. Clearly, for each $1 \leq i \leq k$, each connected component of G_i is a k-outerplanar graphs. The nice tree decomposition of each such k-outerplanar graph G_i can be computed in polynomial time with treewidth $3k + 1$ using the algorithm in [15]. Let S_i be an optimum solution for the MAX-EXP-COVER-1 problem in $G_i = (V, E_i)$ where S_i is obtained in time $O^*(8^{O(k)})$ using the exact algorithm in Sect. 2.1. Let S be the set amongst $\{S_1, \ldots, S_k\}$ which achieves the maximum expected coverage in G. In other words, let $S = \arg\max_{S_i \in \{S_1 \ldots S_k\}} \mathcal{C}(G, V, S_i)$. We use the input graph as a parameter to the coverage function.

Lemma 2. *Let G be an instance of* MAX-EXP-COVER-1. *Let $S \subseteq V$ be the set identified by the procedure described above. The solution S is a $(1 - \frac{1}{k})$ approximation for* MAX-EXP-COVER-1 *problem on input G.*

The procedure described in the proof and Lemma 2 together complete the proof of the main theorem of our paper which is as follows.

Theorem 3. *Given an instance G of* MAX-EXP-COVER-1 *and an $0 < \epsilon < 1$ there is an algorithm running in time $O^*(8^{O(\frac{1}{\epsilon})})$ that outputs a set $S \subseteq V$ which is a $(1 - \epsilon)$ approximation to the optimum, that is $\mathcal{C}(V, S) \geq (1 - \epsilon)\mathcal{C}(V, OPT)$.*

Submodularity of $\mathcal{C}(V, F)$ and the greedy approximation. It is also well-known from [12] that the R-neighbourhood function defined on the set of all subsets of vertices in a graph is a monotone submodular function for each $R \geq 0$. It is well known, for example from [14], that a positive linear combination of submodular functions $f_1 \ldots f_k$ using k non-negative constant $c = \{c_1 \ldots c_k\}$ is a submodular function. In other words, $f'(X) = \sum_{i=1}^{k} c_i \cdot f_i(X)$ is a submodular function. Since $\mathcal{C}(V, F)$ is the expectation of the size (or weighted size) of the R-neighbourhood of F in the graph in $\mathcal{Q}$, it follows that $\mathcal{C}(V, F)$ is a convex combination (a special type of positive linear combination) of submodular functions. Therefore $\mathcal{C}(V, F)$ is a submodular function. Consequently, due to Nemhauser et al. [14] the MAX-EXP-COVER-R problem has a $(1 - \frac{1}{e})$-approximation algorithm and it runs in polynomial time for all those distributions $\mathcal{D}$ for which $\mathcal{C}(V, F)$ can be computed in polynomial time. In particular, for the LRO failure model MAX-EXP-COVER-R has a polynomial time greedy $(1 - \frac{1}{e})$-approximation algorithm as has been mentioned by Hassin et al. in [10].

References

1. Baker, B.S.: Approximation algorithms for NP-complete problems on planar graphs. J. ACM **41**(1), 153–180 (1994)
2. Bienstock, D., Monma, C.L.: On the complexity of embedding planar graphs to minimize certain distance measures. Algorithmica **5**(1), 93–109 (1990)
3. Bodlaender, H.L.: A tourist guide through treewidth. Acta Cybern. **11**, 1–23 (1993)
4. Colbourn, C.J., Xue, G.: A linear time algorithm for computing the most reliable source on a series-parallel graph with unreliable edges. Theor. Comput. Sci. **209**(1), 331–345 (1998)
5. Daskin, M.S.: A maximum expected covering location model: formulation, properties and heuristic solution. Transp. Sci. **17**(1), 48–70 (1983)
6. Ding, W.: Computing the most reliable source on stochastic ring networks. In: 2009 WRI World Congress on Software Engineering, vol. 1, pp. 345–347, May 2009
7. Ding, W.: Extended most reliable source on an unreliable general network. In: 2011 International Conference on Internet Computing and Information Services, pp. 529–533, September 2011
8. Ding, W., Xue, G.: A linear time algorithm for computing a most reliable source on a tree network with faulty nodes. Theor. Comput. Sci. **412**(3), 225–232 (2011). Combinatorial Optimization and Applications
9. Hassin, R., Ravi, R., Salman, F.S.: Tractable cases of facility location on a network with a linear reliability order of links. In: Fiat, A., Sanders, P. (eds.) ESA 2009. LNCS, vol. 5757, pp. 275–276. Springer, Heidelberg (2009). https://doi.org/10.1007/978-3-642-04128-0_24
10. Hassin, R., Ravi, R., Salman, F.S.: Multiple facility location on a network with linear reliability order of edges. J. Comb. Optim. **34**(3), 1–25 (2017)
11. Kloks, T. (ed.): Treewidth: Computations and Approximations. LNCS, vol. 842. Springer, Heidelberg (1994). https://doi.org/10.1007/BFb0045375

12. Lovasz, L.: Matching Theory (North-Holland Mathematics Studies). Elsevier Science Ltd., Oxford (1986)
13. Melachrinoudis, E., Helander, M.E.: A single facility location problem on a tree with unreliable edges. Networks **27**(3), 219–237 (1996)
14. Nemhauser, G.L., Wolsey, L.A., Fisher, M.L.: An analysis of approximations for maximizing submodular set functions. Math. Program. **14**(1), 265–294 (1978)
15. Shmoys, D.B., Williamson, D.P.: The Design of Approximation Algorithms, 1st edn. Cambridge University Press, New York (2011)

On the Decision Trees with Symmetries

Artur Riazanov[1,2]([✉])

[1] St. Petersburg Academic University, Khlopina 8, St. Petersburg, Russia
[2] St. Petersburg Department of V. A. Steklov Institute of Mathematics
of the Russian Academy of Sciences, St. Petersburg, Russia
aariazanov@gmail.com

Abstract. We introduce a propositional proof system based on decision trees utilizing symmetries of formulas. We refer to this proof system as decision trees with symmetries (SDT). SDT can be polynomially simulated by the proof system SR-I introduced by Krishnamurthy [7]; SR-I extends Resolution with the symmetry rule. We show that there are polynomial-size proofs of the functional pigeonhole principle (FPHP_n^{n+1}) and of an encoding of the clique coloring principle ($\text{CLIQUE-COLORING}_{n,k}$). On the other hand we show that any SDT for the pigeonhole principle (PHP_n^{n+1}) has size $2^{\Omega\left(n^{1/3-o(1)}\right)}$ despite that PHP_n^{n+1} has a lot of symmetries. In 1999 Urquhart [11] showed that PHP_n^{n+1} has a polynomial-size SR-I refutation. Thus SDT is strictly weaker than SR-I. The smallest decision tree for PHP_n^{n+1} has size $2^{\Omega(n \log n)}$; we show that there exists an SDT for PHP_n^{n+1} of size $2^{O(\sqrt{n})}$.

1 Introduction

Symmetry is widely used in informal proofs, we use symmetry arguments every time we say "without losing of generality" or just "analogously". In this paper, we will consider the use of symmetries in formal propositional proof systems.

The first application of symmetry in a propositional proof system was introduced by Krishnamurthy in [7] who extended the Resolution by the symmetry rule (we denote the resulting proof system by SR-I). Let φ be the formula which is being refuted. Then symmetry rule allows deriving $\pi(C)$ from C, where C is a clause and π is a permutation of the variables of φ such that the set of the clauses of φ is not changed after the application of the permutation π to its variables. In 1999 Urquhart [11] proved that the pigeonhole principle PHP_n^{n+1} has polynomial-size proof in SR-I which implies that SR-I is strictly stronger than Resolution since PHP_n^{n+1} is hard for Resolution [4]. On the other hand Urquhart proved an exponential lower-bound on the size of SR-I refutation of Tseitin Formulas.

There are also several stronger symmetry based proof systems that were studied. The proof system SR-II was studied by Arai and Urquhart in [1]; this system allows the application of local symmetry rule which allows deriving the clause $\pi(C)$ from a clause C where π is a symmetry of a subset $\Gamma \subseteq \varphi$ (where φ is the refuted formula) and C is derived from Γ. In 2005 Szeider [10] suggested

© Springer International Publishing AG, part of Springer Nature 2018
F. V. Fomin and V. V. Podolskii (Eds.): CSR 2018, LNCS 10846, pp. 282–294, 2018.
https://doi.org/10.1007/978-3-319-90530-3_24

proof systems HR-I and HR-II that differ from SR-I and SR-II by using homomorphisms instead of permutations. Szeider proved exponential lower bounds for complexity HR-I and HR-II as well as exponential separation between HR-I and HR-II.

All mentioned lower bounds were obtained as follows: we take a formula that is hard for resolution and then artificially make it asymmetrical. Thus all known lower bounds for SR-I are proved for formulas that have a small number of symmetries. It is more interesting to have an example of a hard formula that has many "essential" symmetries.

In this paper, we introduce a new kind of symmetry based proof system, decision trees with symmetries (SDT). A decision tree for a CNF formula φ is a binary tree with vertices labeled with variables of φ such that all inner vertices have two outgoing edges one labeled with 0 and another labeled with 1 and for each leaf there exists a clause of φ which is falsified by the assignment corresponding to the path from the root to the leaf. Also, any variable appears only once on every path from the root to a leaf. A decision tree with symmetries has two types of leaves. Leaves of the first type are the same as in a plain decision tree: for each of these leaves, there exists a clause of φ falsified by the assignment corresponding to the path from the root to this leaf. Leaves of the second type do not falsify any clauses but for each such leaf there exists an inner vertex v of the SDT such that the assignment corresponding to the leaf is isomorphic to the assignment corresponding to v i.e. the assignments are the same up to the renaming of the variables according to a symmetry of φ. Similar notions were studied in the context of CSP-solvers. Symmetry Excluding Search Trees were introduced by Backofen and Will in [2]; so-called Group Equivalence trees were defined by Roney-Dugal et al. in [9]. In the mentioned papers symmetries are used in order to reduce the search space for satisfiable formulas, while SDT uses symmetries for unsatisfiable formulas in order to cut off the cases that are symmetric to the already explored ones.

We show that SR-I polynomially simulates SDT. It is well-known that tree-like Resolution is equivalent to decision trees. But it is easy to see that tree-like SR-I is equivalent to tree-like Resolution. And it follows from our result that SDT are strictly stronger than tree-like Resolution. In this paper we prove the following upper bounds for SDT:

- We show that there exists an SDT of size $O(n^3)$ for the functional pigeonhole principle $FPHP_n^{n+1}$ which is known to be hard for the Resolution [3].
- We show that there exists an SDT of size $O(k^2n)$ for the appropriate encoding of the clique coloring principle $CLIQUE\text{-}COLORING_{k,n}$. It follows from this and [8], that SDT is not p-simulated by Cutting Planes. This enhances the similar result of Urquhart that separates SR-I from Cutting Planes [11].

SDT has advantages over decision trees only for formulas with a huge number of symmetries. Namely, we show that an SDT for a formula with s symmetries is at most s times smaller than the smallest decision tree for the same formula.

The main result of the paper is a lower bound on the size of an SDT for the pigeonhole principle. We show that the size of any SDT for PHP_n^{n+1} is

$2^{\Omega(n^{1/3-o(1)})}$. As a corollary, we get that SR-I is strictly stronger than SDT. This result is also interesting since PHP_n^{n+1} has a lot of essential symmetries (these symmetries help in case of SR-I but do not help in the case of SDT). In order to prove this result, we develop a game that can be used for proving lower bounds on SDT.

Due to constraints on the volume of the paper, we omit some of the proofs.

2   Preliminaries

Propositional Formulas. We consider only formulas in conjunctive normal form (CNF). We identify a clause with the set of its literals, i.e. $z \in C$ means that the clause C contains the literal z. If the corresponding clause is empty, we denote it as $\square$. We identify a CNF formula with the set of its clauses. If φ is the conjunction of the empty set of clauses then we assume that φ is identically true. We denote the set of variables of φ by $\mathrm{Var}(\varphi)$.

A *partial assignment* is a mapping $\alpha : X \rightarrow \{0, 1, *\}$, where X is a set of propositional variables. Let $\varphi|_\alpha$ be the result of substitution of a partial assignment $\alpha : \mathrm{Var}(\varphi) \rightarrow \{0, 1, *\}$ to a formula φ. Let $\mathcal{A}_\varphi$ denote the set of partial assignments to variables of a formula φ. A formula φ is said to be *satisfiable* if there exists a full assignment α such that $\varphi|_\alpha = 1$ and *unsatisfiable* otherwise.

The Pigeonhole Principle. PHP_n^m is a CNF formula with variables P_{ij}, $i \in [m]$, $j \in [n]$. P_{ij} states whether the i'th pigeon flies to the j'th hole or not. The formula consists of the following clauses: $\bigvee_{j=1}^n P_{ij}$ for all $i \in [m]$ (every pigeon flies to some hole); $\neg P_{i,k} \vee \neg P_{j,k}$ for all $i \neq j$; $i, j \in [m]$; $k \in [n]$ (no two pigeons fly to the same hole).

We refer to the first type of clauses as *pigeon axioms* and for the second type as *hole axioms*. Note that PHP_n^m *does not* say that a pigeon fly to *exactly* one hole as well as it doesn't say that for every hole some pigeon flies to it.

We define a formula for the pigeonhole principle with the requirement that every pigeon flies exactly into one hole as the conjunction of PHP_n^m add clauses $\neg P_{ij} \vee \neg P_{ik}$ for all $i \in [m]$; $j, k \in [n]$; $j < k$ and call it the *functional pigeonhole principle* (FPHP_n^m).

Clique-Coloring Tautology. For n and k such that $k < n$ define unsatisfiable formula $\mathrm{CLIQUE\text{-}COLORING}_{k,n}(x, y, z) = \mathrm{CLIQUE}_{k,n}(x, z) \wedge \mathrm{COLORING}_{k-1,n}(y, z)$, which says that the graph encoded by variables $\{z_{ij}\}_{i<j\in[n]}$ has a k-vertex clique and a vertex coloring in $k - 1$ colors such that no two vertices with the same color are connected by an edge. $z_{ij} = 1$ iff (i, j) is the edge of the graph. We identify z_{ij} and z_{ji} for convenience. A clique is encoded by nk variables $\{x_{ij}\}_{i\in[k]; j\in[n]}$. $x_{ij} = 1$, iff j is the i'th vertex of the clique. A coloring is encoded by $(k - 1)n$ variables $\{y_{ij}\}_{i\in[k-1]; j\in[n]}$. $y_{ij} = 1$, iff the color i is assigned to the vertex j. The formula has the following clauses:

CL1 $\neg x_{iu} \vee \neg x_{jv} \vee z_{uv}$ for all $i, j \in [k]$; $i \neq j$; $u, v \in [n]$; $u \neq v$ (if both vertices are in the set then they are connected by an edge);

CL2 $\neg x_{iu} \vee \neg x_{iv}$ for all $i \in [k]$; $u, v \in [n]$; $u \neq v$ (there is no more than one vertex with number i);

CL3 $\bigvee_{j=1}^{n} x_{ij}$ for all $i \in [k]$ (there is at least one vertex with number i);

CL4 $\neg x_{iu} \vee \neg x_{ju}$ for all $u \in [n]$; $i, j \in [k]$; $i \neq j$ (each vertex should have a unique number in the clique).

COL1 $\bigvee_{j=1}^{k-1} y_{ji}$ for all $i \in [n]$ (vertex i has to be assigned with a color);

COL2 $\neg y_{ai} \vee \neg y_{bi}$ for all $a \neq b \in [k - 1]$, (vertex can have at most one color);

COL3 $\neg z_{ij} \vee \neg y_{ai} \vee \neg y_{aj}$ for all $i, j \in [n]$; $i \neq j$ and $a \in [k - 1]$ (if two vertices are assigned with the same color, they can not be connected by an edge).

Usually the formula is stated without constraints **CL2** and **COL2** because it's unsatisfiable without them, but we need these constraints for our construction.

Pudlak proved that the formula CLIQUE-COLORING$_{k,n}$ is hard for Resolution and Cutting Planes [8].

Decision Trees. Informally a decision tree is a protocol of (unsuccessful) backtracking search for the satisfying assignment for an unsatisfiable formula φ. Each vertex of a tree corresponds to a splitting based on the value of some variable, thus each vertex corresponds to a partial assignment. For each leaf, an assignment associated with it falsifies some clause of φ.

Formally, *a decision tree* for a formula φ is a binary tree T with root r, such that all its inner vertices have two outgoing edges and the following conditions hold. Each inner vertex v is labeled with a variable $\text{var}(v) \in \text{Var}(\varphi)$ and the edges going out of v are labeled with 0 and 1. We refer to $\text{var}(v)$ as *the splitting variable for v*. Let's denote the end of the edge from v labeled with b as $\text{son}(v, b)$. Each vertex v of T is associated with a partial assignment $\mathcal{S}_v : \text{Var}(\varphi) \to \{0, 1, *\}$ corresponding to the path from the root to v. Each variable can appear *at most once* on every path from the root to a leaf of a decision tree. For the root all variables are left unassigned: $\mathcal{S}_r \equiv *$. If we reason about several decision trees we write $\mathcal{S}_c^T$ instead of $\mathcal{S}_c$ in order to clarify the tree where c belongs.

For each leaf v there exists a clause $C \in \varphi$, such that $C\big|_{\mathcal{S}_v} = 0$. If this condition doesn't hold for some of the leaves we call such tree *a partial decision tree*.

We denote the set of all vertices of T as $\mathcal{V}(T)$. The size of a tree is the number of vertices in it. A *subtree* of a vertex c is the set of its descendants, we denote it by $\text{subtree}(c)$.

It is easy to see that there is a decision tree for an unsatisfiable formula (see for example [6]) i.e. decision trees is a complete proof system.

Symmetries of Formulas and SR-I. Let φ be a CNF formula: $\varphi = \bigwedge_{i=1}^{m} C_i$. Let $\pi : \text{Var}(\varphi) \to \text{Var}(\varphi)$ be a permutation. We extend π to the set of literals as $\pi(\neg x) = \neg \pi(x)$ for $x \in \text{Var}(\varphi)$. We denote the clause obtained by the application of π to all literals of a clause C by $\pi(C)$. Similarly for a CNF formula φ denote $\pi(\varphi) = \bigwedge_{C \in \varphi} \pi(C)$. π is called *a symmetry* of φ if $\varphi = \pi(\varphi)$ as sets of clauses, i.e. application of φ permutes clauses of φ. We denote the set of symmetries of

φ as $\mathrm{Sym}(\varphi)$. Let $\pi(\alpha)$ be the result of the application of symmetry π to an assignment $\alpha : \mathrm{Var}(\varphi) \to \{0, 1, *\}$, namely $\pi(\alpha)(x) = \alpha(\pi(x))$.

Let A and B be some clauses and let x be a propositional variable. *The resolution rule* allows to derive $A \vee B$ from $A \vee x$ and $A \vee \neg x$.

If φ is a CNF formula then *a derivation of a clause R from φ in the system SR-I* [7] is a sequence of clauses $C_1, \ldots, C_k$ such that $C_k = R$ and each of the clauses in the sequence satisfies one of the conditions:

1. $C_i \in \varphi$.
2. C_i is derived from clauses C_j, C_k (where $j, k < i$) by the resolution rule.
3. There exists $j < i$ and $\pi \in \mathrm{Sym}(\varphi)$ such that $C_i = \pi(C_j)$.

A refutation of φ in SR-I is a derivation of $\square$ from φ.

Decision Trees with Symmetries. A partial decision tree T for a formula φ is a *decision tree with symmetries* (SDT) of φ if each of its leaves is labeled with an element of the set $\mathcal{V}(T) \cup \{\bot\}$ and the following conditions hold: (i) If a leaf l is labeled with $\bot$ then there exists $C \in \varphi$ such that $C|_{\mathcal{S}_l} = 0$. (ii) If a leaf l is labeled with inner vertex u then there exists $\pi \in \mathrm{Sym}(\varphi)$ such that $\mathcal{S}_l = \pi(\mathcal{S}_u)$. We refer to these leaves as *symmetry-derived*. If for some of the leaves neither (i) nor (ii) is true, we call the corresponding partial decision tree *a partial* SDT and call these leaves *uncovered*.

Lemma 1. *Let T be an SDT for some formula. Then there exists an order of vertices $v_1, \ldots, v_n$ such that each symmetry-derived leaf v_i is labeled with a vertex v_j such that $j < i$; and if an inner vertex v_k has two sons v_{k_0} and v_{k_1} then $k_0, k_1 < k$.*

Remark 1. Every decision tree is an SDT by definition and thus SDT is the complete proof system, i.e. there is an SDT for any unsatisfiable formula.

Proposition 1. *If T is an SDT for φ of size S then there exists a refutation of φ in SR-I of size at most S.*

Thus SDT is a sound proof system (i.e. if there is an SDT for a formula then it is unsatisfiable) and all lower bounds for SR-I yield lower bounds for SDT.

Lemma 2 *[hanging lemma].* *Let T_1 be a partial SDT for φ with one uncovered leaf c. Let T_2 be an SDT for $\varphi|_{\mathcal{S}_c^{T_1}}$ such that for every symmetry π used in T_2, π' the extension of π to $\mathrm{Var}(\varphi)$ preserving the variables fixed by $\mathcal{S}_c^{T_1}$ is a symmetry of φ. Then T is obtained by hanging T_2 to c is an SDT for φ.*

3   Polynomial Upper Bounds on SDT

SDT for the Functional Pigeonhole Principle. The first thing we need to show is that PHP_n^m (as well as FPHP_n^m) has a lot of symmetries.

Lemma 3. *Let* $\pi_1 \in S_n$ *and* $\pi_2 \in S_m$ *be two permutations. Then* σ : $\mathrm{Var}\,(\mathrm{PHP}_n^m) \to \mathrm{Var}\,(\mathrm{PHP}_n^m)$ *defined by the rule* $\sigma\,(P_{ij}) = P_{\pi_1(i),\pi_2(j)}$ *is a symmetry of* PHP_n^m. *For* $n, m \geq 3$ *there are no other symmetries.*

Theorem 1. *There is an* SDT *of size* $O(n^3)$ *for* FPHP_n^{n+1}.

Lemma 4. *Let* $\varphi = \mathrm{FPHP}_n^m$ *with variables* P_{ij}. *Then for any* $k \leq \min\{n, m\}$ *there exists a partial* SDT T *of size* $O(k \max\{n, m\}^2)$ *with at most one uncovered leaf* c *such that for each* $P_{ij} \in \mathrm{Var}\,(\mathrm{FPHP}_n^m)$:

$$
\mathcal{S}_c(P_{ij}) = \begin{cases} * & i > k \wedge j > k \\ 1 & i = j \leq k \\ 0 & (i \leq k \vee j \leq k) \wedge i \neq j \end{cases}
$$

The assignment $\mathcal{S}_c$ corresponds to the distribution where for every $i \in [k]$ the i'th pigeon flies to the i'th hole, no other pigeons fly to this hole and the i'th pigeon does not fly anywhere else.

The idea of the proof of Lemma 4 is straightforward: we give the explicit construction of the SDT for $k = 1$ and after that hang the tree for $\mathrm{FPHP}_{n-1}^{m-1}$ by the leaf c using Lemma 2.

Proof (of Theorem 1). We use Lemma 4 for FPHP_n^{n+1} and $k = n$ and hang a constant-size full-search tree by the leaf c. □

SDT for CLIQUE-COLORING. We define $\alpha \in \mathcal{A}_{\mathrm{COLORING}_{k-1,n}}$ as $\alpha(z_{ij}) = 1$ for $i, j \in [k]$ and $\alpha(z_{ij}) = *$ otherwise. α corresponds to the fact that $[k]$ is a clique in the graph encoded by z.

Lemma 5. *For any permutations* $\pi_1 \in S_n, \pi_2 \in S_k, \pi_3 \in S_{k-1}$ *a permutation defined as* $x_{ij} \mapsto x_{\pi_2(i),\pi_1(j)}$, $y_{ij} \mapsto y_{\pi_3(i),\pi_1(j)}$, $z_{ij} \mapsto z_{\pi_1(i),\pi_1(j)}$ *is a symmetry of* CLIQUE-COLORING$_{k,n}$.

Using Lemma 4 we prove the following lemma:

Lemma 6. *There exists a partial* SDT T *of size* $O(nk^2)$ *for* CLIQUE$_{k,n}$ *with exactly one uncovered leaf* c *such that* $\forall i, j \in [k]\, \mathcal{S}_c(z_{ij}) = 1$.

Lemma 7. COLORING$_{k-1,n}\big|_\alpha$ *contains* $\psi(\mathrm{FPHP}_{k-1}^k)$ *as a subformula, where* ψ *is the renaming of the variables by the rule* $\psi(P_{ij}) = y_{ji}$.

Theorem 2. *There exists an* SDT *of size* $O(nk^2)$ *for* CLIQUE-COLORING$_{k,n}$.

Proof. Let T_1 be the partial SDT obtained using Lemma 6, let c be the uncovered leaf of T_1. Note that $\alpha(z_{ij}) = 1$ implies that $\mathcal{S}_c(z_{ij}) = 1$.

Let T_2 be an SDT for FPHP_{k-1}^k using Theorem 1, after the renaming of the variables according to ψ (see Lemma 7).

We consider the tree T obtained by hanging T_2 to the leaf c of the tree T_1.

The trees T_1 and T_2 satisfy the prerequisites of Lemma 2, indeed all symmetries of the formula FPHP_{k-1}^k used in the Theorem 1 after the renaming according to ψ are symmetries of $\text{CLIQUE-COLORING}_{k,n}$ involving only variables $\{y_{ij}\}$, since the symmetries used in Theorem 1 only rearrange holes which correspond to colors in the formula $\text{COLORING}_{k-1,n}$. Therefore T is an SDT for $\text{CLIQUE-COLORING}_{k,n}$.   □

Corollary 1. *Cutting Plane does not polynomially simulates* SDT.

4   Symmetries in a Plain Decision Tree

In this section, we establish a connection between decision trees and SDTs for formulas. Although we call SDT "a tree" it is, in fact, a DAG. The following simple result helps us to extract a decision tree from an SDT and analyze it, instead of analyzing an SDT itself.

Let φ be a CNF formula and let T be a decision tree for it. We define a binary relation $\sim$ on the set of assignments $\mathcal{A}_\varphi$. If $\alpha_1, \alpha_2 \in \mathcal{A}_\varphi$ then $\alpha_1 \sim \alpha_2$ iff there exists $\pi \in \text{Sym}(\varphi)$, such that $\pi(\alpha_1) = \alpha_2$. We refer to assignments related by $\sim$ as *isomorphic*. We define $\sim$ on $\mathcal{V}(T)$ by the rule $u \sim v$ iff $\mathcal{S}_u \sim \mathcal{S}_v$. We denote the set of equivalence classes of a set A with respect to $\sim$ as $A/\sim$.

It is clear that $\sim$ is an equivalence relation. The following lemma provides a way to prove lower bounds on the size of SDT using the number of equivalence classes with respect to this relation.

Lemma 8. *Let T be an SDT for $\varphi = \bigwedge_{i=1}^m C_i$. Then there exists a decision tree T_1 for φ such that $|\mathcal{V}(T_1)/\sim| = |\mathcal{V}(T)/\sim|$.*

The idea of the proof is to expand all symmetry derivations i.e. substitute each symmetry-derived leaf with a subtree of its label with the corresponding symmetry applied to all vertices of the subtree. It is straightforward that the number of equivalence classes remains the same after this operation.

Corollary 2. *If the number of equivalence classes with respect to $\sim$ on the set of vertices of any decision tree for φ is at least k, then any SDT for φ has size at least k.*

Proposition 2. *If φ has L symmetries and there exists an SDT T for φ such that $|T| = k$, then there exists a decision tree T' for φ such that $|T'| \le Lk$.*

The similar proposition holds for SR-I: if φ has L symmetries and an SR-I refutation of size k then it has a resolution refutation of size Lk.

5   Complexity of PHP_n^{n+1}

We have shown that there is a polynomial-size SDT for FPHP_n^m. In this section we give an exponential lower bound on the size of an SDT for PHP_n^{n+1}.

The strategy of the proof of a lower bound is as follows: we consider a plain decision tree for PHP_n^{n+1} and show that there are exponentially many vertices v such that $\mathcal{S}_v$ are not isomorphic to each other. Hence by Corollary 2 any SDT for PHP_n^{n+1} has exponential size. We implement this strategy using the following plan:

1. For every set of partial assignments S we define a game of two players Alice and Bob such that if Bob has a winning strategy then every plain decision tree for PHP_n^{n+1} contains a vertex v such that $\mathcal{S}_v \in S$.
2. We construct exponentially many sets S_i such that for every S_i Bob has winning strategy in the game defined in the previous step and for all $i \neq j$, and for every $\alpha \in S_i$ and $\beta \in S_j$, α and β are not isomorphic.

5.1   Game Interpretation

Let φ be an unsatisfiable CNF formula and $S \subseteq \mathcal{A}_\varphi$ be a set of partial assignments. Consider a game of Alice and Bob which proceeds as follows.

In the beginning of the game $\alpha \equiv *$. Then at each round, Alice chooses a variable v that is unassigned by α. After that Bob chooses a value $b \in \{0,1\}$. Then they assign $\alpha(v) := b$ and continue. The game ends if one of the following happens: (i) α falsifies a clause $C \in \varphi$. In this case Alice wins; (ii) α does not falsify a clause of φ and $\alpha \in S$. In this case Bob wins. We denote this game as $\Gamma(\varphi, S)$.

Lemma 9. *If there is a winning strategy for Bob for $\Gamma(\varphi, S)$, then in every decision tree T for φ there exists a vertex v such that $\mathcal{S}_v \in S$.*

5.2   Construction of Non-isomorphic Winning Sets

According to our plan, we have to construct exponentially many disjoint sets of partial assignments such that (1) assignments form different sets are not isomorphic and (2) Bob has winning strategy for every such set. We will define such sets as preimages of a mapping with an image of exponential size. In order to satisfy property (1) we will consider only specific mappings:

Let φ be a CNF formula and $\mathcal{A}_\varphi$ be the set of its partial assignments. A function $\mu : \mathcal{A}_\varphi \to A$ (where A is an arbitrary set) is *an invariant* with respect to the symmetries of φ if for any $\alpha, \beta \in \mathcal{A}_\varphi$, $\alpha \sim \beta$ implies $\mu(\alpha) = \mu(\beta)$.

It is easy to see that the sets $\mu^{-1}(a)$ for different $a \in A$ satisfy property (1) if μ is an invariant function. Now we need to construct an invariant function such that Bob has a winning strategy for $\Gamma\left(\text{PHP}_n^{n+1}, \mu^{-1}(a)\right)$ for an exponential number of elements $a \in A$.

5.3   Naive Invariant

Let us start with a simple and natural invariant on $\mathcal{A}_{\text{PHP}_n^{n+1}}$. We identify the variables of PHP_n^{n+1} with the cells of a matrix $(n+1) \times n$ then every partial assignment is a $(n+1) \times n$ matrix with elements from $\{0, 1, *\}$.

$$
\begin{array}{|ccc|}
1 & 0 & * \\
0 & * & 1 \\
* & 0 & * \\
0 & 1 & *
\end{array}
\quad
\begin{array}{|ccc|}
0 & * & 1 \\
0 & 1 & * \\
1 & * & 0 \\
* & * & 0
\end{array}
\quad
\begin{array}{|ccc|}
1 & 0 & 0 \\
* & 1 & 0 \\
* & * & 1 \\
* & * & *
\end{array}
$$

$$\alpha_1 \qquad\quad \alpha_2 \qquad\quad \alpha_3$$

Fig. 1. Examples of partial assignments of PHP_3^4. By Lemma 3, the application of any symmetry of PHP_n^{n+1} to an assignment α is a permutation of rows and columns of the corresponding table. Note that $\alpha_1 \sim \alpha_2$ (one can permute columns of α_1 according the permutation (23) and permute rows of α_1 according to the permutation (341)) but $\alpha_1 \not\sim \alpha_3$ since α_1 and α_3 have different number of zeroes. $\mu_0(\alpha_1) = \mu_0(\alpha_2) = \{(1,1),(1,0)\}$; $\mu_0(\alpha_3) = \{(2,1),(1,1),(0,1),(0,0)\}$.

For $\alpha \in \mathcal{A}_{\mathrm{PHP}_n^{n+1}}$ and $i \in [n+1]$ we denote $\mathbb{1}_\alpha(i) = \#\{j : \alpha(P_{ij}) = 1\}$ and $\mathbb{0}_\alpha(i) = \#\{j : \alpha(P_{ij}) = 0\}$ where $\#S$ stands for the size of the set S. Consider an invariant μ_0 that maps $\alpha \in \mathcal{A}_{\mathrm{PHP}_n^{n+1}}$ to the set $\mu_0(\alpha) = \{(\mathbb{0}_\alpha(i), \mathbb{1}_\alpha(i)) : i \in [n+1]\}$, i.e. the set of pairs (number of zeroes in a row, number of ones in a row).

Proposition 3. μ_0 is an invariant with respect to symmetries of PHP_n^{n+1}.

Proof. The proof is straightforward since permutation of rows and columns of the matrix of α does not change $\mu_0(\alpha)$ (Fig. 1). □

Notice that we can not implement our plan with this μ_0 since Alice has winning strategies for $\Gamma(\mathrm{PHP}_n^{n+1}, \mu_0^{-1}(a))$ for many distinct a. For example, if $a \supseteq \{(3,0),(1,0)\}$, then Alice can choose variables in the order $P_{1,1}, P_{2,1}, \ldots, P_{n+1,1}, P_{1,2}, \ldots, P_{n+1,2}, \ldots, P_{n+1,n}$ (i.e. hole by hole) and then in every moment for every two rows the difference between the number of assigned values in them does not exceed 1. Hence Bob will never get $(3,0)$ and $(1,0)$ simultaneously.

Thus we construct a strategy using an invariant with a smaller image.

5.4   More Complicated Invariant

Let $T_n = \{(x,y) \in \mathbb{N}^2 : x,y \geq 0; x+y \leq n\}$. For any $g : T_n \to A$ we define the function $I_g : \mathcal{A}_{\mathrm{PHP}_n^{n+1}} \to 2^A$ as follows: $I_g(\alpha) = \{g(x) : x \in \mu_0(\alpha)\}$. It is straightforward that I_g is an invariant with respect to symmetries of PHP_n^{n+1}.

Our goal is to find such $g : T_n \to A$ that for a large number of $x \in 2^A$ Bob has a winning strategy for $\Gamma(\mathrm{PHP}_n^{n+1}, I_g^{-1}(x))$. First of all we define an auxiliary game that is roughly speaking a one-pigeon variant of the game $\Gamma(\mathrm{PHP}_n^{n+1}, I_g^{-1}(S))$.

Auxiliary Game. Let $g : T_n \to A \cup \{\$\}$, $d \in [n]$, $a \in A$. Alice and Bob have one pigeon axiom of PHP_n^{n+1}: $P_{1,1} \vee P_{1,2} \vee \ldots \vee P_{1,n}$. At each round Alice has two options: (i) Assign zero to one of the variables $P_{1,i}$. (ii) Choose a variable

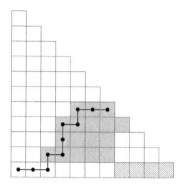

Fig. 2. T_{10} and a trajectory in it. Grey cells corresponds to the same value of the function. With respect to the depicted trajectory the function is 3-robust (although it is not necessary 3-robust in general). Hatched cells are the ones restricted by the condition 2b of robustness. No trajectory can enter these cells.

$P_{1,i}$ and let Bob to assign it. Alice can use the first option at most d times. For $k \in [n]$ let α_k be the resulting partial assignment after k rounds. Then define $\ell_k := g\left(\#\{i \in [n] : \alpha_k(P_{1,i}) = 0\}, \#\{i \in [n] : \alpha_k(P_{1,i}) = 1\}\right)$. The game ends after n rounds and Bob wins if there exists $i \in [n-1]$ such that $\ell_j = \$$ for $j \le i$ and $\ell_j = a$ for $j > i$. Otherwise, Alice wins. We denote this game as $\gamma_g(d, a)$.

Let us give some intuition how this auxiliary game helps in the main game Γ. Assume that we are describing a strategy of Bob in the game $\Gamma\left(\mathrm{PHP}_n^{n+1}, I_g^{-1}(S \cup \{\$\})\right)$, where $S = \{s_1, \ldots, s_k\} \subseteq A$. Roughly speaking, we will implement Bob's winning strategies for $\gamma_g(d, s_1), \ldots, \gamma_g(d, s_k)$ for different pigeons. Alice's moves of type (i) correspond to the situations where assigning 1 to a variable will lead to a conflict with a hole axiom (Fig. 2).

Let $g : T_n \to A \cup \{\$\}$ and let $a \in A$. A *trajectory* of a is a sequence (x_0, y_0), $(x_1, y_1), \ldots, (x_n, y_n) \in T$ such that (i) $x_0 = y_0 = 0$, $(x_{i+1}, y_{i+1}) \in \{(x_i + 1, y_i), (x_i, y_i + 1)\}$; (ii) for each $i \in [n-1]$, $y_n > 0$; (iii) for $i \in [n]$ if $y_i > 0$ then $g(x_i, y_i) = a$.

Now we specify requirements to function g that allow us to construct a winning strategy for Bob in the game $\gamma_g(d, a)$. A function $g : T_n \to A \cup \{\$\}$ is called *d-robust* if the following conditions hold:

1. $g(x, 0) = \$$ for every $x \in \{0, \ldots, n\}$.
2. For each $a \in A$, there exists a trajectory $(x_0, y_0), \ldots, (x_n, y_n)$ of a such that for all $i \in \{0, \ldots, n\}$ and $\delta \in \{0, \ldots, d\}$ (a) if $y_i > 0$ and $x_i + y_i + \delta \le n$ then $g(x_i + \delta, y_i) = a$ (b) if $y_i = 0$ then $x_i + d < n$.

Proposition 4. *If $g : T_n \to A \cup \{\$\}$ is d-robust then Bob has a winning strategy in $\gamma_g(d, a)$ for any $a \in A$.*

Proof. Let a sequence $(x_0, y_0), \ldots, (x_n, y_n)$ be a trajectory of a satisfying the conditions (2a) and (2b) of robustness. Let m be the number of Bob's moves. A winning strategy for Bob is to make moves according to a trajectory $(x_0, y_0), \ldots, (x_n, y_n)$ satisfying the properties (2a) and (2b) of robustness (regardless of Alice's moves). Suppose Z_k is the number of zeroes assigned by Bob after his k'th move (Alice's moves of type (i) do not count) and O_k is the number of ones assigned by Bob after his k'th move. Bob will make moves such that $(Z_k, O_k) = (x_k, y_k)$. By the definition of a trajectory, it is always possible. Since Alice can assign zero by herself no more than d times the real pair of the number assigned zeroes and the number of assigned ones will belong to the set $\{(Z_k + i, O_k) : i \in \{0\} \cup [d]\}$. Using the property (2a) of robustness, if $O_k > 0$ then $g(Z_k, O_k) = g(Z_k + 1, y_k) = \ldots = g(Z_k + d, O_k) = a$. By the properties (1) and (2b) of the robustness if $O_k = 0$ then $g(Z_k, O_k) = g(Z_k + 1, O_k) = \ldots = g(Z_k + d, O_k) = \$$ and $Z_k + d < n$ thus if $O_k = 0$ then $Z_k + O_k + d < n$. Since at the end of the game all variables should be assigned, $O_m > 0$ where m is the number of Bob's moves. Since $O_0 \leq O_1 \leq \ldots \leq O_m$ and $O_m > 0$, $(\ell_0, \ldots, \ell_n) = (\$, \ldots, \$, a, \ldots, a)$. Notice that without the property (2b) the number of zeroes assigned by Alice and Bob can reach n making $\ell_n = \$$.   □

Winning Strategy in the Main Game. For a function $g : T_n \to A \cup \{\$\}$ we denote $m_a(g) = \max\{y \in [n] : \exists x \in [n] : (x, y) \in g^{-1}(a)\}$ for $a \in A$. Then *the rank of g is defined as follows:* $r(g) = \sum_{a \in A} m_a(g)$.

Recall that we plan to describe a strategy of Bob in the game $\Gamma\left(\text{PHP}_n^{n+1}, I_g^{-1}(\{s_1, \ldots, s_k, \$\})\right)$ that implements Bob's winning strategies for $\gamma_g(d, s_i)$ for each of the pigeons. By this moment we did not say what value of d we use. We will use $d = r(g)$; indeed, all Alice's moves in the auxiliary game $\gamma_g(d, s_i)$ of type (i) correspond to conflicts with hole axioms, and hence the number of such moves does not exceed the number of assigned ones. Since Bob implements the strategy from Proposition 4 for every $i \in [k]$, the number of the assigned ones does not exceed $\sum_{i=1}^k m_{s_i}(g) \leq r(g)$.

Theorem 3. *If $g : T_n \to A \cup \{\$\}$ is $r(g)$-robust then for each $Q \subseteq A$ Bob has a winning strategy in $\Gamma\left(\text{PHP}_n^{n+1}, I_g^{-1}(Q \cup \{\$\})\right)$.*

Proof. Let $Q = \{q_1, q_2, \ldots, q_m\}$, let t_j be a trajectory for q_j for $j \in [m]$ that satisfies properties (2a) and (2b) of $r(g)$-robustness. We denote the current assignment by α. Bob maintains the following invariant: for every $i \in [n + 1]$ either $\mathbb{1}_\alpha(i) = 0$ or $\mathbb{1}_\alpha(i) > 0$ and there is $k(i) \in [m]$ such that $g(\mathbb{0}_\alpha(i) - d_\alpha(i), \mathbb{1}_\alpha(i)) \in t_{k(i)}$ where $d_\alpha(i)$ is the number of Bob's forced moves for the pigeon i (i.e. Bob can not assign 1 because of a conflict with a hole axiom). All defined $k(i)$ are distinct. Initially α is the empty assignment thus the invariant holds. Assume that Alice asks for a variable P_{ij}. If $k(i)$ is defined then Bob implements a winning strategy for the auxiliary game $\gamma_g(r(g), q_{k(i)})$ where Alice's moves of type (i) correspond to the forced moves of Bob in the main game. If $k(i)$ is undefined Bob assigns $P_{ij} = 1$ iff the j'th hole is empty and $g(\mathbb{0}_\alpha(i) - d_\alpha(i), 1) \in t_s$ where $s \in [m]$ and $k(i') \neq s$ for all $i' \in [n + 1]$.

Let us prove that Bob wins. Assume that Alice wins. Since we never falsify a hole axiom, in the last move we falsified the i'th pigeon axiom. Notice that in this case $\mathbb{0}_\alpha(i) - d_\alpha(i)$ took all values from $[n - r(g)]$ throughout the game. Then for each $j \in [n - r(g) - 1]$ if a trajectory t_s contains the point $(j, 1)$ then it was occupied by another pigeon at the moment when $\mathbb{0}_\alpha(i) - d_\alpha(i) = j$. Since by the property (2b) of robustness and by the property (i) of the trajectory, every trajectory contains a point from the set $[n - r(g) - 1] \times \{1\}$, all trajectories are occupied. For each pigeon i that occupies a trajectory Bob implements a winning strategy for $\gamma_g(r(g), q_{k(i)})$, hence $q_{k(i)} \in I_g(\alpha)$ and $I_g(\alpha) = Q \cup \{\$\}$ before the last move. Therefore Bob have already won before the considered move. □

Corollary 3. *If $g : T_n \to A \cup \{\$\}$ is a function such that g is $r(g)$-robust then any SDT for PHP_n^{n+1} has size at least $2^{|A|}$.*

Lemma 10. *There exists a family of functions $g_n : T_n \to [k_n] \cup \{\$\}$, each of which is $r(g_n)$-robust and $k_n = \Omega(n^{1/3 - o(1)})$.*

Lemma 10 and Corollary 3 imply

Theorem 4. *The size of any SDT for PHP_n^{n+1} is $2^{\Omega(n^{1/3 - o(1)})}$.*

5.5 An Upper Bound for the Size of SDT for PHP_n^{n+1}

We construct an SDT of size $2^{O(\sqrt{n})}$ for PHP_n^{n+1} which is significantly smaller then the smallest decision tree for PHP_n^{n+1}. In the proof we use the asymptotic of the number of partitions of an integer by Hardy and Ramanujan [5]. This upper bound is non-trivial but does not match with our lower bound. It is interesting problem to find matching bounds.

Theorem 5. *There exists an SDT for PHP_n^{n+1} of size $2^{O(\sqrt{n})}$.*

Acknowledgements. The author thanks Dmitry Itsykson for the problem statement and fruitful discussions and anonymous referees for useful comments and corrections. The research was supported by Russian Science Foundation (project 16-11-10123).

References

1. Arai, N.H., Urquhart, A.: Local symmetries in propositional logic. In: Dyckhoff, R. (ed.) TABLEAUX 2000. LNCS (LNAI), vol. 1847, pp. 40–51. Springer, Heidelberg (2000). https://doi.org/10.1007/10722086_3
2. Backofen, R., Will, S.: Excluding symmetries in constraint-based search. In: Jaffar, J. (ed.) CP 1999. LNCS, vol. 1713, pp. 73–87. Springer, Heidelberg (1999). https://doi.org/10.1007/978-3-540-48085-3_6
3. Ben-Sasson, E., Wigderson, A.: Short proofs are narrow: resolution made simple. In: Proceedings of the Thirty-first Annual ACM Symposium on Theory of Computing, STOC 1999, pp. 517–526. ACM (1999)
4. Haken, A.: The intractability of resolution. Theor. Comput. Sci. **39**, 297–308 (1985)

5. Hardy, G.H., Ramanujan, S.: Asymptotic formulaæ in combinatory analysis. Proc. Lond. Math. Soc. **2**(1), 75–115 (1918)
6. Jukna, S.: Boolean Function Complexity: Advances and Frontiers. Springer Publishing Company, Incorporated, Heidelberg (2012). https://doi.org/10.1007/978-3-642-24508-4
7. Krishnamurthy, B.: Short proofs for tricky formulas. Acta Informatica **22**(3), 253–275 (1985)
8. Pudlák, P.: Lower bounds for resolution and cutting plane proofs and monotone computations. J. Symb. Log. **62**(3), 981–998 (1997)
9. Roney-Dougal, C.M., Gent, I.P., Kelsey, T., Linton, S.: Tractable symmetry breaking using restricted search trees. In: Proceedings of the 16th European Conference on Artificial Intelligence, pp. 211–215. IOS Press (2004)
10. Szeider, S.: The complexity of resolution with generalized symmetry rules. Theory Comput. Syst. **38**(2), 171–188 (2005)
11. Urquhart, A.: The symmetry rule in propositional logic. Discret. Appl. Math. **96**, 177–193 (1999)

On Emptiness and Membership Problems
for Set Automata

A. Rubtsov[1,2] and M. Vyalyi[1,2,3]([✉])

[1] National Research University Higher School of Economics, Moscow, Russia
rubtsov99@gmail.com
[2] Moscow Institute of Physics and Technology, Moscow, Russia
[3] Dorodnicyn Computing Centre, FRC CSC RAS, Moscow, Russia
vyalyi@gmail.com

Abstract. We consider a computational model which is known as set automata.

The set automata are one-way finite automata with an additional storage—the set. There are two kinds of set automata—the deterministic and the nondeterministic ones. We denote them as DSA and NSA respectively. The model was introduced by Kutrib et al. in 2014 in [2,3].

In this paper we characterize algorithmic complexity of the emptiness and membership problems for set automata. More definitely, we prove that both problems are **PSPACE**-complete for both kinds of set automata.

1 Introduction

We consider a computational model which is known as set automata. A set automaton is a one-way finite automaton equipped with an additional storage—the set $\mathbb{S}$—which is accessible through the work tape. On processing of a word, the set automaton can write a word z on the work tape and perform one of the following operations: the operation **in** inserts the word z into the set $\mathbb{S}$, the operation **out** removes the word z from the set $\mathbb{S}$ if $\mathbb{S}$ contains z, and the operation **test** checks whether z belongs to $\mathbb{S}$. After each operation the work tape is erased.

There are two kinds of set automata—the deterministic and the nondeterministic ones. We denote them as DSA and NSA respectively.

If determinism or nondeterminism of an automaton is not significant, we use abbreviation SA, and we refer to the class of languages recognizable by (N)SA as SA. We denote as DSA the class of languages recognizable by DSA.

Set automata were introduced by Kutrib et al. in 2014 in [2,3]. The results of these conference papers are covered by the journal paper [4], so in the sequel we give references to the journal variant.

We recall briefly results from [4] about structural and decidability properties of DSA. They are presented in the tables, see Fig. 1. In the first table we

The study has been funded by the Russian Academic Excellence Project '5-100'. Supported in part by RFBR grants 16–01–00362 and 17–51–10005.

	DSA	CFL	DCFL
$L \overset{?}{=} \varnothing$	+	+	+
$L \overset{?}{\in} REG$	+	−	+
$L \overset{?}{=} R$	+	−	+
$\|L\| \overset{?}{<} \infty$	+	+	+

	DSA	CFL	DCFL
$L_1 \cdot L_2$	−	+	−
$L_1 \cup L_2$	−	+	−
$L_1 \cap L_2$	−	−	−
$\Sigma^* \setminus L$	+	−	+
$L \cup R$	+	+	+
$L \cap R$	+	+	+

Fig. 1. Structural and decidability properties

list decidability problems: emptiness, regularity, equality to a regular language and finiteness. In the tables, R denotes an arbitrary regular language. The second table describes the structural properties: L, L_1 and L_2 are languages from the corresponding classes; we write $+$ in a cell if the class is closed under the operation, otherwise we write $-$.

From Fig. 1 one can see that DSA languages look similar to DCFL. This similarity was extended in [7]. It appears that languages recognizable by NSA form a rational cone, so as CFL. Also, several algorithmic results were obtained in [7]:

- DSA $\subset$ **P** and there are **P**-complete languages in DSA;
- the membership problem is **P**-complete for DSA without ε-loops;
- SA $\subset$ **NP** and there are **NP**-complete languages in SA;
- the emptiness problem is **PSPACE**-hard for DSA.

In this paper we complete the algorithmic characterization of the emptiness and membership problems for set automata. We prove that the membership problem is **PSPACE**-hard for general DSA. We also prove that the emptiness problem for NSA is in **PSPACE**. Thus, it is **PSPACE**-complete. Due to easy reductions, it implies **PSPACE**-completeness for the membership problem.

It is worth to mention a quite similar model presented by Lange and Reinhardt in [6]. We refer to this model as L-R-SA. In this model there are no **in** and **out** operations; in the case of the negative result of the test, the tested word is added to the set after the query; also L-R-SA have no ε-moves in contrast to SA. The results from [6] on computational complexity for L-R-SA are similar to the results in [7]: the membership problem is **P**-complete for L-R-DSA, and **NP**-complete for L-R-NSA. Note that the proofs for the set automata are more sophisticated than for the L-R-SA model. It was shown in [6] that the emptiness problem for L-R-SA is decidable. Actually, the proof can be easily modified to show that the emptiness problem for L-R-SA is in **PSPACE**. Again, the proof presented here for SA is more complicated than for the L-R-SA model.

The rest of the paper is organized as follows. In Sect. 2 we give the definitions of set automata and auxiliary notions needed to analyze their behavior. In Sect. 3 we give a proof of the inclusion SA $\subset$ **PSPACE**. In Sect. 4 we discuss the membership problem and prove **PSPACE**-hardness of the membership problem for DSA.

2   The Definitions

We start with formal definitions. A set automaton M is defined by a tuple

$$M = \langle S, \Sigma, \Gamma, \lhd, \delta, s_0, F \rangle, \text{where}$$

- S is the finite set of states;
- Σ is the finite alphabet of the input tape;
- Γ is the finite alphabet of the work tape;
- $\lhd \notin \Sigma$ is the right endmarker;
- $s_0 \in S$ is the initial state;
- $F \subseteq S$ is the set of accepting states;
- δ is the transition relation:

$$\delta \subseteq S \times (\Sigma \cup \{\varepsilon, \lhd\}) \times \left[(S \times (\Gamma^* \cup \{\mathbf{in}, \mathbf{out}\})) \cup (S \times \{\mathbf{test}\} \times S) \right].$$

In the deterministic case δ is the function

$$\delta \colon S \times (\Sigma \cup \{\varepsilon, \lhd\}) \to \left[(S \times (\Gamma^* \cup \{\mathbf{in}, \mathbf{out}\})) \cup (S \times \{\mathbf{test}\} \times S) \right].$$

As usual, in the deterministic case, if $\delta(s, \varepsilon)$ is defined, then $\delta(s, a)$ is not defined for every $a \in \Sigma$.

A *configuration* of M is a tuple $(s, v, z, \mathbb{S})$ consisting of the state $s \in S$, the unprocessed part of the input tape $v \in \Sigma^*$, the content of the work tape $z \in \Gamma^*$, and the content of the set $\mathbb{S} \subset \Gamma^*$. The transition relation determines the action of M on configurations. We use $\vdash$ notation for this action. It is defined as follows

$$
\begin{aligned}
(s, xv, z, \mathbb{S}) &\vdash (s', v, zz', \mathbb{S}) & &\text{if } (s, x, (s', z')) \in \delta, & z' \in \Gamma^*; & \quad (1) \\
(s, xv, z, \mathbb{S}) &\vdash (s', v, \varepsilon, \mathbb{S} \cup \{z\}) & &\text{if } (s, x, (s', \mathbf{in})) \in \delta; & & \quad (2) \\
(s, xv, z, \mathbb{S}) &\vdash (s', v, \varepsilon, \mathbb{S} \setminus \{z\}) & &\text{if } (s, x, (s', \mathbf{out})) \in \delta; & & \quad (3) \\
(s, xv, z, \mathbb{S}) &\vdash (s_+, v, \varepsilon, \mathbb{S}) & &\text{if } (s, x, (s_+, \mathbf{test}, s_-)) \in \delta, & z \in \mathbb{S}; & \quad (4) \\
(s, xv, z, \mathbb{S}) &\vdash (s_-, v, \varepsilon, \mathbb{S}) & &\text{if } (s, x, (s_+, \mathbf{test}, s_-)) \in \delta, & z \notin \mathbb{S}. & \quad (5)
\end{aligned}
$$

Operations with the set (transitions (2–5) above) are called *query operations*. A word z in a configuration to which a query is applied (the content of the work tape) is called a *query word*.

We call a configuration *accepting* if the state of the configuration is accepting (belongs to the set F) and the word is processed till the endmarker. So the accepting configuration has the form $(s_f, \varepsilon, z, \mathbb{S})$, where $s_f \in F$.

The set automaton accepts a word w if there exists a run from the initial configuration $(q_0, w \lhd, \varepsilon, \varnothing)$ to some accepting one.

Now we introduce the main tool of our analysis: *protocols*.

A protocol is a word $p = \#u_1 \# \mathrm{op}_1 \# u_2 \# \mathrm{op}_2 \# \cdots \# u_n \# \mathrm{op}_n$, where $u_i \in \Gamma^*$ and $\mathrm{op}_i \in \mathbf{Ops} = \{\mathbf{in}, \mathbf{out}, \mathbf{test+}, \mathbf{test-}\}$.

We say that p is *a correct protocol for SA M on an input $w \in L(M)$*, if there exists an accepting run of M on the input w such that M performs the operation op_1 with the query word u_1 on the work tape at first, then performs op_2 with

u_2 on the work tape, and so on. In the case of a test operation, the symbol op_i indicates the result of the test: **test+** or **test−**.

We call p a *correct protocol for SA* M if there exists a word $w \in L(M)$ such that p is a correct protocol for SA M on the input w. And finally, we say that p is a *correct protocol* if there exists an SA M such that p is a correct protocol for M.

We define SA-PROT to be the language of all correct protocols over the alphabet of the work tape $\Gamma = \{a, b\}$.

It is quite clear from the definition that a correct protocol just describes a sequence of operations with the set. Thus, a correct protocol

$$p = \#u_1\#\mathrm{op}_1\#u_2\#\mathrm{op}_2\# \cdots \#u_t\#\mathrm{op}_t \tag{6}$$

determines the sequence $S_1, S_2, \ldots, S_t$ of the set contents: S_i is the set content after processing the prefix $\#u_1\#\mathrm{op}_1\# \cdots \#u_i\#\mathrm{op}_i$, i.e. performing i first operations with the set.

Query blocks (blocks in short) are the parts of the protocol in the form $\#u_k\#\mathrm{op}_k$, $u_k \in \Gamma^*$, op $\in$ **Ops** (a query word followed by an operation with the set). Note that we distinguish different occurrences of the same word $\#u_k\#\mathrm{op}_k$ in a protocol.

This convention simplifies the statement of a protocol correctness criterion. It can be expressed as follows.

We say that a query block p_i *supports* a query block p_j if $u_i = u_j$ and $\mathrm{op}_i = $ **in**, $\mathrm{op}_j = $ **test+** or $\mathrm{op}_i = $ **out**, $\mathrm{op}_j = $ **test−** and there is no query block p_k such that $\mathrm{op}_k \in \{\mathbf{in}, \mathbf{out}\}$, $u_k = u_i$ and $i < k < j$. Note that each **test+**-block is supported by some **in**-block, but blocks with the operation **test−** may have no support in a correct protocol. *Standalone blocks* are blocks in the form $p_j = \#u\#$ op, where op $\in \{\mathbf{out}, \mathbf{test}-\}$, that have no support (if op = **test−**) and there is no block $p_k = \#u\#\mathbf{in}$, where $k < j$.

The following lemma immediately follows from the definitions.

Lemma 1. *A protocol p is correct iff each query block $\#u_i\#$test+ is supported and each query block $\#u_i\#$test− is either supported or standalone.*

3 Computational Complexity of the Emptiness Problem

An instance of the emptiness problem for SA is a description of an SA M. The question is to decide whether $L(M) = \varnothing$.

Theorem 2 ([7]). *The emptiness problem is* **PSPACE**-*hard for DSA.*

Our main contribution is the matching upper bound of computational complexity of the emptiness problem.

Theorem 3. *The emptiness problem for SA is in* **PSPACE**.

In [7] the emptiness problem was related to so-called the regular realizability (NRR) problem for the language SA-PROT of protocols.

The problem NRR(F) for a language F (a parameter of the problem) is to decide on the input nondeterministic finite automaton (NFA) $\mathcal{A}$ whether the intersection $L(\mathcal{A}) \cap F$ is nonempty.

Lemma 4 (Lemma 12 in [7]). *The emptiness problem for SA is equivalent to* NRR(SA-PROT) *with respect to log-space reductions.*

Thus, to prove Theorem 3 it is sufficient to prove the following fact.

Lemma 5. NRR(SA-PROT) $\in$ **PSPACE**.

An idea of the proof is straightforward. We simulate a successful run of an automaton on a protocol in nondeterministic polynomially bounded space. Applying Savitch theorem we get NRR(SA-PROT) $\in$ **PSPACE**.

A possibility of such a simulation is based on a structural result about correct protocols for SA, see Lemma 10 in the next subsection.

We present the proof of Lemma 5 in Subsect. 3.2.

3.1 Successful Automata Runs on a Protocol

Let $\mathcal{A}$ be an NFA with the set of states $Q = Q_\mathcal{A}$ of size n. Suppose that there exists a successful run of the automaton $\mathcal{A}$ on a correct protocol (a word in the language SA-PROT). It implies the positive answer for the instance $\mathcal{A}$ of the regular realizability problem NRR(SA-PROT).

Our goal in this section is to prove an existence of a successful run of $\mathcal{A}$ on another correct protocol that satisfies specific requirements, see Lemma 10 below.

Let p be a correct protocol accepted by $\mathcal{A}$. Fix some successful run of $\mathcal{A}$ on p. The run is partitioned as follows

$$s_0 \xrightarrow{\#u_1\#\,\mathrm{op}_1} s_1 \xrightarrow{\#u_2\#\,\mathrm{op}_2} s_2 \xrightarrow{\#u_3\#\,\mathrm{op}_3} \cdots \xrightarrow{\#u_t\#\,\mathrm{op}_t} s_t, \quad s_t \in F. \qquad (7)$$

Here we assume that $\mathcal{A}$ starts from the initial state s_0, finishes in the accepting state s_t, processes the query block $p_i = \#u_i\#\,\mathrm{op}_i$ from the state s_i and comes to the state s_{i+1} at the end of the processing. We say that (7) is the *partition* of the run and two runs have the same partition if they have the same sequences of states $\{s_i\}$ and operations $\{\mathrm{op}_i\}$.

Let α be a triple (q, q', op), where $q, q' \in Q$ and op is an operation with the set, i.e. op $\in$ **Ops**. The language $R(\alpha) \in \Gamma^*$ consists of all words $u \in \Gamma^*$ such that

$$q' \in \delta_\mathcal{A}(q, \#u\#\,\mathrm{op}),$$

where $\delta_\mathcal{A}$ is the transition relation of the NFA $\mathcal{A}$.

It is obvious that $R(\alpha)$ is a regular language and the number of states of a minimal automaton recognizing it does not exceed n (the number of $\mathcal{A}$'s states).

The total number N of languages $R(\alpha)$ is poly(n) (actually, $O(n^2)$). Each partition of the run (7) determines the sequence $\alpha_i = (s_i, s_{i+1}, \text{op}_i)$; we say that the partition has the type $\boldsymbol{\alpha} = (\alpha_i)_{i=1}^t$.

Suppose that $\boldsymbol{\alpha}$ is a partition type of a correct protocol. Generally, there are many partitions of correct protocols having the type $\boldsymbol{\alpha}$. We are going to choose among them as simple as possible. More exactly, we are going to minimize the maximal size of the set contents determined by correct protocols admitting a partition of the type $\boldsymbol{\alpha}$.

For this purpose we start from a partition (7) of a correct protocol and change some query words u_i in it to keep the partition type and the correctness of the protocol and to make the set contents smaller.

Actually, we will analyze a slightly more general settings. We replace the family of regular languages $R(\alpha)$ indexed by triples $\alpha = (q, q', \text{op})$ by an arbitrary finite family $\mathcal{R} = \{R(1), \ldots, R(N)\}$ of arbitrary[1] languages over the alphabet Γ indexed by integers $1 \le \alpha \le N$; we call them *query languages*.

We extend types from partitions to protocols in a straightforward way: we say that a protocol p (6) has the type $\boldsymbol{\alpha} = (\alpha_1, \ldots, \alpha_t)$, $1 \le \alpha_i \le N$, if $u_i \in R(\alpha_i)$ for each i.

We transform the protocol (6) in two steps to achieve the desired 'simple' protocol.

The first transformation of a correct protocol (6) of a type $\boldsymbol{\alpha}$ gets a correct protocol

$$p' = \#u_1'\#\text{op}_1\#u_2'\cdots\#u_t'\#\text{op}_t \tag{8}$$

of the same type $\boldsymbol{\alpha}$ such that all set contents S_k' determined by the protocol (8) are polynomially bounded in the number N of query languages.

This property of a protocol will be used in our **PSPACE**-algorithm for the emptiness problem for SA.

The formal definition of the transformation $p \to p'$ is somewhat tricky. So we explain the intuition behind the construction.

We are going to preserve all operations with the set in the transformed protocol. Also we are going to make a sequence of the set contents S_k' determined by the transformed protocol as monotone as possible.

In general, it is impossible to make the whole sequence monotone. Thus we select a subset of query words (*stable words* in the sequel) and do not change these words during the transformation.

The rest of query words are *unstable words*. We are going to make the sequence of the set contents monotone on unstable words. It means that an unstable word added to the set is never deleted from it. To satisfy this condition, we make the corresponding $\{\mathbf{out}, \mathbf{test}-\}$-query blocks standalone, i.e. we substitute unstable query words in these blocks by words that are never presented in the set. To satisfy this requirement, the query language $R(i)$ corresponding to the block should be large enough. Below we give a formal definition of large and small languages to satisfy this requirement.

---

[1] The structural Propositions 8 and 9 hold for all query languages. Lemma 10 and the **PSPACE** algorithm below are applied to regular query languages only.

The transformation strategy of **in**-blocks is to insert into the set the only one word from any large query language. In the formal construction below we call these words *critical*.

Now we present formal definitions to realize informal ideas explained above. We define the 'small' languages by an iterative procedure.

Step 0. All query languages $R(i)$ that contain at most N words are declared small.

Step $j + 1$. Let W_j be the union of all query languages that were declared small on steps $0, \ldots, j$. All query languages $R(i)$ that contain at most N words from $\Gamma^* \setminus W_j$ are declared small.

Query languages that are not small are declared *large*.

Let s be the last step on which some language was declared small. Thus W_s is the union of all small languages. The query words from W_s are called *stable*. It is clear from the definition that each unstable query word belongs to a large language.

It is easy to observe that there are relatively few stable query words.

Proposition 6. $|W_s| \leq N^2$.

Proof. The total number of small languages doesn't exceed N. Each small language contributes to the set W_s at most N words. □

To define critical query words we assume that the protocol (6) is correct and has the type $\boldsymbol{\alpha} = (\alpha_1, \ldots, \alpha_t)$. Let $S_1, \ldots, S_t$ be the sequence of the set contents determined by the protocol.

An unstable query word u is *critical* if it is contained in an **in**-block $p_k = \#u\#\mathbf{in}$ indicating insertion into the set an unstable word u from a large language, none of whose unstable words have been inserted into the set earlier. Formally it means that there exists a large language $R(i)$ such that $u \in R(i) \setminus W_s$ and $(R(i) \setminus W_s) \cap S_j = \varnothing$ for all $j < k$.

Proposition 7. *There are at most N critical query words in the correct protocol p (6).*

Proof. The protocol p is correct. Thus the first occurrence of a critical query word u is in an **in**-block $p_k = \#u\#\mathbf{in}$. If a large language $R(i)$ contains u, then $S_k \cap (R(i) \setminus W_s) \neq \varnothing$.

Therefore a large language can certify the critical property for at most one critical query word. But the number of large languages is at most N. □

Look at a $\{\mathbf{test}+, \mathbf{in}\}$-block $p_k = \#u_k\#\,\mathrm{op}_k$ containing an unstable noncritical query word u_k. Note that the query language $R(\alpha_k)$, which is specified by an index α_k, $1 \leq \alpha_k \leq N$, from the type of the protocol (6), is large (otherwise the query word u_k would be stable). The query word u_k is not critical. Thus $(R(\alpha_k) \setminus W_s) \cap S_j \neq \varnothing$ for some $j < k$. The smallest j satisfying this condition indicates the query block describing the first insertion of an unstable query word

u_j from the language $R(\alpha_k)$ into the set. Note that the query word u_j is critical: $(R(\alpha_k) \setminus W_s) \cap S_i = \varnothing$ for $i < j$. Denote u_j as $\tilde{u}_k$. We assign the query word $\tilde{u}_k$ to the block p_k and use this assignment in the construction below.

We are ready to define formally the transformation $p \rightarrow p'$ of protocols assuming correctness of p.

The type of the transformed protocol is the same: $\boldsymbol{\alpha} = (\alpha_1, \ldots, \alpha_t)$. Operations do not change: $\mathrm{op}'_i = \mathrm{op}_i$.

All stable query words do not change. More exactly, if $u_k \in W_s$, then $u'_k = u_k$.

An unstable query word in a $\{\mathbf{test}-, \mathbf{out}\}$-block $\#u_k\# \mathrm{op}_k$ is substituted by a word u'_k from $R(\alpha_k) \setminus (C \cup W_s)$, where C is the set of critical query words. The substitution is possible because there are more than N words in $R(\alpha_k) \setminus W_s$ (the language $R(\alpha_k)$ is large) and there are at most N critical query words due to Proposition 7.

Critical query words in $\{\mathbf{test}+, \mathbf{in}\}$-blocks do not change: if u_k is critical and $\mathrm{op}_k \in \{\mathbf{test}+, \mathbf{in}\}$, then $u'_k = u_k$. (The resulting protocol remains correct due to the previous rule: critical query words are not removed from the set.)

An unstable noncritical query word in a $\{\mathbf{test}+, \mathbf{in}\}$-block $\#u_k\# \mathrm{op}_k$ is substituted by the critical query word $\tilde{u}_k$ assigned to the block. (The choice of $\tilde{u}_k$ guarantees correctness of the resulting protocol.)

Proposition 8. *If the protocol p (6) is correct, then the transformed protocol p' (8) is correct.*

Each set content S'_k determined by the protocol p' contains at most $N^2 + N$ words.

To bound the sizes of the set contents in the modified protocol, note that they can contain only two kinds of query words: stable query words (there are at most N^2 of them) and critical query words (there are at most N of them).

The rest of the proof is technical and omitted here due to space limitations.

Query words in a correct protocol may be very long. To operate with them in polynomial space, we describe them implicitly in the **PSPACE**-algorithm below. All relevant information about a word is a list of query languages containing it. Thus, we divide the whole set of words over the alphabet Γ in the non-intersecting *elementary languages*

$$R_I = \bigcap_{i \in S} R(i) \cap \bigcap_{i \notin S} \overline{R(i)}, \quad I \subseteq \{1, 2, \ldots, N\}. \tag{9}$$

We call I a *type* of a query word u, if $u \in R_I$. Words will be represented by their types in the algorithm below.

Such a representation causes a problem: the set content is represented in this way by types of words and it is unclear how many different words of the same type are in the set.

To avoid this problem, we assume that at most one word of each type is in the set at any moment. To justify the assumption, we need the second transformation of protocols.

To define the second transformation of the protocol (6), let choose two query words u_I, v_I in each elementary language R_I containing at least two words. The transformed protocol

$$p'' = \#u_1''\#\text{op}_1\#u_2''\#\text{op}_2\#\cdots\#u_t''\#\text{op}_t \qquad (10)$$

is produced as follows.

The type of p'' equals the type of p. All operations are the same.

If a word u_i has a type I and the language R_I contains only one word, we do not change u_i. Otherwise we substitute u_i by u_I in the case of $\text{op}_i \in \{\textbf{in}, \textbf{test}+\}$ or by v_I in the case of $\text{op}_i \in \{\textbf{out}, \textbf{test}-\}$.

Proposition 9. *If the initial protocol* (6) *is correct, then the transformed protocol* (10) *is also correct.*

Each set content S_k'' determined by the transformed protocol contains at most one word from any elementary language R_I, $I \subseteq \{1, 2, \ldots, N\}$.

The proof is a straightforward application of Lemma 1. It is skipped due to space limitations.

Now we return to successful runs of an automaton on a correct protocol. Propositions 8 and 9 immediately imply the following lemma.

Lemma 10. *Let $\mathcal{A}$ be an NFA with n states. If $\mathcal{A}$ accepts a correct protocol, then it accepts a correct protocol such that $|S_i| = \text{poly}(n)$ for all i and each set content contains at most one word from each elementary language.*

Proof. Recall that the number N of query languages $R(q, q', \text{op})$ for the NFA $\mathcal{A}$ is $O(n^2)$. So, we just apply the second transformation to the result of the first transformation. Note that the second transformation does not increase the sizes of the set contents. □

3.2 Proof of Lemma 5

Let $\mathcal{A}$ be an input automaton for the emptiness problem. We measure the size of the input by the number of states of this automaton.

As it mentioned above, we simulate a successful run of the automaton on a protocol by a nondeterministic algorithm using polynomially bounded space.

Due to Lemma 10 there exists a correct protocol p and a successful run of the automaton $\mathcal{A}$ on the protocol such that the sizes of set contents during the protocol are upperbounded and the set contains at most one word from each elementary language. The upper bound is polynomial in n and can be derived explicitly from the arguments of the previous subsection. It is easy to verify that $M = 32n^4$ upperbounds the sizes of set contents determined by the protocol from Lemma 10.

A word in the set is represented in the algorithm by a description of an elementary language containing the word. The second condition of Lemma 10 implies correctness of this representation. An elementary language can be described by a set I of triples in the form (q_1, q_2, op), where $q_1, q_2 \in Q(\mathcal{A})$ and op $\in$ $\{\text{in}, \text{out}, \text{test}+, \text{test}-\}$. The language R_I is specified by Eq. (9). Such a representation has polynomial size, namely, $O(n^2)$.

The simulating algorithm stores the description of the set content and the current state of the automaton $\mathcal{A}$. The algorithm nondeterministically guesses the change of this data after reading the next query block of the simulated protocol.

To complete the proof we should devise an algorithm to check correctness of a simulation step. It is specified by indicating the state of $\mathcal{A}$ after reading a query block, an elementary language R_I, containing the query word on this step, where I is the type of a query block to be read, and the description of the set content after performing the operation with the set on this step.

It is easy to check that the state of $\mathcal{A}$ is changed correctly using the description of $\mathcal{A}$.

The language R_I should be nonempty (otherwise the simulation goes wrong). It is well-known that the intersection problem for regular languages is in **PSPACE** [1,5]. Thus, verifying $R_I \neq \varnothing$ can be done in polynomial space. It guarantees the possibility of the query indicated by the current query block.

Test results are easily verified by the description of the set content before a query.

If the current query is **in**, then the set content is changed as follows. Due to Lemma 10 we assume that there is at most one word of each type I in the set. So, if the current query word has a type I and there are no words of the type I in the set, then the type I is included in the set. Otherwise, the set is not changed.

If the current query is **out** and the set includes the current query type I, then two outcomes are possible: either a word of the type I is deleted from the set or the set remains unchanged. The latter is possible iff the elementary language R_I contains at least two words.

Thus, to complete the proof we need the following proposition.

Proposition 11. *There exists a polynomial space algorithm to verify*

$$\left| \bigcap_i R_i \right| \geq 2,$$

where R_i are regular languages represented by NFA.

The proposition is proved by a reduction of the question to the intersection problem for regular languages. It is skipped due to space limitations.

4   The Membership Problem for SA

An instance of the membership problem for SA is a word w and a description of an SA M. The question is to decide whether $w \in L(M)$.

An upper bound of computational complexity of the membership problem can be easily obtained from the following observation.

Proposition 12. *The membership problem for SA is polynomially time reduced to the non-emptiness problem for SA and the membership problem for DSA is polynomially time reduced to the non-emptiness problem for DSA.*

Proof. Let an SA M and a word w be an instance of the membership problem. It is easy to construct in polynomial time an SA M' such that $L(M') = L(M) \cap \{w\}$. Thus $w \in L(M)$ iff $L(M') \neq \varnothing$. This reduction preserves determinism of SA. □

Due to this proposition and Theorem 3 the membership problems for SA and DSA are in **PSPACE** (the class is closed under the complement).

In this section we prove the matching lower bound of computational complexity of the membership problem. More exactly, we will prove that the membership problem for DSA is **PSPACE**-hard. It is in a striking contrast to the results for DSA without ε-loops. Recall that it was proved in [7] that the membership problem for the latter case is in **P**.

Theorem 13. *It is* **PSPACE***-hard to check that $\varepsilon \overset{?}{\in} L(M)$ for DSA M with the unary alphabet of the work tape.*

To prove Theorem 13 we simulate the operation of a deterministic Turing machine M with the 2-element alphabet $\{0, 1\}$ in polynomially bounded space by a DSA A_M with the unary alphabet $\{\#\}$ of the work tape. We describe only ε-transitions of DSA, because the rest of transitions does not affect the answer to the question $\varepsilon \overset{?}{\in} L(M)$.

Denote by N the number of tape cells available for the operation of M. Let Q be the state set of M, $\delta \colon \{0, 1\} \times Q \to \{0, 1\} \times Q \times \{-1, 0, +1\}$ be the transition function of M.

Our goal is to construct a DSA A_M with $\text{poly}(N)$ states simulating the operation of M. A state of A_M carries an information about a position of the head of M and a current rule of the transition function δ. Information about the tape content of M is stored in the set $\mathbb{S}$ as follows.

Let i be the index of a tape cell of M, $1 \le i \le N$, and $x_i \in \{0, 1\}$ be the value stored in this cell. During the simulation process the DSA A_M maintains the relation: $x_i = 1$ is equivalent to $p_i \in \mathbb{S}$, where $p_i = \#^i$.

To insert words p_i and delete them from the set, the states of the DSA A_M include a counter to write the appropriate number of $\#$ on the work tape. The value of the counter is at most N.

As it mentioned above, the head position of M and the current rule of the transition function δ are maintained in the state set of A_M. Thus, the set state of A_M has the form $\{1, 2, \ldots, N\} \times \Delta \times \{0, 1, 2, \ldots, N\} \times C$, where $\Delta = \{0, 1\} \times Q$ is the set of the rules of the transition function δ and C is used to control computation, $|C| = O(1)$ (assuming $N \to \infty$).

The DSA A_M starts a simulation step in the state $(k, (a, q), 0, \mathsf{start})$, where k is the head position of the simulated TM, q is its state, and $x_k = a$. The last component $\mathsf{start} \in C$ indicates the beginning of a simulation step.

Let $\delta(a, q) = (a', q', d)$ be the corresponding rule of the transition function δ.

A simulation step consists of the actions specified below, each action is marked in the last state component to avoid ambiguity:

1. write k symbols # on the work tape, in this action the third state component increased by 1 up to k;
2. update the set, if $a' = 0$, then the DSA performs **out** operation, otherwise it performs **in** operation;
3. change the value of k according to the value of d: the new value $k' = k + d$;
4. write k' symbols # on the work tape, the action is similar to the action 1;
5. perform the test operation, if the result is positive, then change the second component to $(1, q')$, otherwise change the second component to $(0, q')$;
6. change the third component to 0 and the last component to start.

Inspecting these actions, it is easy to see that all information about the configuration of the simulated TM is updated correctly.

Proof (of Theorem 13). The following problem is **PSPACE**-complete: an input is a pair $(M, 1^N)$, one should decide whether $\varepsilon \in L(M)$, here M is an input TM with the alphabet $\{0, 1\}$ that uses at most $2N - 1$ cells on the tape. The hardness of this problem can be proved by a direct reduction of any **PSPACE** language L to it (hardwire a word into the description of M and determine N by a polynomial bound of a space used by the algorithm recognizing L).

We reduce the problem to the question $\varepsilon \overset{?}{\in} L(A_{M,N})$, where $A_{M,N}$ is the DSA with the unary alphabet of the work tape simulating the operation of M on $2N - 1$ memory cells as described above.

It is easy to see that the DSA $A_{M,N}$ can be constructed in time polynomial in the length of description of TM M and N. □

Acknowledgments. The authors are thankful to the first referee for helpful comments.

References

1. Kozen, D.: Lower bounds for natural proof systems. In: Proceedings of the 18th Annual Symposium on Foundations of Computer Science, SFCS 1977, pp. 254–266. IEEE Computer Society, Washington, D.C. (1977)
2. Kutrib, M., Malcher, A., Wendlandt, M.: Deterministic set automata. In: Shur, A.M., Volkov, M.V. (eds.) DLT 2014. LNCS, vol. 8633, pp. 303–314. Springer, Cham (2014). https://doi.org/10.1007/978-3-319-09698-8_27
3. Kutrib, M., Malcher, A., Wendlandt, M.: Regularity and size of set automata. In: Jürgensen, H., Karhumäki, J., Okhotin, A. (eds.) DCFS 2014. LNCS, vol. 8614, pp. 282–293. Springer, Cham (2014). https://doi.org/10.1007/978-3-319-09704-6_25

4. Kutrib, M., Malcher, A., Wendlandt, M.: Set automata. Int. J. Found. Comput. Sci. **27**(02), 187–214 (2016)

5. Lange, K.-J., Rossmanith, P.: The emptiness problem for intersections of regular languages. In: Havel, I.M., Koubek, V. (eds.) MFCS 1992. LNCS, vol. 629, pp. 346–354. Springer, Heidelberg (1992). https://doi.org/10.1007/3-540-55808-X_33

6. Lange, K.-J., Reinhardt, K.: Set automata. In: Bridges, D.S., Calude, C., Gibbons, J., Reeves, S., Witten, L. (eds.) Combinatorics, Complexity, Logic. Proceedings of the DMTCS 1996, pp. 321–329. Springer, Heidelberg (1996)

7. Rubtsov, A.A., Vyalyi, M.N.: On computational complexity of set automata. In: Charlier, É., Leroy, J., Rigo, M. (eds.) DLT 2017. LNCS, vol. 10396, pp. 332–344. Springer, Cham (2017). https://doi.org/10.1007/978-3-319-62809-7_25

On Strong NP-Completeness of Rational Problems

Dominik Wojtczak[⊠]

University of Liverpool, Liverpool, UK
d.wojtczak@liverpool.ac.uk

Abstract. The computational complexity of the partition, 0-1 subset sum, unbounded subset sum, 0-1 knapsack and unbounded knapsack problems and their multiple variants were studied in numerous papers in the past where all the weights and profits were assumed to be integers. We re-examine here the computational complexity of all these problems in the setting where the weights and profits are allowed to be any rational numbers. We show that all of these problems in this setting become strongly NP-complete and, as a result, no pseudo-polynomial algorithm can exist for solving them unless $P = NP$. Despite this result we show that they all still admit a fully polynomial-time approximation scheme.

1 Introduction

The problem of partitioning a given set of items into two parts with equal total weights (that we will refer to as PARTITION) goes back at least to 1897 [14]. A well-known generalisation is the problem of finding a subset with a given total weight (0-1 SUBSET SUM) and the same problem where each item can be picked more than once (UNBOUNDED SUBSET SUM). Finally, these are commonly generalised to the setting where each item also has a profit and the aim is to pick a subset of items with the total profit higher than a given threshold, but at the same time their total weight smaller than a given capacity (0-1 KNAPSACK). A variant of the last problem where each item can be picked more than once is also studied (UNBOUNDED KNAPSACK).

The SUBSET SUM problem has numerous applications: its solutions can be used for designing better lower bounds for scheduling problems (see, e.g., [7,8]) and it appears as a subproblem in numerous combinatorial problems (see, e.g., [18]). At the same time, many industrial problems can be formulated as knapsack problems: cargo loading, cutting stock, capital budgeting, portfolio selection, interbank clearing systems, knapsack cryptosystems, and combinatorial auctions to name a couple of examples (see Chap. 15 in [12] for more details regarding these problems and their solutions).

The decision problems studied in this paper were among the first ones to be shown to be NP-complete [11]. At the same they are considered to be the easiest problems in this class, because they are polynomial time solvable if items' weights and profits are represented using the unary notation (in other words, they are

© Springer International Publishing AG, part of Springer Nature 2018
F. V. Fomin and V. V. Podolskii (Eds.): CSR 2018, LNCS 10846, pp. 308–320, 2018.
https://doi.org/10.1007/978-3-319-90530-3_26

only *weakly* NP-complete). In particular, they can be solved in polynomial time when these numbers are bounded by a fixed constant (and the number of items is unbounded). Furthermore, the optimisation version of all these decision problems admit *fully polynomial-time approximation schemes (FPTAS)*, i.e., we can find a solution with a value at least equal to $(1 - \epsilon)$ times the optimal in time polynomial in the size of the input and $1/\epsilon$ for any $\epsilon > 0$.

To the best of our knowledge, the computational complexity analysis of all these problems was only studied so far under the simplifying assumption that all the input values are integers. However, in most settings where these problems are used, these numbers are likely to be given as rational numbers instead. We were surprised to discover that the computational complexity in such a rational setting was not properly studied before. Indeed, as pointed out in [12]:

> A rather subtle point is the question of rational coefficients. Indeed, most textbooks get rid of this case, where some or all input values are non-integer, by the trivial statement that multiplying with a suitable factor, e.g. with the smallest common multiple of the denominators, if the values are given as fractions or by a suitable power of 10, transforms the data into integers. Clearly, this may transform even a problem of moderate size into a rather unpleasant problem with huge coefficients.

This clearly looks like a fundamental gap in the understanding of the complexity of these computational problems. Allowing the input values to be rational makes a lot of sense in many settings. For example, we encountered this problem when studying an optimal control in multi-mode systems with discrete costs [15,16] and looking at the weighted voting games (see, e.g., [4] where the weights are defined to be rational). An interesting real-life problem is checking whether we have the exact amount when paying, which is important in a situation when no change can be given. While we take decimal monetary systems for granted these days, there were plenty of non-decimal monetary systems in use not so long ago. For example, in the UK between 1717 and 1816 one pound sterling was worth twenty shillings, one shilling was worth twelve pence, and one guinea was worth twenty one shillings.

We show here that allowing the input numbers to be rational makes a significant difference and in fact all these decision problems in such a setting become strongly NP-complete [6], i.e., they are NP-complete even when all their numerical values are at most polynomial in the size of the rest of the input or, equivalently, if all these numerical values are represented in unary. To prove this we will show an NP-completeness of a new variant of the SATISFIABILITY problem and use results regarding distribution of prime numbers. As a direct consequence of our result, there does not exist any pseudo-polynomial algorithms for solving these decision problems unless $P = NP$. At the same time, we will show that they still all admit a fully polynomial-time approximation scheme (see, e.g., [12]). This may seem wrong, because the paper that introduced strong NP-completeness [6] also showed that no strongly NP-hard problem can admit an FPTAS unless $P = NP$. However, the crucial assumption made there is that the objective function is integer valued, which does not hold in our case.

Related Work. The decision problems studied in this paper are so commonly used that they have already been thousands of papers published about them and many of their variants, including multiple algorithms and heuristic for solving them precisely and approximately. There are also two full-length books, [12,21], solely dedicated to these problems.

Several extensions of the classic knapsack problem were shown to be strongly NP-complete. These include partially ordered knapsack [10] (where we need to pick a set of items closed under predecessor), graph partitioning [10] (where we need to partition a graph into m disjoint subgraphs under cost constraints), multiple knapsack problem [2], knapsack problem with conflict graphs [17] (where we restrict which pairs of items can be picked together), and quadratic knapsack problem [5,12] (where the profit of packing an item depends on how well it fits together with the other selected items).

The first FPTAS for the optimisation version of the KNAPSACK problem was established in 1975 by Ibarra and Kim [9] and independently by Babat [1]. Multiple other, more efficient, FPTAS for these problems followed (see, e.g., [12]).

Plan of the Paper. In the next section, we introduce all the used notation as well as formally define all the decision problems that we study in this paper. In Sect. 3, we analyse the amount of space one needs to write down the first n primes in the unary notation as well as a unique representation theorem concerning sums of rational numbers. In Sect. 4, we define a couple of new variants of the well-known satisfiability problem for Boolean formulae in 3-CNF form and show them to be NP-complete. Our main result, concerning the strong NP-hardness of all the decision problems studied in this paper with rational inputs, can be found in Sect. 5 and it builds on the results from Sects. 3 and 4. We briefly discuss the existence of FPTAS for the optimisation version of our decision problems in Sect. 6. Finally, we conclude in Sect. 7.

2   Background

Let $\mathbb{Q}_{\geq 0}$ be the set of non-negative rational numbers. We assume that a non-negative rational number is represented as usual as a pair of its numerator and denominator, both of which are natural numbers that do not have a common divisor greater than 1. A unary representation of a rational number is simply a pair of its numerator and denominator represented in unary. For two natural numbers a and n, let $a \bmod n \in \{0, \ldots, n-1\}$ denote the remainder of dividing a by n. For any two numbers $a, b \in \{1, \ldots, n\}$, we define their addition modulo n, denoted by $\oplus_n$, as follows: $a \oplus_n b = ((a + b - 1) \bmod n) + 1$. Note that we subtract and add 1 in this expression so that the result of this operation belongs to $\{1, \ldots, n\}$. Similarly we define the subtraction modulo n, denoted by $\ominus_n$, as follows: $a \ominus_n b = ((n + a - b - 1) \bmod n) + 1$. We assume that $\oplus_n$ and $\ominus_n$ operators have higher precedence than the usual $+$ and $-$ operators.

We now formally define all the decision problems that we study in this paper.

Definition 1 (SUBSET SUM **problems**). *Assume we are given a list of n items with rational non-negative weights $A = \{w_1, \ldots, w_n\}$ and a target total weight $W \in \mathbb{Q}_{\geq 0}$.*

0-1 SUBSET SUM*: Does there exists a subset B of A such that the total weight of B is equal to W?*

UNBOUNDED SUBSET SUM*: Does there exist a list of non-negative integer quantities $(q_1, \ldots, q_n)$ such that*

$$\sum_{i=1}^{n} q_i \cdot w_i = W?$$

(Intuitively, q_i denotes the number of times the i-th item in A is chosen.)

A natural generalisation of this problem where each item gives us a profit when picked is the well-known knapsack problem.

Definition 2 (KNAPSACK **problems**). *Assume there are n items whose non-negative rational weights and profits are given as a list $L = \{(w_1, v_1), \ldots, (w_n, v_n)\}$. Let the capacity be $W \in \mathbb{Q}_{\geq 0}$ and the profit threshold be $V \in \mathbb{Q}_{\geq 0}$.*

0-1 KNAPSACK*: Is there a subset of L whose total weight does not exceed W and total profit is at least V?*

UNBOUNDED KNAPSACK*: Is there a list of non-negative integers $(q_1, \ldots, q_n)$ such that*

$$\sum_{i=1}^{n} q_i \cdot w_i \leq W \qquad and \qquad \sum_{i=1}^{n} q_i \cdot v_i \geq V?$$

(Intuitively, q_i denotes the number of times the i-th item in A is chosen.)

Finally, a special case of the SUBSET SUM problem is the PARTITION problem.

Definition 3 (PARTITION **problem**). *Assume we are given a list of n items with non-negative rational weights $A = \{w_1, \ldots, w_n\}$.*

Can the set A be partitioned into two sets with equal total weights?

Now let us compare the size of a PARTITION problem instance when represented in binary and unary notation. Let $A = \{w_1, \ldots, w_n\}$ be an instance such that $w_i = \frac{a_i}{b_i}$ where $a_i, b_i \in \mathbb{N}$ for all $i = 1, \ldots, n$. Notice that the size of A is $\Theta\left(\sum_{i=1}^{n} \log(a_i) + \sum_{i=1}^{n} \log(b_i)\right)$ when written down in binary and $\Theta\left(\sum_{i=1}^{n} a_i + \sum_{i=1}^{n} b_i\right)$ in unary. If we now multiply all weights in A by $\prod_{i=1}^{n} b_i$ then we would get an equivalent instance A' with only integer weights. The size of A' would be $\Theta\left(\sum_{i=1}^{n} \log(a_i/b_i \prod_{j=1}^{n} b_j)\right) = \Theta\left(\sum_{i=1}^{n} \log a_i + (n-1) \sum_{i=1}^{n} \log b_i\right)$ when written down in binary and in unary: $\Theta\left(\sum_{i=1}^{n} (a_i/b_i \prod_{j=1}^{n} b_j)\right) = \Omega(\min a_i \cdot (\min b_i)^{n-1})$. Notice that the first expression is polynomial in the size of the original instance while the second one may grow exponentially. A similar analysis shows the same behaviour for all the other decision problems studied in this paper.

3   Prime Suspects

In this section we first show that writing down all the first n prime numbers in the unary notation can be done using space polynomial in n. Let π_i denote the i-th prime number. The following upper bound is known for π_i.

Theorem 1 (inequality (3.13) in [19]).

$$\pi_i < i(\log i + \log \log i) \quad for \quad i \geq 6$$

This estimate gives us the following corollary that will be used in the main result of this paper.

Corollary 1. *The total size of the first n prime numbers, when written down in unary, is $\mathcal{O}(n^2 \log n)$. Furthermore, they can be computed in polynomial time.*

Proof. Let $n \geq 6$, because otherwise the problem is trivial. Thanks to Theorem 1, it suffices to list all natural numbers smaller than $2n \log n$ (because $n(\log n + \log \log n) \leq 2n \log n$) and use the sieve of Eratosthenes to remove all nonprime numbers from this list. It follows that writing down the first n prime numbers requires $\mathcal{O}(n^2 \log n)$ space. The sieve can easily be implemented in polynomial time and, to be precise, in this case $\mathcal{O}(n^2 \log^2 n)$ additions and $\mathcal{O}(n \log n)$ bits of memory would suffice. □

Now we prove a result regarding a unique representation of rational numbers expressed as sums of fractions with prime denominators, which in a way is quite similar to the Chinese remainder theorem.

Lemma 1. *Let $(p_1, \ldots, p_n)$ be a list of n different prime numbers. Let $(a_0, a_1, \ldots, a_n)$ and $(a_0, b_1, \ldots, b_n)$ be two lists of integers such that $|a_i - b_i| < p_i$ holds for all $i = 1, \ldots, n$. We then have*

$$a_0 + \frac{a_1}{p_1} + \ldots + \frac{a_n}{p_n} = b_0 + \frac{b_1}{p_1} + \ldots + \frac{b_n}{p_n}$$

if and only if

$$a_i = b_i \quad for \ all \quad i = 0, \ldots, n.$$

Proof. ($\Leftarrow$) If $a_i = b_i$ for all $i = 0, \ldots, n$ holds then obviously

$$a_0 + \frac{a_1}{p_1} + \ldots + \frac{a_n}{p_n} = b_0 + \frac{b_1}{p_1} + \ldots + \frac{b_n}{p_n}.$$

($\Rightarrow$) We need to consider two cases: $a_0 = b_0$ and $a_0 \neq b_0$. In the first case, suppose that $a_j \neq b_j$ for some $j \in \{1, \ldots, n\}$. If we multiply

$$\frac{a_1 - b_1}{p_1} + \ldots + \frac{a_n - b_n}{p_n} \quad by \quad \prod_{i=1}^{n} p_i$$

then we would get an integer, which is not divisible by p_j (because $0 < |a_j - b_j| < p_j$) and so this expression cannot be equal to 0. Therefore, in this case,

$$a_0 + \frac{a_1}{p_1} + \ldots + \frac{a_n}{p_n} \neq b_0 + \frac{b_1}{p_1} + \ldots + \frac{b_n}{p_n}.$$

In the second case, if $a_i = b_i$ for all $i = 1, \ldots, n$ holds then clearly

$$a_0 + \frac{a_1}{p_1} + \ldots + \frac{a_n}{p_n} \neq b_0 + \frac{b_1}{p_1} + \ldots + \frac{b_n}{p_n}.$$

Otherwise, again suppose that $a_j \neq b_j$ for some $j \in \{1, \ldots, n\}$. If we multiply

$$a_0 - b_0 + \frac{a_1 - b_1}{p_1} + \ldots + \frac{a_n - b_n}{p_n} \quad by \quad \prod_{i=1}^{n} p_i$$

then we would get an integer, which is not divisible by p_j (because $0 < |a_j - b_j| < p_j$) and so this expression cannot be equal to 0. Therefore, again, in this case,

$$a_0 + \frac{a_1}{p_1} + \ldots + \frac{a_n}{p_n} \neq b_0 + \frac{b_1}{p_1} + \ldots + \frac{b_n}{p_n}.$$

$\square$

4   In the Pursuit of Satisfaction

The Boolean satisfiability (SATISFIABILITY) problem for formulae was the first problem to be shown NP-complete by Cook [3] and Levin [13]. Karp [11] showed that SATISFIABILITY is also NP-complete for formulae in the conjunctive normal form where each clause has at most three literals. Of course, the same holds for formulae with exactly three literals in each clause. This is simply because we can introduce a new fresh variable for every missing literal in each clause of the given formula without changing its satisfiability. The set of all formulae with exactly three literals in each clause will denoted by *3-CNF*. Tovey [22] showed that SATISFIABILITY is also NP-complete for 3-CNF formulae in which each variable occurs at most 4 times. We will denote the set of all such formulae by 3-CNF$_{\leq 4}$. Schaefer defined in [20] the ONE-IN-THREE-SAT problem for 3-CNF formulae in which one asks for an truth assignment that makes exactly one literal in each clause true, and showed it to be NP-complete. We define here a new ALL-THE-SAME-SAT problem for 3-CNF formulae, which asks for a valuation that makes exactly the same number of literals true in every clause (this may be zero, i.e., such a valuation may not make the formula true). This problem will be a crucial ingredient in the proof of the main result of this paper.

The first step is to show that ONE-IN-THREE-SAT problem is NP-complete even when restricted to 3-CNF$_{\leq 4}$ formulae.

Theorem 2. *The* ONE-IN-THREE-SAT *problem for 3-CNF$_{\leq 4}$ is NP-complete.*

Proof. Obviously the problem is in NP, because we can simply guess a valuation and check how many literals are true in each clause in linear time.

To prove NP-hardness, we are going to reduce from the SATISFIABILITY problem for 3-CNF$_{\leq 4}$, which is NP-complete [22]. Assume we are given a 3-CNF$_{\leq 4}$ formula

$$\phi = C_1 \wedge C_2 \wedge \ldots \wedge C_m$$

with m clauses $C_1, \ldots, C_m$ and n propositional variables $v_1, \ldots, v_n$, where $C_j = x_j \vee y_j \vee z_j$ for $j = 1, \ldots, m$ and each x_j, y_j, z_j is a literal equal to v_i or $\neg v_i$ for some i. We will construct a 3-CNF$_{\leq 4}$ formula ϕ' with $3m$ clauses and $n + 4m$ propositional variables such that ϕ is satisfiable iff ϕ' is an instance of the ONE-IN-THREE-SAT problem. This will be based on the construction already given in [20].

The formula ϕ' is constructed by replacing each clause in ϕ with three new clauses. Specifically, the j-th clause $C_j = x_j \vee y_j \vee z_j$ is replaced by $C'_j :=$ $(\neg x_j \vee a_j \vee b_j) \wedge (b_j \vee y_j \vee c_j) \wedge (c_j \vee d_j \vee \neg z_j)$ where a_j, b_j, c_j, d_j are four fresh propositional variables. It is quite straightforward to check that only a valuation that makes C_j true can be extended to a valuation that makes exactly one literal true in each of the clauses in C'_j. Notice that such a constructed ϕ' is a 3-CNF$_{\leq 4}$ formula, because this transformation does not increase the number of occurrences of any of the original variables in ϕ and each of the new variables is used at most twice.

Now, if there exists a valuation that makes every clause in ϕ true, then as argued above it can be extended to a valuation that makes exactly one literal true in every clause in ϕ'.

To show the other direction, let ν be a valuation that makes exactly one literal true in every clause in ϕ'. Consider for every $j = 1, \ldots, m$ the projection of ν on the set of variables occurring in the clause C_j. Suppose that such a valuation makes C_j false. It follows that it would not be possible to extend this valuation to a valuation that makes exactly one literal true in every clause in C'_j. However, we already know that ν is such a valuation, so this leads to a contradiction. $\qquad \square$

Although Theorem 2 is of independent interest, all that we need it for is to prove our next theorem.

Theorem 3. *The* ALL-THE-SAME-SAT *problem for 3-CNF$_{\leq 4}$ formulae is* NP-*complete.*

Proof. The ALL-THE-SAME-SAT problem is clearly in NP, because we can simply guess a valuation and check whether it makes exactly the same number of literals in every clause true.

To proof NP-hardness, we reduce from the ONE-IN-THREE-SAT problem for 3-CNF$_{\leq 4}$ formulae (Theorem 2). Let ϕ be any 3-CNF$_{\leq 4}$ formula and let us consider a new formula $\phi' = \phi \wedge (x \vee x \vee \neg x)$, where x is a fresh variable that does not occur in ϕ. Notice that ϕ' is also a 3-CNF$_{\leq 4}$ formula. We claim that ϕ is an instance of ONE-IN-THREE-SAT iff ϕ' is an instance of ALL-THE-SAME-SAT.

($\Rightarrow$) If ν is a valuation that makes exactly one literal in every clause in ϕ true, then extending it by setting $\nu'(x) = \bot$ would make exactly one literal in every clause in ϕ' true.

($\Leftarrow$) Let ν be a valuation that makes the same number of literals in every clause in ϕ' true. This number cannot possibly be 0 or 3, because there is at least one true literal and one false literal in the clause $(x \vee x \vee \neg x)$.

If ν makes exactly one literal in every clause in ϕ' true, then the same holds for ϕ.

If ν makes exactly two literals in every clause true, then consider the valuation ν' such that $\nu'(y) = \neg \nu(y)$ for every propositional variable y in ϕ'. Notice that ν' makes exactly one literal in every clause in ϕ' true, so the same holds for ϕ. □

5 Being Rational Makes You Stronger

In this section, we build on the results obtained in the previous two section and show strong NP-hardness of all the decision problems defined in Sect. 2. As a direct consequence, no pseudo-polynomial algorithm can exist for solving any of these problems unless $P = NP$. Instead of showing strong NP-hardness for each of these problems separably, we will show one "master" reduction for the UNBOUNDED SUBSET SUM problem instead. This reduction will then be reused to show strong NP-hardness of the PARTITION problem, and from these two results the strong NP-hardness of all the other decision problems studied in this paper will follow.

Theorem 4. *The* UNBOUNDED SUBSET SUM *problem with rational weights is strongly NP-complete.*

Proof. For a given instance $A = \{w_1, \ldots, w_n\}$ and target weight W, we know that the quantities q_i, for all $i = 1, \ldots, n$, have to satisfy $q_i \leq W/w_i$. This in fact shows that the problem is in NP, because all the quantities q_i when represented in binary can be written down in polynomial space and can be guessed at the beginning. We can then simply verify whether $\sum_{i=1}^{n} q_i \cdot a_i = W$ holds in polynomial time by adding the rational numbers inside this sum one by one (while representing all the numerators and denominators in binary).

To prove strong NP-hardness, we provide a reduction from the ALL-THE-SAME-SAT problem for 3-CNF$_{\leq 4}$ formulae (which is NP-complete due to Theorem 3). Assume we are given a 3-CNF$_{\leq 4}$ formula

$$\phi = C_1 \wedge C_2 \wedge \ldots \wedge C_m$$

with m clauses $C_1, \ldots, C_m$ and n propositional variables $x_1, \ldots, x_n$, where $C_j = a_j \vee b_j \vee c_j$ for $j = 1, \ldots, m$ and each a_j, b_j, c_j is a literal equal to x_i or $\neg x_i$ for some i. For a literal l, we write that $l \in C_j$ iff l is equal to a_j, b_j or c_j. We will now construct a set of items A of size polynomial in $n + m$ and a polynomial weight W such that A with the total weight W is a positive instance of UNBOUNDED SUBSET SUM iff ϕ is satisfiable.

We first need to construct a list of $n + m$ different prime numbers $(p_1, \ldots, p_{n+m})$ that are all larger than $n + 5$. It suffices to pick $p_i = \pi_{i+n+5}$ for all i, because clearly $\pi_j > j$ for all j. Thanks to Corollary 1, we can list all these primes numbers in the unary notation in time and space polynomial in $n + m$.

The set A will contain one item per each literal. We set the weight of the item corresponding to the literal x_i to

$$1 + \frac{1}{p_i} - \frac{1}{p_{i \oplus_n 1}} + \sum_{\{j | x_i \in C_j\}} \left(\frac{1}{p_{n+j}} - \frac{1}{p_{n+j \oplus_m 1}} \right)$$

and corresponding to the literal $\neg x_i$ to

$$1 + \frac{1}{p_i} - \frac{1}{p_{i \oplus_n 1}} + \sum_{\{j | \neg x_i \in C_j\}} \left(\frac{1}{p_{n+j}} - \frac{1}{p_{n+j \oplus_m 1}} \right).$$

Notice that each of these weights is $\geq 1 - \frac{5}{p_1} > 0$, because each literal occurs at most four times in ϕ, and $p_1 > 5$ is the smallest prime number among p_i-s. At the same time, all of them are also $\leq 1 + \frac{5}{p_1} < 2$. Moreover, they can all be written in unary using polynomial space, because each literal occurs in at most four clauses and so this sum will have at most 11 terms in total. We can then combine all these terms into a single rational number. Its denominator will be at most equal to p_{n+m}^{10}, because p_{n+m} is the largest prime number among p_i-s and the first of these terms is equal to 1. The numerator of this rational number has to be smaller than $2p_{n+m}^{10}$, because we already showed this number to be < 2. So both of them will have size $\mathcal{O}((2n + m)^{10} \log^{10}(2n + m))$ when written down in unary, because $p_{n+m} = \pi_{2n+m+5} < 2(2n + m + 5) \log(2n + m + 5)$ due to Theorem 1. Set A has $2n$ such items and so all its elements' weights can be written down in unary using $\mathcal{O}(n(2n + m)^{10} \log^{10}(2n + m))$ space.

Notice that the total weight of A is equal to

$$2n + \sum_{i=1}^{n} \left(\frac{2}{p_i} - \frac{2}{p_{i \oplus_n 1}} \right) + \sum_{j=1}^{m} \left(\frac{3}{p_{n+j}} - \frac{3}{p_{n+j \oplus_m 1}} \right)$$

because there are $2n$ literals, each variable corresponds to two literals, and each clause contains exactly three literals. As both of the two sums in this expression are telescoping, we get that the total weight is in fact equal to $2n$. We claim that the target weight $W = n$ is achievable by picking items from A (each item possibly multiple times) iff ϕ is a positive instance of ALL-THE-SAME-SAT.

($\Rightarrow$) Let q_i and q_i' be the number of times an item corresponding to, respectively, literal x_i and $\neg x_i$ is chosen so that the total weight of all these items is n.

For $i = 1, \ldots, n$, we define $t_i := q_i + q_i'$. For $j = 1, \ldots, m$, we define t_{n+j} to be the number of times an item corresponding to a literal in C_j is chosen. For example, if $C_j = x_1 \vee \neg x_2 \vee x_5$ then $t_{n+j} = q_1 + q_2' + q_5$. Finally, let $T := \sum_{i=1}^{n} q_i + q_i'$ be the total number of items chosen. Notice that $T \leq W/(1 - \frac{5}{p_1}) < W/(1 - \frac{5}{n+5}) = n + 5$.

Now the total weight of the selected items can be expressed using t_i-s as follows:

$$\sum_{i=1}^{n} t_i + \sum_{i=1}^{n} \frac{t_i - t_{i\ominus_n 1}}{p_i} + \sum_{j=1}^{m} \frac{t_{n+j} - t_{n+j\ominus_m 1}}{p_{n+j}} \qquad (\star)$$

Notice that $|t_i - t_{i\ominus_n 1}| < n + 5$ and $p_i > n + 5$ for all $i = 1, \ldots, n$, and $|t_{n+j} - t_{n+j\ominus_m 1}| < n + 5$ and $p_{n+j} > n + 5$ for all $j = 1, \ldots, m$. It now follows from Lemma 1 that $(\star)$ can be equal to n if and only if $\sum_{i=1}^{n} t_i = n$, and $t_1 = t_2 = \ldots = t_n$, and $t_{n+1} = t_{n+2} = \ldots = t_{n+m}$. The first two facts imply that for all $i = 1, \ldots, n$, we have $t_i = 1$ and so exactly one item corresponding to either x_i or $\neg x_i$ is chosen. The last fact states that in each clause exactly the same number of items corresponding to its literals is chosen. It is now easy to see that the ALL-THE-SAME-SAT condition is satisfied by ϕ for the valuation ν such that, for all $i \in \{1, \ldots, n\}$, we set $\nu(x_i) = \top$ iff $q_i = 1$.

($\Leftarrow$) Let ν be a valuation for which ϕ satisfies the ALL-THE-SAME-SAT condition. We set the quantities q_i and q_i', the number of times an item corresponding to the literal x_i and $\neg x_i$ is picked, as follows. If $\nu(x_i) = \top$ then we set $q_i = 1$ and $q_i' = 0$. If $\nu(x_i) = \bot$ then we set $q_i = 0$ and $q_i' = 1$.

Let us define t_i-s as before. Note that we now have $t_i = 1$ for all $i = 1, \ldots, n$ and $t_{n+1} = t_{n+2} = \ldots = t_{n+m}$, because the ALL-THE-SAME-SAT condition is satisfied by ν. We can now easily see from the expression $(\star)$ that the total weight of the just picked items is equal to n. $\square$

Although the strong NP-hardness complexity of the PARTITION problem does not follow from the statement of Theorem 4, it follows from its proof as follows.

Theorem 5. *The PARTITION problem with rational weights is strongly NP-complete.*

Proof. Just repeat the proof of Theorem 4 without any change. In this case we know *a priori* that $q_i \in \{0, 1\}$, which does not make any difference to the used reasoning. Notice that the target weight W chosen in the reduction is exactly equal to half of the total weights of all the items in A, so the UNBOUNDED SUBSET SUM problem instance constructed can also be considered to be a PARTITION problem instance. $\square$

Now, as the SUBSET SUM problem is a generalisation of the PARTITION problem, we instantly get the following result.

Corollary 2. *The SUBSET SUM problem with rational weights is strongly NP-complete.*

Finally, we observe that the 0-1 KNAPSACK and UNBOUNDED KNAPSACK problems are generalisations of the SUBSET SUM and UNBOUNDED SUBSET SUM problems, respectively. To see this just restrict the weight and profit of each item to be equal to each other as well as require $V = W$. Any such instance is a positive instance of 0-1 KNAPSACK (UNBOUNDED KNAPSACK) if and only if it is a positive instance of SUBSET SUM (respectively, UNBOUNDED SUBSET SUM).

Corollary 3. *The 0-1* KNAPSACK *and* UNBOUNDED KNAPSACK *problems with rational weights are strongly NP-complete.*

6   Approximability

In this section, we briefly discuss the counter-intuitive fact that the optimisation version of all the decision problems defined in Sect. 2 admit a fully polynomial-time approximation scheme (FPTAS) even though we just showed them to be strongly NP-complete. First, let us restate a well-known result concerning this.

Corollary 4 (Corollary 8.6 in [23]). *Let* Π *be an NP-hard optimisation problem satisfying the restrictions of Theorem 8.5 in [23] (first shown in [6]). If* Π *is strongly NP-hard, then* Π *does not admit an FPTAS, assuming* $P \neq NP$.

The crucial assumption made in Theorem 8.5 of [23] is that the objective function is integer valued, which does not hold in our case, so there is no contradiction.

First, let us formally define the optimisation version of some of the decision problems studied. The optimisation version of the 0-1 KNAPSACK problem with capacity W asks for a subset of items with the maximum possible total profit and whose weight does not exceed W. As for the SUBSET SUM problem, its optimisation version asks for a subset of items whose total weight is maximal, but $\leq W$. The optimisation version of the other decision problems from Sect. 2 can also be defined (see, e.g., [12]).

Now, let us formally define what we mean by an approximation algorithm for these problems. We say that an algorithm is a *constant factor approximation algorithm* with a *relative performance* ρ iff, for any problem instance, I, the cost of the solution that it computes, $f(I)$, satisfies:

- for a maximisation problem: $(1 - \rho) \cdot \text{OPT}(I) \leq f(I) \leq \text{OPT}(I)$
- for a minimisation problem: $\text{OPT}(I) \leq f(I) \leq (1 + \rho)\text{OPT}(I)$

where $\text{OPT}(I)$ is the optimal cost for the problem instance I. We are particularly interested in polynomial-time approximation algorithms. A polynomial-time approximation scheme (PTAS) is an algorithm that, for every $\rho > 0$, runs in polynomial-time and has relative performance ρ. Note that the running time of a PTAS may depend in an arbitrary way on ρ. Therefore, one typically strives to find a fully polynomial-time approximation scheme (FPTAS), which is an algorithm that runs in polynomial-time in the size of the input and $1/\rho$.

We will focus here on defining an FPTAS for 0-1 KNAPSACK problem with rational coefficients. An FPTAS for the other optimisation problems considered in this paper can be defined in essentially the same way and thus their details are omitted.

Theorem 6. *The 0-1* KNAPSACK *problem with rational coefficients admits an FPTAS.*

Proof. We claim that we can simply reuse here any FPTAS for the 0-1 KNAP-SACK problem with integer coefficients. Let I be a 0-1 KNAPSACK instance with rational coefficients. We turn I into an instance with integer coefficients only, I', by the usual trick of multiplying all the rational coefficients by the least common multiple of the denominators of all the rational coefficients in I. Let us denote this least common multiple by α. Assuming that all the coefficients in I are represented in binary, then when multiplying them by α (again in binary representation), their size, as argued at the end of Sect. 2, will only increase polynomially. Therefore, the size of I' is just polynomially larger than I. (We should not use the unary notation because then these numbers may grow exponentially.)

Notice that $\alpha \cdot \text{OPT}(I) = \text{OPT}(I')$. In fact, the profit of any subset of items A in I', denoted by $\text{profit}'(A)$, is α times bigger than the profit of this set of items in I, denoted by $\text{profit}(A)$. Let us now run on I' any FPTAS for 0-1 KNAPSACK problem with integer coefficients with relative performance ρ (e.g., [9]). This will return as a solution a subset of items, B, such that $\text{profit}'(B) \geq (1 - \rho)\text{OPT}(I')$. This implies that $\text{profit}(B) \geq (1 - \rho)\text{OPT}(I)$ so the same subset of items B has also the same relative performance ρ on the original instance I. $\square$

7   Conclusions

In this paper we studied how the computational complexity of the PARTITION, 0-1 SUBSET SUM, UNBOUNDED SUBSET SUM, 0-1 KNAPSACK, UNBOUNDED KNAPSACK problems changes when items' weights and profits can be any rational numbers. We showed here, as opposed to the setting where all these values are integers, that all these problems are strongly NP-hard, which means that there does not exists a pseudo-polynomial algorithm for solving them unless $P = NP$. Nevertheless, we also showed that all these problem admit an FPTAS just like in the integer setting. Finally, we just want to point out that if we restrict ourselves to only rational weights and profits with a finite representation as decimal numerals, then these problems are no longer strongly NP-complete. This is because we could then simply multiply all the input numbers by a sufficiently high power of 10 and get an instance, with all integer coefficients, whose size is polynomial in the size of the original instance.

Acknowledgments. We would like to thank the anonymous reviewers whose comments helped to improve this paper. This work was partially supported by the EPSRC through grants EP/M027287/1 (Energy Efficient Control) and EP/P020909/1 (Solving Parity Games in Theory and Practice).

References

1. Babat, L.G.: Linear functions on n-dimensional unit cube. Doklady Akademii Nauk SSSR **221**(4), 761–762 (1975)
2. Chekuri, C., Khanna, S.: A polynomial time approximation scheme for the multiple knapsack problem. SIAM J. Comput. **35**(3), 713–728 (2005)

3. Cook, S.A.: The complexity of theorem-proving procedures. In: Proceedings of the Third Annual ACM Symposium on Theory of Computing, pp. 151–158. ACM (1971)
4. Elkind, E., Goldberg, L.A., Goldberg, P.W., Wooldridge, M.: On the computational complexity of weighted voting games. Ann. Math. Artif. Intell. **56**(2), 109–131 (2009)
5. Gallo, G., Hammer, P.L., Simeone, B.: Quadratic knapsack problems. In: Padberg, M.W. (ed.) Combinatorial Optimization, pp. 132–149. Springer, Heidelberg (1980). https://doi.org/10.1007/BFb0120892
6. Garey, M.R., Johnson, D.S.: "Strong" NP-completeness results: motivation, examples, and implications. J. ACM (JACM) **25**(3), 499–508 (1978)
7. Guéret, C., Prins, C.: A new lower bound for the open-shop problem. Ann. Oper. Res. **92**, 165–183 (1999)
8. Hoogeveen, J.A., Oosterhout, H., van de Velde, S.L.: New lower and upper bounds for scheduling around a small common due date. Oper. Res. **42**(1), 102–110 (1994)
9. Ibarra, O.H., Kim, C.E.: Fast approximation algorithms for the knapsack and sum of subset problems. J. ACM (JACM) **22**(4), 463–468 (1975)
10. Johnson, D.S., Niemi, K.A.: On knapsacks, partitions, and a new dynamic programming technique for trees. Math. Oper. Res. **8**(1), 1–14 (1983)
11. Karp, R.M.: Reducibility among combinatorial problems. In: Miller, R.E., Thatcher, J.W., Bohlinger, J.D. (eds.) Complex. Comput. Comput., pp. 85–103. Springer, Heidelberg (1972)
12. Kellerer, H., Pferschy, U., Pisinger, D.: Knapsack Problems. Springer, Heidelberg (2004). https://doi.org/10.1007/978-3-540-24777-7
13. Levin, L.A.: Universal sequential search problems. Problemy Peredachi Informatsii **9**(3), 115–116 (1973)
14. Mathews, G.B.: On the partition of numbers. Proc. Lond. Math. Soc. **1**(1), 486–490 (1897)
15. Mousa, M.A.A., Schewe, S., Wojtczak, D.: Optimal control for simple linear hybrid systems. In: Proceedings of TIME, pp. 12–20. IEEE Computer Society (2016)
16. Mousa, M.A.A., Schewe, S., Wojtczak, D.: Optimal control for multi-mode systems with discrete costs. In: Abate, A., Geeraerts, G. (eds.) FORMATS 2017. LNCS, vol. 10419, pp. 77–96. Springer, Cham (2017). https://doi.org/10.1007/978-3-319-65765-3_5
17. Pferschy, U., Schauer, J.: The knapsack problem with conflict graphs. J. Graph Algorithms Appl. **13**(2), 233–249 (2009)
18. Pisinger, D.: An exact algorithm for large multiple knapsack problems. Eur. J. Oper. Res. **114**(3), 528–541 (1999)
19. Rosser, J.B., Schoenfeld, L.: Approximate formulas for some functions of prime numbers. Ill. J. Math. **6**(1), 64–94 (1962)
20. Schaefer, T.J.: The complexity of satisfiability problems. In: Proceedings of the Tenth Annual ACM Symposium on Theory of Computing, pp. 216–226. ACM (1978)
21. Silvano, M., Paolo, T.: Knapsack Problems: Algorithms and Computer Implementations. Wiley, Hoboken (1990)
22. Tovey, C.A.: A simplified NP-complete satisfiability problem. Discrete Appl. Math. **8**(1), 85–89 (1984)
23. Vazirani, V.V.: Approximation Algorithms. Springer Science & Business Media, Heidelberg (2013). https://doi.org/10.1007/978-3-662-04565-7

A New Algorithm for Finding Closest Pair of Vectors (Extended Abstract)

Ning Xie$^{(\boxtimes)}$, Shuai Xu$^{(\boxtimes)}$, and Yekun Xu$^{(\boxtimes)}$

Florida International University, Miami, FL 33199, USA
{nxie,sxu010,yxu040}@fiu.edu

Abstract. Given n vectors $x_0, x_1, \ldots, x_{n-1}$ in $\{0,1\}^m$, how to find two vectors whose pairwise Hamming distance is minimum? This problem is known as the *Closest Pair Problem*. If these vectors are generated uniformly at random except two of them are correlated with Pearson-correlation coefficient ρ, then the problem is called the *Light Bulb Problem*. In this work, we propose a novel coding-based scheme for the Close Pair Problem. We design both randomized and deterministic algorithms, which achieve the best known running time when the minimum distance is very small compared to the length of input vectors. When applied to the Light Bulb Problem, our algorithms yields state-of-the-art deterministic running time when the Pearson-correlation coefficient ρ is very large.

1 Introduction

We consider the following classic *Closest Pair Problem*: given n vectors $x_0, x_1, \ldots, x_{n-1}$ in $\{0,1\}^m$, how to find the two vectors with the minimum pairwise distance? Here the distance is the usual Hamming distance: $\text{dist}(x_i, x_j) = |\{k \in [m] : (x_i)_k \neq (x_j)_k\}|$, where $(x_i)_k$ denotes the kth component of vector x_i. Without loss of generality, we assume that $d_{\min} = \text{dist}(x_0, x_1)$ is the unique minimum distance and all other pairwise distances are greater than $d_{\min}$.

The Closest Pair Problem is one of the most fundamental and well-studied problems in many science disciplines, having a wide spectrum of applications in computational finance, DNA detection, weather prediction, etc. For instance, the Closest Pair Problem recently finds the following interesting application in bioinformatics. Scientists wish to find connections between Single Nucleotide Polymorphisms (SNPs) and phenotypic traits. SNPs are one of the most common types of genetic differences among people, with each SNP representing a variation in a single DNA block called *nucleotide* [19]. Screening for most correlated pairs of SNPs has been applied to study such connections [9,13,16,30]. As the number of SNPs in humans is estimated to be around 10 to 11 million, for problem size n of this size, any improvement in running time for solving the Closest Pair Problem would have huge impacts on genetics and computational biology [30].

A full version of the paper is available at https://arxiv.org/abs/1802.09104.
N. Xie, S. Xu and Y. Xu—Research supported in part by NSF grant 1423034.

© Springer International Publishing AG, part of Springer Nature 2018
F. V. Fomin and V. V. Podolskii (Eds.): CSR 2018, LNCS 10846, pp. 321–333, 2018.
https://doi.org/10.1007/978-3-319-90530-3_27

In theoretical computer science, the Closest Pair Problem has a long history in computational geometry, see e.g. [34] for a survey of many classic algorithms for the problem. The naive algorithm for the Closest Pair Problem takes $O(mn^2)$ time. When the dimension m is a constant, either in the Euclidean space or ℓ_p space, the classic divide-and-conquer based algorithm runs in $O(n \log n)$ time [12]. Rabin [33] combined the floor function with randomization to devise a linear time algorithm. In 1995, Khuller and Matias [25] simplified Rabin's algorithm to achieve the same running time $O(n)$ and space complexity $O(n)$. Golin et al. [21] used dynamic perfect hashing to implement a dictionary and obtained the same linear time and space bounds.

When the dimension m is not a constant, the first subquadratic time algorithm for the Closest Pair Problem is due to Alman and Williams [4] for m as large as $\log^{2-o(1)} n$. The algorithm is built on a recently developed framework called *polynomial method* [2,39,40]. In particular, Alman and Williams firstly constructed a probabilistic polynomial of degree $O(\sqrt{n \log 1/\epsilon})$ which can compute the MAJORITY function on n variables with error at most ϵ, then applied the polynomial method to design an algorithm which runs in $n^{2-1/O(s(n) \log^2 s(n))}$ time where $m = s(n) \log n$, and computed the minimum Hamming distance among all red-blue vector pairs through polynomial evaluations. In a more recent work, Alman et al. [3] unified Valiant's fast matrix multiplication approach [37] with that of Alman and Williams' [4]. They constructed probabilistic *polynomial threshold functions* (PTFs) to obtain a simpler algorithm which improved to randomized time $n^{2-1/O(\sqrt{s(n)} \log^{3/2} s(n))}$ or deterministic time $n^{2-1/O(s(n) \log^2 s(n))}$.

The Light Bulb Problem. A special case of the Closest Pair Problem, the so-called *Light Bulb Problem*, was first posed by Valiant in 1988 [38]. In this problem, we are given a set of n vectors in $\{0,1\}^m$ chosen uniformly at random from the Boolean hypercube, except that two of them are non-trivially correlated (specifically, have Pearson-correlation coefficient ρ, which is equivalent to that the expected Hamming distance between the correlated pair is $\frac{1-\rho}{2}m$), the problem then is to find the correlated pair.

Paturi et al. [32] gave the first non-trivial algorithm, which runs in $O(n^{2-\log(1+\rho)})$. The well-known *locality sensitive hashing* scheme of Indyk and Motwani [23] performs slightly worse than Paturi et al.'s hash-based algorithm. More recently, Dubiner [18] proposed a Bucketing Coding algorithm which runs in time $O(n^{\frac{2}{1+\rho}})$. As ρ gets small, all these three algorithms have running time $O(n^{2-O(\rho)})$. Comparing the constants in these three algorithms, Dubiner achieves the best constants, which is $O(n^{2-2\rho})$, in the limit of $\rho \to 0$. Asymptotically the same bound was also achieved by May and Ozerov [27], in which the authors used algorithms that find Hamming closest pairs to improve the running time of decoding random binary linear codes.

In a recent breakthrough result, Valiant [37] presented a fast matrix multiplication based algorithm which finds the "planted" closest pair in time $O(\frac{n^{\frac{5-\omega}{4-\omega}+\epsilon}}{\rho^{2\omega}}) < n^{1.62} \cdot \text{poly}(1/\rho)$ with high probability for any constant $\epsilon, \rho > 0$ and $m > n^{\frac{1}{4-\omega}}/\rho^2$, where $\omega < 2.373$ is the exponent of fast matrix multiplications.

The most striking feature of Valiant's algorithm is that ρ does not appear in the exponent of n in the running time of the algorithm. Karppa et $al.$ [24] further improved Valiant's algorithm to $n^{1.582}$. Both Valiant and Karppa et $al.$ achieved runtime of $n^{2-\Omega(1)}(m/\epsilon)^{O(1)}$ for the Light Bulb Problem, which improved upon previous algorithms that rely on the Locality Sensitive Hashing (LSH) schemes. The LSH methods based algorithm only achieved runtime of $n^{2-O(\epsilon)}$ for the Light Bulb Problem.

We remark that all the above-mentioned algorithms (except May and Oze-rov's work) that achieve state-of-the-art running time are based on either involved probabilistic polynomial constructions or impractical $O(n^\omega)$ fast matrix multiplications[1], or both. Moreover, these algorithms are all randomized in nature while our approach yields simple and practical randomized as well as deterministic algorithms.

1.1 Our Approach

We propose a simple, error-correcting code based scheme for the Closest Pair Problem. Apart from achieving the best running time for certain range of param-eters, we believe that our new approach has the merit of being simple, and hence more likely being practical as well. In particular, neither complicated data struc-ture nor fast matrix multiplication is employed in our algorithms.

Algorithm 1. General Idea of Main Algorithm

    input : A set of n vectors $x_0, \ldots, x_{n-1}$ in $\{0,1\}^m$ and $d_{\min}$
    output: Two vectors and their distance

1 generate a binary code $C \subseteq \{0,1\}^m$
2 pick a random $y \in \{0,1\}^m$
3 **for** $j \leftarrow 0$ **to** $n-1$ **do**
4 | decode $y + x_j$ in C, and denote the resulting vector by $\tilde{x}_j$
5 **end**
6 sort $\tilde{x}_0, \ldots, \tilde{x}_{n-1}$
7 **for** each of the $n-1$ pairs of adjacent vectors in the sorted list **do**
8 | compute the distance between the two $original$ vectors.
9 **end**
10 output the pair of vectors with the minimum distance and their distance

---

The basic idea of our algorithms is very simple. Suppose for concreteness that x_0 and x_1 is the unique pair of vectors that achieve the minimum distance. Our

---

[1] Subqubic fast matrix multiplication algorithms are practical only for Strassen-based ones [11,22]. Even though the recent breakthrough results [20,35,42] achieve asymp-totically faster than Strassen's algorithm [36], however, these algorithms are all based on Coppersmith-Winograd's algorithm [15], and to the best of our knowledge, there are no practical implementations of these trilinear based algorithms.

scheme is inspired by the extreme case when x_0 and x_1 are identical vectors. In this case, a simple *sort and check* approach solves the problem in $O(mn \log n)$ time: sort all n vectors and then compute only the $n - 1$ pairwise distances (instead of all $\binom{n}{2}$ distances) of adjacent vectors in the sorted list. Since the two closest vectors are identical, they must be adjacent in the sorted list and thus the algorithm would compute their distance and find them. This motivates us to view the input vectors as received messages that were encoded by an error correction code and have been transmitted through a noisy channel. As a result, the originally identical vectors are no longer the same, nevertheless are still very close. Directly applying the *sort and check* approach would fail but a natural remedy is to decode these received messages into codewords first. Indeed, if the distance between x_0 and x_1 is small and we are lucky to have a codeword c that is very close to both of them, then a unique decoding algorithm would decode both of these two vectors into c. Now if we "sort" the *decoded* vectors and then "check" the corresponding *original* vectors of each adjacent pair of vectors[2], the algorithm would successfully find the closest pair. How to turn this "good luck" into a working algorithm? Simply try different shift vectors y and view $y + x_i$ as the input vectors, since the Hamming distances are invariant under any shift. The basic idea of our approach is summarized in Algorithm 1.

Figure 1 illustrates the effects "bad" shift vectors and "good" shift vectors on the decoding part of our algorithm.

Figure 2 illustrates what happens if we sort the vectors directly and why sorting decoded vectors works.

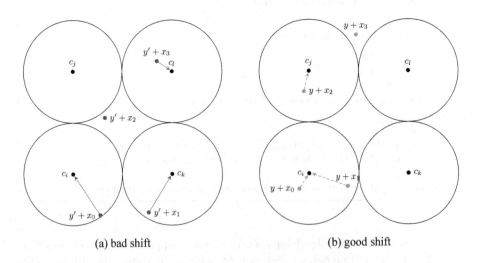

(a) bad shift   (b) good shift

Fig. 1. Decoding with good and bad shift vectors

---

[2] Actually, we only need to "check" when the two adjacent decoded vectors are identical.

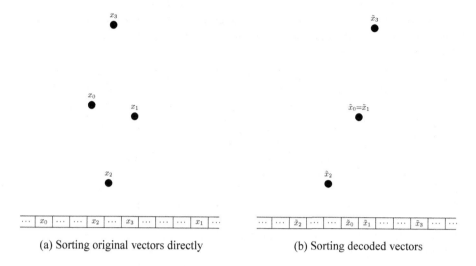

(a) Sorting original vectors directly   (b) Sorting decoded vectors

Fig. 2. Difference between sorting input vectors directly and sorting decoded vectors.

Making the idea of decoding work for larger minimum pairwise distance involves balancing the parameters of the error-correcting code so that it is efficiently decodable as well as having appropriate decoding radius. The decoding radius r should have the following properties. On one hand, r should be small to ensure that there is a codeword c such that only x_0 and x_1 will be decoded into c (therefore x_0 and x_1 will be adjacent in the sorted array and hence will be compared with each other). On the other hand, we would like r to be large so as to maximize the number of "good" shift vectors which enable both x_0 and x_1 decoding to the same codeword. As a result, our algorithms generally perform best when the closest pair distance is very small.

1.2   Our Results

Our simple error-correcting code based algorithm can be applied to solve the Closest Pair Problem and the Light Bulb Problem.

The Closest Pair Problem. Our main result is the following simple randomized algorithm for the Closest Pair Problem.

Theorem 1 (Main). *Let $x_0, x_1, \ldots, x_{n-1}$ in $\{0,1\}^m$ be n binary vectors such that x_0 and x_1 is the unique pair achieving the minimum pairwise distance d_{min} (and the second smallest distance can be as small as $d_{min} + 1$). Suppose we are given the value of d_{min} and let $\delta \overset{\text{def}}{=} d_{min}/m$. Then there is a randomized algorithm running in $O(n \log^2 n \cdot 2^{(1 - \kappa_Z(\delta) - \delta)m} \cdot \text{poly}(m))$ which finds the closest pair x_0 and x_1 with probability at least $1 - 1/n^2$. The running time can be improved to*

$O(n \log^2 n \cdot 2^{(1-\kappa_{GV}(\delta)-\delta)m} \cdot \mathrm{poly}(m))$, *if we are given black-box decoding algorithms for an ensemble of $O(\log m/\epsilon)$ binary error-correcting codes that meet the Gilbert-Varshamov bound.*

Here $\kappa_{GV}(\delta)$ and $\kappa_z(\delta)$ are functions derived from the Gilbert-Varshamov (GV) bound and the Zyablov bound respectively (see full version for details).

The running time of our algorithm depends on—in addition to the number of vectors n—both dimension m and $\delta \overset{\mathrm{def}}{=} d_{\min}/m$. To illustrate its performance we choose two typical vector lengths m, namely those corresponding to the Hamming bound[3] and the Gilbert-Varshamov (GV) bound[4], and list the exponents γ' in the running time of the GV-code version of our algorithm as a function of $d_{\min}$ (in fact δ) in Table 1. Here, we write the running of the algorithm as $\tilde{O}(n^{\gamma'})$, where $\tilde{O}$ suppresses any polylogarithmic factor of n. One can see that our algorithm runs in subquadratic time when δ is small, or equivalently when the Hamming distance between the closest pair is small.

Table 1. Running time of our algorithm when vector length m meets the Hamming bound and GV bound

δ	Hamming bound		GV bound	
	Length of vector $(m/\log n)$	Exponent (γ')	Length of vector $(m/\log n)$	Exponent (γ')
0.01	1.0476	1.0742	1.0879	1.0770
0.025	1.1074	1.1591	1.2029	1.1728
0.05	1.2029	1.2844	1.4013	1.3313
0.075	1.2999	1.4021	1.6242	1.5024
0.1	1.4013	1.5171	1.8832	1.6949
0.125	1.5090	1.6316	2.1909	1.9170

In the setting of $m = c \log n$ for some not too large constant c, Alman *et al.* [3] gave a randomized algorithm which runs in $n^{2-1/O(\sqrt{c}\log^{3/2} c)}$ time for the Closest Pair Problem. As it is very hard to calculate the hidden constant in the exponent of their running time, it is impossible to compare our running time with theirs quantitatively. However, as the running time of Alman *et al.* is of the form $n^{2-g(c)}$ for some function g, it is reasonable to believe that our algorithms run faster when the minimum distance is small enough.

Deterministic algorithm. By checking all shift vectors up to certain Hamming weight, our randomized algorithm can be easily derandomized to yield the following.

---

[3] The Hamming bound, also known as the sphere packing bound, specifies an upper bound on the number codewords a code can have given the block length and the minimum distance of the code.

[4] The GV bound is known to be attainable by the random codes.

Theorem 2. *Let $x_0, x_1, \ldots, x_{n-1}$ in $\{0,1\}^m$ be n binary vectors such that x_0 and x_1 is the unique pair achieving the minimum pairwise distance d_{min} (and the second smallest distance can be as small as $d_{min} + 1$). Suppose we are given the value of d_{min} and let $\delta \stackrel{def}{=} d_{min}/m$. Then there is a deterministic algorithm that finds the closest pair x_0 and x_1 with running time $O(n \log n \cdot 2^{H_2(1-\kappa_Z(\delta))m} \cdot \text{poly}(m))$, where $H_2(\cdot)$ is the binary entropy function. Moreover, if we are given as black box the decoding algorithm of a random Varshamov linear code with block length m and minimum distance $d_{min} + 1$, then the running time is $O(n \log n \cdot 2^{H_2(1-\kappa_{GV}(\delta))m} \cdot \text{poly}(m))$.*

Searching for d_{min}. If we remove the assumption that d_{min} is given, our algorithm can be modified to search for d_{min} first without too much slowdown; more details appear in full version.

Theorem 3. *Let $x_0, x_1, \ldots, x_{n-1}$ in $\{0,1\}^m$ be n binary vectors such that x_0 and x_1 is the unique pair achieving the minimum pairwise distance d_{min}. Then for any $\epsilon > 0$, there is a randomized algorithm runs in $O(\epsilon^{-1} n \log^2 n \cdot 2^{(1-\kappa_Z((1+\epsilon)\delta)-\delta H_2(\frac{1-\epsilon}{2}))m} \cdot \text{poly}(m))$ which finds the d_{min} (and the pair x_0 and x_1) with probability at least $1 - 1/n$, The running time can be improved to $O(\epsilon^{-1} n \log^2 n \cdot 2^{(1-\kappa_{GV}((1+\epsilon)\delta)-\delta H_2(\frac{1-\epsilon}{2}))m} \cdot \text{poly}(m))$, if we are given black-box decoding algorithms for an ensemble of $O(\log m/\epsilon)$ binary error-correcting codes that meet the Gilbert-Varshamov bound.*

Gapped version. Intuitively, if there is a gap between d_{min} and the second minimum distance, the Closest Pair Problem should be easier. This is reminiscent of the case of the ϵ-Approximate NNS Problem versus the NNS Problem. However, as we still need to find the *exact* solution to the Closest Pair Problem, the situation here is different.

Theorem 4 (Gapped version). *Let $x_0, x_1, \ldots, x_{n-1}$ in $\{0,1\}^m$ be n binary vectors such that x_0 and x_1 is the unique pair achieving the minimum pairwise distance d_{min}. Suppose we are given the values of d_{min} as well as the second minimum distance d_2. Let $\delta \stackrel{def}{=} d_{min}/m$ and $\delta' \stackrel{def}{=} d_2/m$. Then there is a randomized algorithm running in $O(n \log^2 n \cdot 2^{(1-\kappa_Z(\delta')-\delta-(1-\delta)H_2(\frac{\delta'-\delta}{2(1-\delta)}))m} \cdot \text{poly}(m))$ which finds the closest pair x_0 and x_1 with probability at least $1 - 1/n^2$. Moreover, the running time can be further improved to $O(n \log^2 n \cdot 2^{(1-\kappa_{GV}(\delta')-\delta-(1-\delta)H_2(\frac{\delta'-\delta}{2(1-\delta)}))m} \cdot \text{poly}(m))$, if we are given the black box access to the decoding algorithm of an (m, K, d)-code which meets the Gilbert-Varshamov bound.*

Our gapped version algorithm uses $d_2/2$ instead of $d_{\text{min}}/2$ as the decoding radius. This, however, does not always give improved running time as illustrated in Fig. 3. In Fig. 3, we set $\delta' = (1 + \epsilon)\delta$ and write the running time as $O(n \log^2 n \cdot 2^{\gamma m} \cdot \text{poly}(m))$ for both the gapped version (the blue line) and the non-gapped version (the green line). One can see that the gapped version performs better only when ϵ is small enough.

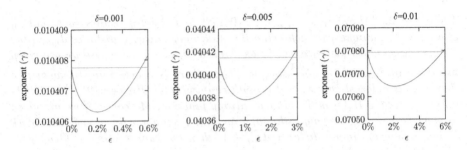

Fig. 3. The range of ϵ in which gapped version outperforms non-gapped version (Color figure online)

The Light Bulb Problem. Applying our algorithms for the Closest Pair Problem to the Light Bulb Problem easily yields the following.

Theorem 5. *There is a randomized algorithm for the Light Bulb Problem which runs in time*

$$O(n \cdot \mathrm{poly}(\log n)) \cdot 2^{(1-\kappa_Z(\frac{1-\rho}{2})-\frac{1-\rho}{2})\frac{4\ln 2 \cdot \log n}{\rho^2}}(1+o(1))$$

and succeeds with probability at least $1 - 1/n^2$. The running time can be further improved to

$$O(n \cdot \mathrm{poly}(\log n)) \cdot 2^{(1-\kappa_{GV}(\frac{1-\rho}{2})-\frac{1-\rho}{2})\frac{4\ln 2 \cdot \log n}{\rho^2}}(1+o(1)),$$

if we are allowed a one-time preprocessing time of $n^{2.773/\rho^2}$ to generate the decoding lookup table of a random Gilbert's $(m, K, (1 - \rho)m/2)$-code. Similar results can also be obtained for deterministic algorithms.

Our deterministic algorithm for the Light Bulb Problem is, to the best of our knowledge, the only deterministic algorithm for the problem. Moreover, we believe that our algorithms are very simple and therefore are likely to outperform other complicated ones for at least not too large input sizes.

1.3 Related Work

The Nearest Neighbor Search problem. The Closest Pair Problem is a special case of the more general *Nearest Neighbor Search* (NNS) problem, defined as follows. Given a set S of n vectors in $\{0, 1\}^m$, and a query point $q \in \{0, 1\}^m$ as input, the problem is to find a point in S which is closest to q. The performance of an NNS algorithm is usually measured by two parameters: the space (which is usually proportional to the preprocessing time) and the query time. It is easy to see that any algorithms for NNS can also be used to solve the Closest Pair problem, as we can try each vector in S as the query vector against the remaining vectors in S, and output the pair with minimum distance.

Most early work on this problem is for fixed dimension. Indeed, when $m = 1$ the problem is easy, as we can just sort the input vectors (which in this case are numbers), then perform a binary search to find the closest vector to the input query. For $m \geq 2$, Clarkson [14] gave an algorithm with query time polynomial in $m \log n$, and space complexity $O(n^{\lceil m/2 \rceil})$. Meiser [28] designed an algorithm which runs in $O(m^5 \log n)$ time and uses $O(n^{m+\epsilon})$ space for arbitrary $\epsilon > 0$. By far, all efficient data structures for NNS have dimension m appear in the exponent of the space complexity, a phenomenon commonly known as the *curse of dimensionality*.

This motivates people to introduce a relaxed version of Nearest Neighbor Search called the ϵ-*Approximate Nearest Neighbor Search* (ϵ-*Approximate* NNS) Problem in the 1990s. The problem now is, for an input query point q, find a point p in S such that the Hamming distance is:

$$\text{dist}(p, q) \leq (1 + \epsilon) \min_{p' \in S} \text{dist}(p', q).$$

We call such a p as an ϵ-approximate nearest neighbor of input query q.

The ϵ-Approximate NNS Problem has been studied extensively in the last two decades. In 1998, Indyk and Motwani [23] used a set of hash functions to store the dataset such that if two points are close enough, they will have a very high probability to be hashed into the same buckets. As a pair of close points have higher probability than a pair of far-apart points to fall into the same bucket, the scheme is called *locality sensitive hashing* (LSH). The query time of LSH is $O(n^{\frac{1}{1+\epsilon}})$, which is sublinear, and the space complexity of LSH is $O(n^{1+\frac{1}{1+\epsilon}})$, which is subquadratic. After Indyk and Motwani introducing the locality sensitive hashing, there have been many improvements on the parameters under different metric spaces, such as ℓ_p metric [5,17,26,29,31]. Recently, Andoni *et al.* [7] gave tight upper and lower bounds of time-space trade-offs for hashing based algorithms for the ϵ-Approximate NNS Problem. This is the first algorithm that achieves sublinear query time and near-linear space, for any $\epsilon > 0$. For many results on the Approximate NNS problem in high dimension, see e.g. [6] for a survey. Some algorithms for the low dimension problem are surveyed in [8].

Recently, Valiant [37] leveraged fast matrix multiplication to obtain a new algorithm for the ϵ-Approximate NNS Problem that is not based on LSH. The general setting of Valiant's results is the following. Suppose there is a set of points S in m-dimensional Euclidean (or Hamming) space, and we are promised that for any $a \in S$ and $b \in S$, $\langle a, b \rangle < \alpha$, except for only one pair which has $\langle a, b \rangle \geq \beta$ (which corresponds to the closest pair, and β is known as the Pearson-correlation coefficient), for some $0 < \alpha < \beta < 1$. Valiant's algorithm finds the closest pair in $n^{\frac{5-\omega}{4-\omega} + \omega \frac{\log \beta}{\log \alpha}} m^{O(1)}$ time, where ω is the exponent for fast matrix multiplication ($\omega < 2.373$). Notice that, if the Pearson-correlation coefficient β is some fixed constant, then when α approaches 0 the running time tends to $n^{\frac{5-\omega}{4-\omega}}$, which is less than $n^{1.62}$. Valiant applied his algorithms to get improved bounds[5] for the Learning Sparse Parities with Noise Problem, the Learning k-Juntas with Noise Problem, the Learning k-Juntas without Noise Problem, and so on. More recently, Karppa et al. [24] improved upon Valiant's algorithm and obtained an algorithm that runs in $n^{\frac{2\omega}{3}+O(\frac{\log \beta}{\log \alpha})}m^{O(1)}$ time.

Note that in general algorithms for the ϵ-Approximate NNS Problem can not be used to solve the Closest Pair Problem, as the latter requires to find the *exact* solution for the closest pair of vectors.

Decoding Random Binary Linear Codes. In 2015, May and Ozerov [27] observed that algorithms for high dimensional Nearest Neighbor Search Problem can be used to speedup the approximate matching part of the *information set decoding* algorithm. They designed a new algorithm for the *Bichromatic Hamming Closest Pair* problem when the two input lists of vectors are pairwise independent, and consequently obtained a decoding algorithm for random binary linear codes with time complexity $2^{0.097n}$. This improved upon the previously best result of Becker et al. [10] which runs in $2^{0.102n}$.

The Bichromatic Hamming Closest Pair problem. In fact, the problem studied in [3,4,27] is the following *Bichromatic Hamming Closest Pair Problem*: we are given n red vectors $R = \{r_0, r_1, \cdots, r_{n-1}\}$ and n blue vectors $B = \{b_0, b_1, \cdots, b_{n-1}\}$ from $\{0,1\}^m$, and the goal is to find a red-blue pair with minimum Hamming distance. It is easy to see that the Closest Pair Problem is reducible to the Bichromatic Hamming Closest Pair Problem via a random reduction. On the other hand, our algorithm for the Closest Pair Problem can also be easily adapted to solve the Bichromatic Hamming Closest Pair Problem as follows. Run the decoding part of our algorithm on both sets R and B to get $\tilde{R} = \{\tilde{r}_0, \tilde{r}_1, \cdots, \tilde{r}_{n-1}\}$ and $\tilde{B} = \{\tilde{b}_0, \tilde{b}_1, \cdots, \tilde{b}_{n-1}\}$, sort $\tilde{R}$ and $\tilde{B}$ separately (without comparing the orginal vectors for adjacent pairs in the sorted lists), then merge the two sorted lists into one, and compute the distance between the original vectors for each red-blue pair of vectors that are compared during the merging process. On the other hand, the Bichromatic Closest Pair Problem is unlikely to have *truly* subquadratic algorithms under some mild conditions. Assuming the Strong Exponential Time Hypothesis (SETH), for any $\epsilon > 0$, there exists a constant c such that when the dimension $m = c \log n$, then there is no $2^{o(m)} \cdot n^{2-\epsilon}$-time algorithm for the Bichromatic Closest Pair Problem [1,4,41].

1.4 Full Version of the Paper

Due to space constraints, we omit the remaining sections from this extended abstract. A full version of this paper is available at https://arxiv.org/abs/1802. 09104.

---

[5] All these results are due to the fact that Valiant's algorithms are much more robust to weak correlations than other algorithms. Our algorithms therefore do not give improved bounds for these learning problems in the general settings.

Acknowledgements. We would like to thank Karthik C.S. and the anonymous referees for their valuable comments.

References

1. Abboud, A., Rubinstein, A., Williams, R.: Distributed PCP theorems for hardness of approximation in P. In: Proceedings of 58th Annual IEEE Symposium on Foundations of Computer Science, pp. 25–36 (2017)
2. Abboud, A., Williams, R., Yu, H.: More applications of the polynomial method to algorithm design. In: Proceedings of 26th ACM-SIAM Symposium on Discrete Algorithms, pp. 218–230 (2015)
3. Alman, J., Chan, T.M., Williams, R.: Polynomial representations of threshold functions and algorithmic applications. In: Proceedings of 57th Annual IEEE Symposium on Foundations of Computer Science, pp. 467–476 (2016)
4. Alman, J., Williams, R.: Probabilistic polynomials and Hamming nearest neighbors. In: Proceedings of 56th Annual IEEE Symposium on Foundations of Computer Science, pp. 136–150 (2015)
5. Andoni, A., Indyk, P.: Near-optimal hashing algorithms for approximate nearest neighbor in high dimensions. Commun. ACM **51**, 117–122 (2008)
6. Andoni, A., Indyk, P.: Nearest neighbors in high-dimensional spaces. In: Goodman, J., O'Rourke, J., Toth, C.D. (eds.) Handbook of Discrete and Computational Geometry, 3rd (edn). Chapman and Hall/CRC (2017)
7. Andoni, A., Laarhoven, T., Razenshteyn, I., Waingarten, E.: Optimal hashing-based time-space trade-offs for approximate near neighbors. In: Proceedings of 28th ACM-SIAM Symposium on Discrete Algorithms, pp. 47–66 (2017)
8. Arya, S., Mount, D.: Computational geometry: proximity and location. In: Mehta, D.P., Sahni, S. (eds) Handbook of Data Structures and Applications. Chapman and Hall/CRC (2005)
9. Aston, C., Ralph, D., Lalo, D., Manjeshwar, S., Gramling, B., DeFreese, D., West, A., Branam, D., Thompson, L., Craft, M., et al.: Oligogenic combinations associated with breast cancer risk in women under 53 years of age. Hum. Genet. **116**(3), 208–221 (2005)
10. Becker, A., Joux, A., May, A., Meurer, A.: Decoding random binary linear codes in $2^{n/20}$: how $1 + 1 = 0$ improves information set decoding. In: EUROCRYPT, pp. 520–536 (2012)
11. Benson, A., Ballard, G.: A framework for practical parallel fast matrix multiplication. In: ACM SIGPLAN Notices, pp. 42–53 (2015)
12. Bentley, J.: Multidimensional divide-and-conquer. Commun. ACM **23**(4), 214–229 (1980)
13. Cho, J., Nicolae, D., Gold, L., Fields, C., LaBuda, M., Rohal, P., Pickles, M., Qin, L., Fu, Y., Mann, J., et al.: Identification of novel susceptibility loci for inflammatory bowel disease on chromosomes 1p, 3q, and 4q: evidence for epistasis between 1p and IBD1. Proc. Natl. Acad. Sci. **95**(13), 7502–7507 (1998)
14. Clarkson, K.: A randomized algorithm for closest-point queries. SIAM J. Comput. **17**(4), 830–847 (1988)
15. Coppersmith, D., Winograd, S.: Matrix multiplication via arithmetic progressions. In: Proceedings of 19th Annual ACM Symposium on the Theory of Computing, pp. 1–6 (1987)
16. Cordell, H.: Detecting gene × gene interactions that underlie human diseases. Nat. Rev. Genet. **10**(6), 392–404 (2009)

17. Datar, M., Immorlica, N., Indyk, P., Mirrokni, V.: Locality-sensitive hashing scheme based on p-stable distributions. In: Proceedings of 20th Symposium on Computational Geometry, pp. 253–262 (2004)
18. Dubiner, M.: Bucketing coding and information theory for the statistical high-dimensional nearest-neighbor problem. IEEE Trans. Inf. Theory **56**(8), 4166–4179 (2008)
19. Frazer, K., Ballinger, D., Cox, D., Hinds, D., Stuve, L., Gibbs, R., Belmont, J., Boudreau, A., Hardenbol, P., Leal, S., et al.: A second generation human haplotype map of over 3.1 million SNPs. Nature **449**(7164), 851–861 (2007)
20. Le Gall, F.: Faster algorithms for rectangular matrix multiplication. In: Proceedings of 53rd Annual IEEE Symposium on Foundations of Computer Science, pp. 514–523 (2012)
21. Golin, M.J., Raman, R., Schwarz, C., Smid, M.: Simple randomized algorithms for closest pair problems. Nord. J. Comput. **2**(1), 3–27 (1995)
22. Huang, J., Rice, L., Matthews, D., van de Geijn, R.: Generating families of practical fast matrix multiplication algorithms. In: 2017 IEEE International on Parallel and Distributed Processing Symposium (IPDPS), pp. 656–667 (2017)
23. Indyk, P., Motwani, R.: Approximate nearest neighbors: towards removing the curse of dimensionality. In: Proceedings of 30th Annual ACM Symposium on the Theory of Computing, pp. 604–613 (1998)
24. Karppa, M., Kaski, P., Kohonen, J.: A faster subquadratic algorithm for finding outlier correlations. In: Proceedings of 27th ACM-SIAM Symposium on Discrete Algorithms, pp. 1288–1305 (2016)
25. Khuller, S., Matias, Y.: A simple randomized sieve algorithm for the closest-pair problem. Inf. Comput. **118**(1), 34–37 (1995)
26. Kushilevitz, E., Ostrovsky, R., Rabani, Y.: Efficient search for approximate nearest neighbor in high dimensional spaces. In: Proceedings of 30th Annual ACM Symposium on the Theory of Computing, pp. 614–623 (1998)
27. May, A., Ozerov, I.: On computing nearest neighbors with applications to decoding of binary linear codes. In: EUROCRYPT, pp. 203–228 (2015)
28. Meiser, S.: Point location in arrangements of hyperplanes. Inf. Comput. **106**(2), 286–303 (1993)
29. Motwani, R., Naor, A., Panigrahi, R.: Lower bounds on locality sensitive hashing. In: Proceedings of 22nd Symposium on Computational Geometry, pp. 154–157 (2006)
30. Musani, S., Shriner, D., Liu, N., Feng, R., Coffey, C., Yi, N., Tiwari, H., Allison, D.: Detection of gene × gene interactions in genome-wide association studies of human population data. Hum. Hered. **63**(2), 67–84 (2007)
31. O'Donnell, R., Wu, Y., Zhou, Y.: Optimal lower bounds for locality-sensitive hashing (except when q is tiny). ACM Trans. Comput. Theory **6**(1), 5 (2014)
32. Paturi, R., Rajasekaran, S., Reif, J.: The light bulb problem. Inf. Comput. **117**, 187–192 (1995)
33. Rabin, M.: Probabilistic algorithms. In: Algorithms and Complexity, pp. 21–30 (1976)
34. Smid, M.: Closest-point problems in computational geometry. In: Sack, J., Urrutia, J. (eds.) Handbook of computational geometry, pp. 877–935. Elsevier Science Publishing (1997)
35. Stothers, A.: On the complexity of matrix multiplication. Ph.D. thesis, The University of Edinburgh (2010)
36. Strassen, V.: Gaussian elimination is not optimal. Numer. Math. **13**(4), 354–356 (1969)

37. Valiant, G.: Finding correlations in subquadratic time, with applications to learning parities and juntas. J. ACM **62**, 13 (2015). Earlier version in FOCS 2012
38. Valiant, L.G.: Functionality in neural nets. In: AAAI, pp. 629–634 (1988)
39. Williams, R.: Faster all-pairs shortest paths via circuit complexity. In: Proceedings of 46th Annual ACM Symposium on the Theory of Computing, pp. 664–673 (2014)
40. Williams, R.: The polynomial method in circuit complexity applied to algorithm design (invited talk). In: Conference on Foundation of Software Technology and Theoretical Computer Science (FSTTCS), pp. 47–60 (2014)
41. Williams, R.: On the difference between closest, furthest, and orthogonal pairs: nearly-linear vs barely-subquadratic complexity. In: Proceedings of 29th ACM-SIAM Symposium on Discrete Algorithms, pp. 1207–1215 (2018)
42. Williams, V.: Multiplying matrices faster than Coppersmith-Winograd. In: Proceedings of 44th Annual ACM Symposium on the Theory of Computing, pp. 887–898 (2012)

Author Index

Printed in the United States
By Bookmasters